2nd International Symposium on Current Progress in Functional Materials (ISCPFM 2017)

Functional Materials for Energy and Environment Applications

IOP Conference Series: Materials Science and Engineering
Volume 496

Bali, Indonesia
8 - 9 November 2017

Editors:

Rosari Saleh
Ivandini T. Anggraningrum
Djoko Triyono

ISBN: 978-1-5108-8628-5
ISSN: 1757-8981

Printed from e-media with permission by:

Curran Associates, Inc.
57 Morehouse Lane
Red Hook, NY 12571

Some format issues inherent in the e-media version may also appear in this print version.

Printed by Curran Associates, Inc. (2019)

For permission requests, please contact the Institute of Physics at the address below.

Institute of Physics
Dirac House, Temple Back
Bristol BS1 6BE UK

Phone: 44 1 17 929 7481
Fax: 44 1 17 920 0979

techtracking@iop.org

Additional copies of this publication are available from:

Curran Associates, Inc.
57 Morehouse Lane
Red Hook, NY 12571 USA
Phone: 845-758-0400
Fax: 845-758-2633
Email: curran@proceedings.com
Web: www.proceedings.com

Table of contents

Volume 496

2nd International Conference on Current Progress in Functional Materials 2017

8–9 November 2017, Bali, Indonesia

Accepted papers received: 6 February 2019
Published online: 22 February 2019

Preface

Papers

IOP Conf. Series: Materials Science and Engineering **496** (2019) 011001 doi:10.1088/1757-899X/496/1/011001

Preface from the editors

The Second International Symposium on Current Progress in Functional Material 2017 (2nd ISCPFM 2017) held in Bali on November 8-9, 2017 with the theme of "Functional Materials for Energy and Environment Applications". The theme was selected based on the society's need for new and modified materials for energy applications and reducing environmental problems. This symposium can become a forum for scientists, to meet, discuss, collaborate and build functional material research networks.

Succeeding the first ISCPFM, which also conducted in Bali in 2016, the second symposium in 2017 held by welcoming prominent scientists from Japan, Taiwan, China and Indonesia as keynote speakers and invited speakers to share their valuable experiences during the plenary session. Participants who attended the symposium were almost ninety. This number increased significantly from last year's event, which showed that the event began to be recognised and noticed by functional material researchers and one day it will become the aim for researchers in disseminating the latest results in functional material research.

This proceeding will contain selected papers, which had been presented and discussed during the 2nd ISCPFM event. Sixty-three papers had accepted for publishing in this proceeding after going through the review process by peer-reviewers. Papers with the subject of nano and 2D materials are the most widely published papers in this proceeding, followed by papers with the subject of metal oxide and composite materials, then several papers with ceramics and biomaterials topics and some papers that reported the results in computational and simulation of functional material.

On this occasion, we would also like to thank our colleagues on the international advisory board and the scientific committee who had assisted in conducting symposium and publishing the proceedings including the abstract selection, the manuscript selection and the reviewer's selection processes. We are also very grateful for the excellent cooperation between the substantive editorial team, the technical editor team, the language editorial team and the supporting editorial team in preparing the camera-ready version of the manuscript files before sending to the publisher. We express our gratitude for the financial support from the Universitas Indonesia for conducting the symposium and publishing the proceedings and to the committees that had arranged the schedule and venue of the symposium so that it could run smoothly. Finally, we would also like to thank the participants for attending the symposium and the authors who contributed their papers to this proceeding.

Depok, December 2018

Editors,
Rosari Saleh
Ivandini T Anggraningrum
Djoko Triyono

IOP Conf. Series: Materials Science and Engineering **496** (2019) 011002 doi:10.1088/1757-899X/496/1/011002

International Advisory Board

Shu Yin (Tohoku University, Japan)
Shinjiang Jessie Lue (Chang Gung University, Taiwan)
Kiyoshi Kanie (Tohoku University, Japan)
Hsiang-Lin Liu (National Taiwan Normal University, Taiwan)

Scientific Committee

Ivandini T Anggraningrum (Universitas Indonesia)
Suminar S Pratapa (Institut Teknologi Sepuluh Nopember)
Wayan Nata Septiadi (Universitas Udayana)
Dede Djuhana (Universitas Indonesia)

Faculty's Conference Proceeding Committee

Djoko Triyono
Lusitra Munisa
Yustin S Dewi
Lisna Arumsari

IOP Conf. Series: Materials Science and Engineering **496** (2019) 011003 doi:10.1088/1757-899X/496/1/011003

Organizing Committee

Advisor	:	Prof. Dr. Ir. Muhammad Anis, M.Met
		Prof. Dr.rer.nat. Rosari Saleh
Honorary Chairman	:	Dr.rer.nat. Abdul Haris
		Prof. Dr. Ridla Bakri, M.Phil
Conference Chair	:	Dr. Djoko Triyono
Co-chair	:	Dr. Tito Latif Indra
Program Chair	:	Dr. Ratna Yuniati
Committee	:	Lusitra Munisa, M.Si
		Yustin S Dewi, S.Si
		Hesni Adila Shabrany, S.Si
		Lisna Arumsari, A.Md
		Nur Afifah, S.Si.
		Ardiansyah Taufik, S.Si.
		Fariz Hafizi Roihan, S. Sn.
		Caya Mikail, A.Md

IOP Conf. Series: Materials Science and Engineering **496** (2019) 011004 doi:10.1088/1757-899X/496/1/011004

ISCPFM 2017

The International Symposium on Current Progress in Functional Materials 2017 (ISCPFM 2017)

held by

FMIPA

Faculty of Mathematics and Natural Sciences
Universitas Indonesia

Supported by

Universitas Indonesia

IOP Conf. Series: Materials Science and Engineering **496** (2019) 011005 doi:10.1088/1757-899X/496/1/011005

Peer review statement

All papers published in this volume of *IOP Conference Series: Materials Science and Engineering* have been peer reviewed through processes administered by the proceedings Editors. Reviews were conducted by expert referees to the professional and scientific standards expected of a proceedings journal published by IOP Publishing.

IOP Conf. Series: Materials Science and Engineering **496** (2019) 012001 doi:10.1088/1757-899X/496/1/012001

Electro-oxidation of palm oil mill effluent using a boron-doped diamond anode

S Muharam[1,2], J Gunlazuardi[1] and T A Ivandini[1]

[1] Department of Chemistry, Faculty of Mathematics and Natural Sciences (FMIPA), Universitas Indonesia, Depok 16424, Indonesia

[2] Department of Chemistry, University Muhammadiyah Sukabumi, Sukabumi 43155, Indonesia

Corresponding author: ivandini.tri@sci.ui.ac.id

Abstract. The palm oil industry produces a large amount of wastewater as palm oil mill effluent (POME) that contains high concentrations of organic compounds, which are difficult to completely remove and consequently result in the degradation of aquatic ecosystems. In this work, palmitic acid was used as a model compound of POME to study its oxidation reactions. The experiment was conducted in a flow cell system with boron-doped diamond as the anode. A voltammetric technique was used to remove palmitic acid while observing the change in chemical oxygen demand concentration to monitor the reaction. The effects of supporting electrolyte, potential, time, and flow rate on the oxidation were studied. The results indicate that degradation of palmitic acid occurs as an indirect effect of electro-oxidation at high potential in the region of oxygen evolution or formation of hydroxyl radicals. Furthermore, under optimum conditions, up to 87.91% of the palmitic acid could be electrochemically removed.

Keywords: palmitic acid, hydroxyl radicals, boron-doped diamond, electro-oxidation

1. Introduction

Indonesia is the largest palm-oil-producing country, accounting for approximately 44% of the total production in the world [1]. Continuous palm oil production produces wastewater in the form of palm oil mill effluent (POME). The direct disposal of POME to local aquatic ecosystems can cause pollution of environment. The contents of organic matter such as palmitic and oleic acids in POME were around 45% and 40%, respectively, and are hence difficult to recover [2]. According to environmental quality standards, POME must be treated and disposed of as clean water.

In the palm oil industry, the biological method is the most widely used to treat POME because several organic compounds in POME can be degraded by microorganisms, which results in less sludge and requires less energy [1]. However, the biological method has some disadvantages such as long treatment times (45–60 days), large area requirements, color removal inefficiency, unstable fluctuations of chemical and physical composition, harmful biogases discharged to the atmosphere, and high operational costs.

Recently, advanced oxidation processes (AOPs) based on hydroxyl radicals have become the focus of research in the development of wastewater treatment technologies. AOPs have been widely applied and have been shown to eliminate hazardous and toxic pollutants successfully, converting them into biodegradable compounds that can be treated by conventional methods [3]. Electrochemical oxidation is a technology based on hydroxyl radical (•OH) discharge, which is more environmentally compatible, energy efficient, versatile, and low cost than other technologies. The most attractive electrode used for electrochemical oxidation to remove organic pollutants in wastewater treatment is boron-doped diamond (BDD). BDD shows excellent characteristics such as low background current, wide potential window, biocompatibility, and high stability in strong acid solutions [4–6]; therefore, BDD can rapidly treat wastewater and oxidize organic compounds to more than 90% chemical oxygen demand (COD) [7,8].

In this research, voltammetric studies of palmitic acid at BDD electrode were carried out, and application of the BDD electrode in wastewater treatment was evaluated using palmitic acid as a model compound of POME.

2. Experimental

The model compound of POME was prepared by dissolving 10 mg of palmitic acid in a mixed solution of ethanol and water (1:1). The mixture was heated at 80 °C while stirring at 500 rpm until the volume of the remaining mixture was reduced to half.

Furthermore, a thin layer of BDD was deposited by chemical vapor deposition using a hot filament on a low-resistivity p-Si substrate (p-Si/BDD). The filament and substrate were kept at around 2500 and 830 °C, respectively, and then the reactive gas containing 1% CH_4 in H_2 was introduced at a flow rate of 5 L/min in the presence of trimethoxy boron to produce 1000 ppm B/C ratio. The formed thin layer of polycrystalline BDD was characterized by cyclic voltammetry using a three-electrode cell. The electrode system comprised a BDD disc 1 cm in diameter as the working electrode, a Pt spiral as the counter electrode, and an Ag/AgCl reference electrode. Cyclic voltammetry was performed in a solution of palmitic acid containing 0.1 M Na_2SO_4 as the supporting electrolyte.

A galvanostatic method was used to perform the electrolysis of palmitic acid in a continuous two-electrode cell, with the BDD electrode as the working electrode and a Pt plate as the counter electrode; the diameter of each electrode was 1 cm (figure 1). Palmitic acid solution (20 mL) was added into the container and continuously circulated through the electro-oxidation cell using a peristaltic pump at a rate of 16–46 mL/min. The potential applied to the cell was in the range of 2000–10,000 mV, and the electrolysis time ranged from 0 to 240 min. Na_2SO_4 (0.1 M) was used as the supporting electrolyte for the electro-oxidation process. The reduction of palmitic acid in the electrolysis process was evaluated and quantified as COD using a potassium dichromate method [9].

3. Results and discussion

3.1. Cyclic voltammetry measurements of the BDD electrode in palmitic acid solution

Figure 2a shows the indirect oxidation of palmitic acid on the surface of BDD by cyclic voltammetry, which occurs in the oxygen evolution area at potentials between +1.626 and +2.0 V (*E* vs. Ag/AgCl). The oxidation current slightly decreased, before becoming constant at the sixth cycle. The deactivation percentage after the sixth cycle was 27.5% (table 1), indicating that the surface of the BDD electrode was contaminated by a thin layer of intermediate substances produced from the indirect oxidation of palmitic acid.

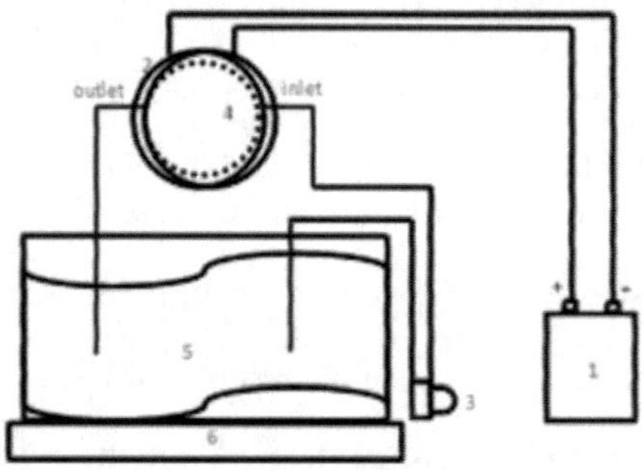

Figure 1. Schematic illustration of a continuous cell: 1, DC current source; 2, continuous-type electrolysis cell; 3, peristaltic pump; 4, BDD and platinum electrodes 1 cm in diameter; 5, sample container; 6, hot plate and stirrer.

Table 1. Peak potential, current density, and deactivation percentage of the peak oxidation of palmitic acid, measured by cyclic voltammetry for different numbers of cycles with a limit potential of +2.0 V (*E* vs. Ag/AgCl).

E_{pa} (V) vs. Ag/AgCl	Cycle	Current density ($\times10^{-}$ A)	Deactivation (%)
1.994	1	9.61	27.5
	2	8.67	
	3	8.24	
	4	7.32	
	5	7.093	
	6	6.97	

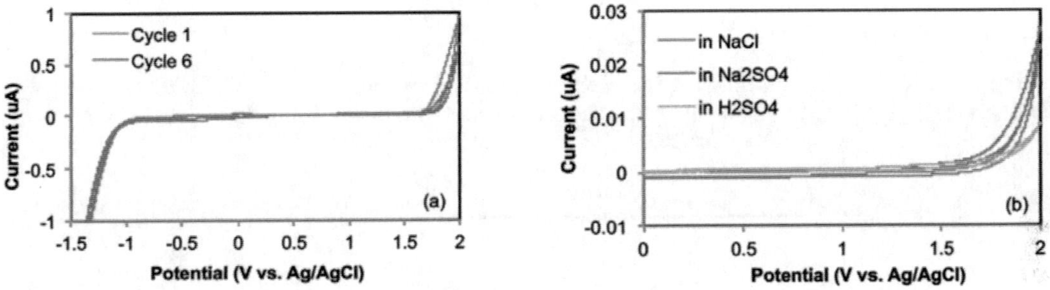

(a) (b)

Figure 2. Cyclic voltammograms obtained with the BDD electrode for palmitic acid (a) in 0.1 M Na_2SO_4 as the supporting electrolyte; showing 1st cycle and 6th cycle, and (b) in different supporting electrolytes. Each concentration of the electrolyte was 0.1 M. Scan rate was 100 mV/s).

Table 1 shows the deactivation percentage of the BDD surface during the electrolysis of palmitic acid solution with 0.1 M Na_2SO_4 as the supporting electrolyte. The oxidation of palmitic acid in this potential region around 2 V occurred through the formation of hydroxyl radical intermediates. The results clearly indicate that the BDD electrode is very stable and can consistently produce active hydroxyl radicals that play a role in palmitic acid decomposition.

In the case of wastewater treatment, the supporting electrolyte is important for improving the conductivity and compatibility of the electrochemical process. A good supporting electrolyte provides low electrochemical potential (high conductivity) and leads to a rapid reaction that destroys organic pollutants in wastewater. Therefore, the energy consumption for wastewater treatment will be reduced drastically Na_2SO_4 was reported to exhibit high peak currents for oxygen evolution, resulting in more hydroxyl radicals being produced by the BDD electrode [10].

Figure 2b shows the cyclic voltammograms of the electrolysis of palmitic acid using the BDD electrode in the presence of different supporting electrolytes. The cyclic voltammograms are similar, but there is a slight increase in the peak current when using H_2SO_4 and Na_2SO_4 as the supporting electrolytes.

3.2. Electro-oxidation of palmitic acid by BDD

3.2.1. Effect of potential. The electrolysis of palmitic acid solution in 0.1 M Na_2SO_4 was performed galvanostatically (I = 1 A). Figure 3 shows the evolution of COD with different potentials. The inset shows the instantaneous current efficiency (ICE) during electro-oxidation at various potentials, which takes place under mass transfer control.

On the basis of our calculations, the value of the mass transfer constant was $2 \times 10^{-}$ m/s, and thus the average limiting current density for the electro-oxidation of palmitic acid was 0.0448 A/cm^2, which is lower than the applied current density, (1 A). Under this condition, the electrolysis was controlled by mass transfer and COD removal occurred at high potential (10,000 mV), yielding around 82.63 % degradation after one hour of electrolysis. At this potential, charge was transferred to the electrode and more hydroxyl radicals were produced.

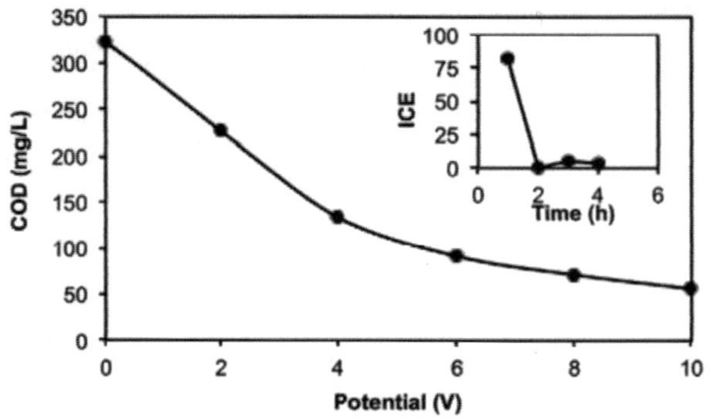

Figure 3. Effect of potential during the electro-oxidation of palmitic acid by BDD (supporting electrolyte: 0.1 M Na_2SO_4; flow rate: 16 mL/min; electro-oxidation time: 1 h). Inset shows the dependence of ICE on the electrolysis time.

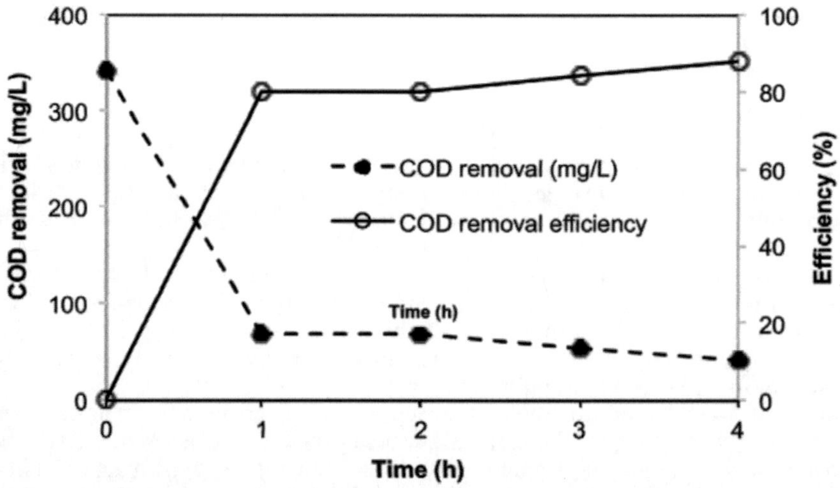

Figure 4. Effect of time on the electro-oxidation of palmitic acid by BDD (supporting electrolyte: 0.1 M Na_2SO_4; flow rate: 16 mL/min; applied potential: 10 V).

Hydroxyl radicals are formed on the surface of BDD as a result of the reaction with water and they are responsible for palmitic acid degradation. Palmitic acid near the surface of the BDD electrode undergoes an oxidation reaction, wherein carbon dioxide is produced. The presence of the supporting electrolyte Na_2SO_4 in the electrolysis process leads to the production of another oxidant, i.e., peroxo disulfate, which increases the rate of the oxidation reaction of palmitic acid, and consequently, the COD value decreases [10].

3.2.2. Effect of time. Figure 4 shows the profile of COD as a function of electrolysis time. The COD value decreased exponentially up to 1 h and became steady after 2–4 h of electrolysis. The profile confirmed that the electro-oxidation of palmitic acid was controlled by a secondary reaction, i.e., the oxygen evolution reaction. Part of the current (energy) was used for side reactions, which was confirmed by the decreasing ICE curve, in the Inset of figure 3 [11].

From these results, it is concluded that the electro-oxidation of palmitic acid by BDD takes place under the control of mass transfer and can be achieved within 1 h by electrolysis.

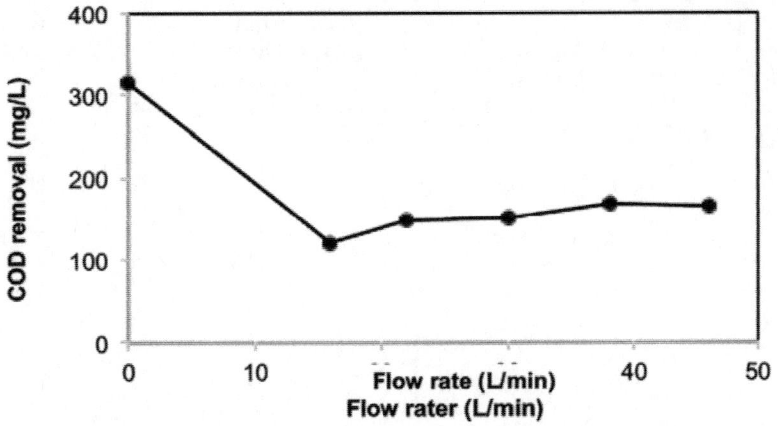

Figure 5. Effect of flow rate on the electro-oxidation of palmitic acid by BDD (supporting electrolyte: 0.1 M Na_2SO_4; applied potential: 10 V; electro-oxidation time: 1 h).

3.2.3. Effect of flow rate. The interaction between the electrode surface and the solution in a flow electrolysis system is different from that in a bulk system; hence, the oxidation product will also be different. In a flow electrolysis system, the diffusion layer is thinner than that in a bulk system, which leads to inter-radical or intermediate reactions [12].

Figure 5 shows the profile of COD as a function of the flow rate of palmitic acid solution in the electro-oxidation cell. Increasing the flow rate caused decreases in the COD removal efficiency. This is probably because, at high flow rate, the interaction of water with the surface of the BDD electrode decreases so the production of hydroxyl radicals (•OH) will also decrease. Under this condition, the hydroxyl radical tends to react with other hydroxyl radicals or intermediates to produce oxidant species weaker than the hydroxyl radical (•OH), such as peroxide, ozone, and organic radical (•R), which will prevent the formation of new hydroxyl radicals.

4. Conclusions

The cyclic voltammetry study showed that indirect oxidation of palmitic acid occurred at a potential of around 1.626 V (*E* vs. Ag/AgCl), i.e., in the region of oxygen evolution or formation of hydroxyl radicals (•OH). The electro-oxidation of palmitic acid by a BDD electrode was controlled by mass transfer at high-applied potential.

Acknowledgements

The author would like to thank the Ministry of Research, Technology and Higher Education Republic of Indonesia (Kemenristekdikti RI) for the grant No.2706/UN2.R3.1/HKP05.00/2017 and Einago (www.einago.com) for the English language review.

References

[1] Rupani P F Singh R P, Ibrahim M H and Esa N 2010 *World Appl. Sci. J.* **11** 70–81.
[2] Primandari P R S, Yaakob Z, Mohammad M and Muhammad A B 2013 *World Appl. Sci. J.* **27** 1482–4.
[3] Wang J L and Xu L J 2012 *Environ. Sci. Technol.* **42** 251–325
[4] Suzuki A, Ivandini T A, Kamiya A, Nomura S, Yamanuki M, Matsumoto K, Fujishima A and Einaga Y 2007 *Sens. Actuators B Chem.* **120** 500-7
[5] Shibano S, Ivandini TA, Terashima C, Nakata K and Einaga Y 2014 *Chem. Lett.* **43** 1292-3
[6] Medina-Sanchez M, Mayorga-Martinez C C, Watanabe T, Ivandini T A, Honda Y, Pino F, Einaga Y and Merkoci A 2016 *Biosens. Bioelec.* **75** 365-74
[7] Ivandini T A and Einaga Y 2017 *Chem. Commun.* **53**, 1338-47
[8] Peralta-Hernandez J M, Mendez-Tovar M, Guerra-Sanchez R, Martine-Huitle C A and Nava J L 2012 *Int. J. Electrochem.* **2012** ID 154316

IOP Conf. Series: Materials Science and Engineering **496** (2019) 012001 doi:10.1088/1757-899X/496/1/012001

[9] Li *et al.* 2009 *Desalination* **239** 139–45
[10] De Souza, R B A and Ruotolo L A M 2013 *Int. J. Electrochem. Sci.* **8** 643–57
[11] Dos Santos E V, Sena S F M, Da Silva J R, Ferro S, De Battisti A, Martinez-Huitle C A 2014 *Environ. Sci. Pollut. Res.* **21** 8466–75
[12] Sumi T, Saitoh T, Natsui K, Yamamoto T, Atobe M, Einaga Y and Nishiyama S 2012 *Angew. Chem. Int. Ed.* **51** 1–5

Characterization of PDPPBTT:PC_{71}BM bulk heterojunction nanostructures synthesized by a porous alumina template-assisted method

D Novialisa, V Fauzia and C Imawan

Department of Physics, Faculty of Mathematics and Natural Sciences (FMIPA)
Universitas Indonesia, Depok 16424, Indonesia

Corresponding author: vivi@sci.ui.ac.id

Abstract. To enhance the charge transfer efficiency of bulk heterojunction materials, a nanostructure comprising blended PDPPBTT and PC_{71}BM prepared via the porous alumina template-assisted method was proposed. Both materials were dissolved in chloroform in a 1:1 ratio at two different concentrations (5 and 10 mg mL^{-1}) and infiltrated into the nanopore template using a centrifuge method. The characterization results showed that one-dimensional nanostructures were successfully prepared. The high concentration solution resulted in nanostructures with large diameters but short lengths, which could absorb light effectively. Moreover, the high concentration solution exhibited low photoluminescence intensity, thus indicating efficient charge transfer from PDPPBTT to PC_{71}BM.

Keywords: organic semiconductor, PDPPBTT, PC_{71}BM, nanostructures, anodized aluminum oxide, hard template

1. Introduction

Organic semiconductors, which are economical and environmentally safe with unlimited availability, are increasingly being used in flexible optoelectronic device applications. They have high optical absorption coefficients. Hence, it is possible to fabricate thin-film devices via a low-cost approach [1]. For organic solar cell applications, a bulk heterojunction (BHJ) is the most preferred structure for active layers. BHJs comprise a mixture of electron donor and acceptor materials, which provides a large interfacial area for exciton dissociation into free electrons [2]. However, the efficiency of organic solar cells is still low compared to that of inorganic solar cells owing to inefficient charge transfer. To enhance the charge transfer efficiency, a nanostructured BHJ was proposed, which is expected to reduce the distance that the photogenerated free carriers have to travel to reach the respective electrodes; thus, the recombination can be greatly reduced [3–5].

In this study, a new low-band gap (1.38 eV) p-type semiconducting polymer, poly{2,2'-(2,5-bis(2-octyldodecyl)-3,6-dioxo-2,3,5,6-tetrahydropyrrolo[3,4-c]pyrrole-1,4diyl) dithieno[3,2-b]thiophene-5,5'-diyl-alt-thiophen-2,5-diyl}(PDPPBTT), was used in combination with a conventional n-type fullerene, [6,6]-phenyl C_{71} butyric acid methyl ester (PC_{71}BM), as the active layer materials in organic solar cells [6–9]. A simple and cost-effective hard template method based on nanoporous anodized aluminum oxide (AAO) was employed to fabricate the nanostructured active layer materials used in this experiment [10,11]. Although several methods exist to infiltrate organic materials into the nanopores of AAO templates, such as immersion and spin coating, in this experiment, a centrifuge method was selected owing to the strong centrifugal force that could push the materials into the nanopores to form better nanostructures [10,12,13]. The high density of the nanowires was proved to

IOP Conf. Series: Materials Science and Engineering **496** (2019) 012002 doi:10.1088/1757-899X/496/1/012002

Figure 1. Schematic of the position of the template inside a modified tube.

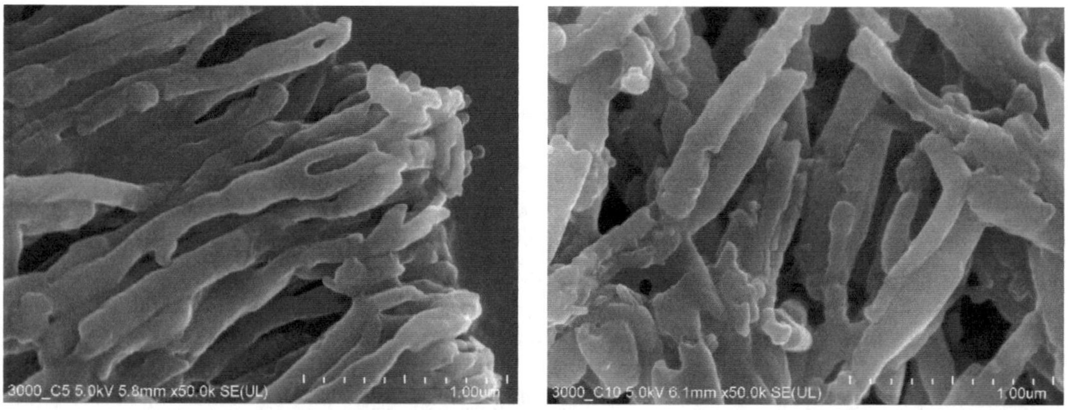

(a) (b)

Figure 2. FESEM results of nanostructures produced at 3000 rpm with different solution concentrations: (a) 5 mg mL^{-1} and (b) 10 mg mL^{-1}.

absorb at lower energies; this may enhance the optical properties by promoting the interchain interaction [13]. The dependences of the morphological, optical, and structural properties of blended PDPPBTT:PC$_{71}$BM materials on solution concentration were investigated in detail.

2. Materials and Methods

PDPPBTT (Mw > 30,000 GPC) and PC$_{71}$BM (Mw = 1030.93 g mol^{-1}) were purchased from Luminescence Technology Corp. and used without further purification. Porous alumina templates (Whatman Anodics, Sigma-Aldrich, St. Louis, USA) with pore diameters in the range of 20–200 nm and a thickness of 60 μm were used for synthesizing bulk heterojunction materials. Before the infiltration process, the template was cleaned by sonication in acetone, ethanol, and deionized water for 15 min each. Then, the template was rinsed with deionized water and dried completely at room temperature.

Mixtures of PDPPBTT and PC$_{71}$BM with a ratio of 1:1 with two different concentrations, 5 mg mL^{-1} (C5) and 10 mg mL^{-1} (C10), were dissolved in chloroform and stirred for 24 h. Then, 100 μL of the solution was dropped onto the AAO template placed inside a modified tube (figure 1) in a centrifuge (Kubota B2420), as reported in a previous study [13]. The template surface was in the vertical position; hence, the centrifugal force horizontally infiltrated the solution into the pores of the template at 3000 rpm.

The UV–Visible absorption spectra of the PDPPBTT:PC$_{71}$BM nanostructure were recorded using Shimadzu UV-3101PC, and the Raman and photoluminescence (PL) spectra were recorded using Renishaw RE 04. For the morphological characterization, the template was removed first by dissolving alumina in 3 M sodium hydroxide (NaOH) solution for 24 h. The morphological characterization was analyzed by field emission scanning electron microscopy (FESEM; Hitachi SU-8030).

3. Results and discussion

The morphology of the PDPPBTT:PC$_{71}$BM nanostructure are shown in figure 2, which shows that one-dimensional nanostructures have been successfully prepared as a result of the pore shape in the template. The resulting structure also suggests that the solution could infiltrate the pores owing to a

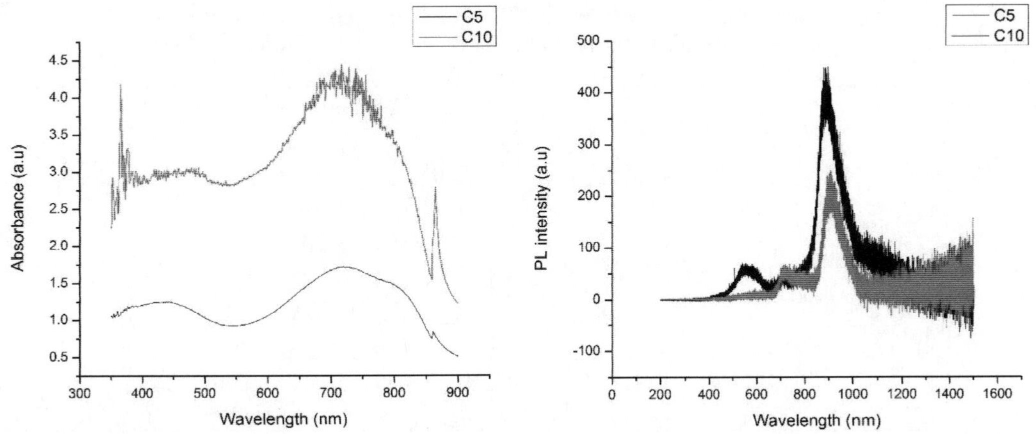

(a) (b)

Figure 3. (a) UV–Vis absorption and (b) photoluminescence spectra of PDPPBTT and $PC_{71}BM$ bulk heterojunction nanostructures prepared with two different solution concentrations

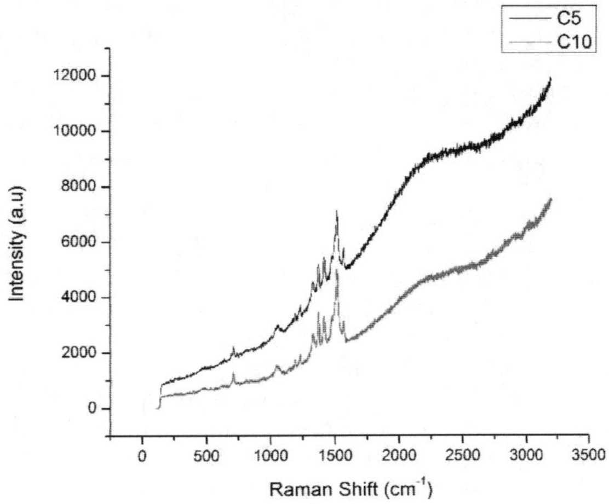

Figure 4. Raman Spectra of PDPPBTT and $PC_{71}BM$ BHJ nanostructures prepared with two different solution concentrations.

combination of centrifugal force and capillary force. However, figure 2 does not clearly show whether a nanorod or nanotube structure was formed, but in a previous study, this method resulted in a nanotube structure owing to the stronger intermolecular force between the liquid and the template surface compared with the cohesion force between the monomers in the solution [10]. The solution with a concentration of 10 mg mL^{-1} produced nanotubes with an average diameter of 0.2 μm, while the solution with a concentration of 5 mg mL^{-1} produced small nanotubes with an average diameter of 0.1 µm. In addition, the low concentration solution produced long nanotubes probably because the solution could penetrate deep into the pores owing to the low surface energy of the solution.

The UV–Vis absorption and PL spectra of PDPPBTT:$PC_{71}BM$ BHJ nanostructures are shown in figure 3. Two wide peaks centered at 420 and 720 nm can be observed in the UV–Visible absorption spectra (figure 3a), which originate from PDPPBTT absorption [6]. Materials with such a broad absorption spectrum are beneficial for solar cell applications owing to their ability to absorb the solar spectrum, of which 45% of the light is in the visible region [14]. High absorption intensity was obtained using the solution with a high concentration, which may be correlated with the morphology shown in figure 2, i.e., thick and short nanostructures can absorb light effectively. Moreover, there is a

Table 1. Raman peaks of PDPPBTT and $PC_{71}BM$ BHJ nanostructures prepared with two different solution concentrations.

Peak	C5 (cm^{-1})	C10 (cm^{-1})	Assignments
V_1	712	713	symmetric C–S stretching
V_2	1060	1055	localized in-plane bending of thiophene C–H ip δ + stretching $V_{C=C}$
V_3	1192	1193	C–H bending
V_4	1230	1230	(CH_2) torsion
V_5	1326	1329	(CH_2) twisting
V_6	1370	1373	in-plane bending of thiophene $V_{C=C}$ + C–H ip δ, Th V_{C-N}
V_7	1416	1413	$\delta_s(CH_3)$
V_8	1517	1517	Pentagonal pinch modes of the fullerenes
V_9	1569	1569	Anti-symmetric stretch modes of the fullerene

slight red shift in the peak from 5 nm to 20 nm as the solution concentration is increased from 5 mg mL^{-1} to 10 mg mL^{-1}; this indicates that there was more aggregation of chains in the high concentration sample than the low concentration sample. Polymer chains could align themselves and increase the conjugation length in the aggregated configuration, and the aligned polymer chains support more efficient electron excitation from π to π* and charge transfer between the DPP and bithiophene (BTT) units of the main chain [15]. The aggregation and chain alignment could enhance the optical properties by promoting the interchain and intrachain interactions [10].

The PL spectra of PDPPBTT:$PC_{71}BM$ BHJ nanostructures are shown in figure 3b. There are two peaks centered at 702 and 900 nm. The peaks may have resulted from the radiative recombination of two main excitation levels that correspond to the two absorbance peaks at short wavelengths. The peaks are also red-shifted from 709 nm to 715 nm and from 889 nm to 906 nm as the solution concentration is increased from 5 mg mL^{-1} to 10 mg mL^{-1}. The low PL intensity for the high concentration sample indicates efficient charge transfer from PDPPBTT to $PC_{71}BM$ [9].

Figure 4 shows the Raman spectra of PDPPBTT:$PC_{71}BM$ BHJ nanostructures. Four peaks are observed at 1192, 1230, 1517, and 1589 cm-1 that correspond to C–H bending for the first two peaks, the pentagonal pinch modes, and Raman active models of the fullerene, respectively [16]. Furthermore, five peaks are observed at 712, 1060, 1326, 1370 and 1416 cm^{-1} that originate from the PDPPBTT molecular motion [16,17]. Table 1 shows assignments of the Raman spectra for symmetric C–S stretching, a localized mode characterized by C–H in-plane bending together with C=C stretching at the thiophene ring, C–H bending of thiophene, C=C stretching in the DPP unit, and C=C stretching and C–C stretching of the BTT core and thiophene ring, respectively.

An upward shift of 10 mg/mL BHJ PDPPBTT:$PC_{71}BM$ for C–S stretching, C–H in-plane bending as well as C=C stretching in the thiophene unit together with C–H bending of the delocalized thiophene segment can be associated with better molecular arrangement seen at higher concentrations and also to a change in electron density from DPP to the acceptor moiety, so that, the C-C bond length of the material increases.

4. Conclusions

One-dimensional PDPPBTT:$PC_{71}BM$ bulk heterojunction nanostructures have been successfully fabricated using porous alumina templates and a centrifuge method. The high concentration solution (10 mg mL^{-1}) produced nanostructures with large diameters and short lengths owing to increased aggregation of the chains, resulting in high absorption intensity. Low photoluminescence intensity for the BHJ nanostructure was also obtained using the high concentration solution, which may be due to efficient charge transfer from PDPPBTT to $PC_{71}BM$.

Acknowledgements

This research was financially supported by PITTA Grant 2017 with contract number 695/UN2.R311/HKP.05.00/2017 from Universitas Indonesia. The authors would like to thank Einago (www.einago.com) for the English language review.

IOP Conf. Series: Materials Science and Engineering **496** (2019) 012002 doi:10.1088/1757-899X/496/1/012002

References

[1] Bagher A M 2014 *International Journal of Renewable and Sustainable Energy* **3** 53–8
[2] Deibel C and Dyakonov V 2010 *Rep. Prog. Phys.* **73** 1–39
[3] Selvakumar U A, Jegathese C R, Kings S and Jaffar M T 2016 *ARPN Journal of Engineering and Applied Science* **11** 1258–61
[4] Moon J S, Jo J and Heeger A J 2012 *Adv. Energy. Mater.* **2** 304–8
[5] Kietzke T 2007 *Advances in Optoelectronics* **2007** 40285
[6] Bronstein H et al. 2011 *J. Am. Chem. Soc.* **133** 3272–5
[7] Luminescence Technology Corporation, Lumtec 2017 Product Catalog. Avaiable at http://plus.webdo.com.tw/userfiles/lumtec/files/2017_Lumtec_Catalog.pdf
[8] Sze S M and Ng K K 2007 *Physics of Semiconductor Devices 3rd edition* (Hoboken: John Wiley & Sons)
[9] Xing S, Wang H, Kong T, Yu J, Jiang J and Wang C 2017 *IEEE J. Photovolt.* **7** 1058–64
[10] Bakar N A, Supangat A and Sulaiman K 2014 *Nanoscale Res. Lett.* **9** 600
[11] Mijangos C, Hernândez R and Martín J 2016 *Prog. Polym. Sci.* **54–55** 148–82
[12] Fakir M S, Supangat A and Sulaiman K 2014 *Nanoscale Res. Lett.* **9** 225
[13] Doris M, Aziz F, Alhummiany H, Bawazeer T, Alsenany N, Mahmoud A, Zakaria R, Sulaiman K and Supangat A 2017 *Nanoscale Res. Lett.* **12** 67
[14] Oku T 2017 *Solar Cells and Energy Materials* (Berlin: Walter de Gruyter GmbH & Co KG)
[15] Banwell C N 1983 *Fundamental of Molecular Spectroscopy* 3rd edition (New York: McGraw-Hill)
[16] Bakar N A, Supangat A and Sulaiman K 2015 *RSC Adv.* **5** 102689–99
[17] Francis C, Fazzi D, Grimm S B, Paulus F, Beck S, Hillebrandt S, Pucci A and Zaumseil J 2017 *J. Mater. Chem. C* **5** 6176–84

Synthesis and characterization of Ni nanoparticles supported on nitrogen-doped mesoporous carbon

A Z Pamungkas, I Abdullah and Y K Krisnandi

Department of Chemistry, Faculty of Mathematics and Natural Sciences (FMIPA), Universitas Indonesia, Depok 16424, Indonesia

Corresponding author: yuni.krisnandi@sci.ui.ac.id

Abstract. Nitrogen-doped mesoporous carbon (N-MC) was synthesized by the soft-templating method using phloroglucinol/formaldehyde and melamine as carbon and nitrogen precursors, respectively, and Pluronic F127 as a template for the generation of mesopores inside the carbon structure. Scanning electron microscopy analysis revealed that the material had a rough surface, while Raman analysis indicated the existence of sp^2-hybridized carbon rather than sp^3, similar to graphite. The increase of the D-band to G-band intensity ratio indicated that N-MC has a more graphite-like structure than pristine mesoporous carbon. X-ray photoelectron spectroscopy analysis confirmed that nitrogen was successfully doped into the carbon framework. The material was then used as a support for Ni nanoparticles (Ni-NPs). The presence of Ni metal in Ni-NP@N-MC was confirmed by X-ray diffraction analysis.

Keywords: mesoporous carbon, Ni nanoparticles, N-doped carbon, soft-templating method

1. Introduction

Industrial activity has steadily increased since the advent of the industrial revolution, leading to ever increasing emission of CO_2 into the atmosphere. Accordingly, the pre-industrial revolution atmospheric CO_2 concentration of 280 ppm has increased to its current level of 380 ppm [1]. This has become a serious environmental problem because of the contribution CO_2 makes to the greenhouse effect, leading to global climate change. Consequently, to reduce this negative effect, it is absolutely vital that we decrease the amount of anthropogenic CO_2 released into the atmosphere.

The current technology used for large-scale CO_2 absorption is based on liquid amine absorption [2, 3]. Amine groups can remove CO_2 from the air owing to their ability to reversibly form carbamate and bicarbonate compounds. However, this method suffers from disadvantages related to high energy consumption, corrosion, and solvent evaporation [4,5]. A potential strategy for overcoming these problems is the immobilization of the working amine compounds onto the surface of support materials instead of using them in the liquid state [6].

Since its initial development, mesoporous carbon has been used in a wide range of applications, including adsorbents [7], sensors [8], molecule carriers [9], and catalyst supports [10]. However, the use of mesoporous carbon in CO_2 adsorption is still limited as CO_2 is only weakly physisorbed on mesoporous carbon. Recently, our group successfully prepared mesoporous carbon modified with amine compounds [11,12]. The amine compounds were immobilized in the pores of the mesoporous carbon, enhancing its capacity to adsorb CO_2. However, the effect of nitrogen compounds with both sp^3 and sp^2 hybridization on the CO_2 sorption properties of mesoporous carbon remains largely unresearched. Accordingly, in this work, a mesoporous carbon framework was doped with nitrogen using melamine as a nitrogen source.

IOP Conf. Series: Materials Science and Engineering **496** (2019) 012003 doi:10.1088/1757-899X/496/1/012003

(a) (b) (c)

Figure 1. SEM images of (a) MC, (b) N-MC, and (c) Ni@N-MC (×5000)

2. Experimental

2.1. Synthesis of mesoporous carbon (MC)
Mesoporous carbon was synthesized using a previously reported procedure [13]. Briefly, phloroglucinol and Pluronic F127 were dissolved in an ethanol/water mixture. Then, 37% HCl was added, and the solution was vigorously stirred for 30 min. Under continuous stirring, formaldehyde (37%) was then added drop-wise to the solution, and then the mixture was stirred for another 30 min before being split into two layers. The bottom layer was removed and the stirring was continued overnight, before being transferred to an autoclave and cured for 24 h in the oven at 100 °C. The material was carbonized under nitrogen flow with a heating rate of 1 °C/min from 100 to 400 °C and 5 °C/min from 400 to 850 °C before kept at 850 °C for 2 h.

2.2. Synthesis of nitrogen-doped mesoporous carbon (N-MC)
In a typical synthesis, 1.25 g Pluronic F127 and 0.625 g phloroglucinol and were dissolved in a 9.7 g ethanol/water (10:9 *w/w*) solution. Once the solid had completely dissolved, 0.08 mL 37% HCl was added drop-wise, and the solution was stirred for 30 min (mixture A). Separately, 0.625 g melamine was dissolved in 26 mL water in a round-bottom flask, which was heated under reflux at 80 °C (mixture B). Then, 1.25 mL 37% formaldehyde was added to mixture A. After 5 min, mixture A was added slowly to mixture B and the solution was stirred for 30 more min until solid precipitate appeared. The solid was then filtered, washed with water, and dried in airy condition. It was then transferred to a 200 mL Teflon-lined stainless steeled autoclave and cured in an oven at 150 °C for 72 h. The material was carbonized under nitrogen flow with a heating rate of 1 °C/min from 100 to 400 °C and 5 °C/min from 400 to 850 °C before being kept at 600 °C for 2 h.

2.3. Impregnation with Ni nanoparticles (Ni@N-MC)
First, 0.5 g N-MC was dispersed in 10 mL water. Second, in separate flask, 1.23 g $Ni(NO_3)_2 \cdot 6H_2O$ was dissolved in 5 mL of water, followed by being added drop-wise to the carbon dispersion. This mixture was stirred for 6 h, then the solid composite was filtered, washed with water, and dried in airy condition. The obtained solid and 20 mL ethylene glycol were transferred into a round-bottom flask for reflux treatment at 170 °C for 6 h. The final solid was then filtered, washed with water and ethanol, and dried in airy condition.

2.4. CO_2 adsorption
CO_2 and N_2 were introduced to the adsorbent at a flow rate of 20 cm^3/min each. The output CO_2 was dissolved in 0.1 M NaOH solution. The amount of CO_2 was then derived by titration against 0.1 M HCl with phenolphtalein and methyl orange as indicator.

3. Results and discussion
The morphologies of MC, N-MC, and Ni@N-MC were observed by scanning electron microscopy (SEM), as shown in figure 1. All the materials present rough surfaces. A grass-like structure is observed for the surface of MC, as has been previously reported [14]. This characteristic surface mor-

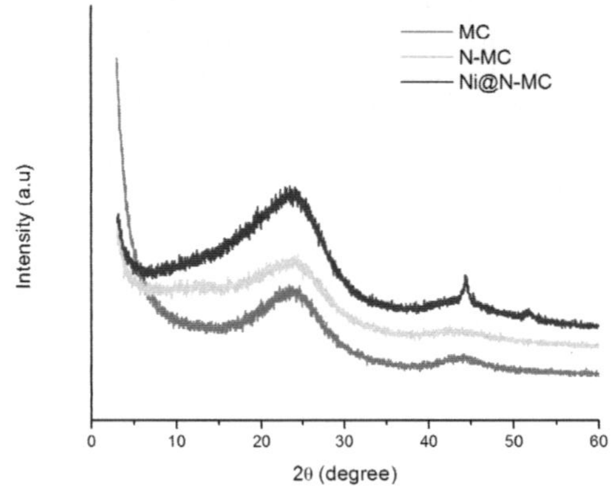

Figure 2. XRD pattern of MC, N-MC, and Ni@N-MC.

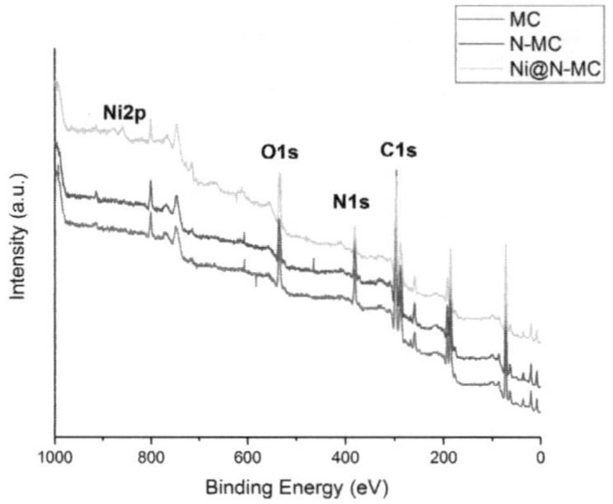

Figure 3. XPS spectra of MC, N-MC, and Ni@N-MC.

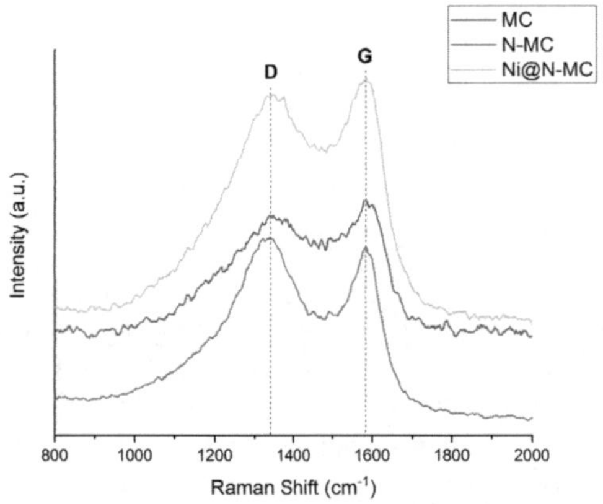

Figure 4. Raman spectra of MC, N-MC, and Ni@N-MC.

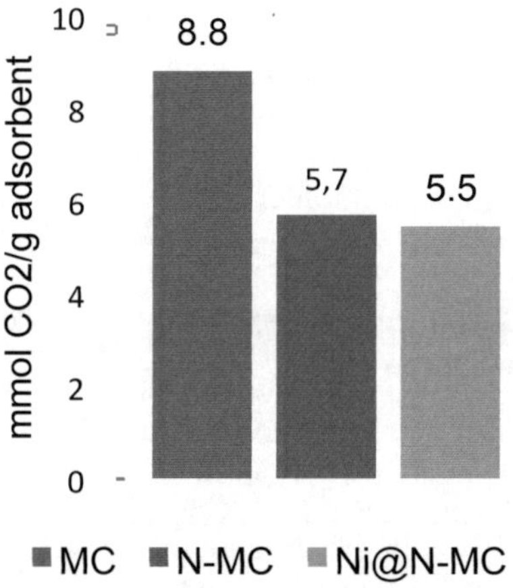

Figure 5. CO_2 adsorption of MC, N-MC, and Ni@N-MC.

phology endows the MC with a high surface area. Both N-MC and Ni@N-MC exist as small particles. There is no observable difference in morphology upon impregnation with Ni nanoparticles.

Figure 2 shows the X-ray diffraction (XRD) patterns for MC, N-MC, and Ni@N-MC. It is clear that all the samples present very similar diffraction patterns. The peaks for $2\theta = 25°$ and $44°$ are attributed to the (002) and (101) planes, respectively, which are characteristic of graphitic carbon [15]. The similarity of the diffraction patterns indicates that there is no significant change in the structure of the material upon nitrogen doping. For Ni@N-MC, there are additional peaks at $2\theta = 45°$ and $52°$ owing to the presence of Ni metal [16]. This confirms the success of the Ni nanoparticle impregnation into the mesoporous carbon. The X-ray photoelectron spectroscopy (XPS) spectra (figure 3) further confirm the presence of Ni in addition to C, N, and O in Ni@N-MC.

The Raman spectra (figure 4) reveal Raman shifts at 1339 and 1584 cm^{-1}, which are the characteristic D-band and G-band, respectively, of graphitic materials. The spectral peaks for all the materials occur at the same frequency, indicating that the Raman spectra are not affected by nitrogen doping. However, the D-band to G-band intensity ratio (I_D/I_G) changes from 1.01 for MC to 0.98 for N-MC and Ni@N-MC. This indicates that nitrogen doping increases the abundance of sp^2 carbon relative to that of sp^3 carbon.

The results for the CO_2 adsorption capacities of the materials are shown in figure 5. Although modification with amine compounds has been shown to increase the CO_2 uptake of mesoporous carbon in previous works [11,12], modification with melamine as the nitrogen source decreases the CO_2 uptake of the N-doped mesoporous carbon. This is most likely due to structural shrinkage of the framework upon nitrogen doping and also the low primary and secondary amine contents, since the nitrogen forms bonding interactions with the carbon framework. Furthermore, the Ni nanoparticles in Ni@N-MC appear to have no significant effect on CO_2 adsorption. Ongoing work is underway in our group to determine the mesoporosity and basicity of these materials, which are important for further application as heterogeneous catalysts for CO_2 fixation reactions.

4. Conclusions

Nitrogen-doped mesoporous carbon was successfully synthesized using a soft-templating method with melamine as nitrogen source. The CO_2 adsorption capacity of N-MC is lower than that of MC. This may be due to structural shrinkage upon nitrogen doping and also the low primary and secondary amine contents. Nevertheless, this material has the potential to be employed as an alternative sorbent for CO_2 capture.

Acknowledgements

The authors gratefully acknowledge Universitas Indonesia for PITTA 2017 No. 702/UN2.R3.1/HKP.05.00/2017 research grant for funding this project and Prof Hisao Yoshida for XRD and SEM measurement at Kyoto University, Japan as well as Mr. Ikhram for XPS measurement at the Keio University and also Einago (www.einago.com) for the English language review.

References

[1] Chatti R, Bansiwal A K, Thote J A, Kumar V, Jadhav P, Lokhande S K, Biniwale R B, Labhsetwar N K and Rayalu S S 2009 *Micropor. Mesopor. Mater.* **121** 84–9
[2] Stowe H M and Hwang G S 2017 *Ind. & Eng. Chem. Res.* **56** 6887–99
[3] Bernhardsen I M and Knuutila H K 2017 *Int. J. Greenh. Gas Con.* **61** 27–48.
[4] Lee S Y, Jang D I, Bae S T and Park S J 2014 Facile synthesis of nitrogen-enriched mesoporous carbon for carbon dioxide capture *Int. J. Hydrog. Energy* **39** 12347–52.
[5] Rabensteiner M, Kinger G, Koller M, Gronald G and Hochenauer C 2014 *Int. J. Greenh. Gas Con.* **27** 1–14.
[6] Ünveren E E, Monkul B Ö, Sarıoğlan Ş, Karademir N and Alper E 2017 *Petroleum* **3** 37–50.
[7] Libbrecht W, Verberckmoes A, Thybaut J W, Van Der Voort P and De Clercq J 2017 *Carbon* **116** 528–46.
[8] Liu L, Zhao H, Shi L, Lan M, Zhang H and Yu C 2017 *Electrochim. Acta* **227** 69–76.
[9] Kiamahalleh M V, Mellati A, Madani S A, Pendleton P, Zhang H and Madani S H 2017 *J. Pharm. Sci.* **106** 1545–52.
[10] Ji T, Chen L, Mu L, Yuan R, Knoblauch M, Bao F S, Shi Y, Wang H and Zhu J 2016 *Chem. Eng. J.* **295** 301–8.
[11] Ardhyarini N and Krisnandi Y 2017 *AIP Conf. Proc.* 1862 030090
[12] Sulistianti I, Krisnandi Y and Moenandar I 2017 *IOP Conf. Ser.: Mater. Sci. Eng.* 188 012041
[13] Górka J, Zawislak A, Choma J and Jaroniec M 2008 *Carbon* **46** 1159–61
[14] Liang C and Dai S 2006 *J. Am. Chem. Soc.* **128** 5316–7
[15] Wang Y, Bai X, Wang F, Qin H, Yin C, Kang S, Li X, Zuo Y and Cui L 2016 *Sci. Rep.* **6** 26673
[16] Sarkar B, Tiwari R, Singha R K, Suman S, Ghosh S, Acharyya S S, Mantri K, Konathala L N S, Pendem C and Bal R 2012 *Catal. Today* **198** 209–14.

The effect of partial substitution of monovalent ion on divalent ion of $La_{0.8}Ca_{0.2-x}Ag_xMnO_3$ (x=0 and x=0.05) compounds in structure, morphology, and purity by sol-gel method

N A Sahara and B Kurniawan

Department of Physics, Faculty of Mathematics and Natural Sciences (FMIPA) Universitas Indonesia, Depok 16424, Indonesia

Corresponding author: bkuru@fisika.ui.ac.id

Abstract. $La_{0.8}Ca_{0.2-x}Ag_xMnO_3$ ceramics with x = 0 and 0.05 has been synthesized using the sol gel method. These materials are one member of perovskite manganites materials which has the general formula $RE_{1-x}AE_xMnO_3$. Material LaCaMnO3 and $La_{0.8}Ca_{0.15}Ag_{0.05}MnO_3$ was characterized by X-ray Diffractometer (XRD), and Scanning Electron Microscopy-Electron Dispersive Spectroscopy (SEM-EDS). According to X-Ray Diffractometer characterization and high score refinement showed that the crystal structure for both materials are orthorhombic. Partial substitution of the Ag^+ in Ca^{2+} did not change the structure of the material but there are changes in lattice parameters. Both materials have a high homogeneity with the larger crystallite average size when doped by Ag, it is seen from the morphology of Scanning Electron Microscope (SEM). EDX characterization using point analysis confirms that in samples with x = 0.05 ($La_{0.8}Ca_{0.15}Ag_{0.05}MnO_3$), the Ag ion have been successfully substituted. The result of the measurement of the composition of the two materials is not much different from the calculation, so it can be concluded that both materials have a good compositional purity.

1. Introduction

In recent years, many researchers have studied the mangantite perovskite with the general formula $RE_{1-x}AE_xMnO_3$, where R is a trivalent rare earth element (Nd, La, Pr, etc.), and A is a divalent alkaline earth metal (Ba, Sr, Ca, etc.) as a result of their rich properties. Partial substitution of divalent ions to trivalent ions can produce a mixture of Mn^{4+}/Mn^{3+} valence, which will give unique phenomena, including metal-insulator transition (T_{MI}), colossal magnetoresistance effects, ferromagnetic-paramagnetic transition (T_c), etc. [1]. The interactions between Mn^{4+} and Mn^{3+} ions are described in the double exchange mechanism, which was first proposed by Zener [2]. A simply, double exchange mechanism is the simultaneous exchange of electrons from Mn^{3+} to Mn^{4+} ions through the oxygen atom to keep the electric charge remains neutral, so the changes in the length and angle of Mn–O–Mn bonds can affect the double exchange interaction [2,3]. Based on literature, modification perovskite manganite material by replacing different dopant elements can affect the microstructure, which results in magnetic and electrical properties [4]. Therefore, the structure of this material is very important to be discussed further.

Recently, engineering of manganite perovskite structures has been carried out by substituting some monovalent ions into divalent ions, which can affect the ratio of Mn^{4+} and Mn^{3+}, shortening the bond distance between Mn–O–Mn and increasing the angle of Mn–O–Mn bond that will affect the double exchange interaction and ihave an impact on the nature of the magnetotransport of the material [4]. So, in this research, we investigated the effect of partial substitution of Ag on Ca-site of the $La_{0.8}Ca_{0.2-x}Ag_xMnO_3$ (x = 0 and 0.05) compounds in structure, morphology and purity.

Tabel 1. Data Ionic Radii

Ion	Ionic radii (nm)
La^{3+}	1.36
Ca^{2+}	1.26
Ag^{+}	1.28
O^{2-}	1.40
Mn^{3+}	0.65

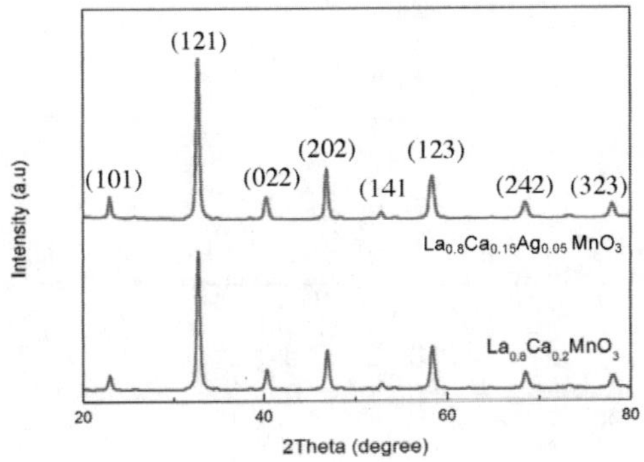

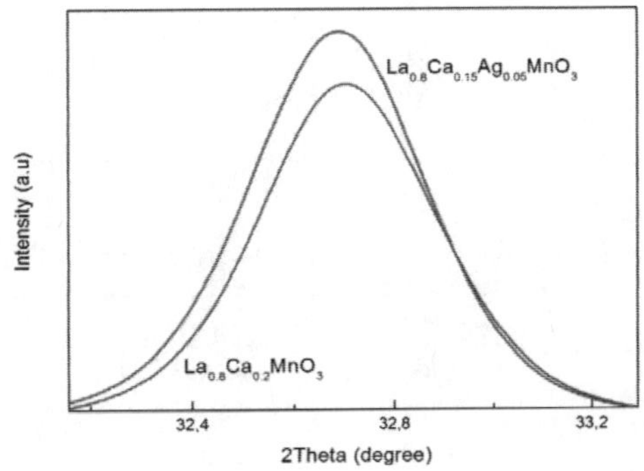

Figure 1. XRD patterns of $La_{0.8}Ca_{0.2}MnO_3$ and $La_{0.8}Ca_{0.15}Ag_{0.05}MnO_3$.

Figure 2. Local XRD patterns around the strongest diffraction peak (121)

2. Experimental

Materials $La_{0.8}Ca_{0.2-x}Ag_xMnO_3$ (x = 0 and x = 0.05) were prepared by a solgel method using the nitrate form of the raw materials. The raw materials are La_2O_3, $Ca(NO_3)_2.6H_2O$, $AgNO_3$, $Mn(NO_3)_2.4H_2O$, and citric acid were weighed according to stoichiometric calculations. All raw materials, except La_2O_3, were dissolved using aquabidest. La_2O_3 was dissolved in aquabidest and nitric acid while stirred using a magnetic stirrer to form lathanum nitrate. After that, the solution of the other raw materials were poured into lanthanum nitrate while continuing to stir and heated with a magnetic stirrer at 80 °C. Then the pH of the mixture solution is adjusted to reach pH of 7 with the addition of ammonium solution. Afterwards, leave the solution of mixture for a few moments until gel is formed. Then the gel was dried for 3 h at 120 °C, then calcined for 5 h at 500 °C. The calcined material is then formed into a pellet with pressure and then sintered for 24 h at 800 °C.

The pellet was characterized using X-Ray Diffractometer from a Rigaku SmartLab with Cu Kα radiation (λ = 1.54178 Å) to determine the crystal structure of the materials. The 2θ range used was 0°–90° and scan rate was 0.02° per second. The XRD pattern was analyzed by Rietvield refinement using high score. The crystallite size D was calculated using the Debye-Scherrer's equation ($D = 0.94\ \lambda/\beta \cos\theta$) with θ and β are the position and the FWHM of highest diffraction peak, respectively and λ is the wavelength of the Cu Kα XRD source. The morphology of the sample surface was characterized by Scanning Electron Microscopy Hitachi SU-3500 with the accelerating voltage used was 20000 Volt and magnification used was 20000 times. SEM observations are carried out by placing a test sample in the form of a pellet above the holder that has been given carbon-type material. The result of EDX characterization were obtained together with SEM measurements using point analysis.

3. Results and discussion

Based on Takehira's research, one of the requirements of successful doping is the dopant ion radius, which is relatively equal to the ion radius to be doped [5]. Table 1 shows the data of ionic radius, which is based on the Shannon research [6]. From the table, it can be seen that the difference of radius of Ca ion and the radius of Ag ion is relatively small (less than ± 5%). It can be confirmed that the silver ion can substitute the Ca ion. The substitution effect on the crystallographic parameter is shown in figure 1 and table 2.

Tabel 2. Rietvield refinement results from XRD data.

Compound	$La_{0.8}Ca_{0.2}MnO_3$	$La_{0.8}Ca_{0.15}Ag_{0.05}MnO_3$
Crystal Structure	Orthorombic	Orthorombic
Space Group	Pnma	Pnma
a (Å)	5.483	5.458
b (Å)	7.760	7.745
c (Å)	5.455	5.501
V ($Å^3$)	232.15	232.61
<D>	22	24
<Mn-O>	1.969	1.966
<Mn-O-Mn>	154.84	156.08
χ^2	1.06	1.09

Tabel 3. Atomic position, wyckoff position and occupancy for $La_{0.8}Ca_{0.2}MnO_3$

Atom	Wyckoff position	Occupancy	x	y	z
La	4c	0.8	0.482	0.25	0.003
Ca	4c	0.2	0.482	0.25	0.003
Mn	4a	1	0	0	0
O1	8d	1	0.226	0.032	0.269
O2	4c	1	0.502	0.25	0.564

Tabel 4. Atomic position, wyckoff position and occupancy for $La_{0.8}Ca_{0.15}Ag_{0.05}MnO_3$

Atom	Wyckoff position	Occupancy	x	y	z
La	4c	0.8	0.482	0.25	0.003
Ca	4c	0.15	0.482	0.25	0.003
Ag	4c	0.05	0.482	0.25	0.003
Mn	4a	1	0	0	0
O1	8d	1	0.226	0.032	0.269
O2	4c	1	0.502	0.25	0.564

Figure 1 shows the XRD diffraction pattern of the $La_{0.8}Ca_{0.2}MnO_3$ and $La_{0.8}Ca_{0.15}Ag_{0.05}MnO_3$ samples sintered at 800 °C. The strongest peaks (121) for both samples are shown in figure 2. In the XRD results, both samples have the same diffraction pattern. There is no change in crystal structure when $La_{0.8}Ca_{0.2}MnO_3$ is doped with monovalent Ag ion. There is no change in structure after doping is indicated by the unchanged position of the wickoff when the material $La_{0.8}Ca_{0.2}MnO_3$ is doped by Ag ions (we can see details in table 3 and table 4). From the result of XRD data processing using the Rietvield refinement method, it is known that both samples $La_{0.8}Ca_{0.2}MnO_3$ and $La_{0.8}Ca_{0.15}Ag_{0.05}MnO_3$ having an orthorombic structure with the position of each atom is shown in table 2. Detailed images of the structure of these two materials are shown in figure 3 and figure 4.

Table 2 presents the result of refinenement XRD data using High Score Plus. The result showed that if Ca ions are substituted by the Ag ion, there is an increase in lattice parameter and cell volume when the concentration is 0.05. This can also be seen in the diffraction pattern in figure 2. The figure shows the diffraction peak shift to a smaller angle for the dopped sample, which indicates an increase in the lattice parameters. The increase in lattice parameters and cell volume may be due to the radius of the Ag^+ ion which is greater than the radius of Ca^{2+} ion (see table 1). These results are the same as the results obtained by Lakshmi and Reddy [7]. Changes in lattice parameters also identify distortions in the crystal structure according to what has been obtained from previous study [8].

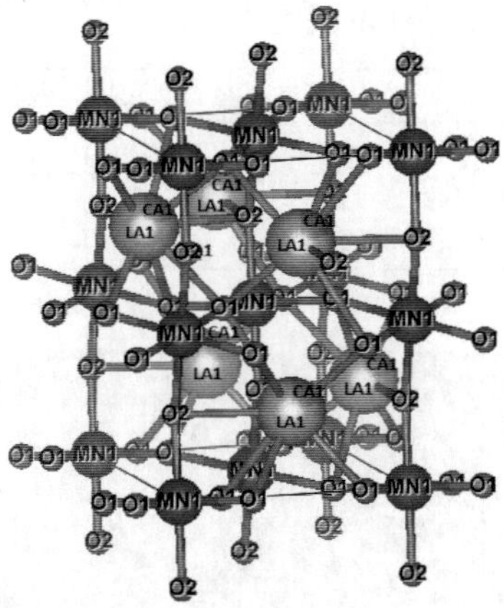

Figure 3. Crystal structure of the $La_{0.8}Ca_{0.2}MnO_3$ material

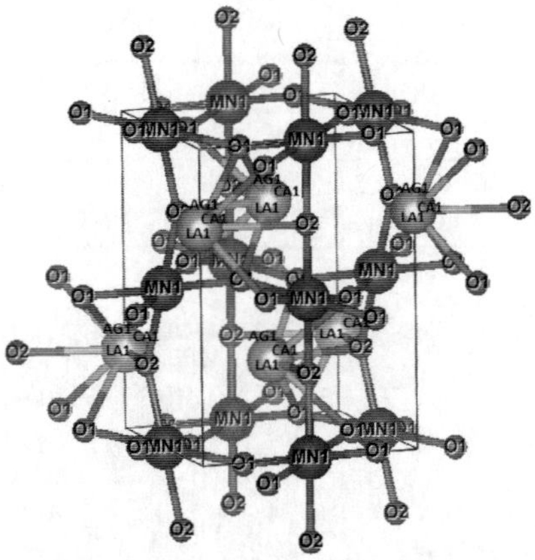

Figure 4. Crystal structure of the $La_{0.8}Ca_{0.15}Ag_{0.05}MnO_3$ material

The presence of distortion in the $La_{0.8}Ca_{0.2}MnO_3$ material is also seen from the FTIR results as previously examined by Shinde *et al.* [9] and Li *et al.* [10]. When Ca doped La on $La_{0.8}Ca_{0.2}MnO_3$ material, there was an absorption band at 600 cm^{-1}, which attributed to the metal oxide stretching vibration that can change the length and angle of the Mn–O–Mn bond, but the vibration of these changes are not detected in the FTIR spectrum which usually appears at 200 cm^{-1} [9,10]. Table 4 shows changes in length and angle of the Mn-O-Mn bonds after being doped by silver. It is possible that there was a change in the absorption band when the $La_{0.8}Ca_{0.2}MnO_3$ material was doped by the Ag ion. However that is only based on the results of the XRD data, so that it cannot be explained in detail.

Changes in bond distance and bond angle of Mn-O-Mn greatly affect the double exchange interaction, which will cause changes in electrical properties of the samples. Based on the bandwidth theory, this interaction has a strong correlation with the conduction bandwidth (W) shown in equation (1):

$$W = \frac{\cos[(\pi-(\theta_{mn-O-Mn}))]}{d_{Mn-O-Mn}} \quad (1)$$

with $d_{Mn-O-Mn}$ represents the length of Mn-O-Mn bond and $\theta_{mn-O-Mn}$ represents the angle of Mn-O-Mn bond [3]. Electron transport will be more difficult with the greater length and angle of the Mn-O-Mn bond, which will affect the weakening of the double exchange interactions [3,8]. From table 2 it can also be seen that when Ca ions are substituted by Ag ions with a concentration of 0.05, the distance of the Mn-O-Mn bond becomes smaller and the bond angle becomes larger (close to 180°) which results in stronger double exchange interaction so electron transport becomes easier.

The surface morphology of $La_{0.8}Ca_{0.2}MnO_3$ and $La_{0.8}Ca_{0.15}Ag_{0.05}MnO_3$ materials are shown in figure 5a and figure 5b. The grain cannot be clearly seen from the SEM results of $La_{0.8}Ca_{0.2}MnO_3$ sample because of the limitation of the magnification of the SEM used, which is only able to enlarge images up to 20,000 times. However logically, substituting 0.05 Ag to the AE site may cause greater grain growth [7]. Increasing grain size due to doping can reduce pores and defects, which can improve the electrical properties of the samples [11]. The average grain size of $La_{0.8}Ca_{0.2}MnO_3$ without and with doping monovalent Ag ions is 105.2 nm and 260 nm, respectively.

The success of material engineering can be seen in figure 6, which is the result of EDX analysis for each material, $La_{0.8}Ca_{0.2}MnO_3$ and $La_{0.8}Ca_{0.15}Ag_{0.05}MnO_3$. The EDX analysis results are important to confirm the purity of the material that has been made. The results show that the Ag ion is substituted the perovskite manganite structure (seen in the red circle in figure 6) so that it can be said that Ag successfully entered $La_{0.8}Ca_{0.2}MnO_3$ material. Table 5 shows the composition obtained from EDX analysis and the results of the calculations. The table shows that the composition of the EDX analysis results for both samples is close to the calculation results, with the EDX analysis composition generally smaller than the calculation results because the composition obtained from EDX analysis

Table 5. Composition of $La_{0.8}Ca_{0.2}MnO_3$ and $La_{0.8}Ca_{0.15}Ag_{0.05}MnO_3$ obtained by EDX analysis and manual calculation.

	At% (EDX analysis)					At% (calculation)				
	La	Ca	Ag	Mn	O	La	Ca	Ag	Mn	O
$La_{0.8}Ca_{0.2}MnO_3$	14.08	3.37	0	15.87	66.68	16	4	0	20	60
$La_{0.8}Ca_{0.15}Ag_{0.05}MnO_3$	13.72	2.68	0.71	17.53	65.36	16	3	1	20	60

(a) (b)

Figure 5. Scanning electron microscope image of (a) $La_{0.8}Ca_{0.2}MnO_3$ and (b) $La_{0.8}Ca_{0.15}Ag_{0.05}MnO_3$.

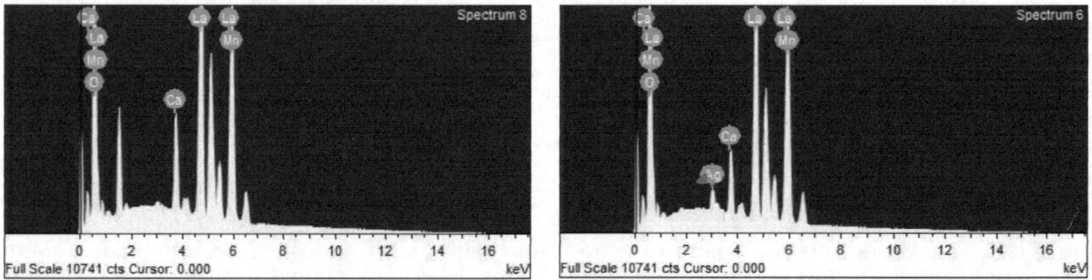

(a) (b)

Figure 6. EDX spectrum of (a) $La_{0.8}Ca_{0.2}MnO_3$ and (b) $La_{0.8}Ca_{0.15}Ag_{0.05}MnO_3$

is semiquantitative. The table also shows that when Ag is doped on $La_{0.8}Ca_{0.2}MnO_3$ material with a concentration of 0.05, the composition of Ag in perovskite material is smaller than the composition of the calculation result. This is due to the possibility that the Ag ion evaporates during the heating processes i.e. calcination process and sintering process. This phenomenon has also been reported before by Lakhsmi and Reddy [7].

4. Conclusions

We have succeeded in synthesizing $La_{0.8}Ca_{0.2-x}Ag_xMnO_3$ with x = 0 and 0.05 using the sol-gel method. XRD data processing using Rietveld refinement produced orthorombic crystal structures for both samples. Substitution of 0.05 Ag into the Ca site did not change the crystal structure but changed the lattice parameters of perovskite manganite material. The monovalen Ag ion substitution into Ca ion can also shorten the bond distance between Mn-O and increase the angle of Mn-O-Mn bond. According to bandwidth theory when the length of Mn-O-Mn bond gets smaller and the angle of the Mn-O-Mn bond increases, the double exchange interaction will be stronger, which will affect electron transport of the material. From the SEM results it can be seen that the grain size increases when $La_{0.8}Ca_{0.2}MnO_3$ is doped with Ag. EDX results indicate that Ag has been substituted in the $La_{0.8}Ca_{0.2}MnO_3$

material. The composition of EDX analysis is close to the results of using manual calculations, and both samples can be said to have high purity in the absence of other elements besides material elements.

Acknowledgements
This work was supported by PITTA Grant 2017 with contract number 633/UN2.R3.1/HKP.05.00/2017

References
[1] Yadav R, Anshul A and Shelke V 2011. *J. Mater. Sci.: Mater. Electron.* **22** 1173–80
[2] Zener C 1951 *Phys. Rev.* **82** 403–5
[3] Manjunatha S O, Rao A, Subhashini and Okram G S 2015 *J. Alloys Compd.* **640**154–61
[4] Keshri S, Biswas S and Wiśniewski P 2016 *J. Alloys Compd.* **656** 245–52
[5] Takehira K 2009 *J. Nat. Gas Chem.* **18** 237–59
[6] Shannon R D 1976 *Acta Cryst. A* **32** 71–767
[7] Lakshmi Y K and Reddy P V 2010 *Solid State Sci.* **12**, 1731–40
[8] Kurniawan B, Winarsih S, Ramadhan M R, Naomi A and Laksmi W 2017 *AIP Conf. Proc.* **1862** 030054
[9] Shinde K P, Oh S S, Baik S K, Kim H S, Sinha B B and Chung K C 2012 *J. Korean Soc.* **61** 2000–4
[10] Li K, Cheng R, Wang S and Zhang Y 1998 *J. Phys.:Condens. Matter* **10** 4315–22
[11] Jin F, Zhang H, Chen X, Liu X and Chen Q 2017 *J. Sol-Gel Sci. Technol.* **82** 193–200

IOP Conf. Series: Materials Science and Engineering **496** (2019) 012005 doi:10.1088/1757-899X/496/1/012005

Chitosan nanoparticles on a natural zeolite as an efficient adsorbent for Congo red

Y Yulizar, T Utari, D O B Apriandanu and Y R P Putri

Department of Chemistry, Faculty of Mathematics and Natural Sciences (FMIPA), Universitas Indonesia, Depok 16424, Indonesia

Corresponding author: yokiy@sci.ui.ac.id

Abstract. Congo red is a toxic synthetic dye that cannot be readily degraded by conventional methods, thus posing a risk to the environment. The increased use of this dye is mainly due to developments in the textile industry. A possible way to reduce the Congo-red waste is by performing the adsorption processes; therefore, in this study, we used a natural zeolite from Bayah and modified it using chitosan nanoparticles (Na-zeolite@chitosan) to increase its adsorption capacity. The obtained Na-zeolite@chitosan material, which was able to efficiently adsorb Congo red, was further characterized by ultraviolet–visible (UV–Vis) and Fourier-transform infrared (FTIR) spectroscopy as well as by transmission electron microscopy (TEM). A maximum of 98.019% of the Congo red was adsorbed at a concentration of 800 ppm and a pH of 5 for 60 min (the adsorption capacity was 0.00428 mmol/g). Our results exhibit that Congo red adsorption on Na-zeolite@chitosan follows the Freundlich adsorption isotherm and can be described using a pseudo-second-order kinetic model.

Keywords: natural zeolite, chitosan, adsorbent, congo red

1. Introduction

Today, textiles have become a major export commodity in several countries; however, their production exhibits a negative effect on the environment because it generates nondegradable, toxic, and stable synthetic dyes [1]. One of these dangerous dyes is Congo red, which can be barely decomposed in nature. This compound exhibits a complex aromatic chemical structure and is therefore physicochemically and thermally stable [2], and it exhibits carcinogenic properties [3].

Liquid dye wastes are treated by chemical, physical, and biological methods such as adsorption, coagulation, advanced oxidation, ozonization, filtration through membranes, and liquid–liquid extraction. Adsorption is an effective, efficient, and economical method for eliminating the contaminants from water [4–6]. Some of the adsorbents used to perform Congo-red adsorption from contaminated water are zeolites, montmorillonites, bentonites, chitosan hydrogels, fly ashes, activated carbon, and clay materials [7–10].

Zeolites exhibit unique ion-exchange and adsorption properties, and they have been extensively used in water-treatment processes, ammonia purification, and the elimination of heavy metals, radioactive substances, organic material, microorganisms, and anions during the desalination of seawater [11–13]. However, they should be modified to increase their adsorption capacity. This has been achieved by modifying them using hexadecyltrimethylammonium bromide (HDTMA-Br) [14], goethite [15], gold nanoparticles [16], hexamethylenediamine (HMDA) [17], and chitosan [18]. Among these modifiers, chitosan exhibits several advantages; it is nontoxic, biodegradable, and antibacterial [19].

Chitosan is a chitin deacetylation biopolymer that can be used as an adsorbent for heavy metals and dyestuff such as Congo red [20]. Other dyestuff adsorbents include cetyltrimethylammonium bromide

(CTAB), which also adsorbs Congo red [2] and chitosan nanoparticles, which can be used as adsorbents for acid orange and acid red [21, 22].

To the best of our knowledge, zeolites that were modified using chitosan nanoparticles have not been used as Congo-red adsorbents until now. In this study, we modified a zeolite with nanochitosan and tested it as a Congo red adsorbent by monitoring several parameters such as pH, reaction time, and dye concentration. We also determined the adsorption isotherm and studied the kinetics of the adsorption process.

2. Experimental

2.1. Materials and methods
Natural zeolite was obtained from Bayah (Lebak, Banten, Indonesia). NaOH, HCl, NaCl, AgNO3, CH3COOH, and Congo red ($C_{32}H_{22}N_6Na_2O_6S_2$) were purchased from Merck (Germany). Chitosan was collected from Bogor, West Java, Indonesia. Double-distilled water was used as a solvent.

2.2. Activation of natural zeolite
Natural zeolite was physically activated by washing in double-distilled water at 70 °C for 1 h. For chemical activation, NaOH and HCl were added to the natural zeolite, and the zeolite cation was uniformed by adding 1 M NaCl. The mixture was further stirred at 70 °C for 6 h and was then precipitated for 12 h. The precipitate was finally dried at 105 °C. The chloride ions of the zeolite were eliminated by washing in double-distilled water.

2.3. Synthesis of chitosan nanoparticles
Synthesis of chitosan nanoparticles was conducted by adding NH_3 and 0.01 M CH_3COOH solution to chitosan powder. The mixture was further stirred for 2.5 h. The formation of chitosan nanoparticles was detected by the appearance of a white colloid.

2.4. Modification of the zeolite with chitosan nanoparticles
The zeolite was mixed with the chitosan-nanoparticle solution. This mixture was then stirred for 2.5 h, centrifuged to form the precipitate, and dried using N_2.

2.5. Na-zeolite@chitosan nanoparticles as adsorbents for Congo red
The obtained material was tested by adding 10 mL of 800 ppm Congo red to 0.1 g Na-zeolite@chitosan nanoparticles at pH 5 and stirring the mixture several times for 60 min. Then, ultraviolet–visible (UV–Vis) spectroscopic experiments were performed between 200 and 800 nm. The kinetic model for Congo red adsorption was determined by monitoring the decrease in absorbance at 499 nm [23]. In addition, Langmuir and Freundlich isotherms were fitted to the experimental data to describe the adsorption processes at the equilibrium point.

2.6. Characterization
A Shimadzu 2600 UV–Vis spectrophotometer was used to obtain the absorption spectra describing the Na-zeolite@chitosan nanoparticle–Congo red adsorption activity. Identification of the functional groups present in the natural zeolite was accomplished by Fourier-transform infrared (FTIR) spectrometry (PerkinElmer) in the range of 4,000–400 cm^{-1}. A JEM 1400 transmission electron microscope was used to determine the structure and size of the Na-zeolite@chitosan nanoparticles.

3. Results and discussion
The uniformity of natural zeolite cation aims to homogenize the cations in zeolite pores. It can facilitate ion exchange in the Congo red adsorption process. The results of the FTIR characterization are presented in figure 1a. Both Na-zeolite@chitosan nanoparticles and Na-zeolite show different peaks at 3,624 cm^{-1}, corresponding to N–H stretching in chitosan. There is also a signal corresponding to the –OH group at 3,422 cm^{-1}. These are the two main functional groups of chitosan. The functional groups $-NH_2$ and –OH give the Na-zeolite@chitosan nanoparticles their adsorption ability. The obtained FTIR spectra show that Na-zeolite has been successfully modified with chitosan nanoparticles.

This result was confirmed by transmission electron microscopy (TEM) characterizations, as shown

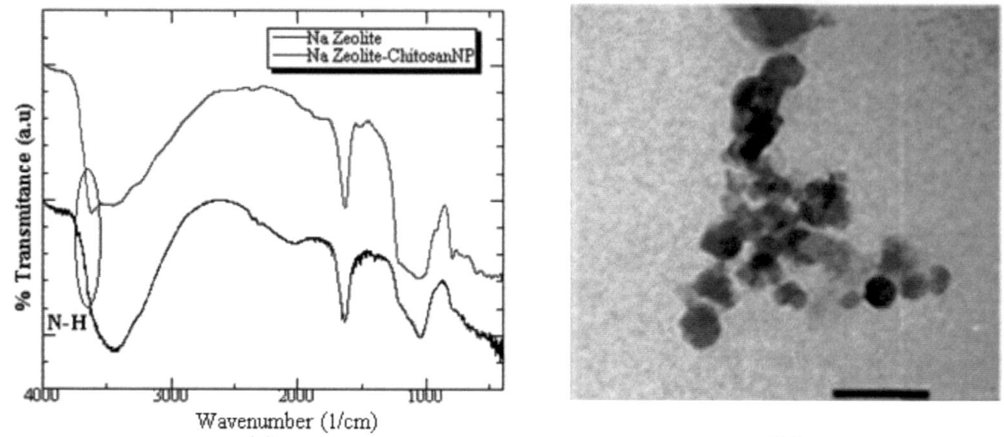

Figure 1. (a) FTIR spectra of Na-zeolite and Na-zeolite@chitosan nanoparticles, (b) TEM image of chitosan nanoparticles.

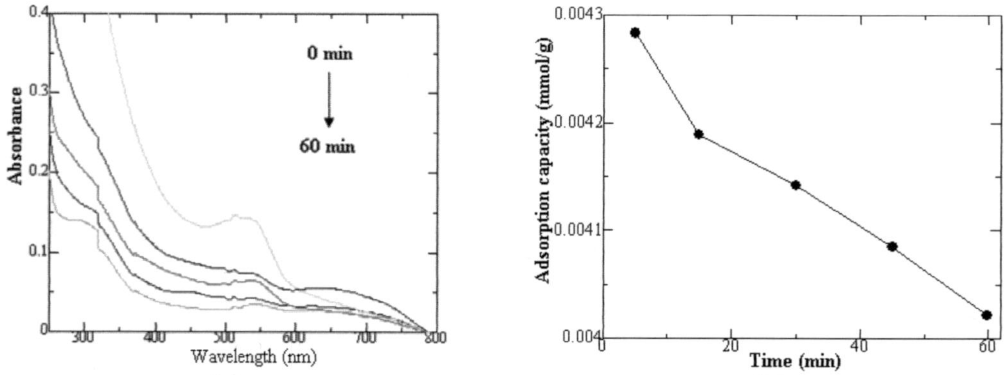

Figure 2. (a) UV–Vis absorption spectra as a function of time for the reduction of Congo red using Na-zeolite@chitosan nanoparticles, (b) Adsorption capacity of the Na-zeolite@chitosan nanoparticles measured during a period of 60 min.

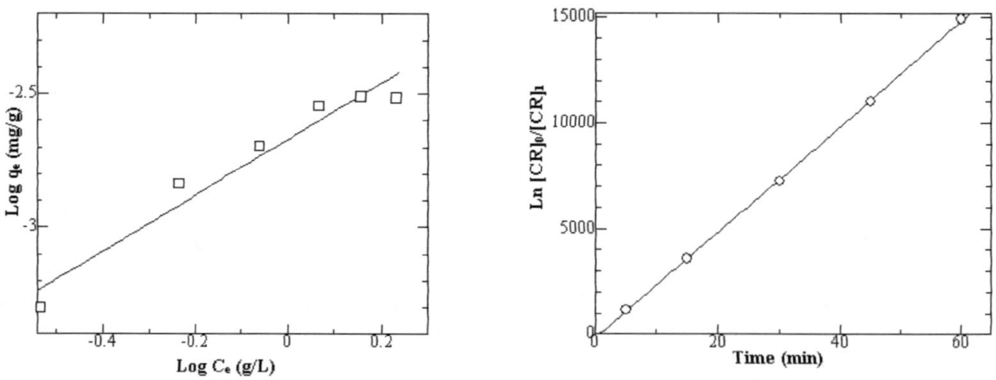

Figure 3. (a) Freundlich's adsorption isotherm, (b) Plot of ln $[CR]_0/[CR]_t$ versus reaction time.

in figure 1b, which were carried out to determine the particle shape and size. Figure 1b shows the shapes and sizes of the Na-zeolite@chitosan nanoparticles. As can be seen, they have an irregular and almost spherical shape, and their size varies from 20 to 40 nm. These results show that the synthesized adsorbent is composed of chitosan nanoparticles. The presence of these nanosized structures leads to an increase in the number of active sites, thereby improving the adsorption properties of the material.

Congo red was adsorbed on the Na-zeolite@chitosan nanoparticle material, and the adsorption activity was monitored by following the decrease in absorbance of the peak observed at 499 nm. The relationship between UV–Vis absorption spectra and time is shown in figure 2a. As can be seen, absorbance decreases at longer times, indicating the adsorption of Congo red. The Na-zeolite@chitosan nanoparticles absorbed 98.019% of Congo red at pH 5 after 60 min. The adsorption capacity was 0.00428 mmol/g at the optimum time of 5 min, as shown in figure 2b, and the adsorption process occurred because of electrostatic interactions between the SO_3^- group of Congo red and the NH_3^+ group of nanochitosan.

Adsorption isotherms were calculated in order to determine the adsorption mechanism for the studied system. We found that Congo-red adsorption on Na-zeolite@chitosan followed Freundlich's adsorption isotherm with an R^2 value of 0.976. The corresponding graphic is shown in figure 3a. The kinetic model for Congo red adsorption on the Na-zeolite@chitosan material was determined by plotting ln $[CR]_0/[CR]_t$ versus the reaction time, as shown in figure 3b. The results show that Congo-red adsorption follows a pseudo-second-order behavior with an R^2 value of 0.999.

4. Conclusions

A Na-zeolite@chitosan nanoparticle material with a particle size of 20–40 nm was successfully synthesized. The modified zeolite was applied as a Congo-red adsorbent, exhibiting an adsorption ability of 0.00428 mmol/g and achieving 98.019% Congo-red adsorption (for a concentration of 800 ppm) at pH 5 after 60 min. Congo-red adsorption can be described by Freundlich's isotherm and a pseudo-second-order kinetic model.

Acknowledgements

The authors would like to thank Universitas Indonesia for funding this research through PITTA Grant with contract number 700/UN2.R3.1/HKP.05.00/2017 and Einago (www.einago.com) for the English language review.

References

[1] Gupta V K, Mittal A, Khrisnan L and Gajbe V 2004 *Sep. Purif. Technol.* **40** 87–96
[2] Chatterjee S, Lee D S, Lee M W and Wooa S H 2009 *Bioresour. Technol.* **100** 2803–9
[3] Mall I D, Srivastava V C, Agarwal N K and Mishra I M 2005 *Chemosphere* **61** 492–501
[4] Jiuhui Q 2008 *J. Environ. Sci.* **20** 1–13
[5] Sen T K, Afroze S and Ang H M 2011 *Water Air Soil Pollut.* **218** 499–515
[6] Yao Z, Wang L and Qi J 2009 *Clean: Soil Air Water* **37** 642–8
[7] Li C, Dong Y, Wu D, Peng L and Kong H 2011 *Appl. Clay Sci.* **52** 353–7
[8] Yermiyahu Z, Lapides I and Yariv S 2003 *Clay Miner.* **38** 483–500
[9] Bulut E, Ozacar M and Sengil I A 2008 *J. Hazard. Mater.* **154** 613–22
[10] Vimonses V, Lei S, Jin B, Chow C W K and Saint C 2009 *Chem. Eng. J.* **148** 354–64
[11] Shoumkova A 2011 *Bulletin of the Australian Institute of High Energetic Materials* **2** 10–70
[12] Zhenyan L, Yunzhuang L and Jiaxin Z 1998 *Sol. Energy Mater. Sol. Cells* **52** 45–53
[13] Wang S and Peng Y 2010 *Chem. Eng. J.* **156** 11–24
[14] Naghash A and Ejhieh A N 2015 *J. Ind. Eng. Chem.* **31** 185–91
[15] Kugbe J, Matsue N and Henmi T 2009 *J. Hazard. Mater.* **164** 929–35
[16] Yulizar Y, Kadja G T M and Yudiana A 2016 *AIP Conf. Proc.* **1729** 020043
[17] Alver E and Metin A 2012 *Eng. Chem. J.* **200–202** 59–67
[18] Zhi J, Wang Y and Luo G 2005 *React. Funct. Polym.* **65** 249–57
[19] Miao J, Chen G H and Gao C J 2009 *Desalination* **181** 173–83
[20] Wang L and Wang A 2007 *J. Hazard Mater.* **147** 979–85
[21] Cheung W H, Szeto Y S and McKay G 2009 *Bioresour. Technol.* **100** 1143–8
[22] Du W L, Xu Z R, Han X Y, Xu Y L and Miao Z G 2008 *J. Hazard Mater.* **153** 152–6
[23] Bhagwat U O, Wu J J, Asiri A M and Anandan S 2017 *J. Photochem. Photobiol. A: Chem.* **346** 559–69

IOP Conf. Series: Materials Science and Engineering **496** (2019) 012006 doi:10.1088/1757-899X/496/1/012006

Radiochromic film label made from chitosan and *Taraxacum officinale* leaf extract

P Widodo, Y Priasetyono, M Harahap, A Listyarini, D Djuhana and C Imawan

Department of Physics, Faculty of Mathematics and Natural Sciences (FMIPA) Universitas Indonesia, Depok 16424, Indonesia

Corresponding author: cuk.imawan@sci.ui.ac.id

Abstract. Monitoring the dose of gamma-ray radiation is important in the healthcare field. A simple indicator for dose monitoring is provided by radiochromic labels. This paper reports the preparation and characterization of radiochromic labels prepared from the natural dye of *Taraxacum officinale* (TO) leaf extract mixed with a chitosan matrix. The radiochromic label was fabricated in the form of a colored plastic biodegradable film. The film structure was characterized by Fourier transform infrared spectroscopy. The color of the film label was quantitatively measured with a UV-Vis spectrophotometer at 433 and 630 nm. The ability of the labels to detect gamma-ray radiation was evaluated for a dose range from 0 to 30 kGy. Gamma-ray radiation caused changes in the color intensity of the film labels. Spectral analysis revealed that the absorbance increased when the film label was exposed to gamma-ray radiation. The results of these experiments show that the radiochromic film label made from TO leaf extract in a chitosan matrix can be used as a radiochromic indicator to determine the dose of gamma radiation and is a good candidate for use in the sterilization of health equipment.

Keywords: Radiochromic film label, Taraxacum officinale, chitosan

1. Introduction

In the industrial field, ionizing radiation is used to improve the quality of a product because it can change the physical, chemical, and biological properties of irradiated objects. Ionizing radiation is used to sterilize health products [1], inhibit budding, promote disinfestation of insects and pathogens, prolong shelf life, and sterilize foodstuffs and agricultural products [2] and modify materials by inducing polymerization [3].

The absorption dose plays an important role in the irradiation process; thus, its monitoring is essential [4]. The "dose" refers to the energy absorbed per unit mass [5]. Various dosing tools have been developed, including radiochromic films, chemical solutions, calorimeters, fluorescent systems, organic crystals, inorganic crystals, semiconductors and diamonds [6].

A radiochromic film label comprises two parts: a dye and a matrix. Several studies have reported the use of dyes as a radiation indicator; examples include leuco brilliant blue cyanide [7], calcein [8], leucocrystal violet [9], and leucomalachite green [10]. The matrix is the material used to immobilize the dyes. Materials used as a matrix include polyvinyl alcohol, films of PVA [8], polyvinyl butyryl, films of PVB [9], solutions, ethanol as a solvent [11], and gels, agarose as gelling agent [12].

In the present study, we report the preparation of a radiochromic film label made from *Taraxacum officinale* (TO) leaf extract immobilized with a chitosan matrix. The TO extract is a biodegradable and nontoxic natural dye. Chitosan is a copolymer derived from the diacetylene chitin process; it is also biodegradable and nontoxic and can form a film [13].

2. Materials and methods

2.1. Materials
The materials used in these experiments were TO leaves extract, ethanol (J.T. Baker), chitosan in powder form (Pusat Penelitian Bioteknologi dan Bioindustri (PPBB), Bogor, Indonesia; degree of deacytelation >80%), and glacial acetic acid (Merck).

2.2. TO leaf extraction
TO leaves were washed with tap water and then dried in an oven at 40 °C for 72 h. The dried leaves then milled using dry milling until they became powder. TO leaf extraction was performed by mixing 40 g of leaf powder with 200 mL of 100% ethanol, and then stirred using a magnetic stirrer at 300 rpm for 24 h at room temperature. The solution was filtered using Whatman No. 1 filter paper. The pH of the resultant extract was 6.2.

2.3. Film label preparation
The chitosan matrix was prepared by dissolving 4 g of chitosan powder into 200 mL of 1% (w/v) acetic acid and stirring the resultant mixture at 300 rpm for 3 h at 55 °C using a magnetic stirrer. The chitosan–TO film label was prepared by mixing 10 mL of TO extract solution with 50 mL of chitosan solution and stirring the resultant mixture for 15 min at room temperature using a magnetic stirrer. The chitosan–TO solution was poured onto a 15 × 15 cm^2 acrylic board and left for 48 h at room temperature to form the film.

2.4. Collection of UV-Vis absorption spectra
Absorption spectra were recorded for the TO extracts, chitosan film label, nonirradiated chitosan–TO film, and the chitosan–TO film label, which were tested for exposure to gamma-ray radiation over the dose range 3–30 kGy. The measurements were performed using a UV-Vis spectrophotometer (Thermo Genesys 10S) in the wavelength range 200–800 nm.

2.5. Film-label irradiation tests
The chitosan–TO film-label response to exposure to gamma-ray radiation was examined using a Gammacell Irradiator 220 (Nordon, Canada), which had a dose rate of 5.7 kGy/h. The dose rate of the Co-60 source in the Gammacell Irradiator 220 was calibrated using a Fricke dosimeter. The chitosan–TO test label, which was 2 × 3 cm^2 in area, was placed on a special holder. The film label was then exposed to gamma-ray radiation with doses of 3, 6, 12, 15, 21, 27, and 30 kGy at chamber temperature.

2.6. FTIR absorption spectroscopy
Fourier transform infrared (FTIR) absorption spectra were collected for the TO extract solution, chitosan–TO film label, and the chitosan–TO film label irradiated at a dose of 30 kGy. Before the spectra were recorded, the TO extract solution was evaporated until it exhibited an opaque color; subsequently, the solution was dripped onto the sample holder. Neither the nonirradiated chitosan–TO film label nor the chitosan–TO film label irradiated at a dose of 30 kGy were subjected to a special treatment prior to measurement. Measurements were conducted using a double-beam spectrophotometer (Thermo Nicolet iS5) in the attenuated total reflectance (ATR) Fourier transform mode. The FTIR absorption spectra were recorded for 500–4000 cm^{-1}.

3. Results and discussion

3.1. Absorption spectra of TO extracts and chitosan film labels
The TO extract solution had a pH of 6.2 and was dark green. Figure 1 shows the absorbance spectra of the TO extract solution, chitosan film label, and the chitosan–TO film label. The absorbance spectrum

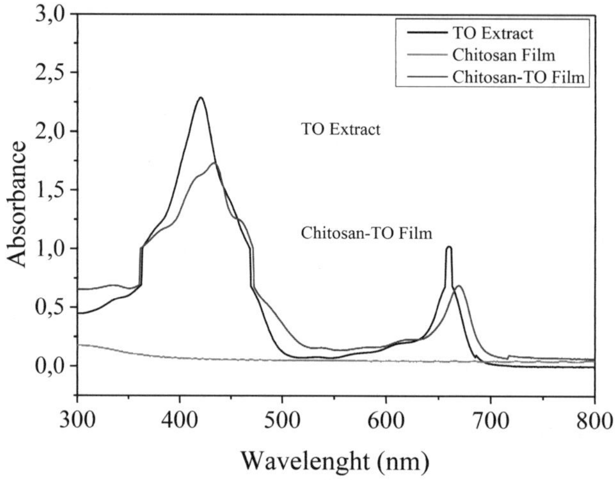

Figure 1. Absorbance spectra of the TO extract, chitosan film, and the chitosan–TO film.

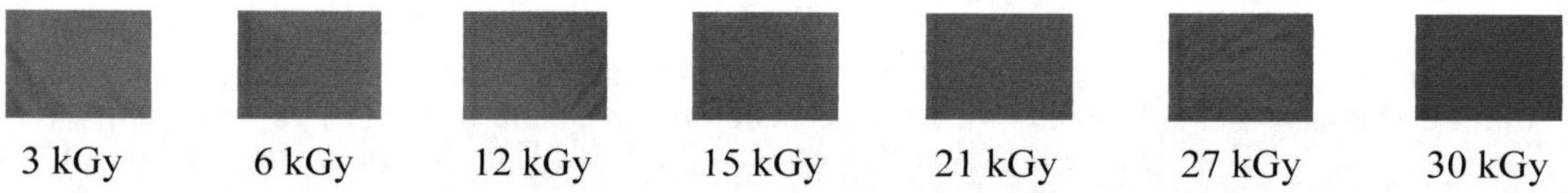

Figure 2. The color changes of chitosan–TO film labels after irradiation.

of the TO extract solution exhibited maximum absorption peaks at 420 and 660 nm. The peak at about 420 nm corresponds to the absorption of purple light and the reflection of yellow-green light, whereas the peak around 660 nm corresponds to the absorption of red light and the reflection of green-blue light. The combination of green-yellow and green-blue colors reflection produces a dark-green color, as shown in figure 1. The chitosan film label has a flat absorption spectrum because of its transparent and colorless nature. The absorbance spectrum of the chitosan–TO label exhibited two absorption peaks at wavelengths of 433 and 670 nm. The absorption peak of the chitosan–TO film label shifted compared with the absorbance peak of the TO extract. The peak at about 433 nm corresponds to blue absorption and yellow reflection, whereas the peak around 670 nm corresponds to red absorption and green-blue reflection. The combination of yellow and green-blue colors reflection produces a lighter green color of the chitosan–TO film label compared with that of the TO extract.

3.2. Absorption spectra of chitosan–TO film labels at various radiation doses

The chitosan–TO film labels respond to exposure to gamma-ray radiation. The color of the chitosan–TO film labels darkened in proportion to the increased absorbed radiation dose. The color changes of chitosan label colors irradiated with doses of 3–30 kGy are shown in figure 2. The film labels irradiated with a dose of 30 kGy exhibit a darker green color compared with other labels irradiated with lower doses.

The chitosan–TO film label response is also shown in the irradiated-film-label absorption spectra. The UV-Vis absorbance spectra of the chitosan–TO film irradiated at doses from 3 to 30 kGy show absorption peaks at 433 nm and 670 nm (figure 3). No absorption peak shift is observed in the absorption spectra. The absorption peak increases inversely proportional to increasing radiation dose. The higher the absorption peak, the darker the color appears. Soliman *et al.* [14] reported similar results wherein the film color intensity darkened in proportion to increasing radiation dose.

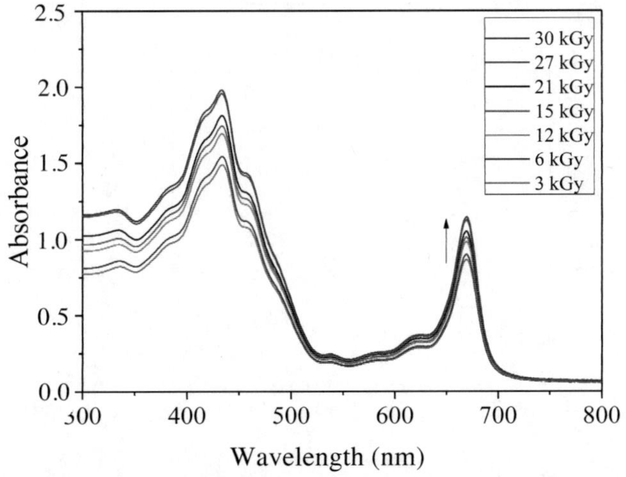

Figure 3. UV-Vis absorption spectra of irradiated chitosan–TO-based film labels.

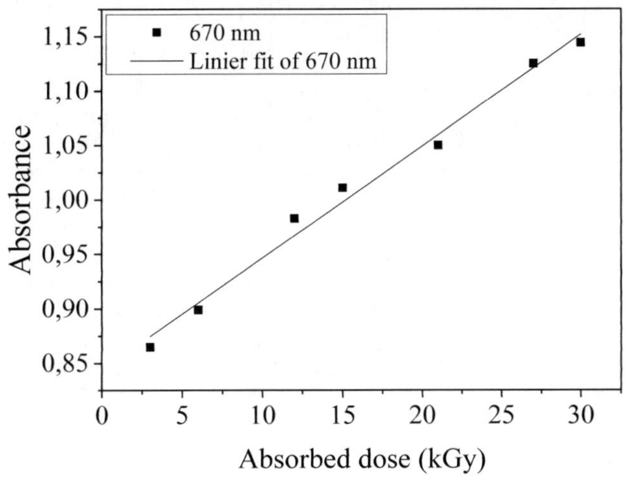

Figure 4. Absorbance of chitosan–TO film labels at a wavelength 670 nm after the film labels were subjected to radiation doses from 3 to 30 kGy.

3.3. Response of the chitosan–TO film label to radiation

The absorption peak representing the response of the chitosan–TO film label to the radiation dose absorbed by the film label were obtained at a wavelength of 670 nm. Figure 4 shows the absorbance response of the chitosan–TO film label to the radiation dose absorbed by the film label in the 3–30 kGy range. The absorbance point value on the graph increases nonlinearly in proportion to the increased radiation dose absorbed by this film label. The points on the graph can be approximated using the linear function described in equation (1):

$$Absorbance = 0.843 + 0.010(Absorbed\ dose(kGy)) \quad (1)$$

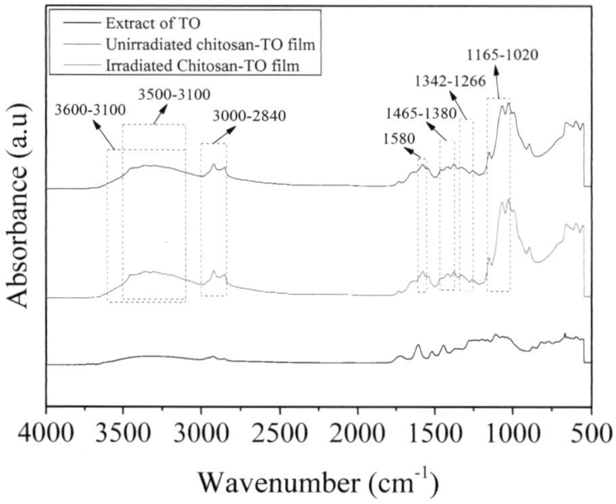

Figure 5. FTIR spectra of the TO extract, a nonirradiated chitosan–TO film label, and a chitosan–TO film label irradiated at 30 kGy.

Equation (1) has a coefficient of determination value of 0.986, which shows that the absorbance value is 0.986 dependent on the absorbed dose value.

3.4. FTIR spectra of the chitosan–TO film label

The chemical structure of the chitosan–TO film label before and after irradiation was characterized via FTIR spectroscopy. Figure 5 shows the FTIR spectrum of the TO extract, the nonirradiated chitosan–TO film label and the chitosan–TO film label irradiated at a dose of 30 kGy. There is no TO extract peak observed in the FTIR spectrum of the chitosan–TO film labels because the volume of TO extract in the chitosan–TO film labels is much smaller than the chitosan matrix volume. In the spectra of the nonirradiated chitosan–TO film labels and the chitosan–TO film labels irradiated at 30 kGy, the O–H stretching vibration is observed in the wavenumber range 3600–3100 cm^{-1}, which overlaps with an N–H stretching vibration at 3500–3100 cm^{-1} [15]. In the range of 3000–2840 cm^{-1}, the stretching vibration of C–H is experienced [16,17]. The N–H bending absorption band also appears at 1580 cm^{-1} [15]. The C–H bending, C–N stretching, and C–O stretching bands appear in the wavenumber ranges of 1465–1380 cm^{-1}, 1342–1266 cm^{-1}, and 1165– 1020 cm^{-1}, respectively. No substantial change was observed in the spectrum of the chitosan–TO film label before and after irradiation at a dose of 30 kGy.

4. Conclusions

Radiochromic film labels prepared from TO leaf extract on a chitosan matrix were used as radiochromic indicator to determine the dose of gamma radiation in the dose range from 3 to 30 kGy. The color of the chitosan–TO film label developed was dark green. The intensity of the color of the chitosan–TO film labels darkened in proportion with the radiation doses. Changes in the color intensity of chitosan–TO film labels were confirmed by increases in the intensity of the absorption peaks at 433 nm and 670 nm in proportion to the radiation doses. The absorbance peaks can be approximated using equation (1) with a coefficient of determination of 0.986 for chitosan-TO film labels.

Acknowledgements

This research was completed with a funding by the PITTA (Publikasi Internasional Terideks untuk Tugas Akhir) Grant in 2017 from Universitas Indonesia. The authors would like to thank Einago (www.einago.com) for the English language review.

References

[1] Porto K M B G, Napolitano C M and Borrely S I 2018 *Radiat. Phys. Chem.* **142** 23–8
[2] Ehlermann D A E 2016 *Radiat. Phys. Chem.* **129** 10–12

[3] Lee H, Ryu J, Kim M, Im S S, Kim I S and Sohn D 2016 *Radiat. Phys. Chem.* **125** 160–4
[4] Kuntz F and Strasser A 2016 *Radiat. Phys. Chem.* **129** 46–9
[5] Canadian Nuclear Safety Commission (CNSC) 2011 *Introduction to Dosimetry* (Ottawa: CNSC)
[6] McLaughlin W L and Desrosiers M F 1995 *Radiat. Phys. Chem.* **46** 1163–74
[7] Ghafar S M and El-Ahdal M A 2015 *Dyes. Pigm.* **114** 273–7
[8] Aldweri F M, Rabaeh K A and Kefaia N A 2017 *Radiat. Phys. Chem.* **139** 1–4
[9] Soliman Y S and Abdel-Fattah A A 2013 *Radiat. Meas.* **49** 1–6
[10] Soliman Y S, Basfar A A and Msalam R I 2014 *Radiat. Meas.* **62** 45–51
[11] K A Rabaeh, A A Moussa, A A Basfar, R I Msalam 2013 *J. Ejmp.* 05. 006
[12] J I Shieh, K Y Cheng, H L Shyu, Y C Yu, L L Hsieh 2014 *J. Radphyschem.* 03. 003
[13] Muxika A, Etxabide A, Uranga J, Guerrero P and de la Caba K 2017 *Int. J. Biol. Macromol.* **105** 1358–68
[14] Soliman Y S, Bayomi A M M, Abdel-Fattah A A and Abdel-Khalek A A 2014 *Sens. Actuators B* **200** 109–16
[15] Takara E A, Marchese J and Ochoa N A 2015 *Carbohydr. Polym.* **132** 25–30
[16] Habiba U, Siddique T A, Talebian S, Lee J J L, Salleh A, Ang B C and Afifi A M 2017 *J. Carbohydr. Polym.* **117** 32–9
[17] Ji X, Li B, Yuan B and Guo M 2017 *Carbohydr. Polym.* **176** 273–80

Development study of Diffusive Gradient in Thin Film (DGT) method with binding agent Fe–Ca complex for phosphate measurement

W Wahyuni and A Saefumillah

Department of Chemistry, Faculty of Mathematics and Natural Sciences (FMIPA), Universitas Indonesia, Depok 16424, Indonesia

Corresponding author: asep-s@ui.ac.id

Abstract The inclusion of phosphorus as a phosphate compound into the aquatic system causes environmental problems, for which monitoring is required in particular on waste treatment. The Diffusive Gradient in Thin Film (DGT) technique with ferrihydrite as binding gel is commonly used for in situ phosphate measurement. Studies on Fe–Mn binary oxide adsorbent have been performed but are only able to absorb phosphate optimally at pH 3–5. In addition to ferric salts, calcium salt is an effective chemical reagent in absorbing phosphate. This research develops new technology by combining both reagent into Fe–Ca complex. The result of this study show that Fe–Ca complex has an antagonistic effect in absorbing phosphate at molar ratio of Fe–Ca 1: 1.5 to 1: 5 and has a synergistic effect when the molar ratio of Fe is increased from 1: 1 to 5: 1. The Optimum absorption occurs in the molar ratio of Fe–Ca complex 3: 1. The measurement of phosphate by spectrophotometric method of molybdenum blue known that Fe–Ca complex has wide working area that is in pH range 2.0–10.0 with optimum contact time 24 hours. Based on the results obtained Fe–Ca complex can be one alternative binding gel on phosphate measurement with DGT technique.

Keywords: diffusive Gradient in Thin Film, Fe–Ca complex

1. Introduction

Phosphorus is an important macronutrient for plants, animals and a crucial component of DNA, RNA, ATP, and phospholipids [1]. Phosphorus contained in natural waters mostly found as phosphates. The compounds are categorised as orthophosphates, condensed phosphates and phosphate which bound organically. Various anthropogenic and natural origins -namely, agricultural activity, drain water, waste from farming, and the degradation of organic compounds- lead to the diverse phosphate forms. Accelerated eutrophication may occur as a result of the phosphorus content of detergents or fertilizer in natural waters. Thus, the combined effects of excessive concentrations of phosphate in natural waters may reduce oxygen concentrations, a change in watercolour, increased turbidity and sedimentation, and a decrease in biological activity [2],

Some quantitative methods are available to determine the phosphate content in the natural water by colorimetric, inorganic phosphate by acid hydrolysis and total phosphate by per sulfate digestion [3]. In consequence of the phosphate's dynamic interaction in natural water, during storage the concentration of each individual species might be changed. The method applied to determine the in situ of the dissolved phosphate in natural water is Diffusive Gradient in Thin film (DGT) with ferrihydrite as binding agent. [4].

A simultaneous oxidation and co-precipitation technique were employed in the synthesize process of Fe-Mn binary oxide absorbent with Fe/Mn molar ratio of 6:1 to remove the phosphate. Some tests were performed to examine the kinetic and equilibrium of adsorption in batch mode. The maximal adsorption capacity was found to be 36mg.g at pH 5.6. It was discovered that adsorption of phosphate was greatly depended on pH [5].

Precipitants of Ferric–calcium complex is a cost effective option for phosphate removal, the study show that removal phosphate using Ferric–calcium complex (Fe–Ca Complex) increased up to 29.77% compared with the traditional technology using ferric salt as sole precipitant. The aim of study is to develop the alternative-binding agent with Fe–Ca Complex to applied on DGT Technique [6].

2. Material and methods

2.1. Material

N,N Methylenebisacrylamide preparation, Ferrihydrite slurry preparation, Fe–Ca complex slurry preparation, Diffusive gel preparation, Ferrihydrite binding gel preparation, Fe–Ca complex binding gel preparation

2.2. Method

2.2.1. Performance test of binding agent to optimize ratio concentration of Fe–Ca Complex. All variant of Fe–Ca complex binding gel were loaded with a known amount of phosphate by placing them in 10 ml KH_2PO_4 (100 μg/L of phosphate) solution for 8 hours. Phosphate on binding gel was eluting with 10 ml of H_2SO_4 0.25 M for 16 hours. The concentration of phosphate is usually quantified by spectrophotometric determination using phosphomolybdenum method.

2.2.2. Performance test of binding agent to optimize weight of slurry. All variant of Fe–Ca complex binding gel were loaded with a known amount of phosphate by placing them in 10 mL KH_2PO_4 (100 μg/L of phosphate) solution for 8 hours. Phosphate on binding gel was eluting with 10 mL of H_2SO_4 0.25 M for 16 hours.

2.2.3. Preparation of DGT devices, diffusive gel and binding gel. DGT devices, diffusive gel and binding gel cut with diameter 25 mm using DGT cutter, and the filter membrane soaked on aqua demineralization. Binding gel was placed first on moulding, and then followed for the diffusive gel and membrane filters. Front cap was pressed down tightly until a good seal was formed.

2.2.4. DGT response as a function of contact time. DGT was exposed to a phosphate solution of 5 ppm for different time periods 4 – 36 hours. Then, to eluate phosphate from binding layer, the layer was placed in a clean plastic tube, after peeled off the diffusive gel, and added with 10 mL 0.025 M H_2SO_4.

2.2.5. DGT response as a function of pH. DGT was exposed to a phosphate solution of 5 ppm for variantion of pH 2–10. Then, the same procedure with section 2.2.4 was performed.

3. Results and discussion

3.1. Performance test for binding agent

Manufacture of diffusive gels and bindings gel based on radical polymerization reactions free. Acrylamide as monomer,N, N'-methylenebisacrylamide as a crosslinker, ammonium per sulfate as initiator, and TEMED as catalyst (figure 1). The concentration of crosslinkers used is 0.05% [7].

Performance test of binding gel is to ensure that binding gel have ability for phosphate adsorption. Fe–Ca complex was made from combination of $Fe_2(SO_4)_3$0.1 M and $Ca(OH)_2$ 0.1 M solution. Determinations of ratio concentration of Fe–Ca complex were conducted to study Fe-Ca ratio influence on the adsorption of phosphate. The synergic effect was defined if the phosphate adsorption, using ferric-calcium complex as binding agent, is higher than that using ferrihydrite as binding agent. On the contrary, antagonistic effect was defined when the phosphate adsorption using ferrihydrite was higher.

In experiment, the total concentration of the mixture was 0.10 M, it shows that when the concentration of $Ca(OH)_2$ 0.1 M the solution does not contain $Fe_2(SO_4)_3$ otherwise when the concentration of $Fe_2(SO_4)_3$ 0.1 M then the solution does not contain $Ca(OH)_2$. According the data on

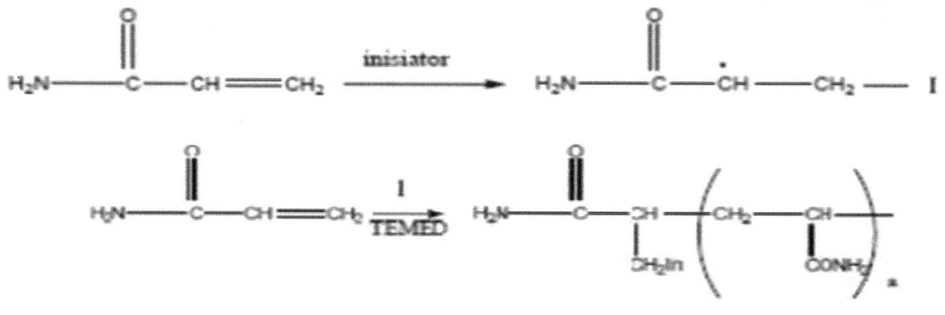

Figure 1. Polyacrylamide polymerization reaction [7].

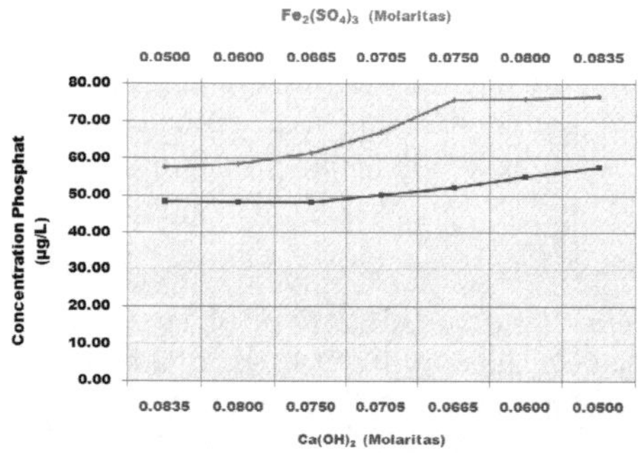

Figure 2. Variation concentration of $Fe_2(SO_4)_3$ and $Ca(OH)_2$ versus phosphate Concentration

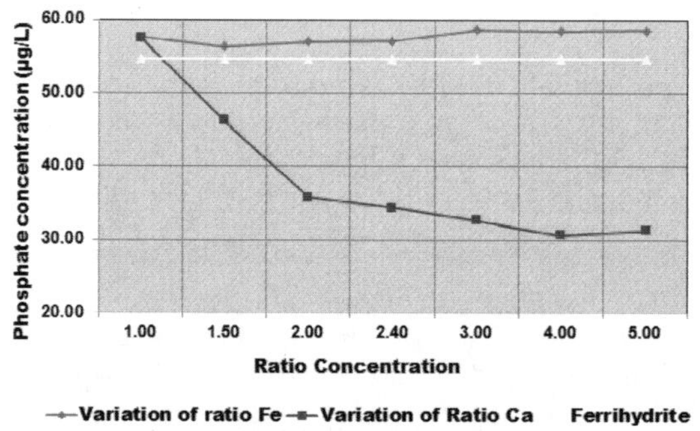

Figure 3. Optimization graph of the Fe–Ca complex ratio concentration test

figure 2 when the volume of $Fe_2(SO_4)_3$ reaches 100%, the phosphate absorption around 75 µg /L but when the volume of $Ca(OH)_2$ reaches 100%, the phosphate absorption only around 52 µg/L. The ratio of the optimum concentration occurs when the solution of $Fe_2(SO_4)_3$ is in the concentration of 0.08 M and $Ca(OH)_2$ 0.02 M. Through the above graph there is a tendency of high phosphate absorption when the binding gel is dominated by $Fe_2(SO_4)_3$, and phosphate uptake a significant decrease when the presence of $Ca(OH)_2$ predominates. Along with testing of ferrihydrite composite binding gel-calcium sulphate was also absorbed by binding gel of ferrihydrite obtained from $Fe(NO_3)_3$ 0.1 M solution, based on the test known that phosphate absorption by ferrihydrite binding gel ranges from 70–72 µg / L.

According the data shown in figure 3 the effect of Fe–Ca complex binding agent on the phosphate adsorption compared ferrihydrite binding agent was antagonistic when the proportion of Ferric was below 50%, while the effect of Fe–Ca complex binding agent on the phosphate adsorption compared ferrihydrite binding agent was synergist when the proportion of Ferric was over 50%. The data on graph show the optimal ratio of Fe–Ca complex was 3:1 or for instance, total volume consist of 75% Ferric solution and 25% calcium solution.

The data from step 1 used for variation weight for applied at binding agent. For all result compared with binding agent of ferrihydrite. The conclusion is same with the step one, If the phosphate adsorption using weight test of ferric–calcium complex as binding agent is higher than phosphate adsorption using ferrihydrite as binding agent, it is defined as synergistic effect. On the contrary, If the phosphate adsorption using weight test of ferric–calcium complex as binding agent is less than phosphate adsorption using ferrihydrite as binding agent, it is defined as antagonistic effect. According the data shown in figure 4 the effect of Fe–Ca complex binding agent on the phosphate adsorption compared ferrihydrite binding agent was antagonistic when the proportion of slurry Fe–Ca complex ratio 3:1 below 1.5 g while the effect of Fe–Ca complex binding agent on the phosphate adsorption compared ferrihydrite binding agent was synergist when the proportion of slurry Fe–Ca complex ratio 3:1 over 1.5 g. The data on graph show the optimal weight of Fe–Ca complex ratio 3:1 was 2 g.

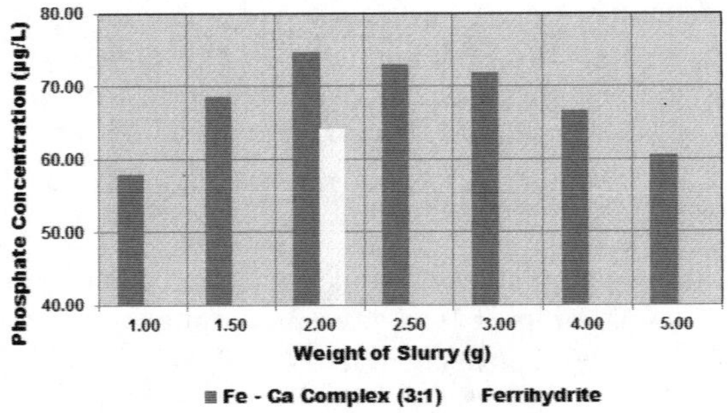

Figure 4. Variation concentration of $Fe_2(SO_4)_3$ and $Ca(OH)_2$ versus phosphate Concentration

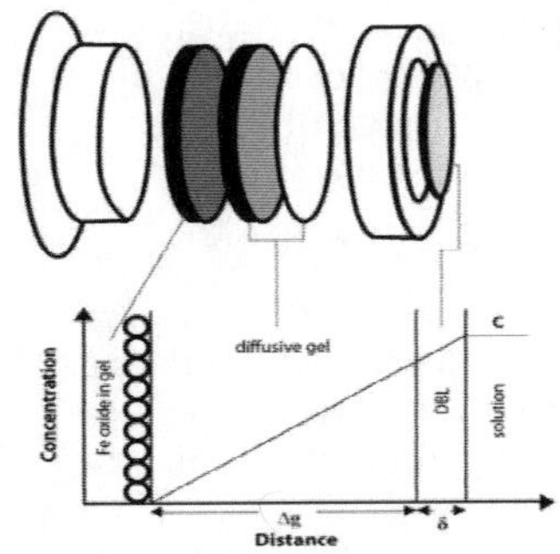

Figure 5. Optimization graph of the Fe–Ca complex ratio concentration test

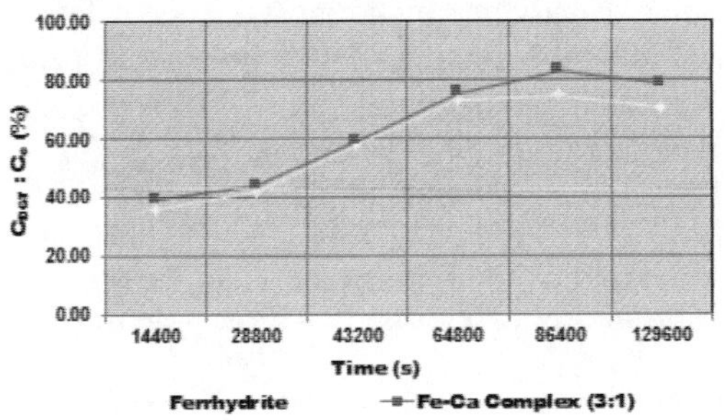

Figure 6. Effect contact time to DGT response

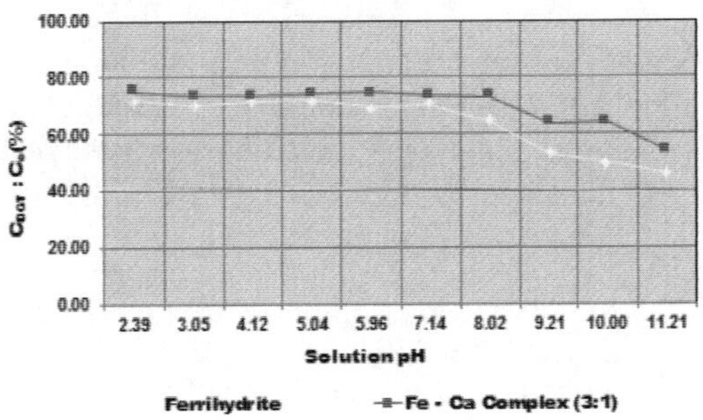

Figure 7. Effect pH to DGT response

3.2. Performance test for DGT device

The DGT method is built upon a common instrument, in which the solute is accumulated on a binder through a hydrogel, serves as a well-defined diffusion layer. The method depends on the formation of concentration gradient of a steady-state form, from solution to the binder. The solution's concentration was measured by Fick's first diffusion law and the solutes' mass accumulated on the binder in a known deployment time was calculated.

For phosphate adsorption, Δr thickness of a binding gel layer is impregnated in a binder. This binding gel is isolated from the solution using Δg thickness of an ion-permeable diffusive gel membrane (figure 5). Diffusive Boundary Layer (DBL) in a thickness of *d* lies between the diffusive gel layer and the bulk solution, where only molecular diffusion plays a role in transporting the ions. The DGT instrument is disseminated into the aqua system to determine the reactive phosphate in situ's concentration. Investigation effect contact times of DGT with Fe–Ca complex ratio 3:1 were conducted to estimate the maximum deployment time for test in real condition of environment. According the data on figure 6 show the DGT device using Fe–Ca complex as binding agent can be used to measure phosphate. The data show phosphate adsorption of DGT device using ferrihydrite as binding agent was increased until 18 hours (64800 s) and stable until 36 hours (129600 s) for DGT. While phosphate adsorption of DGT device using Fe–Ca complex ratio 3:1 as binding agent was increased until 24 hours (86400 s) and stable until 36 hours (129600 s). Contact time of DGT with Fe–Ca complex ratio 3:1 as binding agent is longer than DGT with ferrihydrite as binding agent but the capacity of DGT with Fe–Ca complex as binding agent is bigger than DGT with ferrihydrite as binding agent.

The diffusion rate depends on both the concentration gradient and the solution pH. While the pH of natural water is around pH 4–9, industrial effluents may produce higher pH. Therefore, the pH dependence of the DGT was examined. The DGT was placed inside the phosphate solution with range pH 2–10. pH solution arrange by NaOH 1M or H_2SO_4 1M solutions. According the data on figure 7 show the DGT device using Fe–Ca complex as binding agent can be used to measure phosphate in large of pH range. The data show phosphate adsorption of DGT device using ferrihydrite as binding agent was stable over 60% in range pH 2–8, on pH 9–11 adsorption of phosphate decreased below 50%. While phosphate adsorption of DGT device using Fe–Ca complex ratio 3:1 as binding agent was stable over 60% in range pH 2-10, on pH 11 adsorption of phosphate decreased below 60%. The data show that the adsorption phosphate with DGT using Fe–Ca complex ratio 3:1 as binding agent was optimal in pH 2-10.

4. Conclusions

Fe–Ca complex can act as binding agent, and effective for applied as binding agent on DGT technique. Adsorption phosphate using DGT technique with Fe–Ca complex was gave the strongest synergistic effect at ratio 3:1 with mass of slurry 2 g. Phosphate adsorption using DGT device with Fe–Ca complex ratio 3:1 is optimize at 24 h contact and that the adsorption phosphate with DGT using Fe–Ca complex ratio 3:1 as binding agent was optimal in pH 2-10.

References

[1] Tirta A P 2016 *Studi Lepasnya Fosfat dari Sedimen Danau ke Porewater Menggunakan Perangkat Diffusive Gradient In Thin Film* (Depok: Universitas Indonesia) Master Thesis
[2] Ding S, Xu D, Sun Q, Yin H and Zhang C 2010 *Environ. Sci. Technol.* **44** 8169–74
[3] Pichette C, Zhang H and Sauvé S 2009 *Aquaculture* **286** 198–202
[4] Zhang H, Davison W, Gadi R and Kobayashi T 1998 *Analytica Chimica Acta* **370** 29–38
[5] Svehla G 1979 *Vogel's Textbook of Macro and Semimicro Qualitative Inorganic Analysis* 3 ed (New York: Longman Inc.)
[6] Qiu L, Zheng P, Zhang M, Yu X and Abbas G 2015 Phosphorus removal using ferric-calcium complex as precipitant: Parameters optimization and phosphorus-recycling potential. *Chem, Eng. J.* **268** 230–5.
[7] Schlesinger W H and Bernhardt E S 2013 *Biogeochemistry: An Analysis of Global Change* (New York: Elsevier)

IOP Conf. Series: Materials Science and Engineering **496** (2019) 012008 doi:10.1088/1757-899X/496/1/012008

Microwaved assisted synthesis $Ag_2O/TiO_2/CeO_2$ for highly efficient photocatalyst

A Taufik[1,2], S P Prakoso[2] and R Saleh[1,2]

[1] Department of Physics, Faculty of Mathematics and Natural Sciences (FMIPA) Universitas Indonesia, Depok 16424, Indonesia

[2] Integrated Laboratory of Energy and Environment, Faculty of Mathematics and Natural Sciences (FMIPA) Universitas Indonesia, 16424 Depok, Indonesia

Corresponding author: rosari.saleh@gmail.com

Abstract. Photocatalyst $Ag_2O/TiO_2/CeO_2$ composites were prepared using the microwave assisted method. The concentration of CeO_2 was varied from 25 weight percentage (wt.%) to 75 (wt.%), and the physical properties of the composites were studied by X-ray diffraction, thermogravimetric analysis and different thermal analysis, and UV-Vis spectroscopy. The composites were confirmed by the existence of the cubic phase from Ag_2O and CeO_2 and the anatase structure of TiO_2 nanoparticles. The thermal degradation analysis confirmed the weight lost at 400 C followed by the heat released around 200°C due to the phase transformation from Ag_2O into Ag nanoparticles. The photocatalytic activity of $Ag_2O/TiO_2/CeO_2$ composites were investigate to remove methylene blue from the aqueous solutions under alkaline condition. The degradation of methylene blue was analyzed by investigating the absorbance around 633 nm. Furthermore, the photocatalytic activity results show that the $Ag_2O/TiO_2/CeO_2$ composites exhibit good photocatalytic performance for removing methylene blue from the aqueous solution, indicating that the $Ag_2O/TiO_2/CeO_2$ composites are promising materials for waste water treatment.

1. Introduction

Recently, it is considerably important to modify semiconductor photocatalysts to improve the photocatalytic efficiency [1]. The conventional photocatalyst such as TiO_2 has been known as a promising photocatalyst due to its outstanding photocatalytic performance [2]. However, TiO_2 nanoparticles still have limited efficiency due to the high recombination rate of electrons and holes. Moreover, the single phase TiO_2 can only be activated under UV light irradiation, only requiring 5 % of total solar irradiation [3]. Therefore, it is necessary to find a photocatalyst that has high efficiency and wide irradiation range to activate the photocatalyst.

Combining TiO_2 with other materials, the low band gap semiconductor can commonly extend the activation energy from photocatalyst from UV light irradiation to visible light irradiation [4]. Recently, Ag_2O (silver oxide) nanoparticles has become suitable for supporting the photocatalyst to extend the activation of photocatalyst into visible light range [5]. Several studies have studied the combination of TiO_2 and Ag_2O nanoparticles in the form of Ag_2O/TiO_2 composites and shown that the photocatalytic performance under visible light irradiations has increased [6]. The enhanced photocatalytic performance is due to the p-n junction formation from Ag_2O p-type and TiO_2 n-type semiconductors. The p-n junction formation is helpful in facilitating electron hole migration, efficiently retarding electron hole recombination [7].

Recently, cerium oxide (CeO_2) has also attracted considerable attention in the photocatalytic process due to several unique properties such as the following: CeO_2 is a rare earth metal oxide with good chemical stability. The mutation of the oxidation states of Ce^{4+} and Ce^{3+} in this metal oxide could create oxygen vacancies. The band gap of CeO_2 can be narrowed by adjusting the level of Ce^{4+} ions, which will effectively increase photocatalytic activity under visible light irradiation [8]. Therefore, combining Ag_2O/TiO_2 composites with CeO_2 nanoparticles produces novel materials that are promising for photocatalytic applications.

Therefore, this study evaluates the formation of $Ag_2O/TiO_2/CeO_2$ composites with different CeO_2 loading weight percentages as photocatalysts for waste water removal. In this study, $Ag_2O/TiO_2/CeO_2$ nanoparticles are investigated by X-ray diffraction (XRD), UV-Vis absorbance, differential thermal analysis, and thermogravimetric analysis. The photocatalytic activities were investigated for methylene blue solution through observation of its degradation. The influence of pH, initial dye concentration, catalyst dosage, and the addition of the scavengers were investigated to understand the photocatalytic behavior of $Ag_2O/TiO_2/CeO_2$ composites.

2. Experimental

2.1. Materials
Cerium sulfate pentahydrate [$Ce(SO)_4.5H_2O$, 99% Merck], silver nitrate ($AgNO_3$ 99% Merck), titanium dioxide (TiO_2, 99% Merck), sodium hydroxide (NaOH, 99% Merck) were analytical grade and purchased from Merck without further purifications.

2.2. CeO_2 nanoparticles synthesis
The CeO_2 nanoparticles were synthesized using the sol-gel method. In the typical process, $CeSO_4{\cdot}5H_2O$ was dissolved in water (solution A) and stirred for several minutes. After $CeSO_4{\cdot}5H_2O$ being completely dissolved, then the solution is added with NaOH solution. The mixture was then kept stirred using a magnetic stirrer with a constant temperature of 80 °C for 4h. The precipitate was obtained from the results of the centrifugation process followed by drying in a vacuum, then proceed with calcination at a temperature of 800 °C.

2.3. Synthesis $Ag_2O/TiO_2/CeO_2$ composites
$Ag_2O/TiO_2/CeO_2$ were synthesized using the microwave-assisted method. There are three prepared solution, $AgNO_3$ (silver nitrate) solution (which is dissolved in 40 mL water), SDS (sodium dodecyl sulfate) solution (which is dissolved with 80 mL water), and 10 mM NaOH solution (which is dissolved in 40 mL water). TiO_2 and CeO_2 nanoparticles were dispersed in $AgNO_3$solution with a magnetic stirrer. While still stirring, the SDS solution was then added to the mixture. After that, NaOH solution was added and stirring continued for 10 minutes. After the solution reaches homogeneity, the solution is irradiated with microwave for 2 minutes at 800 W. The precipitate was obtained from centrifugation then it was rinsed with ethanol and DI water, and dried for 5 hours 120°C under air condition.

2.4. Characterization
The structural properties of the samples were characterized by XRD Rigaku Miniflex 600 (30 kV, 15 mA). UV-Vis spectrometer Hitachi UH5300 was used to investigate the optical properties of the samples in the range of 200 nm to 800 nm. The differential thermal analysis (DTA) and thermogravimetric analysis (TGA) (Rigaku Thermo Plus EVO2 TG-8121) was perfomed to characterize the thermal properties of the samples. It operated from room temperature to 500 °C with alumina as a reference.

2.5. Photocatalytic measurement
Photocatalytic performance of $Ag_2O/TiO_2/CeO_2$ were tested under visible light irradiation to remove methylene blue from the solution using 40 W Xe lamp as a irradiation source. The methylene blue solution was set at 20 mg/L and 0.03 g catalyst dosage. To ensure the solution has reached the adsorption-desorption equilibrium before irradiation, the solution is left for 30 minutes in the dark. At the specified time interval, the methylene blue solution was observed by UV-Vis spectroscopy to examine the degradation process. Degradation can be obtained by the following equation:

$$\text{Degradation: } \frac{C_t}{C_0} \qquad (1)$$

where C_0 represents the initial dye concentraton and C_t represents the dye concentration after time t.

Table 1. Lattice parameters and grain size of $Ag_2O/TiO_2/CeO_2$ composites with different molar ratios.

Sample	Lattice parameters					Grain Size (nm)			
	Ag_2O	Ag	CeO_2	TiO_2					
	a=b=c	a=b=c	a=b=c	a=b	c	Ag_2O	Ag	CeO_2	TiO_2
25 wt.% CeO_2	4.7238	4.0894	5.4030	3.7844	9.5171	15	12	13	39
50 wt.% CeO_2	4.7238	4.0894	5.4030	3.7844	9.5171	14	11	12	33
75 wt.% CeO_2	4.7238	4.0894	5.4030	3.7844	9.5171	11	11	12	24

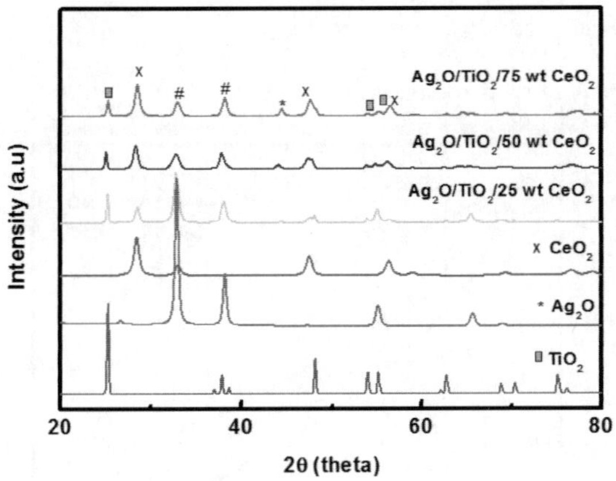

Figure 1. XRD spectra of $Ag_2O/TiO_2/CeO_2$ composites with different molar ratio.

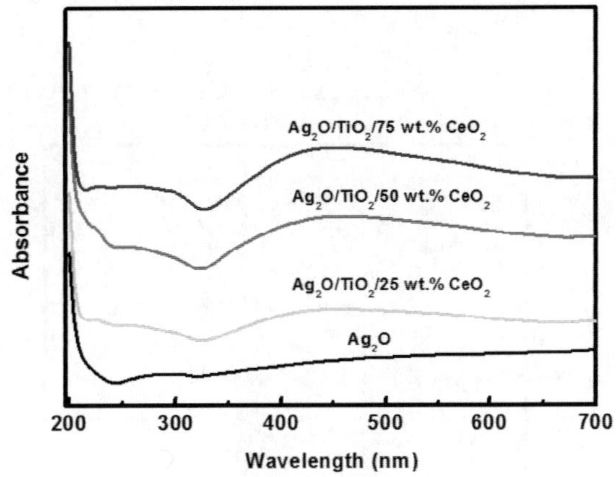

Figure 2. UV-Vis absorbance spectra of $Ag_2O/TiO_2/CeO_2$ composites with different molar ratio.

3. Results and discussion

The structural properties of $Ag_2O/TiO_2/CeO_2$ with different CeO_2 weight percentages investigated through XRD pattern are plotted in figure 1. As comparison, the XRD spectra of Ag_2O, TiO_2, and CeO_2 are also plotted in the figure. The figure shows that the XRD spectra Ag_2O revealed several diffraction peaks at $2\theta \approx 32.79°, 38.02°, 54.98°$, and $65.61°$, showing the diffraction planes of (111), (200), (220), and (311), respectively, from cubic structure of Ag_2O nanoparticles. In addition, the XRD spectra of TiO_2 shows several diffraction peaks at $2\theta \approx 25.30°, 37.60°, 48.02°, 53.55°, 54.93°, 62.53°, 68.75°$, and $70.59°$, and $65.61°$ that correspond to the existence of (101), (004), (200), (105), (211), (204), (116), and (220) planes, respectively, of the anatase structure of TiO_2. The XRD spectra of CeO_2 revealed several peaks at $2\theta \approx 28.21°, 32.82°, 47.33°$, and $56.08°$ that correspond to the existence of (111), (200), (220), and (311) planes, respectively, of cubic structures from CeO_2 nanoparticles. The formations of $Ag_2O/TiO_2/CeO_2$ composites with different CeO_2 loading concentrations confirm the existence of Ag_2O, TiO_2, and CeO_2 structure that indicate the successfully synthesized composites. The peak intensity of CeO_2 also increased with the increased CeO_2 contents. However, with increased CeO_2 content, a new peak at $2\theta \approx 44.10°$ appeared that corresponds to the existence of Ag nanoparticles. The lattice parameters and grain size of the samples were investigated using the Rietveld refinement method and the Debye Scherrer equations, and the results are presented in table 1. The results of the table show that the corresponding lattice parameters of Ag_2O, TiO_2, and CeO_2 remain unchanged after being formed in the $Ag_2O/TiO_2/CeO_2$ composites. This indicates that the formation of $Ag_2O/TiO_2/CeO_2$ does not change the structural properties of Ag_2O, TiO_2, and CeO_2 nanoparticles.

The optical properties of the samples investigated by UV-Vis absorbance spectra are plotted in figure 2. As comparison, the UV-Vis absorbance spectra of Ag_2O are also plotted in the figure.

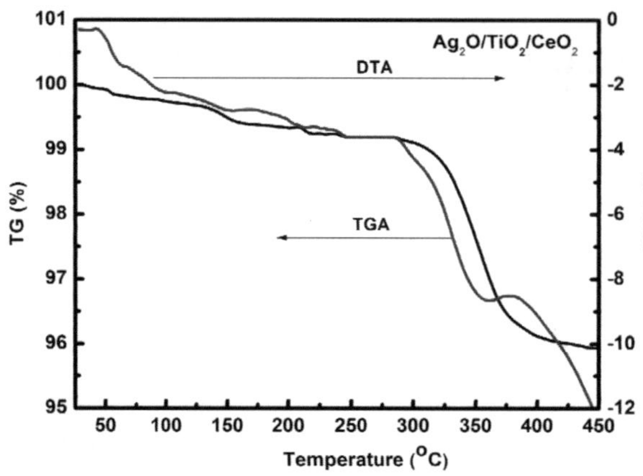

Figure 3. TGA/DTA analysis of $Ag_2O/TiO_2/CeO_2$ composites.

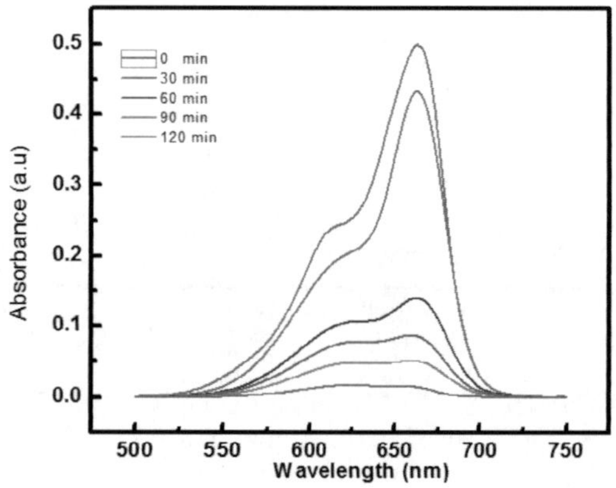

Figure 4. Absorbance spectra of methylene blue after photocatalytic reactions.

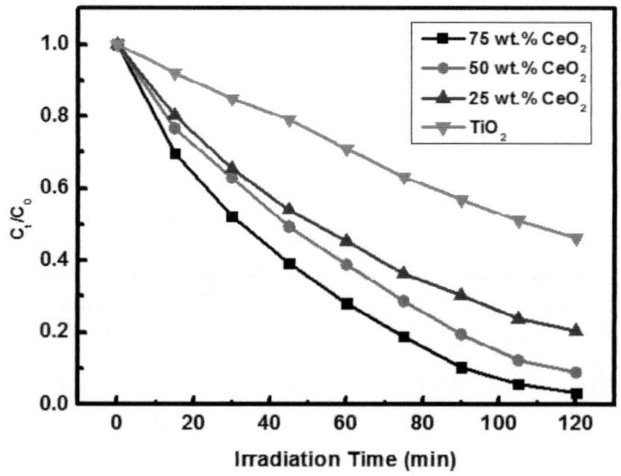

Figure 5. Photocatalytic performance of $Ag_2O/TiO_2/CeO_2$ composites with different molar ratios.

The figure shows that Ag_2O nanoparticles composites show low absorbance ability under visible light irradiations. However, after forming the $Ag_2O/TiO_2/CeO_2$ composites, the broad absorbance under visible light irradiations had a maximum intensity at 400 nm that could be due to the existence of surface plasmon resonance (SPR), i.e., the collective oscillations of conduction electrons in presence of electromagnetic radiation, from the silver nanoparticles. In addition, nanoparticles of noble metal (Au, Ag and Cu) exhibit interesting optical properties that also come from the SPR of the metal nanoparticles.

Thermogravimetric analysis and differential thermal analysis were performed to identify the thermal properties of the samples. The TGA analysis of the $Ag_2O/TiO_2/CeO_2$ composites is plotted in figure 3, and the figure shows that the $Ag_2O/TiO_2/CeO_2$composites pose good stability from room temperature until 400 °C. However, after 400 °C, thermal degradation occurs. The thermal degradation at 400 °C is due to the phase transformation from Ag_2O into Ag nanoparticles, and the differential thermal analysis of the $Ag_2O/TiO_2/CeO_2$ composite confirms that the thermal degradation at 400 °C is an exothermic process.

The methylene blue degradation through photocatalytic process is plotted in figure 4. The figure shows that the methylene blue has a maximum intensity at 630 nm and gradually decreases with increased irradiation time. This indicates that the methylene blue concentration has been reduced after photocatalytic treatment using $Ag_2O/TiO_2/CeO_2$ composites. The photocatalytic performances with different molar ratio are also plotted in Figure 5. The figure shows that all samples have good

Table 2. Degradation percentage and rate constant of methylene blue degradation using $Ag_2O/TiO_2/CeO_2$ composites.

Samples	Degradation (%)	Rate constant (min^{-1})
TiO_2	54	0.00651
25 wt.% CeO_2	80	0.01322
25 wt.% CeO_2	91	0.01994
25 wt.% CeO_2	97	0.02777

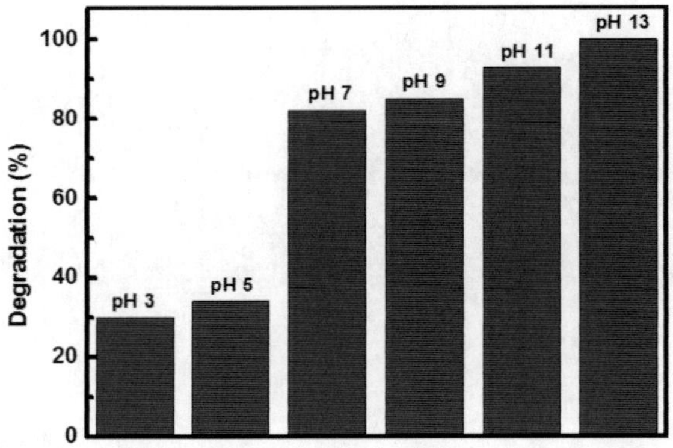

Figure 6. The influence of pH solution on the photocatalytic performance.

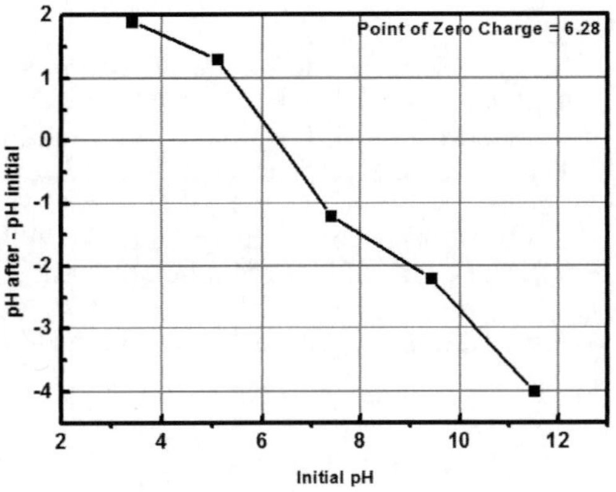

Figure 7. Point of Zero Charge (PZC) measurement.

photocatalytic performance for degrading methylene blue from the aqueous solution. The composites with 25 wt.% CeO_2 has can degrade methylene blue by approximately 80%. The photocatalytic activity further increases with the increased CeO_2 weight percentages from 50 wt.% and 75 wt.% with the degradation ability being approximately 91% and 97%, respectively. The degradation rate of the methylene blue degradation followed pseudo-first order kinetic reactions with a rate constant, as shown in table 2. The results in the table show that the composite samples with 75 wt.% CeO_2 exhibit faster degradation ability.

The photocatalytic performance of $Ag_2O/TiO_2/CeO_2$ composites was performed in methylene blue degradation. The influence of pH solution is an important aspect in the photocatalytic process, and the pH of methylene blue solution varied between 3, 5, 7, 9, 11, and 13. The result is plotted in figure 6. The figure shows that it is very difficult under an acidic condition to degrade methylene blue from the aqueous solution. However, after increasing the pH solution from 7 to 13, the photocatalytic performance significantly increased. The figure also shows that the pH solution plays a significant role in the photocatalytic process. Several researchers suggest that the surface charge between methylene blue and photocatalyst is important in the photocatalytic process. Methylene blue is the cationic dye, when it dissolved in water, it will become positively charged. To know the surface charge of the photocatalyst, point of zero charge measurement were performed, and the point of zero charge result is plotted in figure 7. The figure shows that under acidic conditions (pH 3—5), the surface charge of photocatalyst is positively charged and the point of zero charge of the photocatalyst is 6.28. After pH reaches 6.4, the surface of photocatalyst become negatively charged. The low photocatalytic performance under acidic condition is due to the repulsive effect from the positive charge of the photocatalyst and the cationic surface of methylene blue. However, under a pH above the point of zero charge, the attraction between the negatively charged photocatalyst and methylene blue provides better contact between photocatalyst and methylene blue.

4. Conclusions

$Ag_2O/TiO_2/CeO_2$ composites with different CeO_2 loading were successfully synthesized through the microwave-assisted method. The structural analysis revealed that the composites samples pose a heterostructure consisting of a cubic structure from Ag_2O and CeO_2 and an anatase structure from TiO_2. Moreover, the existence of noble metal silver could be detected by the existence of SPR in the UV-Vis absorbance spectra at 400 nm. The photocatalytic activity revealed that $Ag_2O/TiO_2/CeO_2$ composites with 75 wt.% CeO_2 shows the highest photocatalytic performance, and the surface of the catalyst is positively charged above pH 6.28, implying that methylene blue (cationic dye) is efficiently degraded under alkaline conditions.

Acknowledgements

The authors would like to thank Einago (www.einago.com) for the English language review.

Reference

[1] Jiang L, Wang Y and Feng C 2012 *Procedia Eng.* **45** 993–7
[2] Kudhier M A, Sabry R S and Al-Haidarie Y K 2018 *Mater. Sci. Semicond. Process.* **73** 35–41
[3] Moradi M, Jun M B G, Blackburn A and Herring R A 2018 *Appl. Surf. Sci.* **427** 791–9
[4] Taufik A, Albert A and Saleh R 2017 *J. Photochem. Photobiol. A* **344** 149–62
[5] Ren H T and Yang Q 2017 *Appl. Surf. Sci.* **396** 530–8
[6] Liu B, Mu L, Han B, Zhang J and Shi H 2017 *Appl. Surf. Sci.* **396** 1596–603
[7] Sarkar D, Ghosh C K, Mukherjee S and Chattopadhyay K K 2012 *ACS Appl. Mater. Interfaces* **5** 331–7
[8] Jiang H, Li M, Liu J, Li X, Tian L and Chen P 2018 *Ceram. Int.* **44** 2709–17

A lanthanide complex for selective sensing of Fe(III) ions

A Zulys and P Amanda

Department of Chemistry, Faculty of Mathematics and Natural Sciences (FMIPA), Universitas Indonesia, Depok 16424, Indonesia

Corresponding author : zulys@ui.ac.id

Abstract. The lanthanide complex $[Eu(Tpy–NH_2)_2Cl_2]Cl$ (**1**), where $Tpy–NH_2$ is 4-(p-aminophenyl)-2,2:6,2"-terpyridine, was synthesized. Its photophysical properties, e.g., Fourier Transform Infrared (FTIR) spectra, UV–visible absorption spectra, and luminescence properties, in the presence of various metal ions were studied. $Tpy-NH_2$ acts as chromophore, or light-harvesting photo sensitizing antenna moiety, for Eu^{3+} luminescence. The bright luminescence in **1** suggests that favourable photo-induced energy transfer occurs from the $Tpy-NH_2$ antenna to the Eu^{3+} to fill the long-lived emissive excited states of Eu^{3+} f–f transitions. The luminescence spectral profile of complex **1** showed a characteristic sharp emission band maximum for $^5D_0 \rightarrow ^7F_2$ f–f transitions at 614 nm originating from Eu^{3+} ions. The luminescence studies revealed that this material possesses a highly selective fluorescence quenching response to Fe^{3+} ions in liquid suspension. The dose-dependent fluorescent quenching showed good linearity, and a Fe^{3+} detection limit of 0.655 μM was calculated for complex **1**. This material can potentially be used as a fluorosensor for sensing Fe^{3+} ions; its luminescence quenching response can be explained in terms of a competitive absorption mechanism.

Keywords: lanthanide complex, europium, quenching, fluorescent chemosensors, Fe^{3+} ions

1. Introduction

Metal ions determination has gained increasing attention over recent decades. Many metal ions play important roles in biological systems and contamination of metal ion constitutes a noticeable hazard toward the environmental system[1–5]. Consequently, a highly selective and sensitive scheme and preparation of metal ions is urgently required. Some techniques, namely HPLC, MS and AAS have been introduced for analytical purpose. Unfortunately, a long-time process as well as sophisticated equipment are necessary [6]. In recent years, fluorescent chemosensors have raised a great interest and has been synthesized with updated systems [7]. Fluorescence spectroscopy is commonly applied due to it carries a noticeable sensitivity, easy usage, and affordable instrumentation properties [4]. Specific fluorescent chemosensors development to produce an effective measurement for metal ion analytes becomes one of the crucial aspect in supramolecular chemistry, due to the basic role of the said metals in biological, environmental, and chemical processes [8].

Much effort has been made toward the creation and preparation of the highly selective, highly sensitive, low detection limits and fast response sensors [9]. Lately, as an emerging type of multifunctional luminescent material, lanthanide (Ln)-based coordination compounds have attracted significant attention owing to their typical optical feature, namely high stokes shifts, long lifetimes, and narrow line-like emission bands characteristic. These superiority have resulted in in-depth experiments toward the photosensitized and electrochemically driven luminescent materials and luminescent biolabels progress. The unfavourable absorptivity of lanthanide ions (due to its parity-forbidden characteristic of the inner-shelf f-f transition) is immediately resolved by the addition of chelating chromophores in large extinction coefficients. The fluorophores, e.g., anthracene, coumarin, benzaldehyde hydrazone, and naphthalimide were notified to be beneficial as fluorescent ion probes [6]. Terpyridine and its derivatives were greatly employed due to their typical photophysical characteristic and high binding abilities toward varied types of metal ions. [10]. Terpyridines (Tpy) are heterocyclic ligands that posses a great binding affinities toward the trivalent metals [11–13]. In this case, the study of the complex of Tpy lanthanides is of interest owing to their numerous applications in chemistry and physics. The system of ligand to-lanthanide offers two main superiorities: ligand overcomes the issue with small light absorption of lanthanide ion and lanthanide ions complex prevent the luminescence quenching [14–16]. In this paper, we demonstrate the detection of transition metal ions utilizing the luminescence of a lanthanide complex.

2. Experimental

2.1. Materials and methods

All solvents and reagents were used as-obtained from commercial suppliers. 2-acetylpyridine, 4-nitrobenzaldehyde, KOH, NH_4OH, ethanol p.a, methanol p.a, $NaHCO_3$, $SnCl_2.2H_2O$, Na_2SO_4, Eu_2O_3, HCl, and DMF of analytical grade were procured from Sigma-Aldrich and Merck. $EuCl_3.6H_2O$ was obtained by dissolving Eu_2O_3 in HCl solution. The metal ion solutions were prepared from MCl_2 in DMF, where M = Fe^{3+}, Co^{2+}, Ni^{2+}, Cu^{2+}, or Zn^{2+}. FTIR spectra (400–4000 cm^{-1}) were measured using a Shimadzu 8400-S and the KBr pellet method. UV–visible spectra were measured on a Shimadzu UV-2450 at 298 K. Photoluminescent spectra in liquid suspensions were measured on a Hitachi F-2000 FL spectrophotometer.

2.2. Synthesis and characterization

2.2.1. Synthesis of Tpy-NH_2. Tpy–NH_2 ligand was prepared according to literature procedures and then thoroughly characterized. 4-(4-nitrophenyl)-2,2:6,2"-terpyridine (Tpy–NO_2) was synthesized by the Kröhnke method using an aldol condensation mechanism [17]. The nitro group of Tpy–NO_2 was consequently reduced using $SnCl_2$/ethanol under N_2 to form 4-(4-aminophenyl)-2,2:6,2-terpyridine (Tpy–NH_2) [13]. 1H NMR (500 MHz, $CDCl_3$) δ (ppm/TMS): 8.73(d,2H), 8.66 (d,2H), 8.04(d,2H), 7.89 (d,2H),7.37 (d,2H),6.82 (d,2H), 6.79 (d,2H).UV/Vis (DMF,298K), λmax/nm: 289, 328. FTIR (KBr pellet,cm^{-1}): 3300–3000 (s), 1650(s), 1594(vs), 800(m).

*2.2.2. Synthesis of complex [Eu(Tpy-NH_2)$_2Cl_2$]Cl (**1**).* [Eu(Tpy-NH2)2Cl2]Cl (1) was prepared using a generalized synthetic method obtained from the literature [18]. A warm methanolic solution of Tpy–NH_2 (0.162 g, 0.5 mmol) was mixed dropwise in a methanolic solution of $EuCl_3.6H_2O$ (0.091 g, 0.25 mmol) dissolved in a minimum amount of methanol and refluxed for 24 h. The reaction mixture was then evaporated under reduced pressure. The brown solid crude product was collected and successively washed with diethyl ether. It was then recrystallized from ethanol to obtain the desired pure product [Eu(Tpy-NH_2)$_2Cl_2$]Cl (1); yield 0.112 g.

2.3. Luminescence sensing of metal ions

The fluorescence response during interaction with different metal ions was tested in DMF solution containing MCl_2 (concentration of [M] = 10^{-3} mol/L).

3. Results and discussion

3.1 Synthesis of the complex

The complex [Eu(Tpy–NH_2)$_2Cl_2$]Cl (1) was prepared in 58.7% yield by reacting a methanolic solution of $EuCl_3.6H_2O$ with a methanolic solution of Tpy–NH_2 at a 1:2 molar ratio. 4-(4-nitrophenyl)-2,2:6, 2"-terpyridine (Tpy–NO_2) was prepared according to the described procedures. Reduction of Tpy–NO_2 was conducted with $SnCl_2$/ethanol under N_2 to transform the NO_2 group to an NH_2 group. Complex **1** was observed to be stable in both solid and solution forms, as well as under ambient circumstances. The Tpy–NH_2 ligand and the prepared scomplex **1** were completely determined using ^{1}H NMR, FTIR, UV–visible spectroscopy, and emission studies.

IOP Conf. Series: Materials Science and Engineering **496** (2019) 012009 doi:10.1088/1757-899X/496/1/012009

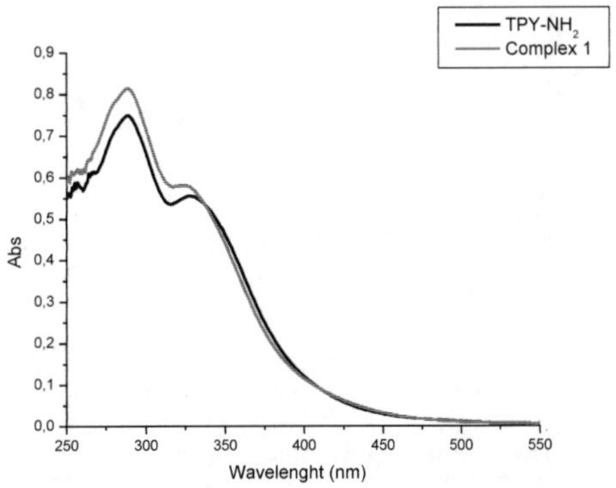

Figure 1. UV–visible absorption spectra of Tpy–NH_2 and complex **1** in DMF at 298 K.

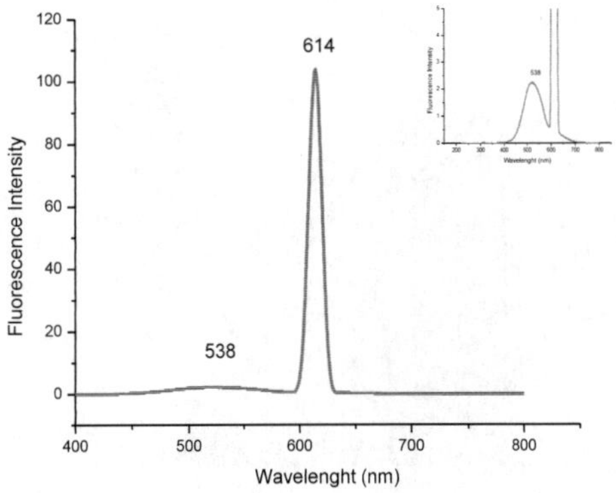

Figure 2. Luminescence spectra for complex **1** at 298 K; λ_{ex} = 289 nm.

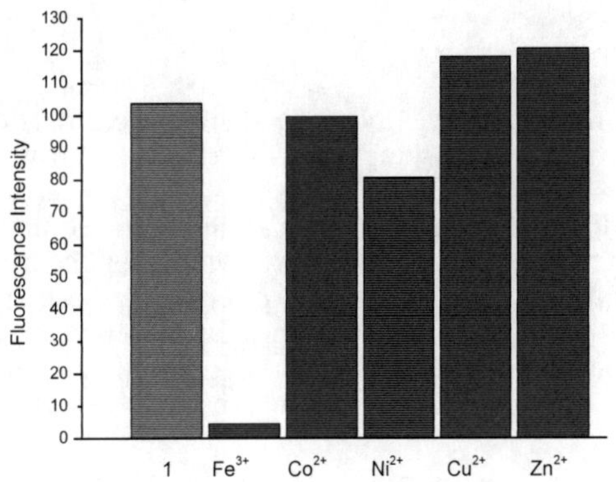

Figure 3. Selectivity profile of complex **1** in DMF containing different metal ions (10^{-3} mol/L) at λ_{ex} = 289 nm.

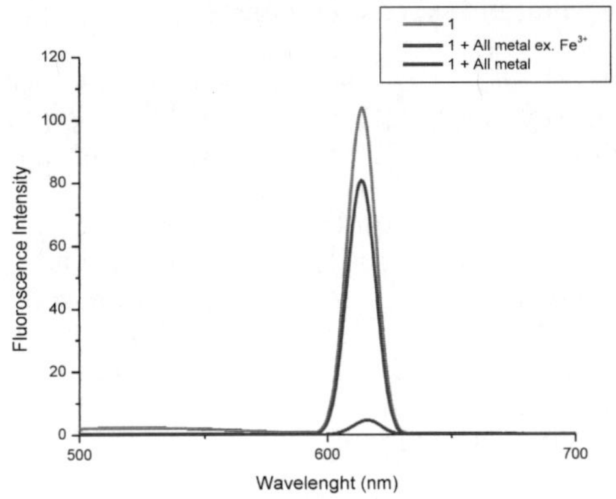

Figure 4. Emission spectra of complex **1** in the anti-interference experiment.

3.2 Photophysical properties

The Tpy–NH_2 ligand in DMF suggested noticeable peaks located at 289 and 328 nm in the UV–visible spectrum owing to ligand-centered π-π* and n-π* transitions (figure 1). Complex **1** showed a similiar electronic spectral signature to that of the ligand. Owing to the poor absorption property, lanthanides are not directly excited; instead, they will excite via a chelated organic ligands. Hence, the absorption spectrum of the complex indicates the ligands' absorption spectrum rather than the lanthanide itself (figure 1).

Figure 2 shows the luminescence spectral profile of complex **1** with a characteristic sharp and narrow emission band maximum for $^5D_0 \rightarrow ^7F_2$ f–f transitions at 614 nm originating from the Eu^{3+} ion.

3.3 Luminescence sensing of metal ions

The selectivity of complex **1** was tested by the adition DMF solution containing MCl_2 at [M] = 10^{-3} mol/L. Figure 3 shows the corresponding bar diagram of complex **1** in the presence and absence of metal ions. The figure shows that Fe^{3+} exhibited a greater indication to quench the fluorescence of complex **1**, as compared with all the other metal ions studied.

An anti-interference experiment (figure 4) was also performed and indicated that the presence of other metal ions (i.e., Co^{2+}, Ni^{2+}, Cu^{2+}, and Zn^{2+}) would not disturb the selective sensing of Fe^{3+}. The

IOP Conf. Series: Materials Science and Engineering **496** (2019) 012009 doi:10.1088/1757-899X/496/1/012009

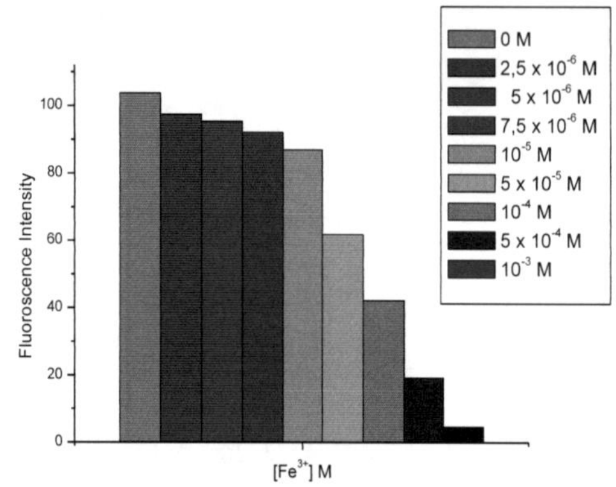

Figure 5. Variation in fluorescence intensity of **1** at different Fe^{3+} ion concentrations; λ_{ex} = 289 nm.

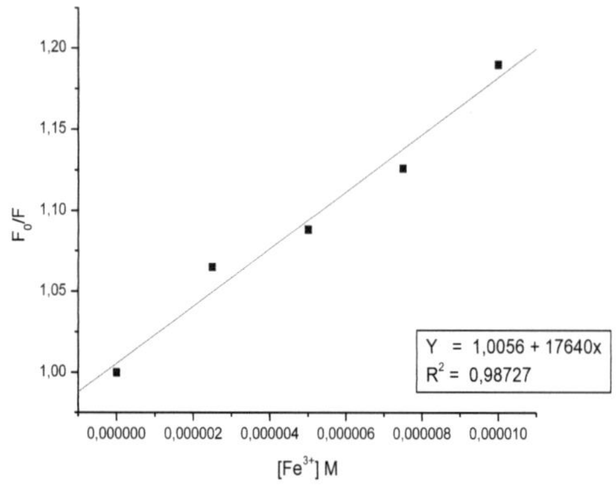

Figure 6. F_0/F vs. [Fe^{3+}] graph for the Stern–Volmer equation.

addition of several metal ions lead to some reduction in intensity of fluorescence and the emission was quenched upon the addition of Fe^{3+}, indicating the high detection selectivity of complex **1** toward Fe^{3+} ions.

The sensitivity of complex **1** toward Fe^{3+} ions was also determined. Figure 5 shows the fluorescence intensity at various Fe^{3+} concentrations. The fluorescence intensity of complex **1** was quenched little by little as the Fe^{3+} ions concentration increased from 0 to 10^{-3} M. The quenched intensity was corresponded linearly against the Fe^{3+} concentration from 0 to 10^{-5} M, with a linear coefficient (R^2) between fluorescence intensity and Fe^{3+} ion concentration of 0.98645. This suggests that complex **1** was highly sensitive toward Fe^{3+}. In fact, the efficiency of quenching of the Fe^{3+} ions was determined by the Stern–Volmer equation:

$$\frac{F_0}{F} = 1 + K_{sv}\,[Q], \tag{1}$$

where F_0 and F are the fluorescence intensities of complex **1** without and with the presence of Fe^{3+} ions, respectively; K_{sv} is the Stern–Volmer quenching constant, and [Q] is the Fe^{3+} ions concentration. Figure 6 presents the relationship between F_0/F and [Fe^{3+}]. The figure suggests that Fe^{3+} concentrations (0–10^{-5} M) and F_0/F values were corresponded linearly (R^2 = 0.98727, with an intercept of 1.0056). The Stern–Volmer constant was achieved by the correlation rearrangement based on equation (1) to give a value of 19.88 × 10^3 M.

The limit of detection (LOD) of **1** was calculated using the formula 3σ/b, where b is the slope of the linear relationship. As such, the LOD was estimated to be very low (0.655 μM).

3.4 Underlying mechanism of luminescence quenching

Lastly, we continued to investigate the underlying mechanism of the inspected fluorescence quenching in regards to the Fe^{3+} ion content. With the addition of metal ions, the emission spectrum of **1** showed no changes in wavelength, indicating that no molecules were chemically altered during the process.

As described above, the luminescence intensity of complex **1** depended greatly on the antenna effect, which induced the efficiency energy from the ligand to the lanthanide ion. With this in mind, we consider that the luminescence quenching of **1** could be due to the competitive absorption of the excitation energy of **1** and Fe^{3+} ions. To verify this potential mechanism, the UV–visible spectra of the solution metal ions were obtained. The results showed that the absorption band of Fe^{3+} ions strongly overlapped with the excitation of **1**. Therefore, we propose that a competitive absorption mechanism may cause the selective quenching response of **1** toward Fe^{3+} ions. This quencing mechanism is well known as the Foster mechanism or Foster/fluorescence resonance energy transfer (FRET).

4. Conclusions

To summarize the research, the lanthanide complex [Eu(Tpy–NH$_2$)$_2$Cl$_2$]Cl (**1**), where Tpy–NH$_2$ is 4-(p-aminophenyl)-2,2:6,2"-terpyridine, was synthesized. The complex 1 showed a selective fluorescence quenching response toward Fe3+ ion in the absence of other metal ions interference. This can be clarified in regards to the competitive absorption of wavelength energy excitation between Fe3+ and 1. This study suggests that 1 is a promising luminescence sensor of Fe3+ions.

Acknowledgements

This research was funded by PITTA Grant from Universitas Indonesia in 2017 with contract number 2068/UN2.R3.1/PPM.00/2017. The author would like to thank Einago (www.einago.com) for the English language review.

References

[1] Taher M A, Asadollahzadeh H and Fazelirad H 2015 *Anal. Methods* **7** 6726–31
[2] Neupane L N, Oh E T, Park H J, and Lee K H 2016 *Anal. Chem.* **88** 3333–40
[3] Hui X, Xiwen Z and Huiling D 2011 *Chin. J. Chem.* **29** 2165–8
[4] Feng X J, Tian P Z, Xu Z, Chen S F and Wong M S 2013 *J. Org. Chem.* **78** 11318–25
[5] Neupane L N, Park J Y, Park J H and Lee K H 2013 *Org. Lett.* **15** 254–7
[6] Jeong Y and Yoon J 2012 *Inorg. Chim. Acta* **381** 2–14
[7] Formica M, Fusi V, Giorgi L and Micheloni M 2012 *Coord. Chem. Rev.* **256** 170–92
[8] Chang K C, Luo L Y, Diau E W G and Chung W S 2008 *Tetrahedron Lett.* **49** 5013–6.
[9] P. Wang, Li Z, Lv G C, Zhou H P, Hou C, Sun W Y and Tian Y P 2012 *Inorg. Chem. Commun.* **18** 87–91
[10] Yuan J L and Wang G L 2006 *Trends Anal. Chem.* **25** 490–500
[11] Hassan E A, Hassan M L, Moorefield C N and Newkome G R 2015 *Carbohydr. Polym.* **116** 2–8.
[12] Hassan M L, Moorefield C M, Elbatal H S, Newkome G R, Modarelli D A and Romano N C 2012 *Mater. Sci. Eng. B Solid State Mater. Adv. Technol.* **177** 350–8.
[13] Bo S, Liu X and Zhen Z 2008 *J. Lumin.* **128** 1725–30.
[14] Werts M H V 2005 *Sci. Prog.* **88** 101–31.
[15] Xu S, Pei W, Ren N, Huang L and Huang W 2014 *Coord. Chem. Rev.* **273–274** 201–12
[16] Armelao L, Quici S, Barigelletti F, Accorsi G, Bottaro G, Cavazzini M and Tondello E 2010 *Coord. Chem. Rev.* **254** 487–505
[17] Andres P R, Lunkwitz R, Pabst G R, Böhn K, Wouters D, Schmatloch S and Schubert U S 2003 *Eur. J. Org. Chem.* **2003** 3769–76
[18] Ghosh S, Abbas Z, Dasari S and Patra A K 2017 *J. Lumin.* **187** 46–52

IOP Conf. Series: Materials Science and Engineering **496** (2019) 012010 doi:10.1088/1757-899X/496/1/012010

Investigation of electrical conductivity and dielectric properties of Sr_2FeTiO_6

N Masta, D Triyono and H Laysandra

Department of Physics, Faculty of Mathematics and Natural Sciences (FMIPA), Universitas Indonesia, Depok 16424, Indonesia

Corresponding author: djoko.triyono@sci.ui.ac.id

Abstract. In this study, we use impedance spectroscopy in the frequency range of 5 kHz to 1 MHz and in the temperature range of 300–550 K in order to study the electrical conductivity and dielectric properties of Sr_2FeTiO_6. The dielectric constants show the classical type of the paraelectric state, with no pseudo relaxation peak. DC conductivity increases with increasing temperature according to the Arrhenius law, indicating a thermally activated process. The activation energy is evaluated based on the DC conductivity and using the Arrhenius equation. The relaxation mechanism is the same at various temperatures, as confirmed by the scaling of the imaginary part of the electric modulus.

Keywords: Sr_2FeTiO_6, conductivity, dielectric, impedance spectroscopy

1. Introduction

Double perovskite $A_2B'B''O_6$ offers a variety of properties depending on the combination of B′ and B″. It has numerous applications, such as solid oxide fuel cells (SOFCs), capacitors, sensors, and spintronic devices [1]. It also has promising high-temperature thermoelectric properties [2]. Double perovskite Sr_2FeTiO_6, which is a promising material for sensors and SOFCs, is composed of $SrTiO_3$ and $SrFeO_3$ [3].

Single perovskite $SrTiO_3$ is known for its quantum paraelectric properties, which are temperature-independent at very low temperatures ($T \sim 3$ K) [4]. $SrTiO_3$ is typically used as a resistance-based gas sensor. Many combinations of B′ and B″ in perovskite $SrTiO_3$ have been studied in the search for enhanced properties of this n-type thermoelectric material. Doping $SrTiO_3$ with Fe decreases the band gap and increases the electron, hole, and oxygen-vacancy densities, leading to mixed ionic-electronic behavior [3], which makes this material suitable as an ideal gas sensor.

Based on previous research, the paraelectric state of Sr2FeTiO6 occurs at T >16 K. Lekshmi *et al.* [5] reported the dielectric properties and electrical conductivity of Sr2FeTiO6, for the temperature range of 200–300 K. In order to study the high-temperature properties of this material, we investigate its dielectric properties and electrical conductivity over the temperature range of 303–398 K.

2. Experimental

The Sr_2FeTiO_6 sample was prepared using the sol gel method. The experimental details are available in [6]. For IS characterization, the sample was pelleted and coated by carbon on both surfaces. The impedance was measured using an RLC meter over a temperature range of 303–523 K and a frequency range from 5 kHz to 1 MHz. Next, the sample was connected to a conducting probe and modeled as a parallel-plate capacitor. The RLC meter with a frequency range from 1 kHz to 1 MHz was used to measure the phase angle and the impedance of the samples as a function of temperature (293–523 K).

(a) (b)

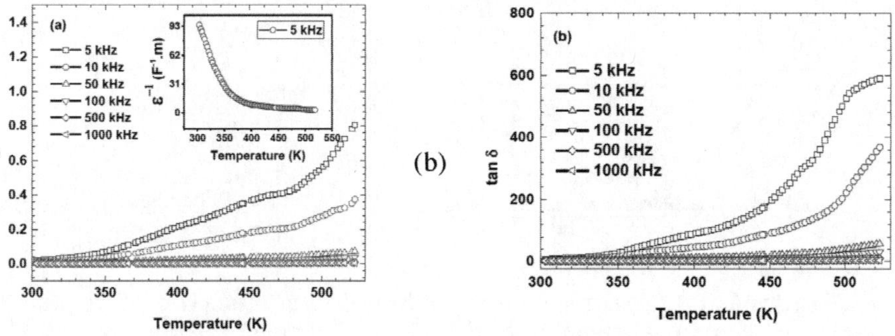

Figure 1. (a) Dielectric constant as a function of temperature for Sr_2FeTiO_6. The inset shows the inverse dielectric constant as a function of temperature. (b) Tangent loss as a function of temperature for Sr_2FeTiO_6. The inset shows a Nyquist plot for Sr_2FeTiO_6 at 523 K as a function of frequency.

The real and imaginary parts of the permittivity, AC conductivity, and electric modulus are determined as follows [7]. The real part of the permittivity is

$$\varepsilon' = \frac{1}{\omega C_0}\left(\frac{Z''}{Z'^2+Z''^2}\right), \tag{1}$$

The imaginary part of the permittivity is

$$\varepsilon'' = \frac{1}{\omega C_0}\left(\frac{Z'}{Z'^2+Z''^2}\right), \tag{2}$$

with Z' is the real parts of the sample impedance, and Z'' is the imaginary part of it, $\omega = 2\pi f$ and $C_0 = \frac{A\varepsilon_0}{t}$ is the vacuum capacitance.
The complex dielectric constant is

$$\varepsilon = \varepsilon' + \varepsilon''. \tag{3}$$

The AC conductivity is

$$\sigma_{AC} = \varepsilon_0 \varepsilon'' \omega, \tag{4}$$

where $\varepsilon_0 = 8.854\times10^{-12}$ F/m is the permittivity in vacuum and ε'' is the imaginary part of the permittivity. The data are fit using the Jonscher power law (JPL):

$$\sigma_{AC} = \sigma_{DC} + A\omega^n, \tag{5}$$

where σ_{DC} is the DC conductivity, A is the temperature and material preexponential factor, the hopping frequency is $\omega = 2\pi f$, the frequency exponent is n and σ_{ac} is the AC conductivity, The value of n $(0 < n \leq 1)$ gives the degrees of freedom of the lattice and ions. The modulus is evaluated using

$$M' = j\omega C_0 Z'', \tag{6}$$
$$M'' = j\omega C_0 Z'. \tag{7}$$

3. Results and discussion

3.1. Dielectric properties

Figure 1a shows the dielectric constant of Sr_2FeTiO_6 as a function of temperature, which behaves as a classical dielectric in the range of 303–523 K. The paraelectric state, which is manifested by a linear

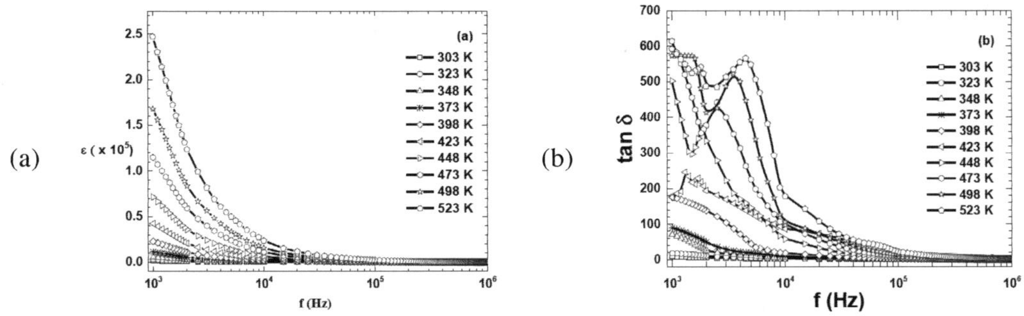

Figure 2. (a) Dielectric constant of Sr_2FeTiO_6 as a function of frequency.
(b) Loss tangent of Sr_2FeTiO_6 as a function of frequency.

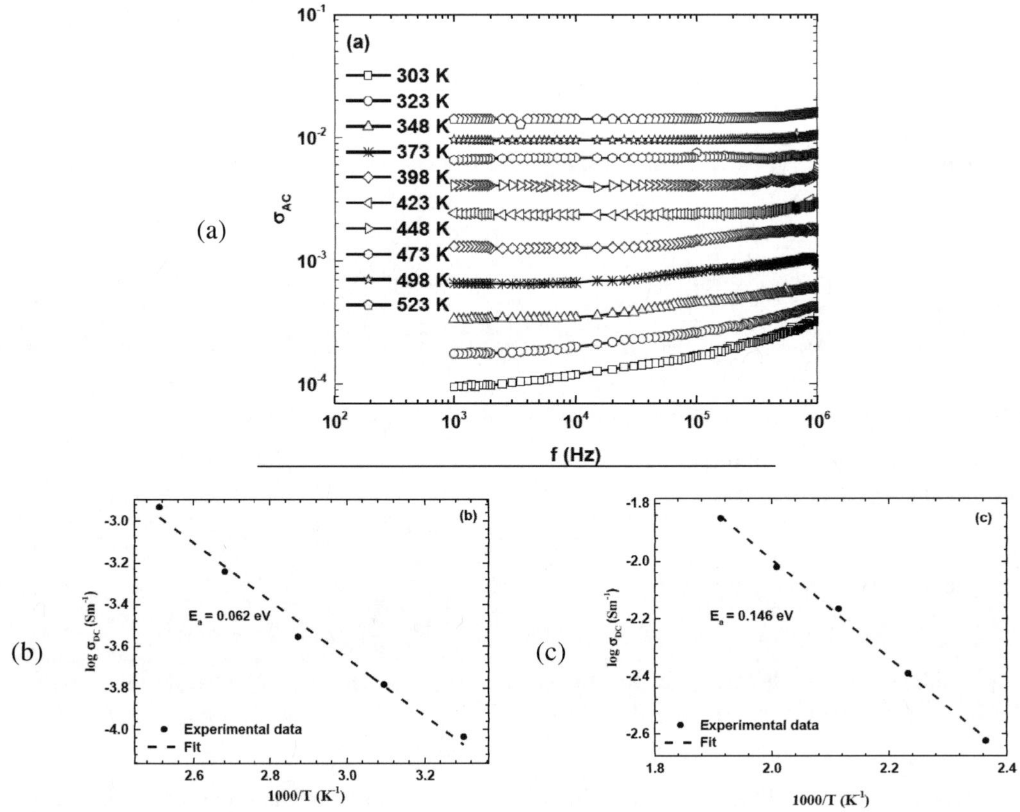

Figure 3. (a) AC conductivity as a function of frequency for various temperature, (b) DC conductivity as a function of inverse temperature at 303–398 K, (c) DC conductivity variation as a function of inverse temperature from 398 to 523 K.

increase in the dielectric constant with temperature, shows no relaxation peak. The dielectric type is confirmed by the shape of the inverse dielectric constant in each temperature (see the inset in figure 1a) and loss tangent as a function of temperature (figure 1b).

Figure 2a shows the dielectric constant of Sr_2FeTiO_6 as a function of frequency. The dielectric constant is large at low frequencies and becomes small and temperature-independent at high frequencies. A similar result was reported by Hzez [8], Pradhan [9], and Phoka [10]. This phenomenon is attributed to space-charge polarization.

IOP Conf. Series: Materials Science and Engineering **496** (2019) 012010 doi:10.1088/1757-899X/496/1/012010

Table 1. DC conductivity and JPL exponent *n* for various temperatures.

Temperature (K)	DC conductivity (S/m)	*n*
303	9.22×10^{-5}	0.50
323	1.65×10^{-4}	0.45
348	2.79×10^{-4}	0.29
373	5.74×10^{-4}	0.30
398	1.17×10^{-3}	0.34
423	2.38×10^{-3}	1.48
448	4.08×10^{-3}	1.75
473	6.83×10^{-3}	2.46
498	9.56×10^{-3}	1.83
523	1.41×10^{-2}	1.42

Figure 2b shows the loss tangent as a function of frequency. The dielectric loss ($\tan\delta$) exhibits a relaxation peak between 10^3 and 10^4 Hz at high temperatures (423–523 K). These peaks are attributed to the grain-boundary effect, which is a manifestation of the electrode-polarization effect [11,12]. The relaxation phenomenon only occurs at high temperatures because the grain-boundary effect is thermally activated.

3.2. AC conductivity

Figure 3a presents the logarithm of the AC conductivity in each frequency as determined by the Jonscher Power Law (JPL) in equation (5). At lower frequencies, the AC conductivity is relatively constant and so is actually a DC conductivity (see table 1). This phenomenon is attributed to Maxwell–Wagner space-charge polarization in the bulk and to the polarization process at the interfaces (electrode space charge) [13]. In the high-frequency region, the conductivity increases according to the power law $A\omega^n$, which puts it in the category of AC conductivity.

Given the temperature dependence of the electric parameters, the JPL gives n. At lower temperatures (303–398 K), the JPL is fully satisfied ($0 < n \leq 1$), which indicates a translational motion by hopping [14]. However, above 398 K, $n > 1$, which denotes localized hopping [14].

Figures 3b and figure 3c show the DC conductivity as a function of temperature. The DC conductivity was determined using the JPL. The activation energies determined by fitting to the JPL are 0.062 and 0.146 eV for translational hopping (figure 3b) and localized hopping (figure 3c), respectively.

3.3. Electric modulus

Figure 4a shows the real part M' of the modulus as a function of frequency. The electric modulus is related to electric-field relaxation in materials when the displacement field is constant. Specifically, the electric modulus explains the relaxation process in the real part of the permittivity. M' shows a tendency to increase with increasing frequency, the the value of M towards a constant value ($M \sim \infty$) at high frequency.

Figure 4b shows the imaginary part M'' of the modulus as a function of frequency. M'' exhibits a broad peak that shifts to a higher frequency as the temperature increases, which indicates that the relaxation rate increases with temperature. The region before the maximum imaginary part of the modulus (M''_{max}) reflects the long path of the charge carrier, and the region after the maximum frequency reflects the short path of the mobile charge carrier. The inset in figure 4b shows the scaling profile of M''. The fact that the curves overlap means that the relaxation at each temperature is based on the same mechanism.

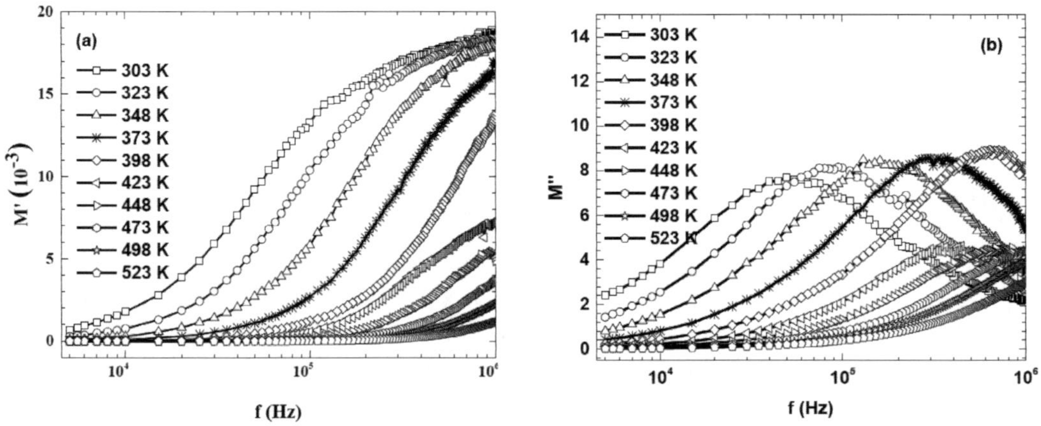

(a) (b)

Figure 4. (a) Real part M' of the modulus as a function of frequency. (b) Imaginary part M'' of the modulus as a function of frequency. The inset shows the scaling profile of M''

4. Conclusions

Sr_2FeTiO_6 was prepared using the sol gel method. The dielectric constant of this material as a function of temperature follows the classical dielectric behavior in the paraelectric state. The AC conductivity fully satisfies the JPL in the temperature range of 303–398 K. The thermal-energy-activated mechanism is confirmed by the increase in DC conductivity as the temperature increases and by the fit to the Arrhenius law.

Acknowledgements

The authors would like thanks to PITTA Grant from Universitas Indonesia for financial support with contract number 2256/UN2.R3.1/HKP05.00/2018 and Einago (www.einago.com) for the English language review.

References

[1] Vasala S and Karppinen M 2015 *Prog. Solid State Chem.* **43** 1–36.
[2] Saxena M and Maiti T 2017 *Ceram. Int.* 44 2732–7
[3] Jung W C and Tuller H L 2009 *Solid State Ionics* **180** 843–7
[4] Tkach A, Amaral J S, Vitor S A B and Vilarinho V M 2017 *J. Eur. Ceram. Soc.* **37** 2391–2397.
[5] Lekshmi P N, Pillai S S, Suresh K G, Santhosh P N and Varma M R 2012 *J. Alloys Compd.* **522** 90–95.
[6] Masta N, Triyono D and Laysandra H 2017 *AIP Conf. Proc.* **1862** 030036
[7] Singh A, Deka B, Ravi S and Pamu D 2017 *Ceram. Int.* **43** 10468–77.
[8] Hzez W, Rahmouni H, Dhahri E and Khirouni K 2017 *J. Alloys Compd.* **725** 342–8.
[9] Pradhan A K, Mandal P R, Bera K, Saha S and Nath T K 2017 *Physica B Condens. Matter.* **525** 1–6.
[10] Phokha S, Hunpratup S, Pinitsoontorn S, Putasaeng B, Rujirawat S and Maensiri S 2015 *Mater. Res. Bull.* **67** 118–25.
[11] Szwagierczak D, Kulawik J and Witek K 2015 *Mater. Sci. Semicon. Process.* **38** 329–35.
[12] Kumar D A, Selvasekarapandian S, Nithya H, Sakunthala A and Hema M 2010 *Physica B Condens. Matter* **405** 3803–7.
[13] Mahato D K, Dutta A and Sinha T P 2012 *Solid State Sci.* **14** 21–5.
[14] Mekap A, Das P R and Choudhary R N P 2014 *J. Electron. Mater.* **43** 3527–33.

A boron-doped diamond electrode decorated with hemoglobin-modified platinum nanoparticles as a biosensor for acrylamide detection

R Wulandari[1,2], T A Ivandini[1] and E Saefudin[1]

[1] Department of Chemistry, Faculty of Mathematics and Natural Sciences (FMIPA), Universitas Indonesia, Depok 16424, Indonesia

[2] Department of Chemical Engineering, Faculty of Engineering, Universitas Serang Raya, Serang 42116, Indonesia

Corresponding author: ivandini.tri@sci.ui.ac.id

Abstract. Acrylamide (AA) is a neurotoxin and potential carcinogen. It has been found in various thermally processed foods, e.g., potato chips and biscuits. Thus, simple, rapid, and sensitive methods for AA detection are needed to ensure food safety. Herein, the fabrication of a highly stable AA biosensor is presented. A boron-doped diamond (BDD) was modified by Pt and hemoglobin. In the first step, platinum nanoparticles (Pt NPs) were chemically seeded onto the BDD surface using $NaBH_4$ as a reducing agent. The electrochemical overgrowth of these Pt NP seeds was conducted at a constant potential of −0.2 V in a 1 mM Pt solution. Then, rapid thermal annealing (RTA) of the BDD/Pt NP composite was conducted at 700 °C under N_2 atmosphere to enhance its stability. After RTA, BDD/Pt NP was electrochemically activated between −0.5 and 1.5 V. Then, further overgrowth was performed using a deposition voltage of −0.2 V to renew the BDD/Pt NP surface. Finally, 0.15-mM hemoglobin was used to modify BDD/Pt NP. The characterization of the resulting surface was performed using scanning electron microscopy. The biosensor exhibited an optimal response (limit of detection = 0.012 nM) at pH 4.9 in a 0.2-M acetate buffer solution.

Keywords: BDD, platinum, hemoglobin, acrylamide, biosensor

1. Introduction

Recently, multiple studies have focused on food safety, most of which are concerned with the monitoring of harmful substances, e.g., acrylamide (AA), in food products and ingredients. AA is a neurotoxin and a potential carcinogen [1] that is materialized in reactions between reducing sugars, e.g., glucose and asparagine. The Maillard reaction mechanism is responsible for its formation in high-content starch food during cooking process at high temperatures (>120 °C) [2]. The standard methods employed for the detection and determination of AA levels in food products are GC-MS/MS [3], LC-MS/MS [4], and HPLC-MS [5]. However, these methods are laborious and expensive [1].

Accordingly, inexpensive biosensors for AA that can be easily produced and used have recently been developed. Garabagiau and Mihailescu [1] developed a simple hemoglobin (Hb)/Au nanoparticle (NP)-modified indium tin oxide glass electrode for the amperometric detection of AA [1]. Krajewska *et al.* [6] developed an electrochemical biosensor for AA, which comprised Hb that was deposited on single-walled carbon nanotubes. Li *et al.* [7] developed a facile electrocatalytic Hb redox system using flower-like Au NPs supported on a boron-doped diamond (BDD) surface. Umam *et al.* [8] developed a biosensor comprising a Hb/Au NP-modified BDD electrode for the detection of AA with a limit of detection (LOD) of 5.1 μM. However, this biosensor was found to be unstable.

Many researchers have focused their attention on enhancing the electron transfer properties of Hb using nanostructured metals and alloys, such as Ag, Au, Pt, Pd, Ir, Rh, Ag/Au, Pt/Ag, and Pt/Pd, as mediators or promoters [9,10]. However, electrochemical deposition of metal NPs on BDD substrates was found to show poor stability due to its sp^3 coordination [11,12]. In addition, their electrochemical activities vary with different substrates and deposition methods [13]. On the other hand, Gao *et al.* [10] reported wet-chemical seeding method to deposit highly stable Pt NPs on diamonds.

In this study, the wet-chemical seeding method was employed to prepare the Pt NPs-modified BDD [10]. This work presents highly stable and active Pt NPs on BDD, which is coated with Hb molecules to promote electron transfer from the electrolyte to the electrode, for the determination of AA via the formation of its Hb adduct.

2. Experimental

2.1 Materials and instruments

BDD was obtained from Keio University, Japan, with the specifications of 0.1% B/C ration and 1000-µm thicknesses $H_2PtCl_6{\cdot}6H_2O$ was purchased from Wako, Japan. Human Hb H7379 and AA were purchased from Sigma-Aldrich. All reagents were used as received without further purification.

Electrochemical experiments were conducted using an eDAQ potentiostat (Bioelectrochemistry Main Laboratory, Universitas Indonesia). An electrochemical containing of Pt and Ag/AgCl was used as the counter and reference electrodes, while the BDD modified with Pt NPs and Hb as the working electrode. Scanning electron microscopy (SEM) images were recorded using an EVO MA 10 instrument (Carl Zeiss Microscopy Gmbh, Germany).

2.2 BDD electrode modification with Pt NPs

The deposition of Pt NPs on BDD was performed with 1-M $NaBH_4$ in 0.1-M NaOH and 1-mM H_2PtCl_6 solution. First, 10 µL of the $NaBH_4$ solution was dropped onto a BDD electrode. Then, 40 µL of the 1-mM H_2PtCl_6 solution was added onto the sample for a reaction with adsorbed $NaBH_4$, generating Pt NPs. The electrode was then washed with water and dried under N_2 flow. Once the seeding process completed, the electrochemical overgrowth of Pt seeds was performed in 5 mL of 1.0-mM H_2PtCl_6 solution for 15 min under a fixed potential of –0.2 V. The Pt NPs-modified BDD was then subjected to rapid thermal annealing (RTA) at 700 °C for 5 min under N_2 atmosphere before activated by cyclic voltammetry in the potential range of –0.5 to +1.5 V at 200 mV/s scan rate for 100 cycles. Subsequently, forth overgrowth was conducted at –0.2 V deposition voltage for 15 min to renew the Pt NP surface.

2.3 Modification of Pt/BDD electrodes with Hb

A 10-µL drop of 0.2-M acetate buffer solution (ABS) solution (pH 4.9) containing 0.15-mM Hb was placed onto a Pt/BDD electrode for 24 h to obtain a Hb/Pt NP-modified BDD electrode. The electrode was rinsed with Milli-Q water and dried in a stream of N_2 before use. The modified electrodes thus obtained were kept in air at 4 °C.

2.4 Detection of AA standard

AA solutions of various concentrations (0.01, 0.02, 0.03, 0.04, and 0.05 nM) were prepared in 0.2 M ABS (pH 4.9). Samples were then analysed using cyclic voltammetry with the Hb/Pt NP/BDD electrode between -0.5 V and 1.5 V at a scan rate of 100 mV/s.

3. Results and discussion

3.1 BDD electrode modification process

BDD provides positively charged C–H dipole layers on the surface [10], therefore, enables $NaBH_4$ adsorption in the process of metal particles wet chemical seeding. The major functional group in $NaBH_4$ plays a role in the adsorption process is non-polarized BH_4^-. Accordingly, an attractive force is generated, which enhances the adsorption of $NaBH_4$ [14]. $NaBH_4$ and H_2PtCl_6 chemical reaction generates Pt NPs, the reaction is highly familiar and commonly used [15–17]:

$$NaBH_4 + 3H_2O + H_2PtCl_6 \rightarrow Pt + H_3BO_3 + 5HCl + NaCl + 2H_2\uparrow$$

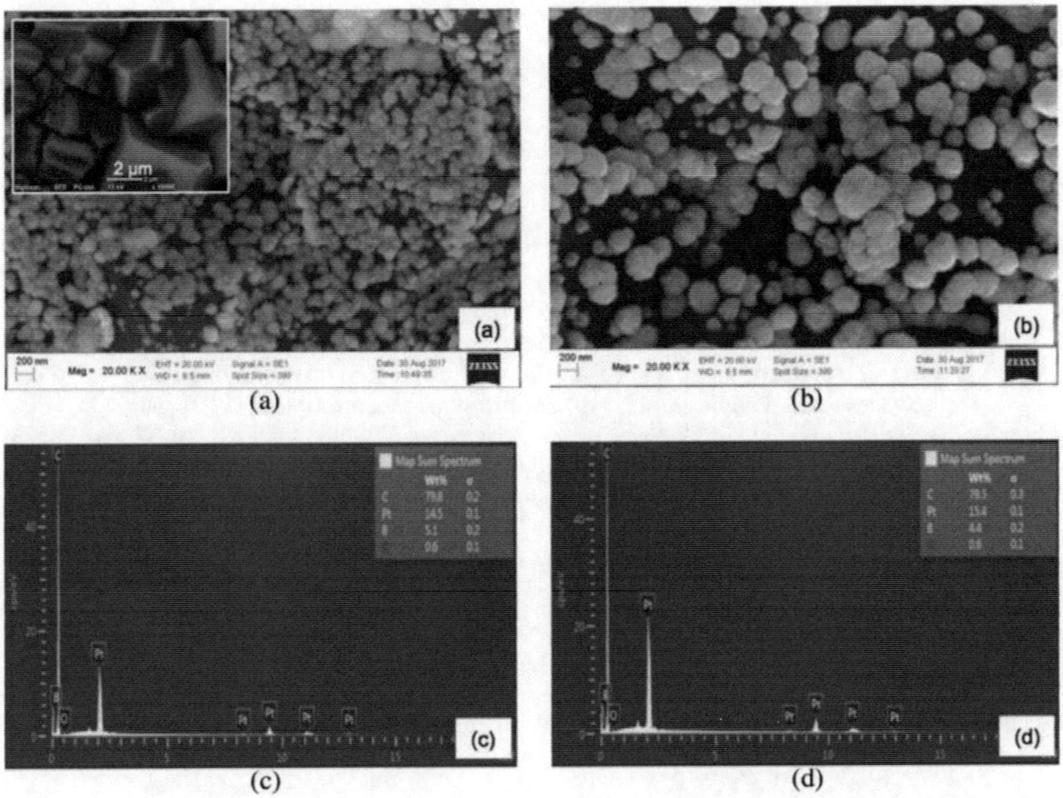

Figure 1. Scanning electron microscopy (SEM) images of Pt/BDD (a) before) and (b) after rapid thermal annealing (RTA) and their related EDX traces (d and e, respectively). Inset of (a) shows the surface of the unmodified BDD.

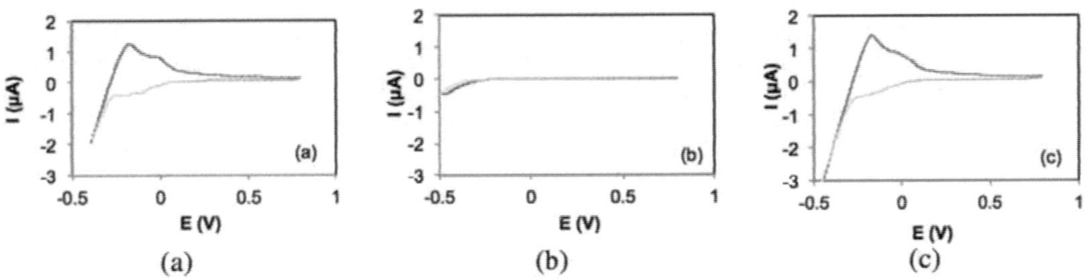

Figure 2. Cyclic voltammograms for Pt nanoparticle (NP)-coated BDD electrodes in 0.1-M H_2SO_4 at a scan rate of 100 mV/s. (a) As-deposited scenario, (b) scenario after RTA, and (c) scenario after RTA and electrochemical activation.

The key to a successfully seeded surface is the adsorption of $NaBH_4$ before the Pt NPs growth, leading to the supersaturation of the surface. It is also possible to generate similar supersaturation can via pulse deposition at a constant potential of –0.2 V. In this way, the nucleation density on the BDD electrode surface is enhanced [14].

RTA helps to form an epitaxial lattice structure between BDD and Pt NPs. Accordingly, RTA of Pt NP-coated BDD was conducted at 700 °C under N_2 atmosphere for 5 min. Figure 1a and figure 1b show the SEM images of Pt/BDD before and after RTA, respectively. The SEM of unmodified BDD

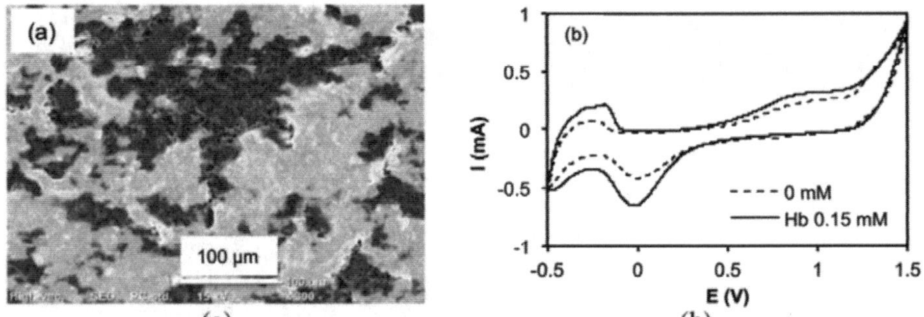

(a) (b)

Figure 3. (a) SEM image and (b) cyclic voltammogram response of Pt/BDD electrodes toward hemoglobin (Hb) at different concentration of 0.15 mM).

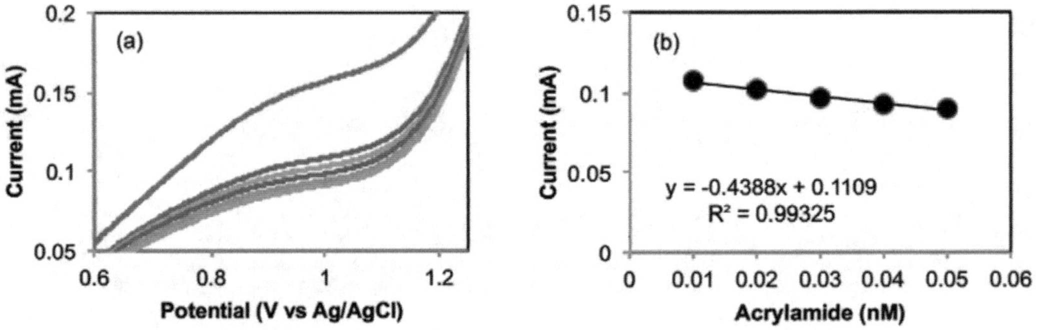

(a) (b)

Figure 4. (a) Cyclic voltammograms showing the response of the Hb/Pt/BDD electrodes toward AA at different concentrations (0, 0.01, 0.02, 0.03, 0.04, and 0.05 nM). (b) Linearity of the response of the acrylamide (AA) biosensor to AA concentration.

was shown in the inset of figure 1a for comparison. The SEM images show that Pt distributed more homogeneously on the BDD surface, afrer RTA. Figures 1c and figure 1d show that the EDX spectra of Pt/BDD before RTA and that after RTA have almost identical elemental composition.

Figure 2 shows the cyclic voltammetry (CV) results for Pt NP-coated BDD electrodes in 0.1-M H_2SO_4 before RTA, after RTA, and after electrochemical activation. Before RTA, a pair couple of peaks were noticeably observed for hydrogen adsorption/desorption, suggesting that the Pt NPs are electrochemically active. Upon completion of the RTA, the peaks disappeared and the background current similar to that of bare BDD was the only one observed, showing a reactivity loss by Pt NPs toward hydrogen adsorption/desorption. This phenomenon could be caused by the passivation film formation, for the carbon dissolves in Pt at high temperature and precipitate on metal surface upon cooling process [18].

To acquire an active Pt NPs, surface reactivation electrochemically was performed by cycling in the 0.1-M H_2SO_4 solution. The strategy is to oxidize and reduce the Pt NPs surface repeatedly, that is refreshed the electrode surface [19]. As shown in figure 2, after RTA and electrochemical activation the hydrogen adsorption/desorption peaks reappeared after this treatment.

3.2 Response of the AA biosensor

AA forms an adduct with Hb as a result of the reaction with the NH_2 group of its N-terminal valine. The formations of this adduct causes changes in the Hb structure [4,5]. This change generally decreases the current responses as the accessibility to the redox-active centre of Hb at the electrode surface was disturbed. Thus, Hb–AA adduct formation alters the electroactivity of Hb, generating the response of the biosensor. Figure 3a shows an SEM image of Hb/Pt/BDD, and figure 3b shows the CV measurements for the Pt/BDD with and without immobilization of Hb.

IOP Conf. Series: Materials Science and Engineering **496** (2019) 012011 doi:10.1088/1757-899X/496/1/012011

The Hb/Pt/BDD electrode exhibits a quasi-reversible electrochemical reaction ($Hb–Fe^{3+}/Hb–Fe^{2+}$). As shown in figure 4, the current peak in the cyclic voltammogram decreases in parallel with an increase in the Hb–AA adduct concentration at the electrode surface. As a result, this biosensor exhibits an electrochemical response toward AA in the linear range 0.01–0.05 nM. The results confirm that an AA biosensor with a considerably low detection limit (0.012 nM) was successfully fabricated.

4. Conclusions

An AA biosensor was successfully fabricated via the Hb modification of Pt NPs attached to a BDD electrode. The resultant device exhibited relatively rapid response and a considerably low detection limit (0.012 nM).

Acknowledgements

This research is partly funded by PITTA Grant 2017 from Universitas Indonesia with contract number 1822/UN2.R3.1/PPM.00.01/2017. The authors would like to thank Einago (www.einago.com) for the English language review.

References

[1] Garabagiu S and Mihailescu G 2011 *J. Electroanal. Chem.* **659** 196-200
[2] Batra B, Lata S and Pundir C S 2013 *Bioprocess. Biosyst. Eng.* **36** 1591-9
[3] Notardonato I, Avino P, Centola A, Cinelli G and Russo M V 2013 *Anal. Bioanal. Chem.* **405** 6137-41
[4] Bortolomeazzi R, Munari M, Anese M and Verardo G 2012 *Food Chem.* **135** 2687-93
[5] Zargar B, Sahraie N R and Khoshnam F 2009 *Anal. Lett.* **42** 1407-17
[6] Krajewska A, Radecki J and Radecka H 2008 *Sensors* **8** 5832-44
[7] Li M, Zhao G, Geng R and Hu H 2008 *J. Bioelechem.* **74** 217-21
[8] Umam K, Saepudin E and Ivandini T A 2016 IOP *Conf. Ser.: Mater. Sci. Eng.* **188** 012006.
[9] Sun W, Cao L, Deng Y, Gong S, Shi F, Li G and Sun Z 2013 *Anal. Chim. Acta* **781** 41-7
[10] Gao F, Yang N, Smirnov W, Obloh H and Nebel C E 2013 *Electrochim. Acta* **90** 445-51
[11] Stobiecka A, Radecka H and Radecki J 2007 *Biosens. Bioelectron.* **22** 2165-70
[12] Hoffmann R, Kriele A, Obloh H, Hees J, Wolfer M, Smirnov W, Yang N and Nebel C E 2010 *Appl. Phys. Lett.* **97** 052103
[13] Kim P, Joo J B, Kim W, Kim J, Song I K and Yi J 2006 *J. Power Sources* **160** 987-90
[14] La-Torre-Riveros L, Abel-Tatis E, Méndez-Torres A E, Tryk D A, Prelas M and Cabrera C R 2011 *J. Nanopart. Res.* **13** 2997-3009
[15] La-Torre-Riveros L, Guzman-Blas R, Méndez-Torres A E, Prelas M, Tryk D A and Cabrera C R 2012 *ACS Appl. Mater. Interfaces* **4** 1134-47
[16] Hamilton J C and Blakely J M 1978 *J. Vacuum Sci. Technol.* **15** 559-62
[17] Yamamoto K, Kolb D M, Kotz R and Lehmpfuhl G 1979 *J. Electroanal. Chem.* **96** 233-9

IOP Conf. Series: Materials Science and Engineering **496** (2019) 012012 doi:10.1088/1757-899X/496/1/012012

Dynamic mechanical properties of enhanced filler dispersion polyethylene-glycol/q-SiO_2 composites

N A Fauziyah, T A Fadly, A R Hilmi, M Mashuri and S Pratapa

Department of Physics, Faculty of Sciences, Institute of Technology Sepuluh Nopember, Surabaya 60111 Indonesia

Corresponding author: suminar_pratapa@physics.its.ac.id

Abstract. In this work, polyethylene-glycol (PEG)/q-SiO_2 (silica quartz) composites were prepared by dispersing q-SiO_2 in various fractions, i.e., 0, 20, and 40 wt.%. We elaborated the silica dispersion in the polymer matrix from the previously reported solid method (reported in [1]) and a liquid method. In the latter, the dispersion of q-SiO_2 in the matrix was done by a stirring process in the liquefied polyethylene glycol with temperature control. The filler content and mixing method were selected to optimize the dynamic mechanical properties of composites, especially, the shear moduli (G') from the Dynamic Mechanical Analyzer (DMA) instrument. The maximum shear modulus value was found with 40 wt.% q-SiO_2 by the liquid method, which showed 46% increment as compared to 40 wt.% q-SiO_2 by the solid method. We showed in this work that the dynamic mechanical properties are tightly related with the dispersion of the filler in the matrix. The quantitative analysis of filler dispersion was done using *PixelProfile* software and dispersibility characterization from SEM-EDX images. We analyzed each image in three regions by RGB (*Red-Green-Blue spectrum*). The lower the dispersibility index ($I_{dispersibility}$), the better the filler dispersion. By applying the liquid method, the dispersibility indices were 0.357 for 20 wt.% and 0.705 for 40 wt.%, better than solid method.

1. Introduction

In the past several decades, polymer composites have been very rapidly developed since their worldwide applications like in aerospace, packaging, cosmetics, automobile and some industrial applications [1–5]. The composites can be designed according to the usage. Some important aspects that affect the properties of composites are (1) type of filler, (2) filler concentration, and (3) filler morphology. Furthermore, the filler dispersion will have a critical role in the composites' properties, besides the properties of its structural components. The effectiveness of filler particles dispersion surely influences the final composites performance such as strength, stiffness, modulus, electrical/thermal conductivity, etc. [6]. Therefore, a well-defined method for improving the filler dispersion is required.

Several attempts have been introduced to increase the distribution of filler in composites, such as ultrasonic dispersion [7], melting blending [8], microemulsion [9], and the assistance of surfactants [10]. Therefore, one should choose the most effective method to optimize the filler distribution in the composite. In most studies, however, the distribution of fillers in composites was presented qualitatively. For instance, Jamil *et al.* [11] observed the nanofiller in polyetherimide (PEI) with various modified montmorillonite (f-Mt) using EDX mapping and qualitatively analyze the distribution of the filler. Moreover, the study of non-covalent functionalization of carboxylated ionic liquid (IL) for improving filler dispersion has been studied by Jiang *et al.* [12] using TEM images to analyze the distribution of of carboxylated multi-walled carbon nanotubes composites (MWCNT). Interestingly, a quantitative approach has been proposed by Glaskova *et al.* [6] where a mathematical model was used to describe the quality of filler dispersions. The model worked well only in nanocomposites systems, and therefore imperative to examine it in other systems.

In this work, we examined the filler distribution in composite systems, which involved the melt of a commercial polyethylene-glycol (PEG) as a matrix. The liquid method may give rise to silica dispersion and, consequently, improve the dynamic mechanical properties of the composites, especially, the shear moduli (*G'*) as compared to the previously reported solid method. All compounds acquired by melt blending were characterized through their morphological images to obtain the quantitative information about the silica dispersion in the polymer matrix.

2. Methods

2.1. Composites preparation

In this research, quartz powder is obtained from purified silica sand. The silica sand comes from Tanah Laut, Pelaihari, South Kalimantan. Polyethylene glycol (PEG) 4000 was purchased from Merck Schuchardt OHG (Hohenbrunn, Germany) and used as the matrix. The purifying process has been reported elsewhere [1,13–15]. The phases present in the quartz powder (q-SiO_2) was identified using X-Ray Diffractometer (XRD). There are two ways of making composite samples in this study: liquid processes and solid processes. Making composite with a liquid process is carried out using q-SiO_2 powder (20 wt.% and 40 wt.%) which acts as a filler, was mixed into the melted PEG then casted into the pellet shape (SQ 20-L and SQ 40-L). Meanwhile, the solid process is done using q-SiO_2 powder that was mixed with PEG then formed into pellets (SQ 20-S and SQ 40-S) [1,14].

2.2. Scanning electron microscopy

To determine the dispersion of the q-SiO_2 filler in the PEG matrix, the microstructure images were taken by Scanning Electron Microscopy (SEM) from all SQ pellets while the element mapping data were collected from the same samples using the embedded EDX. The observations were set in the same condition to ensure compatible relative intensity ratios. The intensity of the SEM-EDX images was obtained using *PixelProfile* software. The dispersion of the filler was determined using equation (1) as follow [16] :

$$I_{dispersibility} = \frac{S_{intensity}}{\bar{x}_{intensity}} \quad (1)$$

where $S_{intensity}$ is the standard deviation of intensity, and $\bar{x}_{intensity}$ indicates the average intensity.

2.3. Dynamic mechanical behavior measurement

All samples were formed in (5 mm × 5 mm) sizes and 1 mm thick in according to the sample holder on the DMA testing for shear mode. The detailed sample preparation has been explained elsewhere [1,14,17]. DMA/SDTA861· Mettler Toledo instrument was the thermo mechanical analyzer. In this experiment, all samples were measured at a temperature range of 25 °C up to 80 °C. Storage modulus (*G'*) was the output of DMA, which becomes the parameter related to the homogeneity of the silica filler in the PEG matrix.

3. Results and discussion

Figure 1 provides the results obtained from the DMA measurement as a function of temperature for all PEG/q-SiO_2 composites, i.e., at different silica weight fraction and preparation method (liquid and solid methods). The pure PEG plot is also presented and shows the lowest storage modulus at room temperature belongs to pristine PEG among all samples. The figure also shows a significant increase in such *G'* values from the SQ composites by both liquid and solid methods. *G'* for the 40 wt.% q-SiO_2 by liquid method sample exhibits a 46% increase as compared to the 40 wt.% q-SiO_2 by solid method sample. Meanwhile, the increase for the 20 wt.% was approximately 25%. It is obvious that the liquid method was more effective in increasing silica dispersion than the solid one. The correlation between silica dispersion and storage modulus will be discussed in detail later.

Furthermore, as *T* increases, the pure PEG and the composite samples exhibit a different decreasing trend. However, no significant difference trend between the two methods was evident. Faster drop in storage moduli values was observed in the SQ composites with 20 wt.% filler compared with 40 wt.%, generally for liquid and solid method. Drop rate of moduli in the glassy region (start from room temperature to around 55 °C) can be interpreted regarding polymer transition processes, where the large-scale chain slippage occurs and then initiated to melt. The presence of hard and rigid filler inhibited the movement of the polymer chains in the process of polymer chains transition and material

Table 1. Reinforcement factor G'/G_o as a function of silica weight fractions of the SQ composites.

Silica wt%	Solid	Liquid
20	7.19	8.25
40	7.85	10.88

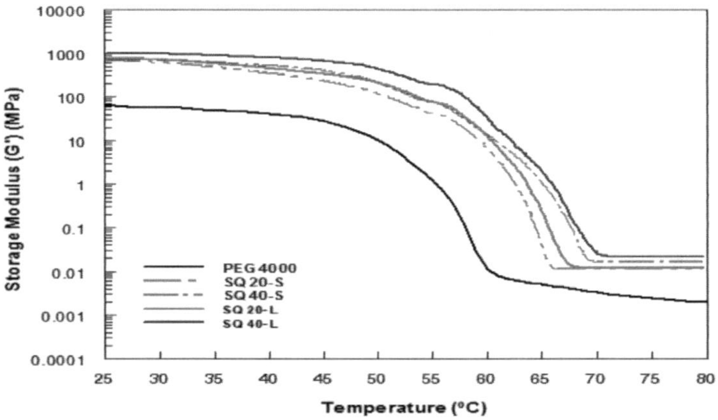

Figure 1. Shear storage moduli (G') of SQ with various silica content of 0, 20, and 40 wt% by solid and liquid method.

melt. Also, the melting temperature (T_m) shifts to the higher values as the filler was added, i.e., 47° and 51 °C for SQ 20-S and SQ 40-S, 48° and 51°C for SQ 20-L and SQ 40-L respectively. The limited PEG chain movement due to the presence of silica filler acts as the barrier which results in the shift of T_m [18]. Further analysis shows that the reinforcement effect can be observed clearly at a higher temperature. At $T > 65$ °C, PEG has entered the liquid region which is indicated a very low moduli values as compared to that at room temperature. The G' value in the liquid region increases with silica content. This effect can be explained by the reinforcement factor from silica (see table 1). The evolution of G'/G_0 (where G' is the storage modulus value in room temperature, and G_0 is the storage modulus value of pure PEG) presents the significant deviation from the dynamic response when the increasing silica weight fraction as filler. For example, in 20 wt.%, q-SiO_2 added in the composite, almost 4% of the total composite volume is filled by the silica particles, and 19 vol% for 40 wt.% q-SiO_2 From previous research [19], this G'/G_0 is related to the filler connectivity threshold which is observed between the polymer chain and the filler. As a comparison, SQ 40-L has the higher reinforcement factor value than SQ 40-S. It means that, for the high reinforcement, the filler/polymer interaction in the liquid process is more dominant than the solid one.

The correlation of dynamic results above was supported by the microstructure observation. The SEM-EDX images of the four blends and the associated elemental mappings are presented in figure 2. The red color represents silicon element, while green and blue are for oxygen and carbon respectively. Qualitatively, one can observe that the liquid process improves the silica dispersion in the PEG matrix. The silica agglomeration was observed for SQ 40-S (see figure 2c). Silica distribution in SQ 20-L and SQ 40-L are more homogenous than in SQ 20-S and SQ 40-S. This observation proves that the liquid method restrains the silica agglomeration in the PEG matrix. Liquid process combined with stirring in the polymer melted matrix will bring up the electrostatic force which plays an important role in enhancing the distribution of silica filler [20]. The electrostatic force occurred due to the interaction between the silica surface and PEG during stirring process. Meanwhile, the temperature treatment was the driving force for the homogenization process. By considering figure 1 and figure 2, consequently, the excellent dispersion of silica in PEG matrix can increase G' values of the composites.

IOP Conf. Series: Materials Science and Engineering **496** (2019) 012012 doi:10.1088/1757-899X/496/1/012012

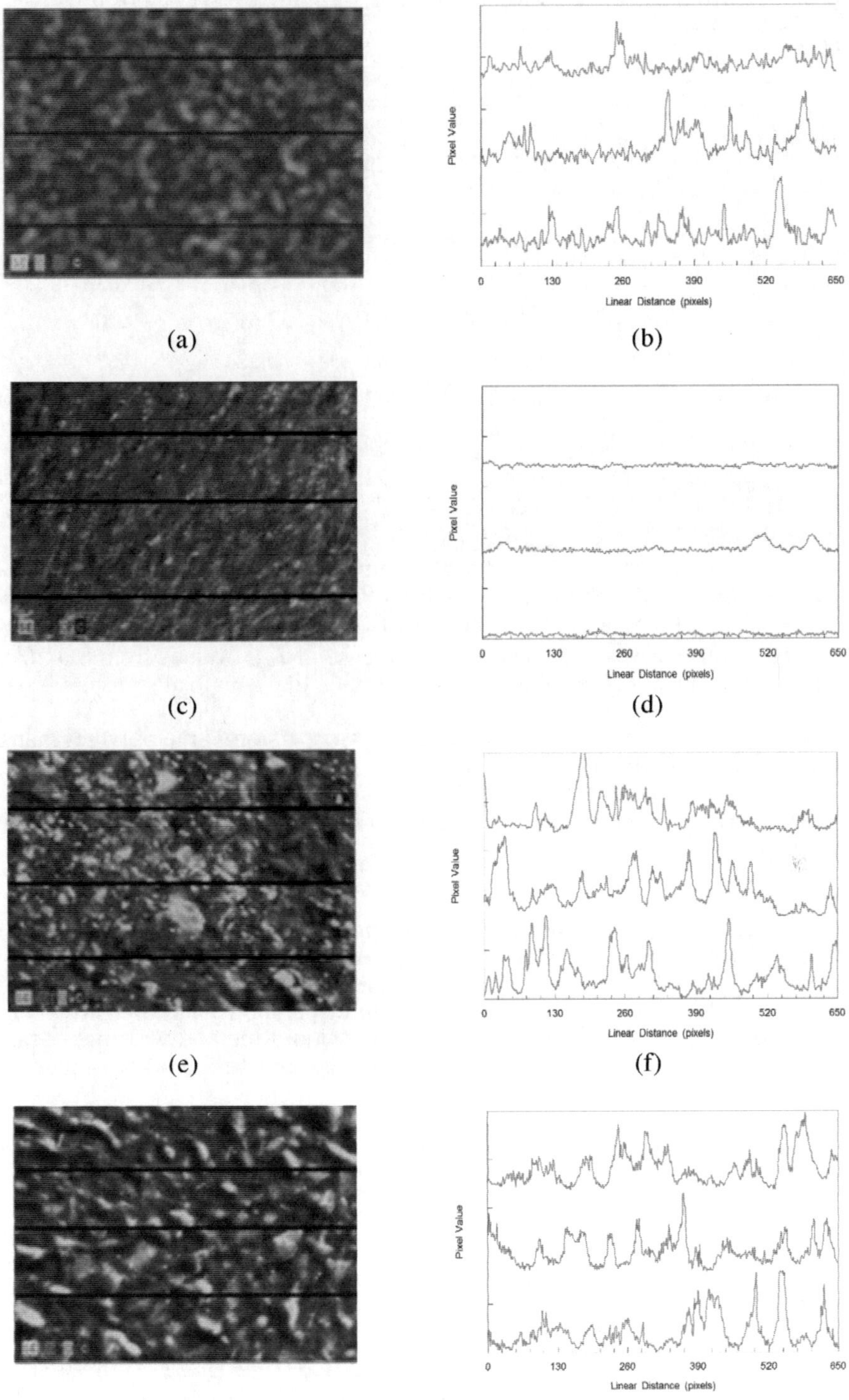

Figure 2. SEM-EDX and *PixelProfile* spectrum of (a) 20 wt.%-S (b) 20 wt.%-L (c) 40 wt.%-S (d) 40 wt.%-L composites.

Table 2. The dispersibility index ($\bar{I}_{dispersibility}$) of the SQ composites

		Solid		Liquid	
		20	40	20	40
$\bar{x}_{intensity}$	1	12.813	58.429	79.993	70.483
	2	17.268	77.019	93.246	63.299
	3	12.438	74.613	89.711	61.413
$S_{intensity}$	1	4.415	45.881	38.346	47.734
	2	11.250	54.134	38.346	40.436
	3	4.211	52.789	33.617	49.004
$I_{dispersibility}$	1	0.345	0.785	0.285	0.677
	2	0.652	0.703	0.411	0.639
	3	0.339	0.707	0.374	0.798
$\bar{I}_{dispersibility}$		0.445	0.732	0.374	0.705

Quantitatively, the SEM-EDX images from figure 2 have been analyzed using the *PixelProfile* software. Table 1 shows the values of $\bar{I}_{dispersibility}$ from all composites. The liquid process yields a lower $\bar{I}_{dispersibility}$ value as compared to the solid process. This result confirmed the SEM-EDX images. The lower the $\bar{I}_{dispersibility}$, the better the dispersion (see table 2). The SEM-EDX images by the solid process show large silica agglomerations, which well confirmed by the increase of $\bar{I}_{dispersibility}$ values from SQ 20-S and SQ 40-S. Thus, the agglomeration is responsible for determining the composite properties, especially its storage moduli.

A liquid process can be alternative for increasing filler dispersion in the polymer matrix [8]. In this process, the interaction between PEG chains and silica particles leads to the decreasing of polymer mobility. Additionally, the existence of electrostatic force is also responsible for enhancing the silica dispersion. This result can be used for further studies on how to improve the dispersion of the filler in polymer composites.

4. Conclusions

This work demonstrated how filler dispersion could be improved in the PEG/silica composites by the liquid method. The quantitative parameter, $I_{dispersibility}$, can be used to measure the degree of dispersibility of the filler. Lower $I_{dispersibility}$ implies better filler distribution. The 20 wt.% of SQ-L sample exhibited the lowest $I_{dispersibility}$ value at 0.357. We have shown that particle dispersibility has a positive correlation with storage shear modulus. The storage shear modulus of the liquid method was higher than that of the solid method, i.e., by 25% for the SQ 20-L sample and by 46% for the SQ 40-L sample.

Acknowledgements

The research supported by Ministry of Research, Technology and Higher Education of the Republic of Indonesia and LPPM ITS through program PMDSU 2017 No.489/PKS/ITS/2017 awarded to NAF and SP.

References

[1] Fadly T A, Fauziyah N A, Rosyidy A, Mashuri and Pratapa S 2017 *AIP Conf. Proc.* **1788** 030033
[2] Zhong B, Jia Z, Luo Y and Jia D 2015 *Compos. Sci. Technol.* **117** 46–53
[3] Liu Z, Yuan X, Beck A J and Jones F R 2011 *Compos. Sci. Technol.* **71** 1529–34
[4] Dziadek M, Stodolak-Zych E and Cholewa-Kowalska K 2017 *Mater. Sci. Eng. C* **71** 1175–91
[5] Bai Y, Post N L, Lesko J J and Keller T 2008 *Thermochim. Acta* **469** 28–35
[6] Glaskova T, Zarrelli M, Borisova A, Timchenko K, Aniskevich A and Giordano M 2011 *Compos. Sci. Technol.* **71** 1543–9
[7] Martin-Gallego M, Verdejo R, Khayet M, de Zarate J M O, Essalhi M and Lopez-Manchado M A 2011 *Nanoscale Res. Lett.* **6** 610

[8] Moniruzzaman M and Winey K I 2006 *Macromolecules* **39** 5194–205
[9] Banos A, Corson N, Daudé É, Gaudou B and Coyrehourcq R S 2017 *Agent-based Spatial Simulation with Netlogo* vol 2 ed A Banos, C Lang and N Marilleau (London: ISTE Press and Elsevier) pp 63–84
[10] Gong X, Liu J, Baskaran S, Voise R D and Young J S 2000 *Chem. Mater.* **12** 1049–52
[11] Jamil A, Ching O P and Shariff A M 2017 *Appl. Clay Sci.* **143** 115–24
[12] Jiang G, Song S, Zhai Y, Feng C and Zhang Y 2016 *Compos. Sci. Technol.* **123** 171–8
[13] Fauziyah N A, Musyarofah M, Triwikantoro T, Mashuri M, Firdaus S and Pratapa S 2015 *Adv. Mater. Res.* **1112** 385–8
[14] Fauziyah N A, Fadly T A, Rosyidi A, Mashuri and Pratapa S 2017 *J. Phys.: Conf. Ser.* **817** 012020
[15] Muchlis K, Fauziyah N A, Soontaranon S, Limpirat W and Pratapa S 2017 *J. Phys.: Conf. Ser.* **795** 012007
[16] Kim J, Oh J, Lee K Y, Jung I and Park M 2017 *Compos. Part B Eng.* **114** 445–56
[17] Pratapa S, Wahyuni T, Fauziyah N A, Apriliyana G A, Mashuri M and Firdaus S 2017 *J. Mech. Sci. Technol.* **31** 3653–6
[18] Menard K P 2008 *Dynamic Mechanical Analysis* 2nd ed (Boca Raton: CRC Press)
[19] Jouault N, Dalmas F, Boué F and Jestin J 2012 *Polymer* **53** 761–75
[20] Cheng Y, Lee L, Zhang W and Wang C-H 2014 *Ind. Eng. Chem. Res.* **53** 14166–74

IOP Conf. Series: Materials Science and Engineering **496** (2019) 012013 doi:10.1088/1757-899X/496/1/012013

Improved separation of Ce, La, and Nd from a concentrate of rare-earth hydroxide via fractional precipitation

I Haritsah[1], W A Adi[2], M V Purwani[3] and A Manaf[1*]

[1] Materials Science Graduate Program, Department of Physics, Faculty of Mathematics and Natural Sciences (FMIPA), Universitas Indonesia, Depok 16424, Indonesia

[2] Centre for Science and Technology of Advanced Materials, National Nuclear Energy Agency (BATAN), Kawasan Puspiptek Serpong, Banten 15314, Indonesia

[3] Center for Science and Technology of Accelerator, National Nuclear Energy Agency (BATAN), Jl. Babarsari Caturtunggal, Sleman, Yogyakarta 55281, Indonesia

Corresponding author: azwar@ui.ac.id

Abstract. In this report, we describe the selective and effective separation of Ce, La, and Nd rare-earth (RE) elements from the concentrate of a RE hydroxide (REOH) feedstock. The separation steps included dissolving the feedstock into a nitric acid solution and oxidizing the solution with potassium bromate ($KBrO_3$) as an oxidizing agent to enable precipitation under the addition of sodium hydroxide (NaOH). The yellow precipitates consisted of a concentrate of Ce precipitates, and the filtrate contained La and Nd. The adjustment of the pH of the filtrate to 8.5 with ammonia solution led to the precipitation of Nd and an La concentrate in the filtrate. The developed method was demonstrated to be selective and adequate for the qualification separation of Ce, La, and Nd. This study demonstrates the identification and analysis of REOH from PTBGN-BATAN, Indonesia, and can be used as a reference for further processing at both the laboratory and industrial scale.

1. Introduction

Rare-earth (RE) elements have low natural abundance, and their reserves are likely to be exhausted. With the development of materials technology, demand for RE elements (REE), which are mostly used in high-technology industries, has increased. RE oxides play important roles in the modern production of materials such as those used in optical, electronic, magnetic, and LED applications [1,2]. In short, REE are expected to play a central role in the development of a huge number of technologies. The pursuit of this aim necessitates the development of separation technologies for various worldwide natural resources [3,4]. Currently, more than 100 RE-bearing minerals are known, but only four are used for the commercial production of RE oxides: monazite ($(RE)PO_4$), bastnasite ($(RE)CO_3F$), xenotime (YPO_4), and RE-bearing clay [5,6].

In recent years, investigators have developed and investigated various separation methods such as solvent extraction, ion exchange, chromatographic extraction, and fractional precipitation. However, the most common method is solvent extraction because of its high efficiency and simple operation. Moreover, various solvents, including di-(2-ethylhexyl) phosphoric acid [7], 2-ethylhexylphosphonic mono-2-ethylhexyl ester [8], bis-2,4,4-tri-methylpentyl phosphinic acid [9], tributylphosphate [10], bis(2-ethylhexyl) (2-ethylhexyl)phosphonate [11], di(1-methyl-heptyl)methyl phosphonate [12], and primary amine N1923 [13,14], have been used to separate Th and other REs. In the process of element

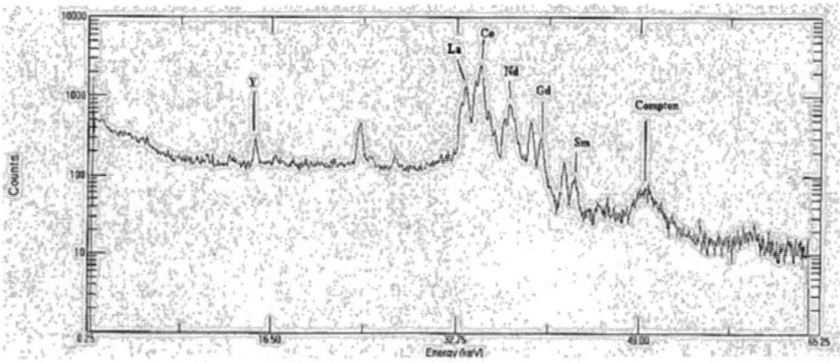

Figure 1. XRF analysis result during reprocessing and reduction of the chlorine and phosphate contents.

separation in monazite, researchers in India noted the potential benefit of Cyanex 923 in separating La, U, and Th [15]. Furthermore, a mixture of TBP and D2EHPA in a nitric acid medium has been used to successfully purify Th, U, and REs by supercritical fluid extraction with carbon dioxide [7]. TEHP and TiAP have also been used to separate Th, U, and REE of monazite in a nitric acid medium [16]. The separation of Th and REE in monazite has also been carried out using phosphorylated calixarene in a nitric acid medium [17].

In this paper, we report a selective and effective method for separating Ce, La, and Nd from a concentrate of rare-earth hydroxide (REOH) by fractional precipitation and our identification and analysis of monazite sand acquired from PTBGN-BATAN.

2. Experimental

REOH was supplied by PTBGN-BATAN; all other reagents were acquired from PSTA-Batan Yogyakarta. A 100-g sample of REOH was dissolved in 280 mL of a nitric acid solution for 30 min at 150 °C. Afterwards, the RE nitrate was dissolved with oxalic acid 15% in 5 L of demineralized water; the precipitate that formed was then filtered. Furthermore, the RE oxalate was added to demineralized water containing sodium hydroxide, and the resulting mixture was stirred and heated at 140 °C for 1 h. The REOH precipitate was filtered, washed with hot water until pH 7, and then dried at 110 °C. The REOH concentrate was dissolved in concentrated nitric acid, potassium bromate was added, and the solution was heated at 100 °C. After the oxidation process, 15% ammonia solution was added to the RE nitrate solution to adjust the pH to 4. The precipitate formed was filtered and analyzed. The filtrate was added to 15% aqueous ammonia solution, and the pH was adjusted to 8.5. Oxalic acid (15%) solution was then added to the pH 8.5 solution until precipitate formation ceased. The precipitation process was performed three times. The resulting precipitates were characterized by X-ray fluorescence (XRF) spectroscopy, Fourier transform infrared (FT-IR) spectroscopy, and X-ray diffraction (XRD) analysis.

3. Results and discussion

The REOH obtained from monazite through the alkaline-based processed contains Ce, La, and Nd. To increase the added value of REOH, each element contained therein, especially Ce, La, and Nd, should be separated and obtained as oxides. In this work, Ce, La, and Nd were obtained as CeO_2, La_2O_3, and $Nd(OH)_3$ concentrate.

3.1. Reprocessing and reducing chlorine and phosphate contents

Figure 1 shows that the intensities of the peaks of the three dominant elements of La, Ce, and Nd at 33.44 keV, 34.72 keV, and 37.36 keV, respectively, in the XRF spectrum do not substantially change. This lack of change demonstrates that the process of reducing the chlorine and phosphate contents does not substantially affect the main elements.

3.2. Fractional precipitation

After the oxidation process, the RE nitrate solution was adjusted to pH 4 by addition of 15% ammonia solution. The precipitates were filtered and characterized by XRF. The XRF spectrum in figure 2

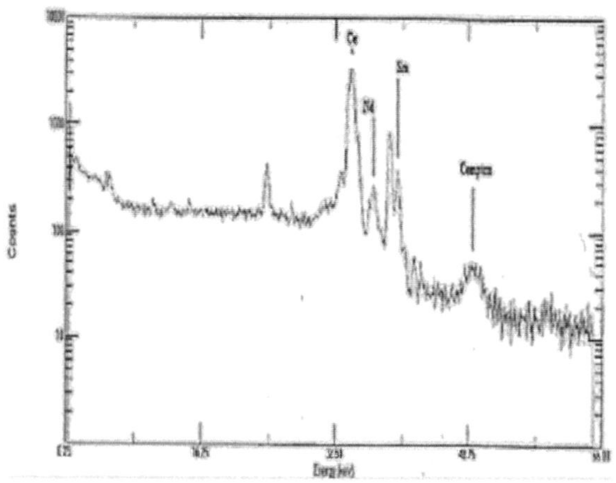

Figure 2. XRF spectrum of the Ce-dominant precipitate formed at pH 4

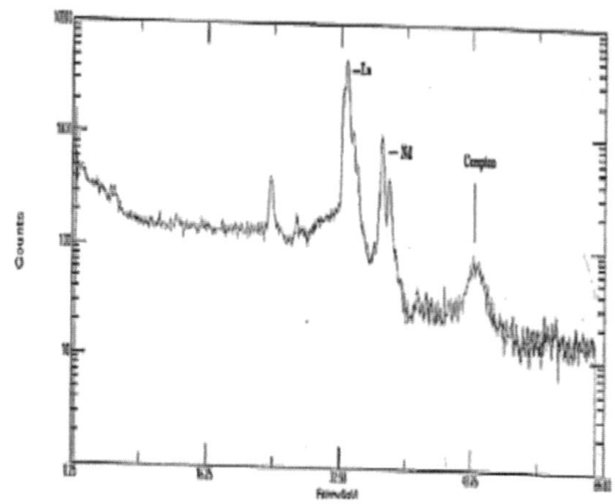

Figure 3. XRF spectrum of the of La-dominant precipitate formed by oxalate addition

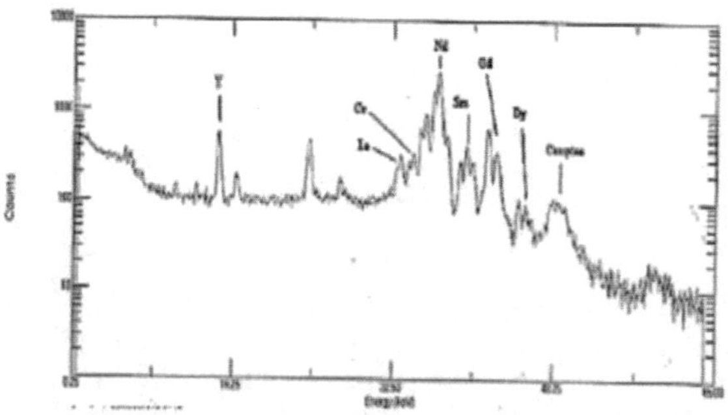

Figure 4. XRF spectrum of the Nd-dominant precipitate formed at pH 8.5

shows that the Ce peak at 34.72 keV was the most intense. The filtrate was further adjusted to pH 8.5 by addition of more 15% ammonia solution; the obtained precipitate was collected by filtration and characterized by XRF. Figure 3 shows that the most intense peak in the spectrum is that of the $Nd(OH)_3$ concentrate at 37.36 keV. The reconstituted filtrate was mixed with oxalic acid until no other precipitates were formed; in the corresponding XRF spectrum in figure 4, the most intense peak is that of $La(OH)_3$ at 33.44 keV.

3.3. Identification of cerium oxide

Figure 5 shows the FT-IR spectrum of the prepared CeO_2 precipitate. The spectrum of the CeO_2 precipitate shows several absorption bands in the 400–4000 cm^{-1} range, most notably those at 505–511 cm^{-1}, 1350–1400 cm^{-1}, 1632 cm^{-1}, and 3100–3400 cm^{-1}. The absorption band at 505–511 cm^{-1} associated with CeO_2 stretching vibration, and at 1632 cm^{-1} with H_2O (H–O–H) molecular bending frequency. The broad absorption band in the range of 3100–3400 cm^{-1} represents water that is hydrated and physically adsorbed in the sample.

The XRD patterns used to identify the samples coded Ce-1, Ce-2, and Ce-3 are shown in figures 6, figure 7, and figure 8, respectively. Figure 6 indicates that sample Ce-1 contains cerianite and neodymium oxide phases. Figures 7 and figure 8 indicate that samples Ce-2 and Ce-3 contain only cerianite.

IOP Conf. Series: Materials Science and Engineering **496** (2019) 012013 doi:10.1088/1757-899X/496/1/012013

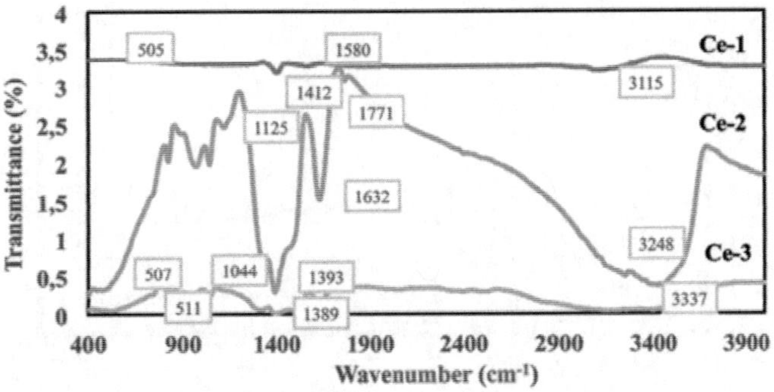

Figure 5. FTIR spectrum of the Ce fractional precipitate

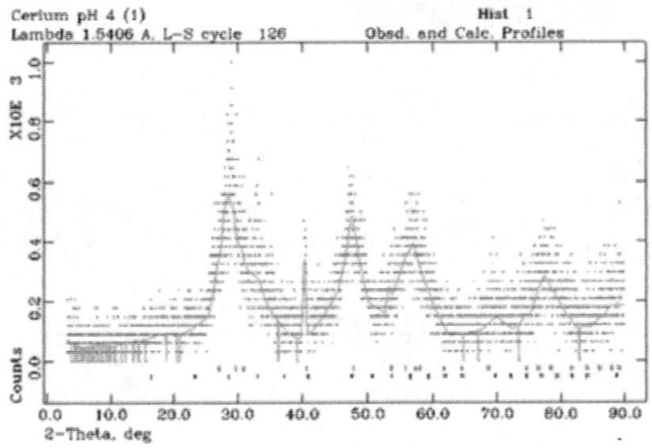

Figure 6. XRD pattern of sample Ce-1 obtained by fractional precipitation of the cerium sample

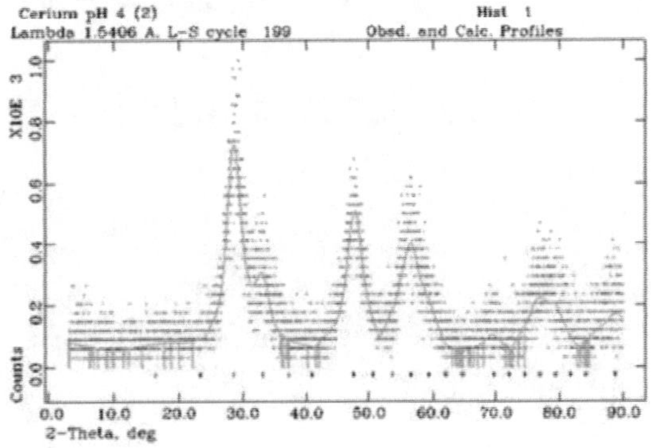

Figure 7. XRD pattern of sample Ce-2 obtained by fractional precipitation of the cerium sample

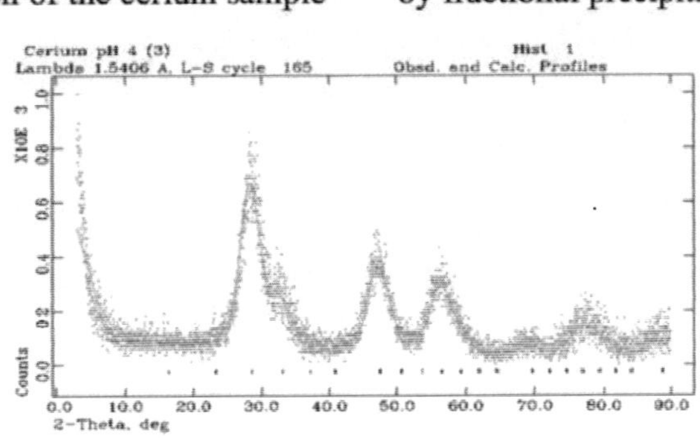

Figure 8. XRD pattern of sample Ce-3 obtained by fractional precipitation of the cerium sample

3.4. Identification of lanthanum oxide

Figure 9 shows the FT-IR spectra of the prepared La_2O_3 precipitates. The absorption peak at 3100–3400 cm^{-1} is related to vibrational modes of water molecules on the sample surface, namely H–O–H bending and the O–H stretching modes that may contribute during the spectra recording process. The weak broad absorption bands at 1320–1400 cm^{-1} are correspond to the mode of asymmetric stretching vibration of CO_3^{2-} groups. The absorption at 802–810 cm^{-1} is assigned to bending vibration modes of out-of-plane and in-plane of water and CO_2 adsorbed onto the $La(OH)_3$ surface. The sharp peak observed at approximately at 500–507 cm^{-1} is attributed to a stretching vibration of La–O bonds.

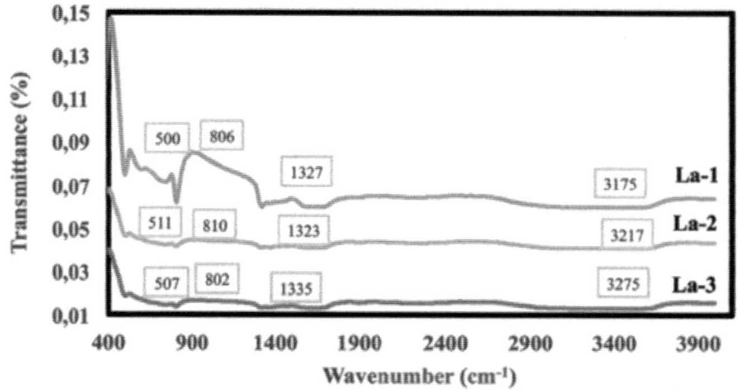

Figure 9. FT-IR spectra of the lanthanum fractional precipitates

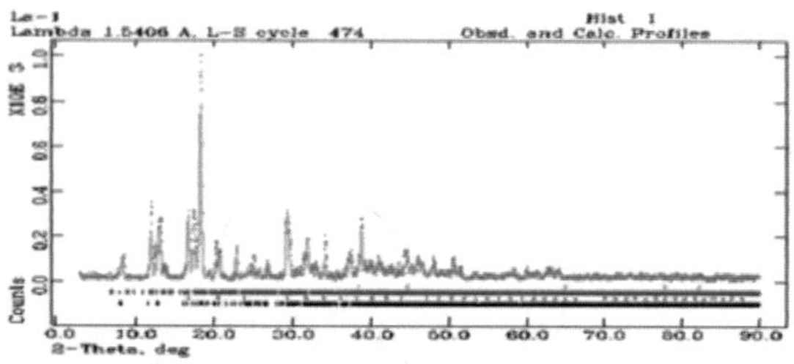

Figure 10. XRD pattern of sample La-1 obtained by fractional precipitation of the lanthanum sample

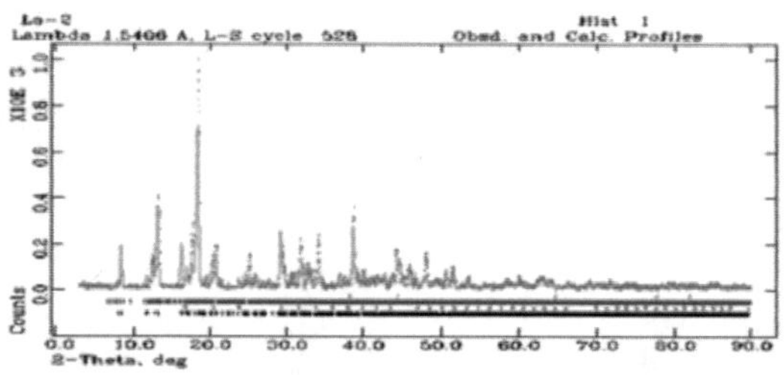

Figure 11. XRD pattern of sample La-2 obtained by fractional precipitation of the lanthanum sample

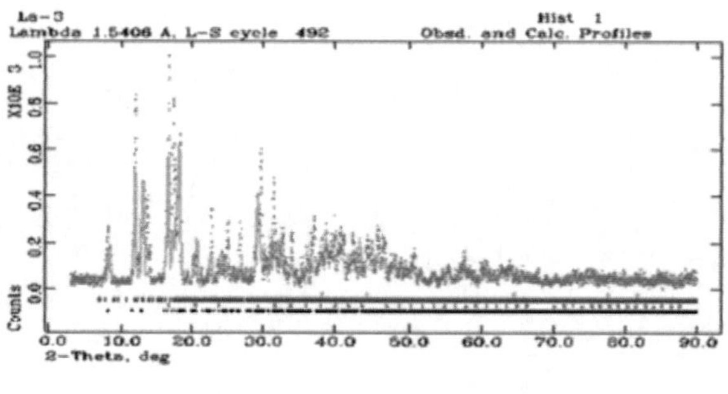

Figure 12. XRD pattern of sample La-3 obtained by fractional precipitation of the lanthanum sample

The XRD patterns used to identify the phases in samples La-1, La-2, and La-3 are shown in figure 10, figure 11, and figure 12, respectively. The results indicate that the phases present in the samples are lanthanum oxalate hydrate, aluminum, yttrium oxide, yttrium tris(enanthate) dihydrate, and lanthanum tris(2-nepthanoate) trihydrate.

3.5. Identification of neodymium oxide

Figure 13 shows the FT-IR spectra of the prepared Nd_2O_3 precipitates. The peak at 1327 cm^{-1} is ascribed to carbonate groups. The peak at 3100–3400 cm^{-1} corresponds to the stretching vibration of the O–H bond of absorbed water. The peak at approximately 820 cm^{-1} is due to neodymium oxide.

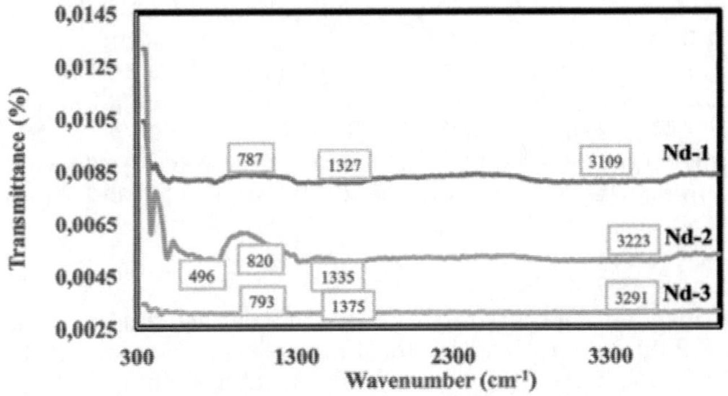

Figure 13. Spectrum FTIR of fractional precipitation of Neodymium

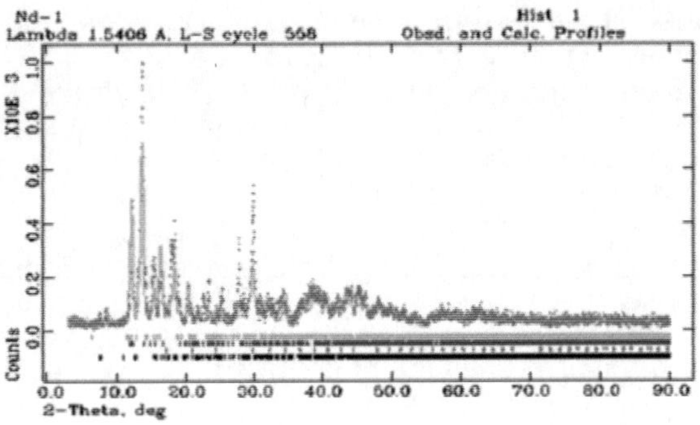

Figure 14. XRD pattern of sample Nd-1 obtained by fractional precipitation of the neodymium sample

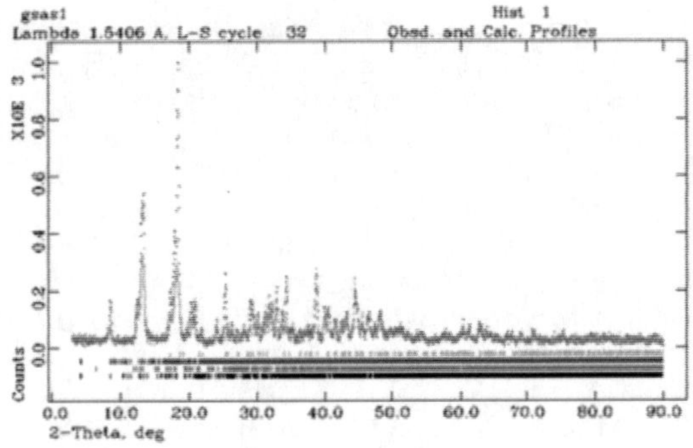

Figure 15. XRD pattern of sample Nd-2 obtained by fractional precipitation of the neodymium sample

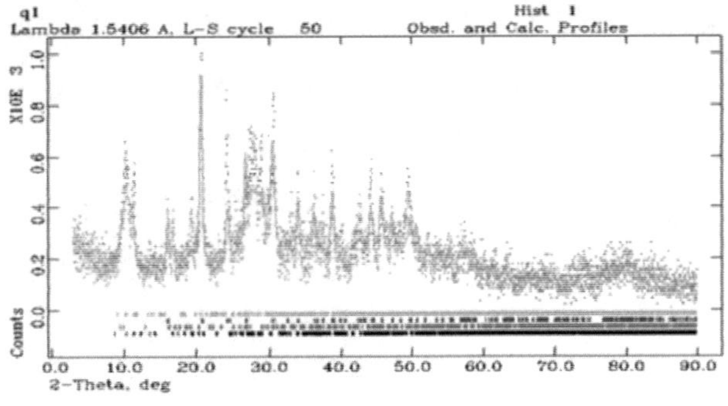

Figure 16. XRD pattern of sample Nd-3 obtained by fractional precipitation of the neodymium sample

The XRD phase analysis of the samples coded Nd-1, Nd-2, and Nd-3 samples are shown in figure 14, figure 15, and figure 16, respectively. Figures 14 and figure 15 indicate that the phases in the Nd-1 and Nd-2 samples were lanthanum oxide, neodymium oxalate hydrate, neodymium squarde hydrate, and neodymium aconite hydrate. Figure 16 indicates that the phases in sample Nd-3 are potassium neodymium silicate, trineodymium silicotetramolybdate, kazonite, and francosite.

4. Conclusions

In this study, we identified and analyze the composition of REOH from PTBGN-BATAN, Indonesia. The XRF spectra showed three dominant elements of La, Ce, and Nd at 33.44 keV, 34.72 keV, and 37.36 keV, respectively. The FT-IR spectra showed absorption bands at 497–510 cm^{-1}, 484–492 cm^{-1}, and 810 cm^{-1} that correspond to stretching and vibration of CeO_2, a stretching vibration of La–O bonds, and to neodymium oxide. The results of both FTIR and XRF analyses were confirmed by XRD analysis, which showed that the Ce-, La-, and Nd-based precipitates consisted primarily of cerianite and neodymium oxide, lanthanum oxalate hydrate and yttrium oxide, and various neodymium oxide phases, respectively.

Acknowledgements

We are thankful for the financial support provided by Universitas Indonesia under a research grant program of International Publication Indexed for Final Project Student 2017 (PITTA 2017) no. 629/UN2.R3.1/HKP.05.00/2017. The authors would like to thank Einago (www.einago.com) for the English language review.

References

[1] Du Y P, Zhang Y W, Yan Z G, Sun L D and Yan C H 2009 *J. Am. Chem. Soc.* **131** 16364–5
[2] Shen B G, Sun J R, Hu F X, Zhang H W and Cheng Z H 2011 *Adv. Mater.* **21** 4545–64
[3] Binnemans K, Jones P T, Blanpain B, Van Gerven T and Pontikes Y 2015 *J. Clean Prod.* **99** 17–38
[4] Abhilash S S, Meshram P and Pandey B D 2016 *Hydrometallurgy* **160** 47–59
[5] Xie F, Zhang T A, Dreisinger D and Doyle F 2014 *Miner. Eng.* **56** 10–28
[6] Zhu Z W, Pranolo Y and Cheng C Y 2015 *Miner. Eng.* **77** 185–96
[7] Samsonov M D, Trofimov T I, Kulyako Y M, Vinokurov S E, Malikov D A, Batorshin G S and Myasoedov B F 2015 *Radiochem.* **57** 343–47
[8] Singh D K, Singh H, Gupta C K and Mathur J N 2001 *J. Radioanal. Nucl. Chem.* **250** 123–8
[9] Tong H, Wang Y L, Liao W P and Li D Q 2013 *Sep. Purif. Technol.* **118** 487–91
[10] Maiorov V G and Nikolaev A I 2006 *Russ. J. Appl. Chem.* **79** 1196–9
[11] Wang Y L, Li Y L, Liao W P and Li D 2013 *J. Radioanal. Nucl. Chem.* **298** 1651–7
[12] Zhu, T 1991 *Hydrometallurgy* **27** 231–45
[13] Li D Q, Ji E R, Xu W and Ni J Z 1987 *Chin. J. Appl. Chem.* **4** 36–41
[14] ZuoY, Chen J, and Li D Q 2008 *Sep. Purif. Technol.* **63** 684–90
[15] Gupta B, Malik P and Deep A 2002 *J. Radioanal. Nucl. Chem.* **251** 451–6
[16] Biswas S, Rupawate V H, Hareendran K N, Parashar H and Roy S B 2015 Thorium Energy Conf. October 12-15, 2015 Mumbai ed Kumar N *et al.* (Trombay: Bhabha Atomic Research Centre) NCPth-2014/54
[17] Lu Y C, Wei H Q, Zhang Z F, Li Y L, Wu G L and Liao W P 2016 Hydrometallurgy **163** 192–7

IOP Conf. Series: Materials Science and Engineering **496** (2019) 012014 doi:10.1088/1757-899X/496/1/012014

Effect of rice husk pretreatment on the conversion of cellulose to levulinic acid over the Mn_3O_4/ZSM-5 catalyst

I Novitasari, D U C Rahayu and Y K Krisnandi

Department of Chemistry, Faculty of Mathematics and Natural Sciences (FMIPA), Universitas Indonesia, Depok 16424, Indonesia

Corresponding author: yuni.krisnandi@sci.ui.ac.id

Abstract. A considerable amount of low-cost wasted rice husk that has not been optimally utilized is available in Indonesia. Almost half of the rice husk, which comprises a biomass resource, is composed of cellulose. Cellulose can be converted to platform chemicals such as levulinic acid. However, cellulose is typically bonded strongly to lignin by β-(1,4)-glycosidic linkages. In this study, we examine the use of four rice husk pretreatment methods for the conversion of cellulose to levulinic acid, namely, dewaxing followed by delignification to remove lignin from biomass, dewaxing followed by ultrasonication in an ethanol solution, ultrasonication using a phosphoric acid solution, and only dewaxing. Meanwhile, hierarchical ZSM-5 impregnated by manganese oxide was used as a catalyst in the conversion of cellulose to levulinic acid at 130 °C. The conversion products were characterized using high-performance liquid chromatography. The dewaxing product after delignification afforded the highest yield of 21.71%, followed by the product ultrasonicated using phosphoric acid (8.81% yield). Hence, ultrasonication using phosphoric acid demonstrates promise for producing platform chemicals by the use of low amounts of chemical compounds. The results of this study can provide useful information regarding the utilization of rice husk waste as a substrate for converting biomass to levulinic acid.

Keywords: cellulose, delignification, hierarchical Mn_3O_4/ZSM-5, levulinic acid, rice husk

1. Introduction

In 2015, 75.36 million tons of dried milled grain was used for the production of rice in Indonesia [1]. Each grain milling process can afford 20–30% of rice husks [2]. The abundant amount of low-cost rice husk waste has not been optimally utilized. Several studies have reported the utilization of rice husk as a biomass resource for raw materials, energy use, or fuel for industries. Cellulose is one of the biomass sources that can be utilized from rice husks.

Rice husk is a biomass where almost half of its composition comprises cellulose, which can be converted into platform chemicals such as levulinic acid. Currently, levulinic acid is still produced on a low scale because of the high price of raw materials and the low yield of the product. In addition, cellulose is typically bonded strongly to lignin by β-(1,4)-glycosidic linkages. Pretreatment was required to remove the lignin from the biomass, which was carried out using four methods. Several methods have been widely used to produce levulinic acid [3]. One method involves the conversion of compounds using a catalyst; in this case, hierarchical ZSM-5 impregnated by manganese oxide was used as the catalyst for the conversion of cellulose to levulinic acid at 130 °C. The product obtained from the conversion was characterized using high-performance liquid chromatography (HPLC). The results of this study will provide useful information about the utilization of rice husk waste as a substrate for the conversion of biomass to levulinic acid.

2. Experimental

2.1. Materials

Rice husk, ethanol, Aqua Dest, Aqua Bidest, levulinic acid (97%), sulfuric acid (96% p.a.), $NaClO_2$ (technical grade), phosphoric acid (85% p.a.), hydrogen peroxide (30% p.a.), n-hexane (technical grade), $Mn(NO_3)_2{\cdot}4H_2O$ (p.a.), sodium hydroxide (10%), and glacial acetic acid were used in this experiment. Hierarchical ZSM-5 was provided by the Solid Inorganic Framework Research Group, Department of Chemistry, Universitas Indonesia. For analysis, an HPLC Ultimate 3000 system with a pump, an autosampler, an Rs column compartment, a UV diode-array detector, and a Rezex™ ROA-organic acid H+ column (8%), with 0.005 N H_2SO_4 as the eluent.

2.2. Synthesis of Mn_3O_4/ZSM-5

A 2% w/w solution of Mn(II) was added to 1 g of ZSM-5, and the mixture was stirred for 24 h at room temperature until a paste was formed, followed by drying overnight at 60 °C, followed by calcination at 550°C for 3 h. The synthesized zeolites were characterized using X-ray diffraction (XRD), Fourier transform infrared (FTIR) spectroscopy, and atomic absorption spectrophotometry.

2.3. Dewaxing

First, rice husk (~10 g of 100 μm granules) was dispersed into 160 mL of n-hexane and 80 mL of ethanol in a round flask and then connected to a Soxhlet apparatus. The solution's temperature was set at 80°C for 6.5 h. This process was carried out twice, and the residue was cooled for 12 h.

2.4. Delignification using NaOH

First, the sample was dissolved in 10% NaOH [ratio of the sample to NaOH was 1:25 (w/v)], followed by heating to 55 °C for 90 min. The residue was then filtered and neutralized using Aqua Dest. This process was carried out twice.

2.5. Ultrasonication using ethanol

The sample from the dewaxing process was added to an ethanol solution and blended twice using an ultrasonicator for 3 min. Then, it was filtered, and the residue was cooled to room temperature.

2.6. Ultrasonication using phosphorous acid

The sample from dewaxing was added to a 40% phosphorous acid solution, followed by the addition of 0.25 mL of 30% hydrogen peroxide, followed by ultrasonication for 3 min. The Mn_3O_4 catalyst was added to the mixture and subjected to ultrasonication for 3 min.

2.7. Conversion of cellulose to levulinic acid

The reaction was carried out in a reactor placed in an oil bath. First, 0.5 g of pretreated rice husk, 10 mL of 40% H_3PO_4, and 0.25 mL of 30% H_2O_2 were added in a reactor, followed by the addition of 0.05 g of the Mn_3O_4/ZSM-5 catalysts and heating for 8 h at 130°C. The reaction product was then cooled in an ice tub. The products from the conversion process were identified using HPLC. The flow rate of the eluent was set to 0.5 mL/min with a column temperature of 40°C and a pressure of 360–370 psi. Three UV lamp detectors were used at wavelengths of 210 nm, 220 nm, and 285 nm, respectively.

3. Results and discussion

3.1. Pretreatment of rice husk

First, rice husk was ground to produce a small-sized sample for increasing homogeneity and permitting facile dissociation of the lignin and cellulose polymer chains. Therefore, it is easier to separate lignin and cellulose. Next, pretreatment began with a dewaxing process, which involves the removal of the extractive substances present in the rice husk. These substances can become impurities in the conversion reaction. Among others, extractive substances include wax, fat, fatty acids, terpenes, and steroids [3]. Polar and nonpolar solvent mixtures, such as ethanol and *n*-hexane, were used in the dewaxing process, where polar compounds were dissolved in ethanol and nonpolar solvents were dissolved in *n*-hexane. Several tests from the American Society for Testing and Materials methods and lignin determination method were employed to examine the chemical components in rice husks, such as water, lignin, holocellulose, and α-cellulose (table 1) [4].

Table 1. Chemical components in the rice husk sample.

Components	Percentage amount (%)
Water	3.46
Extractive component	1.40
Lignin	36.01
Holocellulose	51.43
α-Cellulose	29.40

Table 2. Analysis of the lignin components in the rice husk sample.

Sample	Lignin content (%)
Dewaxing and ultrasonication using phosphoric acid	36.03
Delignification using NaOH	19.55
Ultrasonication using an ethanol solution	35.04

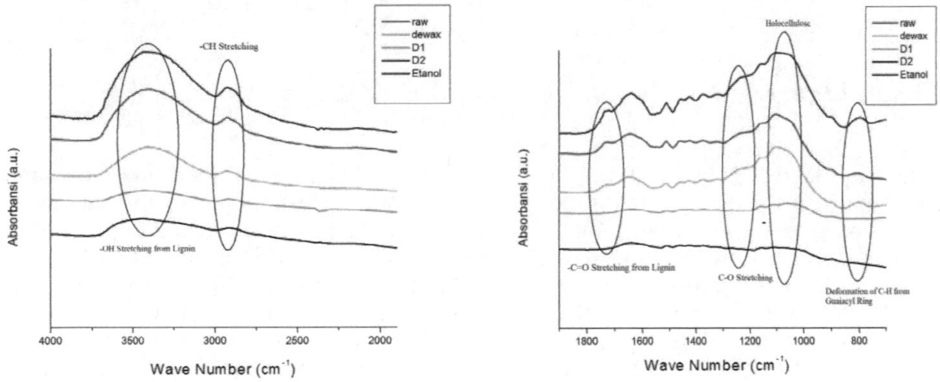

(a) (b)

Figure 1. (a) FTIR spectra of rice husks before and (b) after pretreatment

Delignification involves the removal of lignin from lignocellulosic material to obtain cellulose with a considerable purity and decreased crystallinity, which induces the decomposition of cellulose into glucose [5]. Delignification using sodium hydroxide was carried out, with the highest amount of lignin removed (45.70%) because NaOH as a base can hydrolyze the structures of lignin, both crystalline and amorphous parts, and separate lignin and hemicellulose. Based on the analysis of the lignin content before and after delignification, the lignin content decreased (table 2).

In the second method (i.e., dewaxing followed by ultrasonication using phosphoric acid), phosphoric acid was used because of its use as a solvent in the conversion reaction. All compounds in the reaction were subjected to ultrasonication; hence, lignin compounds are not reduced. Dewaxing followed by ultrasonication using ethanol was basically the same method as the previous one, but ethanol was used as the solvent during the ultrasonication of rice husk, followed by the dissolution of the product in the phosphoric acid solution for the conversion reaction. The lignin component obtained by this treatment did not significantly decrease. In some studies, an organic solvent was used as a mixture for the delignification of lignocellulosic biomass [5,6]. These two processes aimed to decrease the bond between the lignin and cellulose to enhance the reactivity of the cellulose conversion into levulinic acid.

Some rice husks before and after pretreatment were characterized by FTIR (figure 1). The peak intensity of the hydroxyl group (–OH) decreased at 3000–3500 cm^{-1}, indicative of a change in the hydroxyl group possibly caused by the reduced phenolic compounds in lignin. The C–H stretching peaks of the methoxy group were observed at 2930 cm^{-1}, which further confirmed that C–H bonds in cellulose were covered by lignin. In addition, the absorption peak observed at 1730 cm^{-1} decreased, which showed a decrease in the stretching vibration of the C=O group present in the lignin compound.

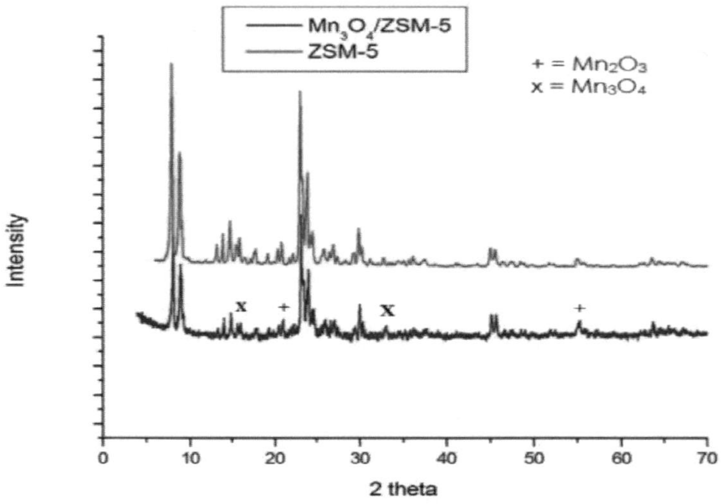

Figure 2. XRD patterns of hierarchical ZSM-5 and Mn_3O_4/ZSM-5.

The peak intensity decreased at 1450 cm^{-1}. Rojith and Singh [7] have reported that the decrease in the peak intensity indicates a change in the bending vibration, corresponding to the methoxy group in lignin. In addition, the differences of intensity at 1240 cm^{-1} as a C–O vibration correspond to the phenolic group of lignin. Kahar [8] has also reported a decrease in the intensity of the peak corresponding to the deformation of the C–H group from the guaiacyl ring present in the lignin structure at 815 cm^{-1}. NaOH pretreatment can also damage the hemicellulose covering the cellulose. Based on the result obtained from the analysis of lignin component and FTIR characterization, delignification using NaOH affords the best result for the removal of the lignin component from rice husks.

3.2. Catalytic properties

Hierarchical zeolite ZSM-5 that was used in the conversion reaction was synthesized according to the Wang method with some modification [9]. The synthetic product was characterized by XRD to determine the crystals formed, followed by energy-dispersive X-ray spectroscopy (EDX) analysis to determine the ratio of Si to Al. Specific peaks for ZSM-5 (i.e., two peaks at 2θ = 7–10° and three peaks at 2θ = 22–25°) were observed (figure 2) [10]. Based on the EDX analysis, the ratio of Si/Al was 32.9580. The synthesized ZSM-5 was then impregnated using manganese metal, which was bound to ZSM-5 as an oxide and did not change the ZSM-5 crystal structure. These results indicated that the catalyst is successfully prepared. By XPS characterization, manganese oxide is bound to ZSM-5 in the Mn_3O_4 form.

3.3. Conversion of cellulose to levulinic acid

Cellulose conversion occurs in three major stages. The first stage involves the dissociation of intramolecular and intermolecular hydrogen bonds. The second stage involves the dissociation of the β-(1,4)-glycosidic linkage via a Fenton-like reaction over Mn metal in equation (1). The last stage involves the conversion of glucose to levulinic acid in equation (2).

$$Mn^{2+} + H_2O_2 \rightarrow Mn^{3+} + HO^- + HO^{\bullet} \quad (1)$$

$$Mn^{3+} + H_2O_2 \rightarrow Mn^{2+} + H_2O^{\bullet} + H^{+\bullet} \quad (2)$$

The percentage yields of levulinic acid based on the different pretreatment methods are summarized in figure 3. Sample delignification using NaOH afforded the highest yield (figure 3). Hence, alkali solutions afford the best result for the removal of lignin from cellulose. Then, dewaxing

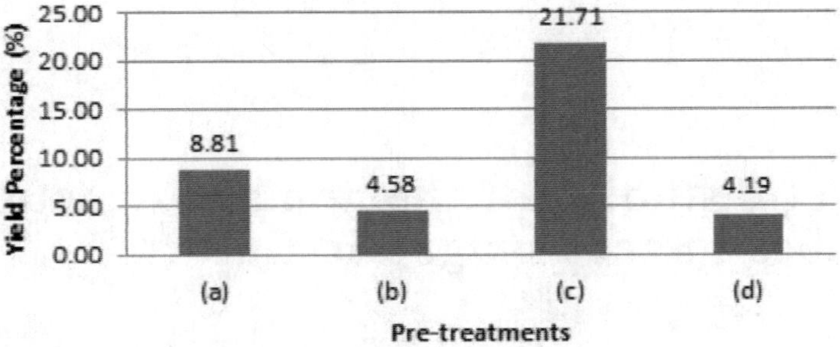

Figure 3. Percentage yields of levulinic acid based on different pretreatment methods: (a) ultrasonication with H_3PO_4, (b) ultrasonication with ethanol, (c) delignification with NaOH, and (**d**) dewaxing only.

followed by ultrasonication in phosphoric acid provided a new perspective for the removal of lignin from cellulose. It might not afford the best result, but it demonstrates promise for producing platform chemicals using low amounts of chemicals.

Glucose was observed in the chromatogram (not shown) as the substrate from the hydrolyzed cellulose. Besides levulinic acid, the reaction also afforded formic acid as a by-product and hydroxymethylfurfural (HMF) as the intermediate product. A different UV spectrum was observed for HMF compared with the other products. The results obtained from this conversion reaction can be a useful resource for producing other chemical compounds.

4. Conclusions

Based on the results obtained in this study, pretreatment by delignification using NaOH afforded the highest yield of levulinic acid (21.7%). However, pretreatment by ultrasonication using H_3PO_4 also demonstrated promise (8.81%) because of the low amounts of chemicals used, the same amount of cellulose as the raw sample, and the high efficiency for producing platform chemicals.

Acknowledgements

Authors wishing to acknowledge financial funding and support from Ministry of Research Technology and Higher Education Republic of Indonesia (Kemenristekdikti RI) through PUPT BOPTN research grant with contract number 2719/UN2.R3.1/HKP05.00/2017 and also Einago (www.einago.com) for the English language review.

References

[1] Statistics Indonesia 2015 *Rice Production in Indonesia* (Jakarta: Biro Pusat Statistik)
[2] Hsu H W and Luh B S 1980 *Rice Hull in Rice Product and Utilization* ed B S Luh (New York: Avi Publishing Co Inc.) pp 736–40
[3] Basu P 2013 *Biomass Gasification, Pyrolysis and Torrefaction* ed P Basu (Boston: Academic Press) pp 47–86
[4] Dence CW 1992 *Methods in Lignin Chemistry* ed S Y Lin and C W Dence (Heidelberg: Springer-Verlag) pp 33–61
[5] Mosier N, Wyman C, Dale B, Elander R, Lee Y Y, Holtzapple M and Ladisch M 2005 *Bioresour. Technol.* **96** 673–86
[6] Chapple C, Ladisch M and Meilan R 2007 *Nat. Biotechnol.* **25** 746–8
[7] Rojith G and Singh I S B 2013 *Journal of Environmental Research Development* **7** 1333–9
[8] Kahar P 2014 *Environmental Biotechnology: New Approaches and Prospective Applications* ed M Petre (London: InTechOpen Ltd.) pp 65–87
[9] Wang L, Zhang Z, Yin C, Shang Z and Xiao F S 2010 *Micro. Meso. Mater.* **131** 58–67
[10] Treacy M M J and Higgins J B 2001 *Collection of Simulated XRD Powders Patterns for Zeolites* (London: Elsevier)

IOP Conf. Series: Materials Science and Engineering **496** (2019) 012015 doi:10.1088/1757-899X/496/1/012015

Micromagnetic studies of the domain structure of $BaFe_{12}O_{19}$ cubes in zero magnetic field

D Djuhana[1*], C Kurniawan[2], and J W Utomo[1]

[1]Department of Physics, Faculty of Mathematics and Natural Sciences (FMIPA) Universitas Indonesia, Depok 16424, Indonesia

[2]Center for Physics Research, The Indonesian Institute of Sciences (LIPI), Kawasan Puspitek No. 441-442 Serpong, Banten 15314, Indonesia

Corresponding author: dede.djuhana@sci.ui.ac.id

Abstract. This article reports an investigation of the domain structure of cube-shaped $BaFe_{12}O_{19}$ models in zero magnetic field using the public-domain OOMMF micromagnetic simulator to model cubes with sides varying from 100 to 900 nm. A Bloch wall appears in $BaFe_{12}O_{19}$ cubes larger than about 510 nm and a single-domain structure appeared below 510 nm. The transition from single- to multiple- domain structure in the cube-shaped models occurred near the size predicted by analytical solutions of Kittel's equation. Analysis of the magnetization energy showed that the exchange energy increased with increasing size once the Bloch wall was formed. We also calculated the width of the Bloch walls from the full width at half maximum of the transverse component of the magnetization vector. The domain wall width increased, as the modeled cubes grew larger.

1. Introduction

Barium hexaferrite ($BaFe_{12}O_{19}$) is a very popular magnetic material owing to its high magnetocrystalline anisotropy, excellent chemical stability, and high Curie temperature (T_c) [1-3]. $BaFe_{12}O_{19}$ is used widely in permanent magnets, microwave absorbers, and memory storage materials [4-7]. In addition to experimental methods, numerical simulations of micromagnetic phenomena are an effective tool to understand magnetic domain structure in ferromagnetic materials [8,9]. We have previously studied the transition from single-domain (SD) structures to multi-domain (MD) structures in $BaFe_{12}O_{19}$ for hexagonal and cylindrical models. The transition occurred around diameters of 430 and 410 nm for the hexagonal and cylindrical models, respectively [10]. However, the two models differed greatly from theoretical predictions, by around 18 %. This disagreement indicates that a better theoretical model is needed to predict the domain structure in $BaFe_{12}O_{19}$, especially the size at which the transition from SD to MD structures occurs.

In this article, we report an investigation of the domain structure of microscale cubes of $BaFe_{12}O_{19}$ using public-domain micromagnetic simulation software. A transition from SD to MD structure appeared when the size of the cube passed through a critical value. We also calculated the width of the domain wall and the magnetization energy in terms of the side length of the cube. Finally, we analytically solved for the critical transition side length and domain wall width using Kittel's equation for comparison with the numerical results.

2. Simulation method

Investigations of the domain structure of $BaFe_{12}O_{19}$ were performed in the OOMMF public-domain micromagnetic software [11] using the Landau–Lifshitz–Gilbert (LLG) [12] equation:

$$\frac{d\vec{m}}{dt} = -\gamma\vec{m}\times\vec{H}_{eff} + \alpha\vec{m}\times\frac{d\vec{m}}{dt} \quad (1)$$

IOP Conf. Series: Materials Science and Engineering **496** (2019) 012015 doi:10.1088/1757-899X/496/1/012015

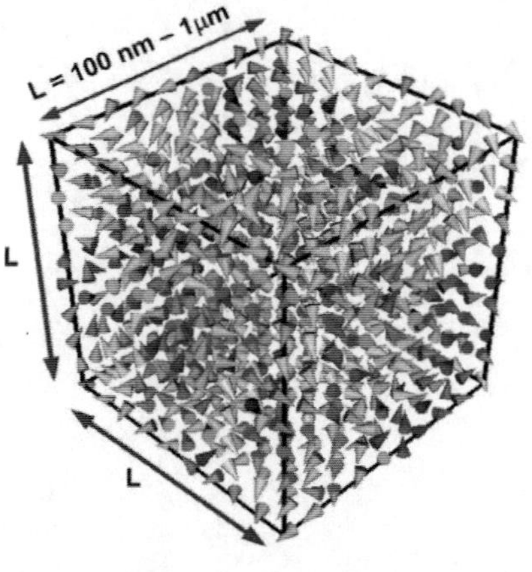

Figure 1. Geometry and dimensions of $BaFe_{12}O_{19}$ cube-shaped model with random initial magnetization. The cube side was varied from 100 to 900 nm in this simulations.

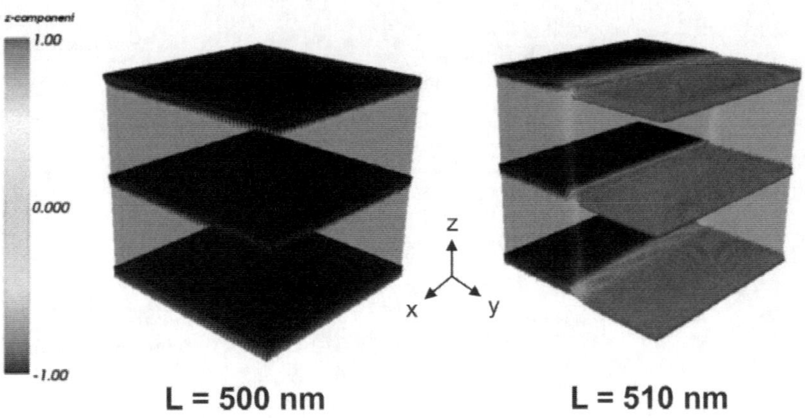

(a) (b)

Figure 2. $BaFe_{12}O_{19}$ cubes show (a) SD structure with sides of 500–nm and (b) MD structure with a Bloch wall between domains with sides larger than 510 nm. The colour in this plot represents magnetization in the *z* direction.

where γ is the gyromagnetic ratio 2.21×10^5 m/A·s, α is the damping factor, $\vec{m}$ is the magnetization vector, and $\vec{H}_{eff}$ is the effective field vector. The sides of the model $BaFe_{12}O_{19}$ cubes were varied from 100 to 1 μm, and the simulation cell size was $10 \times 10 \times 10$ nm^3. This cell volume was chosen because the exchange length of $BaFe_{12}O_{19}$ is 20.8 nm [13]; hence, the calculations will be adequate at this resolution. The following material parameters of $BaFe_{12}O_{19}$ were used: the magnetization saturation $M_s = 2.75 \times 10^5$ A/m, exchange constant $A = 2 \times 10^{-11}$ J/m, anisotropy constant $K = 96 \times 10^3$ J/m^3, and damping factor $\alpha = 0.01$ [14]. The simulation was performed in a zero magnetic field with the spin configuration initially random, as depicted in figure 1. The domain structure was observed once the magnetization system reached the equilibrium state after about 10 ns.

3. Results and discussion

After the system reached equilibrium, we observed SD structure in cubes with sides smaller than 510 nm, and MD structure in cubes with sides larger than 510 nm. A Bloch wall formed in cubes with sides larger than 510 nm. Figure 2 shows the domain structure in cubes with 500-nm and 510-nm sides, respectively. The transition from SD to MD structure indicates a critical size for this

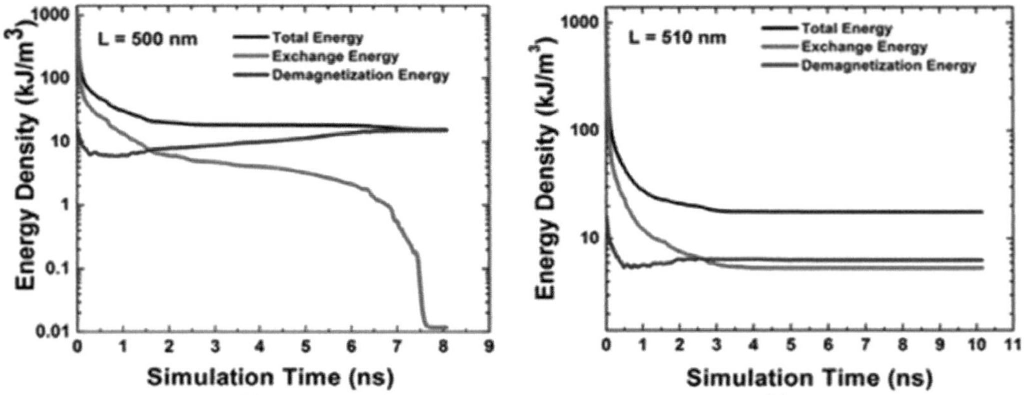

(a) (b)

Figure 3. The magnetization energies of $BaFe_{12}O_{19}$ cubes with side lengths (a) L = 500 nm and (b) L = 510 nm. The demagnetization energy is dominant over the exchange energy with SD structure, while the exchange energy shows increasing once MD structure appears.

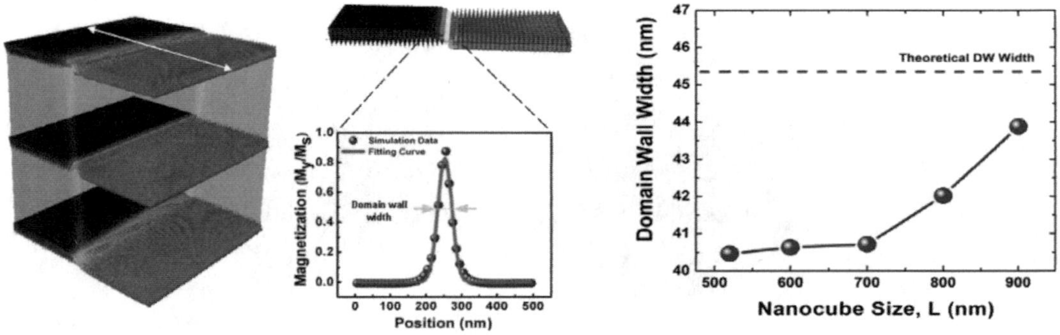

(a) (b)

Figure 4. (a) Determination of the width of Bloch wall in $BaFe_{12}O_{19}$ domain structure from the FWHM of the transverse magnetization (M_y) and (b) width $BaFe_{12}O_{19}$ Bloch wall related to the side length of a $BaFe_{12}O_{19}$ cube. The dashed line plots the domain wall width obtained by theoretical calculations.

ferromagnetic material. For comparison, we also calculated the critical size of cube-shaped $BaFe_{12}O_{19}$ structures using Kittel's formula $D_c = (2\times 9\sigma_w)/\mu_0 M_s^2$ with the domain wall energy $\sigma_w = 2\sqrt{AK_1}$. The predicted critical size is around 525 nm with 3% error, as reported previously [15-17]. The difference between this prediction and our numerical results is caused by the discretization involved in the micromagnetic simulation calculations.

We also analyzed magnetization energies of demagnetization energy and exchange energy. For the SD case, the demagnetization energy was higher than the exchange energy at equilibrium. The exchange energy therefore approaches the demagnetization energy when multiple domains appear. On the other hand, the exchange energy tends to increase for cubes larger than the critical size. Magnetization energy profiles for the cubes with 500-nm and 510-nm sides are shown in figure 3.

In further calculations, we determined the width of the $BaFe_{12}O_{19}$ Bloch wall using the full width at half maximum (FWHM) of the transverse magnetization component (M_y). The FWHM was calculated using Gaussian fitting of the transverse magnetization, as shown in figure 4a. The width of the $BaFe_{12}O_{19}$ Bloch wall in cubes with size lengths 510 to 900 nm is presented in figure 4b. The width of the Bloch wall increased as the cube side length increased. To understand this relation, we calculated the width of the domain wall based on an approximation of Kittel's equation $W_{DW} \approx \pi\sqrt{A/K_1}$ and found a value of 45.35 nm [17,18]. Comparing these values, the width of the $BaFe_{12}O_{19}$ Bloch wall obtained from the FWHM value is close to the theoretical prediction.

4. Conclusions

We have systematically investigated the domain structure of cubic $BaFe_{12}O_{19}$ using micromagnetic simulations with no external magnetic field. A Bloch wall structure appears in $BaFe_{12}O_{19}$ cubes with side lengths larger than 510 nm, and the cubes exhibit SD structure when the side length is less than 510 nm. The exchange energy tends to increase with side length when the micromagnetic structure has multiple domains. We have also calculated the width of the Bloch wall from the FWHM of the transverse magnetization component. The domain wall is wider in larger cubes of $BaFe_{12}O_{19}$. Theoretical calculations were also used to predict the transition regime and the domain wall width using Kittel's equation. Interestingly, the micromagnetic simulation results are close to the theoretical predictions. The results of the above numerical simulations suggest that micromagnetic domain structures are influenced by energy contributions in a ferromagnetic system, which are determined in part by the scale of the material.

Acknowledgements

This work is fully supported by PUPT Grant year 2017 funded by Ministry of Research, Technology and Higher Education (Kemenristekdikti) Republic of Indonesia with contract number 2700/UN2.R3.1/HKP05.00/2017. The authors would like to thank Einago (www.einago.com) for the English language review.

References

[1] Park J, Hong Y K, Kim S G, Kim S, Liyanage L S I, Lee J, Lee W, Abo G S, Hur K H and An S Y 2014 *J. Magn. Magn. Mater.* **355** 1–6
[2] Manikandan M and Venkateswaran C 2014 *J. Magn. Magn. Mater.* **358-359** 82–6
[3] Liu Y, Drew M G B and Liu Y 2012 *Mater. Chem. Phys.* **134** 266–72
[4] Ahmed M A, Mansour S F and Ismael H 2015 *J. Magn. Magn. Mater.* **378** 376–88
[5] Wang Y, Huang Y and Wang Q 2012 *J. Magn. Magn. Mater.* **324** 3024–8
[6] Vinnik D A, Zherebtsov D A, Mashkovtseva L S, Nemrava S, Semisalova A S, Galimov D M, Gudkova S A, Chumanov I V, Isaenko L I and Niewa R 2015 *J. Alloys Compd.* **628** 480–4
[7] Bahadur A, Saeed A, Iqbal S, Shoaib M, Ahmad I, Rahman M S, Bashir M I, Yaseen M and Hussain W 2017 *Ceram. Int.* **43** 7346–50
[8] Gilbert T L 2004 *IEEE Trans. Magn.* **40** 3443–9
[9] Kronmüller H 2007 *Handbook of Magnetism and Advanced Magnetic Materials* vol 2, ed H Kronmüller and S Parkin (Hoboken: John Wiley & Sons) pp 677–715
[10] Djuhana D, Oktri D C C and Kim D H 2014 *Adv. Mater. Res.* **896** 414–7
[11] Donahue M J and Porter D G 2008 *OOMMF User's Guide, Version 1.0* (Gaithersburg: National Institute of Standards and Technology)
[12] Miltat J E and Donahue M J 2007 *Handbook of Magnetism and Advanced Magnetic Materials* vol 2, ed H Kronmuller and S Parkin (John Wiley & Sons, Ltd.) pp 716–36
[13] Song F, Shen X, Liu M and Xiang J 2012 *J. Solid State Chem.* **185** 31–6
[14] Biskeborn B and Jubert P O 2010 *IEEE Trans. Magn.* **46** 880–5
[15] Goto K, Ito M and Sakurai T 1980 *Jpn. J. Appl. Phys.* **19** 1339–46
[16] Hirayama T, Ru Q, Tanji T and Tonomura A 1993 *Appl. Phys. Lett.* **63** 418–20
[17] Kittel C 2005 *Introduction to Solid State Physics* 8th ed (Hoboken: Wiley)
[18] Kittel C 1949 *Rev. Mod. Phys.* **21** 541–83

Modeling of the temperature and magnetic field dependence on the density of states of G4-DNA molecules

R Rahman and E Yudiarsah

Department of Physics, Faculty of Mathematics and Natural Sciences (FMIPA) Universitas Indonesia, Depok 16424, Indonesia

Corresponding author: e.yudiarsah@sci.ui.ac.id

Abstract. In this article, we investigate the charge-transport properties of G4-DNA molecules that are 32 base pairs long using the tight-binding Hamiltonian and Green function in numerical calculations. The transport properties are studied by calculating the density of states (DOS) of the G4-DNA model under the influence of various external magnetic field strengths and temperatures. We found that a shift in phase coherence lowers the DOS as the magnetic field increases. At zero temperature, the DOS spectrum is split into two bands of electron energies with a specific gap of zero values between them. Increasing the temperature decreases the DOS at all electron energies and narrows the energy gap in the DOS spectra when the frequency of the molecule's twisting motion is 0.77 meV.

Keywords: G4-DNA, magnetic field, temperature, DOS

1. Introduction

Nanotechnology can be used to improve the performance of electronic devices [1]. Biomolecular electronics are a promising application of nanotechnology, including the study of future electronic devices based on DNA molecules. DNA molecules can self-assemble into coherent structures, and they conduct electrical current, so that DNA and other chains of nucleic acids can form relatively large molecules for use in electronic devices in a bottom-up fashion. Nanowire structures, for instance, can be fabricated from chains of nucleic acids [2]. Interbase π–π coupling interactions mediate the charge-transport process in DNA molecules [3].

In addition to the standard double-stranded DNA (dsDNA) molecule that consists of adenine (A), cytosine (C), guanine (G), and thymine (T) bases, DNA chains can also be constructed by a variety of guanine-rich structures, such as guanine quadruplex (G4) structures that stack to form G4-DNA molecules [4]. The G4 structure consists of four G bases that are arranged in a square. G4 structures are stacked with separation of about 3.25 Å and are twisted with respect to the stack's symmetrical axis such that two adjacent G4 structures form an angle of about 30° [4]. A procedure for fabricating G4-DNA nanowires from poly(dG)-*n*(dC) was proposed by Borovok *et al.* [5]. The charge-transport capabilities of these G4-DNA nanowires open the possibility of using G4-DNA nanowires in electronic devices [5,6]. The four strands of a G4-DNA molecule allow more charge transport than the double strands of a dsDNA molecule do [7,8]. The counter ions present in a G4-DNA molecule, such as K^+, Na^+, and Mg^{2+}, offer stability in the π stack, which also improves the electronic properties of the structure [9,10].

In this study, we modeled G4-DNA molecules in order to study the molecule's charge-transport properties in the presence of varying applied magnetic field strengths and temperatures. The model system's charge-transport properties are modeled in terms of the density of states (DOS) at each electron energy, calculated using Green's function.

IOP Conf. Series: Materials Science and Engineering **496** (2019) 012016 doi:10.1088/1757-899X/496/1/012016

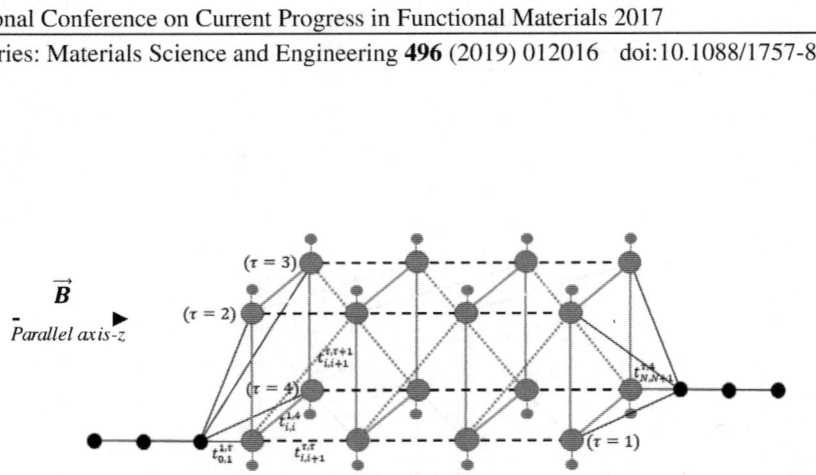

Figure 1: A schematic diagram of the G4-DNA molecule. The blue and red colors represent the guanine bases and the backbone, respectively. The black color represents the electrodes at the ends of the molecule.

2. Theoretical model

We used the tight-binding approach for numerical calculations. The DOS is calculated using Green's function. The charge transport that occurs within the molecule can be modeled using the tight-binding Hamiltonian. The tight-binding Hamiltonian of G4-DNA as depicted in figure 1, can be given by the following equation:

$$\begin{aligned} H_{\mathrm{DNA}} = \sum_{i=1}^{L}\Big[\sum_{\tau=1}^{4}\Big\{\left(\varepsilon_i^{\tau} + \varphi_i^{\tau}\right)|i,\tau\rangle\langle i,\tau| + t_{i,i+1}^{\tau,\tau}|i,\tau\rangle\langle i+1,\tau|\Big\} \\ +\sum_{\tau=1}^{4}\Big\{B_i^{\tau,q}|i,\tau\rangle\langle i,q| + t_{i,i}^{\tau,q}|i,\tau\rangle\langle i,q|\Big\} \\ +\sum_{\tau=1}^{4}\Big\{t_{i,i}^{\tau,\tau+1}|i,\tau\rangle\langle i,\tau+1| + t_{i,i+1}^{\tau,\tau+1}|i,\tau\rangle\langle i+1,\tau+1|\Big\}\Big] + h.c. \end{aligned} \quad (1)$$

In equation (1), *h.c.* stands for Hermitian conjugate. The applied magnetic field induces a phase change in the electronic state, which affects the electron-hopping constant t_{ij} [11]. The electron-hopping constant can be calculated as a function of the applied magnetic field using the Peierls phase factor [12]. If $B \neq 0$, the electron-hopping constant is given by the following equation:

$$t_{ij} = t_0 \exp\left(-i\frac{e}{\hbar}\int_i^j \vec{A}.d\vec{l}\right). \quad (2)$$

In equation (2), t_0 is the electron-hopping constant when the magnetic field is zero, $\vec{A}$ is the vector potential, and $d\vec{l}$ is the electron displacement.

The DOS of the system in an external magnetic field can be calculated using the following equation:

$$\mathrm{DOS}(\varepsilon) = -\frac{1}{\pi}\sum_{\vec{k}} \mathrm{Im}\,\mathrm{Tr}\left[G^R(\vec{k},\varepsilon)\right], \quad (3)$$

where Im indicates that the imaginary part of the trace (Tr) of the Green function matrix $G^R(\vec{k},\varepsilon)$ is summed. The Green function matrix is given by the following equation:

$$[G^R(\vec{k})] = \left[(\varepsilon(\vec{k}) + i\eta)[I] - H_{\mathrm{DNA}} - \Sigma_l^R - \Sigma_r^R\right]^{-1}, \quad (4)$$

where $\varepsilon(\vec{k})$, η, H_{DNA}, and Σ_l^R denote the electron energy as a function of the wave vector $\vec{k}$, an infinitesimal number, the Hamiltonian of the DNA molecule, and the retarded self-energy of the electrode, respectively.

The influence of temperature generates a phonon activity that changes the values of guanine-base on-sites and the electron-hopping constant. The effect of environmental temperature on the electron-hopping constant is modeled as in reference [3] by considering that the twisting angles of the bases follow a Gaussian distribution with zero mean and standard deviation $\sqrt{T/f}$, where T is the temperature and f is a parameter that depends on the twisting frequency of the bases.

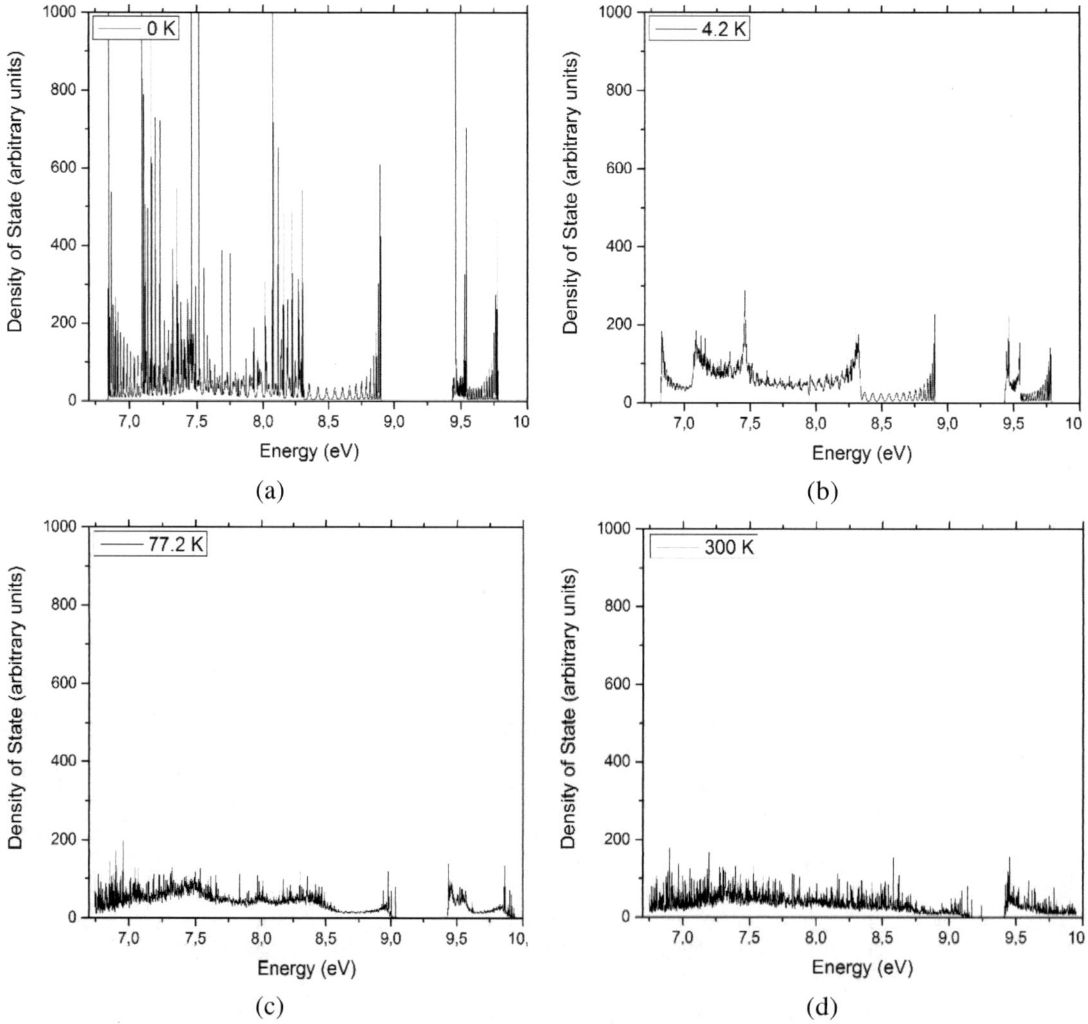

Figure 2. DOS as a function of electron energy under temperatures of (a) 0 K, (b) 4.2 K, (c) 77.2 K, and (d) 300 K for $\omega=0.77$ meV.

3. Results and discussion

The simulated DOS values are plotted for a range of temperatures and magnetic field strengths in figure 2 and figure 3, respectively. Figure 2 plots the DOS of the G4-DNA molecule against electron energy under temperatures in the range 0–300 K with a twisting frequency of f = 0.77 meV and B = 0 T. Temperature destabilizes the molecules, leading to thermal fluctuations that are expressed as phonons. In figure 2, two bands of nonzero DOS values appear across the spectrum of electron energy with an energy gap between them. At 0 K, the DOS is very high for all electron energies, and the DOS spectrum is not affected by the frequency of the twisting motion, since the standard of deviation $\sqrt{T/f}$ is zero if temperature is zero. A temperature of 300 K at f = 0.77 meV causes the electron-occupation state to change and generates many new phonon-induced states in the charge-transport process. As a result, the DOS decreases in certain energy ranges and the energy bands become irregularly shaped. The DOS energy bands also tend to widen, causing the energy gap to shrink. This result is consistent with the findings of Joe *et al.* [11] and Suhendro *et al.* [13] that nonzero temperatures tend to decrease the charge-transport capability of DNA molecules.

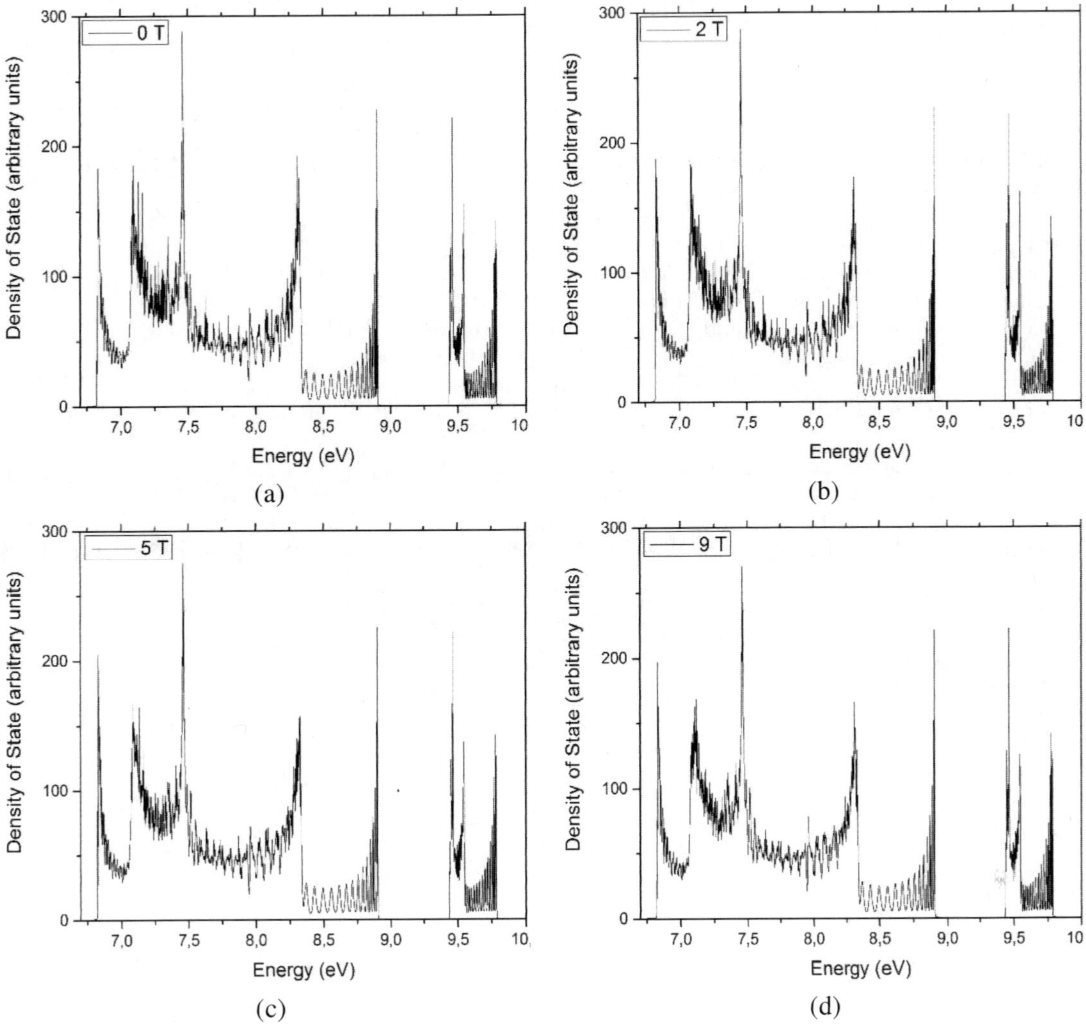

Figure 3. DOS as a function of electron energy in an applied magnetic field of (a) 0 T, (b) 2 T, (c) 5 T, and (d) 9 T for T=4.2 K with f=0.77 meV.

Next, a magnetic field is applied along the axis marked in figure 1. The applied magnetic field affects the physical properties of the system, leading to a shift in the phase coherence of the molecule [11,14]. Figure 3 shows plots of the calculated DOS as a function of electron energy with the applied magnetic fields varying from 0 to 9 T at a temperature of 4.2 K and f = 0.77 meV. A change in the electron state over the whole electron energy range corresponds to a strong applied magnetic field. The magnetic field shifts the electronic interference depending on the coupling configuration of the G4-DNA molecule. These results agree with those of Kang *et al.* [9], who found that an applied magnetic field changes the electronic state in a G4-DNA molecule.

4. Conclusions

The tight-binding Hamiltonian was used to model the influences of temperature and external magnetic field on a G4-DNA molecule. High temperatures deteriorate the electron state and create a new state occupied by phonons with irregularly shaped energy bands. This replacement tends to widen the energy bands, shrinking the gap energy for 0.77 meV. An external magnetic field affects the DOS by causing a shift in phase coherence that changes the molecule's electronic state over the whole energy range that we tested.

Acknowledgements

This research was funded by PITTA Grant year of 2017 with the following contract number 649/U2.R3.1/HKP.05.00/2017. The authors wishing to thank Lab Cisco Department of Physics, FMIPA Universitas Indonesia for allowing the usage of the laboratory and Einago (www.einago.com) for the English language reviews.

References

[1] Wang F, Willner B and Willner I 2013 *Curr. Opin. Biotechnol.* **24** 562–74
[2] Endo M and Sugiyama H 2009 *Chembiochem.* **10**, 2420–43
[3] Endres R G, Cox D L and Singh R R 2004 *Rev. Mod. Phys.* **76** 195–214
[4] Guo A M and Xiong S J 2009 *Phys. Rev. B* **80** 035115
[5] Borovok N, Molotsky T, Ghabboun J, Porath D and Kotlyar A 2008 *Anal. Biochem.* **374** 71–8
[6] Zhang Y Q, Zhang W B, Liu C R, Zhang P, Balaeff A and Beratan D N 2016 *Surf. Sci.* **652** 33–8
[7] Calzolari A, Di Felice R, Molinari E and Garbesi A 2002 *Appl. Phys. Lett.* **80** 3331-3
[8] Woiczikowski P B, Kubař T, Gutiérrez R, Cuniberti G and Elstner M 2010 *J. Chem. Phys.* **133** 035103
[9] Kang D W, Sun M L, Zuo Z W, Wang H X, Lv S J, Li X Z and Li L B 2016 *Phys. Lett. A* **380** 977–82
[10] Guo A M, Yang Z, Zhu H J and Xiong S J 2010 *J. Phys. Condens. Matter* **22** 065102
[11] Joe Y S, Lee S H, Hedin E R and Kim Y D 2013 *J. Nanosci. Nanotechnol.* **13** 3889–96
[12] Peierls R 1933 *Z. Phys. A* **80** 763–91
[13] Suhendro D K, Yudiarsah E and Saleh R 2010 *Physica B* **405** 4806–11
[14] Kang D, Jiang H, Sun Z, Qu Z and Xie S 2011 *J. Phys. Condens. Matter* **23** 55302

IOP Conf. Series: Materials Science and Engineering **496** (2019) 012017 doi:10.1088/1757-899X/496/1/012017

Dietary nano-calcium phosphate and hydrolyzed collagen for the inhibition of osteoporosis in calcium-deficient bones

R Sapundani[1], A L Juwono[1], D S Soejoko[1] and D A Astuti[2]

[1] Department of Physics, Faculty of Mathematics and Natural Sciences (FMIPA) Universitas Indonesia, Depok 16424, Indonesia

[2] Department of Nutrition and Feed Technology, Faculty of Animal Science, Institut Pertanian Bogor (IPB), Kampus Dramaga, Bogor 16680, Indonesia

Corresponding author: ariadne.laksmidevi@ui.ac.id

Abstract. Several studies exhibit that high calcium intake prevents osteoporosis-related bone loss. In this study, the dietary effects of nano-tricalcium phosphate (n-CaP) and hydrolyzed collagen (HC) against bone loss at the peak of the bone growth stage are examined, and the main objectives are to prevent osteoporosis and to induce bone formation. The Sprague Dawley mice (Rattus norvegicus) is observed to be calcium deficient (the average calcium content of serum is ≤ 8%). The mice were randomized into four groups (P1: normal purified diet, P2: HC purified diet, P3: n-CaP purified diet, and P4: n-CaP and HC purified diet). After four weeks, all the groups were euthanized, and their bone formation was analyzed. UV–Vis analysis revealed that the calcium content values were 8.20 (P1), 8.92 (P2), 9.2 (P3), and 10.35 (P4) mg/dL, respectively. The corresponding values obtained by the atomic absorption spectrophotometric analysis are 72.29 (P1), 74.25 (P2), 74.80 (P3), and 76.04 (P4) w/w% respectively. To assess the spinal calcification, the infrared spectra were recorded. The splitting factors were 2.5 (P1), 2.6 (P2), 2.5 (P3), and 2.6 (P4), indicating better absorption of n-CaP for P4. Hence, n-CaP and HC-purified diets are suitable to ensure bone growth.

Keywords: Osteoporosis, nano-calcium phosphate, hydrolyzed collagen

1. Introduction

Bone is a highly specialized form of connective tissue, which is characterized by its hardness, resilience, growth mechanisms, and self-regeneration capability after an injury via the bone healing process. Mature bones predominantly comprise the bone matrix, bone cells, and bone marrow. The bone matrix provides mechanical strength and serves as the storage area of minerals in the human body. The bone matrix is the intercellular substance of the bone tissue, comprising collage fibers, ground substances (composed of water, glucoaminoglycans, proteoglycan and glycoproteins), non-collagenous proteins, and inorganic bone salts. Bone modeling and remodeling ensures that the old or damaged bones are replaced by new bones. These processes are regulated by various osteotropic cytokines and hormones, which affect the bone cell activity for maintaining the coupling between bone resorption and formation. The imbalance of these interactions causes abnormal turnover cycles, leads to diseases such as osteoporosis (caused by an imbalance between new bone formation and old bone resorption). Osteoporosis-related fractures can decrease the quality of life. Several studies related to the recovery of bone fractures, such as autografts, allografts, and alloplastic materials, still exhibit limitations [1–3].

Table 1. Formulated diet for rats

Ingredient	**D**	**P1**	**P2**	**P3**	**P4**
			w/w%		
Rice Flour	25.00	25.00	22.00	25.00	22.00
Casein or HC (*)	18.00	18.00	30.00	18.00	30.00
Corn Oil	3.50	3.30	2.30	3.30	3.30
Sugar Flour	49.00	49.00	40.00	49.00	40.00
DL-Methionine	0.30	0.30	0.30	0.30	0.30
CMC	3.00	3.00	3.00	3.00	3.00
$Ca_3(PO_4)_2$	**0.00**	**0.20**	2.00	**0.00**	0.00
Nano-$Ca_3(PO_4)_2$	**0.00**	0.00	**0.00**	0.20	**0.20**
Vitamin Mix	0.50	0.50	0.50	0.50	0.50
NaCl	0.20	0.20	0.20	0.20	0.20
Mineral Mix (def.Ca)	0.50	0.50	0.50	0.50	0.50
TOTAL	**100**	**100**	**100**	**100**	**100**

Individual studies have reported the implications of high-calcium and special-protein diets on the recovery of bone formations. In this study, the effect of dietary nano-tricalcium phosphate (n-CaP) and hydrolyzed collagen (HC) on the increased bone density of calcium-deficient mice at the peak phase of bone growth was examined. The main objectives of this study included the prevention of osteoporosis and the induction of bone formation. At the end of the treatment, the model animals were euthanized for their serum and skeletal parts (e.g. femur and spines).

2. Materials and methods

2.1. Animals and diet preparation

White female Sprague-Dawley mice (Rattus norvegicus) at their peak phase of bone growth were utilized to examine the effects of n-CaP and HC intake [4,5]. The model animals were approved by the Ethics Committee of the Faculty of Medicine, University of Indonesia with the protocol number 17-08-0841. The formulated diets for mice were denoted as D for tricalcium phosphate deficient, P1 for normal tricalcium phosphate purified diet, P2 for HC purified diet, P3 for n-CaP purified diet, and P4 for n-CaP + HC purified diet (table 1). The raw material used for the diet comprised the purified ingredients of natural diet, which was purchased from the market. HC was obtained from cow bones, and the nanoparticle size was achieved by a high milling process (PI 700i Mixer/Mill machine) at 1000 rpm for 3 h. The purified diet-making process was conducted as follows: all the ingredients (w/w%) were mixed in water. The mixture was further stirred at a medium speed until a homogeneous dough was obtained. The dough was shaped into pellets and heated at 60 °C.

2.2. Collection and bone processing

At the end of the treatment (1 month), the animals were sacrificed. The femur and spines were cleaned from the surrounding soft tissue. Further, the bone samples were baked at 60 °C for 24 h, followed by soaking in hydrazine for 7 days to eliminate the organic substances. Subsequently, immersion of bone samples were carried out for 1 h in alcohol then the sample was rinsed three times with distilled water. For the purpose of further analysis, the specimen were gone through a drying process then weighed, and crushed

2.3. Material characterization

The calcium and phosphate content of blood serum was analyzed by ultraviolet–visible (UV–Vis) on a Beckman DU-640 spectrophotometer. The bone samples were characterized by atomic absorption spectrophotometry (AAS) to measure the mineral contents (Ca, P, and Mg) of the bones. To assess the

IOP Conf. Series: Materials Science and Engineering **496** (2019) 012017 doi:10.1088/1757-899X/496/1/012017

Table 2. UV–Vis spectral analysis results.

Code	Ca	P (mg/dL)	Mg
P1	8.20	4.36	2.71
P2	8.92	1.23	2.36
P3	9.02	0.54	2.94
P4	10.35	0.75	1.88

Table 3. Mineral levels of the right femur sample by AAS (%).

Code	Ca	P (w/w%)	Mg
P1	72.29	13.39	4.01
P2	74.25	13.38	4.14
P3	74.80	16.96	2.02
P4	76.04	12.09	2.77

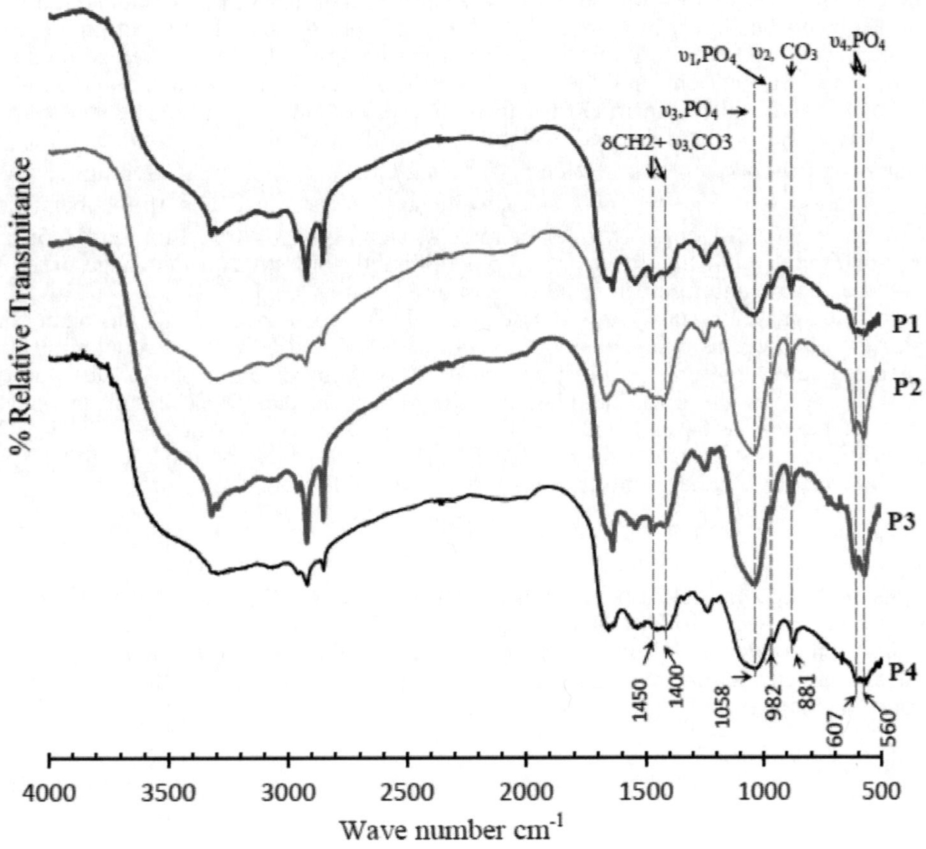

Figure 1. FTIR spectra of spine

chemical structure of the spine, Fourier transform infrared (FTIR) spectra were recorded on a Perkin Elmer system, which aided in the identification of the phosphate and carbonate groups of bones as well as for the determination of the splitting factor, which mainly corresponded to the υ4 in case of the PO_4^{3-} bending vibrations.

3. Results and discussion

The UV–Vis spectral analysis results is summarized in table 2. The blood serum comprised Ca, P, and Mg (table 2). Clearly, the content of Ca in P3 and P4 sera were greater than that in the P1 and P2 sera, indicating that the particle size of n-CaP positively affected the absorption of particles in the bloodstream.

Table 4. Splitting factor

No.	Treatment	Splitting Factor
1.	P1	2.4
2.	P2	2.6
3.	P3	2.7
4.	P4	2.5

The Ca, P, and Mg contents in bones that were analyzed by AAS were summarized in table 3. As observed, the contents of Ca in P4, P3, and P2 were greater than that in P1. This result indicated that the n-CaP and HC diets positively influence of the bone resorption. The FTIR spectrum of rat spines is depicted in figure 1. IR spectroscopy measurement of the rat spinal samples revealed the presence of phosphate groups (560 to 607 cm^{-1} and 982 to 1058 cm^{-1}) and carbonate groups (~881 cm^{-1} and 1400 cm^{-1} to 1450 cm^{-1}). This result reconfirmed the formation of carbonate apatite. The spectrum revealed the presence of three phosphate absorption bands and carbonate absorption bands. The phosphate absorption band comprised symmetric stretching (υ1, 982 cm^{-1}), asymmetric stretching (υ3, 1058 cm^{-1}), and bending vibration (υ4, 560 and 607 cm^{-1}), whereas the carbonate absorption band comprised symmetric stretching (υ2, 881 cm^{-1}), two degenerated stretching bands (υ3, 1400 and 1450 cm^{-1}), and bending (v4, ~650 cm^{-1}). Both the groups were predominantly observed in the range of 1650 to 400 cm^{-1}, which can be observed in figure 1, in cases of spinal samples [6,7].

The calcium phosphate crystals were characterized by the presence of v4 phosphate absorption bands, which were observed as two peaks with maxima at ~564 and 602 cm^{-1}. A continuous band with maxima at approximately 559 cm^{-1}, corresponding to amorphous calcium phosphate, was observed. The splitting factors of all the bone samples are summarized in table 4. The splitting factors were obtained from the peak at ~564 and 602 cm^{-1} [8]. Although the splitting factor of P4 was less than those of P2 and P3, it was still greater than that of P1. This result probably indicated that a special diet containing n-CaP and HC, either simultaneously or individually, positively affected the increase in bone density.

4. Conclusions

Serum analysis revealed that calcium phosphate nanoparticles were easily absorbed into the blood vessels. Furthermore, AAS and FTIR analysis of bones revealed that nano-tricalcium phosphate (n-CaP) and hydrolyzed collagen (HC) intake positively affected the increase in the bone mass and bone structure. Thus, nano-tricalcium phosphate and hydrolyzed collagen diet at the peak stage of bone growth can prevent osteoporosis.

Acknowledgements

This research was funded by PITTA grant year of 2017 No: 624/UN2.R3.1/HKP.05.00/2017. The authors would like to thank Einago (www.einago.com) for the English language review.

References

[1] Sommerfeldt D W and Rubin C T 2001 *Eur. Spine J.* **10** S86–95
[2] Rodan G A and Martin T J 2000 *Science* **289** 1508–14
[3] Russell G, Mueller G, Shipman C and Croucher P 2001 *Novartis Found Symp.* **232** 251–67
[4] Brecevich A T, Kiely P D, Yoon B V, Ngunyen J T, Cammisa F P and Abjornson C 2017 *Spine J.* **17** 855–62
[5] Bagi C M, Berryman E and Moalli M R 2011 *Comp. Med.* **61** 76–85
[6] Petibois C, Wehbe K, Belbachir K, Noreen R and Déléris G 2009 *Acta Phys. Pol. A* **115** 507–512
[7] Granados-Correa F, Bonifacio-Martinez B and Serrano-Gómez J 2010 *Rev. Int. Contam. Ambient.* **26** 129–34
[8] Surovell T A and Stiner MC 2001 Journal of Archaeological Science **28** 633–42.

Synthesis and characterization of ZnO nanorods prepared using microwave-assisted hydrothermal method

M Ridwan[1], **V Fauzia**[1,] **and L Roza**[2]

[1]Department of Physics, Faculty of Mathematics and Natural Sciences (FMIPA) Universitas Indonesia, Depok 16424, Indonesia

[2] Physics Education Program, Faculty of Teacher Training and Education Science, Universitas Muhammadiyah Prof. Dr. Hamka, Jakarta 13830, Indonesia

Corresponding author: vivi@sci.ui.ac.id

Abstract. ZnO is one of the most studied semiconductor materials for many applications; however, the synthesis of one-dimensional ZnO nanostructures by a simple and low-cost method in a short time is still a big challenge. The hydrothermal method is a simple way to produce nanostructures but the process is usually slow. However, the reaction can be accelerated using microwaves. This study aims to grow ZnO nanorods were on the glass substrates using ultrasonic spray pyrolysis and microwave-assisted hydrothermal methods. Our goal was to investigate the effect of the concentration of the growth solution containing equimolar amounts of hexamethyelenentetramine and zinc nitrate tetrahydrate (0.05, 0.1, and 0.15 M) on the morphology and structural and optical properties of the ZnO nanorods. Scanning electron microscopy, X-ray diffraction, and ultraviolet (UV)-visible spectroscopy studies demonstrated that an increase in the concentration of the growth solution results in an increase in the lattice parameters, unit-cell volume, crystallite size, density, and diameter of the nanorods. In addition, increasing the precursor concentrations improves the optical absorbance in the UV region and leads to a slight increase in the bandgap energy (from 3.20 to 3.22 eV).

Keywords: semiconductor, ZnO, nanorods, hydrothermal, microwave

1. Introduction

Zinc oxide (ZnO) is one of the most widely studied semiconductor materials for various applications, such as light-emitting diodes [1], photodetectors [2], photocatalysis [3], and sensors. Several forms of ZnO nanostructures, such as nanodots, nanorods, nanowires, nanobelts, nanotubes, and nanobridges have been prepared. Among these, nanorods have been extensively studied owing to its high surface area and ease of preparation. Although great progress has been made, synthesizing one-dimensional ZnO nanostructures using a simple and low-cost method in a short time is still a big challenge. Physical methods, such as chemical vapor deposition, sputtering, and pulsed laser deposition, require complex systems involving high temperatures and a vacuum]. On the other hand, chemical deposition processes, such as hydrothermal methods, are regarded as more practical techniques [4] because they do not require complicated instruments or involve complicated processes. However, long reaction times are needed to grow ZnO structures, and therefore, microwave-assisted hydrothermal methods have been recently introduced for this purpose as they enable high growth rates and lead to high-quality nanostructures [5]. Herein, we synthesized ZnO nanorods on a glass substrate by a two-step process involving the seeding and growth of the nanostructures using microwave-assisted

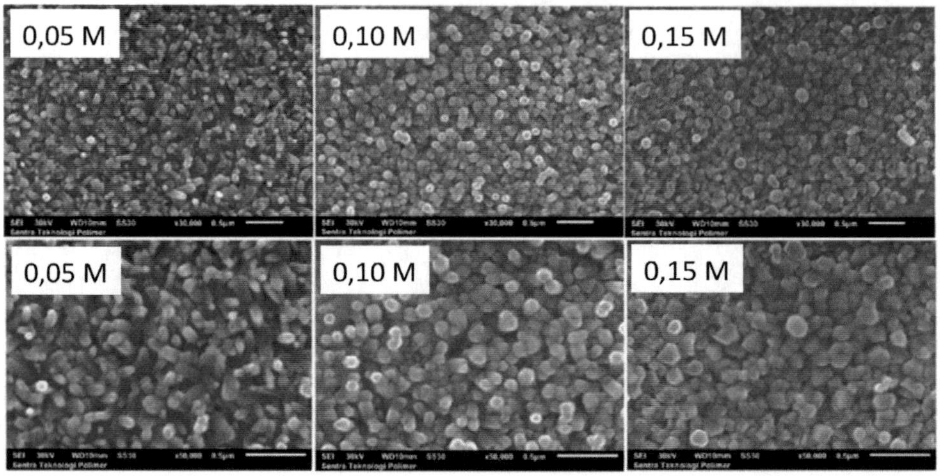

Figure 1. SEM images of the ZnO nanorods prepared herein.

hydrothermal method. Our goal was to investigate the influence of the precursor solution on the morphological, microstructural, and optical properties of the ZnO nanorods.

2. Experiments

In general, the synthesis of ZnO nanorods consists of a seeding process via ultrasonic spray pyrolysis and a growth process via microwave-assisted hydrothermal method. Herein, the seed solution consisted of 0.2 M zinc acetate dihydrate ($Zn(CH_3COO)_2.2H_2O$) in deionized water, which were placed in a container in the nebulizer instrument and vibrated using 1.65 MHz ultrasonic waves. The fine sprays were directed onto the heated glass substrates for 15 min. at 450 °C on a hot plate. The substrates were then annealed at the same temperature for 60 min. The growth solution consisted of equimolar amounts of zinc nitrate tetrahydrate ($Zn(NO_3)_2.4H_2O$) and hexamethylenetetramine ($C_6H_{12}N_4$) at three different concentrations, namely, 0.05, 0.10, and 0.15 M. The glass substrate containing the ZnO seeds was soaked into a glass bottle containing the growth solutions and heated by 240-W microwaves for ten minutes. After completion, the substrates were removed; rinsed using deionized water, and dried with hot air.

Structural characterization was performed using an X-ray diffractometer (Rigaku SmartLab 3kV with CuKα radiation 1.541862 Å), morphological studies were carried out using a JSM-6510LA scanning electron microscope, and the optical properties were determined by recording the absorption spectra on a Thermo Fisher Scientific Genesys 10S Ultraviolet–Visible spectrometer.

3. Results and discussion

3.1. Morphological properties

The surface morphology of the obtained ZnO nanorods is shown in figure 1, where hexagonal nanostructures with different diameters can be observed. In general, the SEM images show that increasing the concentration of the growth solution increases the diameter of the ZnO nanorods. The diameters of the ZnO structures prepared at 0.05 M concentration, were relatively small (89–183 nm) and inhomogeneous, whereas those of the samples prepared at 0.1 and 0.15 M were larger, namely, 91–191 and 118–216 nm, respectively. In addition, the ZnO nanorods were short and relatively transparent. In our previous study, we obtained white ZnO nanorods with a height of more than 1.5 μm [6]. This difference may be due to the low microwave power used herein (240 W). In previous studies, ZnO nanorods longer than 1 μm were prepared using a 750-W microwave power for 20 minutes [7], a 1600-W power for 45 minutes [8], or 1100 W for 20 seconds [9].

The water-based growth solution is basically a microwave absorbent material. The electrical components of electromagnetic waves cause heating through two mechanisms, namely, dipole

Table 1. Microstructural data of the ZnO nanorods.

Concentration	2θ	a=b (Å)	c (Å)	Volume (Å³)	Interplanar spacing (Å)	FWHM	Crystallite size (nm)
0.05 M	34.36	3.249	5.206	47.672	2.61	0.307	26.8
0.10 M	34.34	3.251	5.208	47.678	2.61	0.273	33.8
0.15 M	34.38	3.251	5.208	47.681	2.61	0.212	42.6

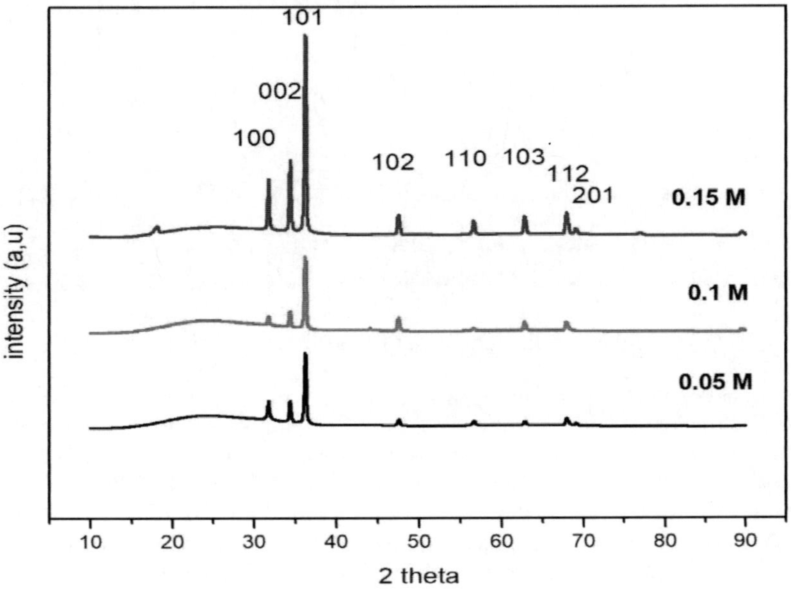

Figure 2. XRD patterns of the ZnO nanorods

polarization and ionic conduction [4]. The electric dipoles in the water molecules align themselves according to the alternating external electric field of the microwaves. The conversion efficiency of electromagnetic wave radiation into heat depends mainly on the dielectric constant of the solution [4]. Microwave radiation can increase the temperature of the solution homogeneously and quickly, which makes it an efficient method for rapid reaction processes.

3.2. Microstructural properties

The X-ray diffraction patterns of ZnO nanorods prepared at three different growth-solution concentrations are shown in figure 2. Based on the reference pattern (ICDD 98-005-7478), the peaks at 2θ 31.72°, 34.36°, 36.20°, 47.48°, 56.54°, 62.78°, 67.88°, and 69.03° correspond, respectively, to the (100), (002), (101), (102), (110), (103), (112), and (201) crystal planes of ZnO. These results show that the material is pure polycrystalline ZnO (without any other phases) and exhibits a hexagonal wurtzite crystal structure. The preferred orientation was quantitatively determined by calculating the value of the Texture Coefficient (TC) [10,11], which showed that the (002) plane had the highest value in all the samples, namely, 1.38, 1.43, and 2.07, for the samples prepared at 0.05, 0.1, and 0.15 M growth-solution concentrations, respectively. These results agree with the morphology images, which show that the nanorods generally grow perpendicular to the substrate (c-axis). Furthermore, it appears that the TC value of (002) increases with increasing growth-solution concentration. This is also in accordance with the morphological observation results in which the nanorod grows more homogenously perpendicular to the substrate. The lattice parameters, cell-unit volumes, interplanar spacing, and crystallite size of the ZnO nanorods in the (002) plane were calculated using the *Highscore* software, as summarized in table 1.

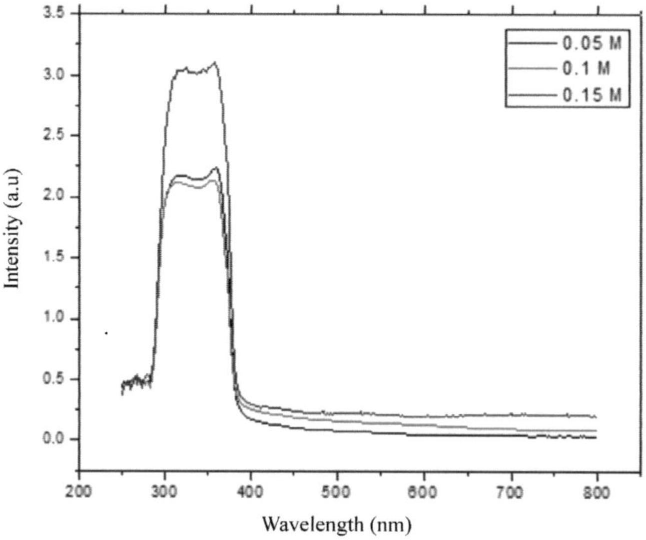

Figure 3. Absorbance spectra of the obtained ZnO nanorods.

Table 1 shows that the lattice parameters and cell-unit volumes of all the samples have values smaller than the standard values under bulk conditions. This may be due to the presence of crystal defects, in the form of oxygen and/or zinc vacancies, which reduce the average value of the lattice parameters and the unit-cell volume. Additionally, increasing the growth-solution concentration slightly increases the value of the lattice parameters a = b from 3.249 to 3.251 Å, as well as the c value from 5.206 to 5.208 Å. In addition, the crystallite size significantly increases from 26.8 to 33.8 and 42.6 nm, which indicates an increase in the crystallinity of the ZnO nanorods formed [12].

3.3. Optical properties

Figure 3 shows the optical absorbance spectra for ZnO nanorods prepared at three different growth-solution concentrations. It is clearly seen that the ZnO nanorods exhibit a high absorption in the wavelength region of 290–375 nm. This has to do with the band gap of ZnO, so the material absorbs photons in the ultraviolet region and these are then used to excite electrons from the valence band to the conduction band. ZnO nanorods prepared at 0.15 M concentration exhibit a higher absorbance than the samples prepared at 0.05 and 0.1 M. This may be associated with the diameter of the nanorods (from SEM results), where the bigger nanorods could be more effective for light absorption.

Figure 3 also shows a sharp curve decline at a wavelength of 375 nm, known as the absorption edge. As the concentration of the growth solution increases, this absorption edge shifts to higher wavelengths. The slope of this curve can be used to determine the band gap of the semiconductor using the Tauc equation, as follows [13]:

$$(\alpha h v)^2 = A(hv - Eg) \tag{1}$$

where h, ν, and E_g are the Planck constant, frequency, and band gap energy, respectively, and α is the absorption coefficient, which is proportional to the absorbance intensity. To obtain the band gap, $(\alpha hv)^2$ is plotted against hv. The band gap value is then determined by extrapolating the linear curve when $(\alpha hv)^2 = 0$, or from the intersection point on the horizontal axis. The band gaps of ZnO nanorods prepared at 0.05, 0.1, and 0.15 M were 3.20, 3.21, and 3.22 eV, respectively. These bandgap values are lower than the ZnO bulk value of 3.37 eV, which is probably due to the presence of crystal defects, such as oxygen and zinc vacancies, that form a new energy level, narrowing the band gap [8]. In addition, it is seen that increasing the growth-solution concentration could slightly increase the band gap.

4. Conclusions

Herein, ZnO nanorods were synthesized on glass substrates by the microwaved-assisted hydrothermal method using varying growth-solution concentrations. Based on our characterization results, we can conclude that increasing the concentration of the growth solution results in increases in the diameter of the obtained ZnO nanorods, the lattice parameters, the volume of the unit cell, and the crystallite size (from 26.8 to 42.6 nm). These changes also lead to an increase in the absorbance intensity in the ultraviolet region as well as to a slight increase in the band gap (from 3.20 to 3.22 eV), which may due to a decrease in the number of oxygen and zinc vacancies.

Acknowledgements

This research was financially supported by PITTA Grant 2017 from Universitas Indonesia with contract number 694/UN2.R31/HKP.05.00/2017. The authors would like to thank Einago (www.einago.com) for the English language review.

References

[1] Voss T and Waldvogel S R 2017 *Mater. Sci. Semicon. Proc.* **69** 52–6
[2] Husham M, Hamidon M N, Paiman S, Abuelsamen A A, Farhat O F and Al-Dulaimi A A 2017 *Sens. Actuators A Phys.* **263** 166–73
[3] Samadi M, Zirak M, Naseri A, Khorashadizade E and Moshfegh A Z 2016 *Thin Solid Films* **605** 2–19
[4] Mirzaei A and Neri G 2016 *Sens. Actuators B Chem.* **237** 749–75
[5] Soleimanzadeh R, Mousavi M S S, Mehrfar A, Esfahani Z K, Kolahdouz M and Zhang K 2015 *J. Cryst. Growth* **426** 228–33
[6] Lestari A, Iwan S, Djuhana D, Imawan C, Harmoko A and Fauzia V 2016 *AIP Conf. Proc.* **1729** 020027
[7] Ahmed F, Arshi N, Anwar M S, Danish R and Koo B H 2013 *Thin Solid Films* **547** 168–72
[8] Ocakoglu K, Mansour S A, Yildirimcan S, Al-Ghamdi A A, El-Tantawy F and Yakuphanoglu F 2015 *Spectrochim. Acta A Mol. Biomol. Spectrosc.* **148** 362–8
[9] Tan S T, Tan C H, Chong WY, Yap C C, Umar A A, Titian G R, Lee H B, Lim K S, Yahaya M and Mat Salleh M 2016 *Sens. Actuators B Chem.* **227** 304–12
[10] Wang Y, Tang W and Zhang L 2015 *J. Mater. Sci. Technol.* **31** 175–81.
[11] Fauzia V, Yusnidar M N, Lalasari L H, Subhan A and Umar A A 2017 *J. Alloys Compd.* **720** 79–85
[12] Rajbongshi B M and Samdarshi S K 2014 *Mater. Sci. Eng. B Adv.* **182** 21–8.
[13] Ma Q, Lv X, Wang Y and Chen J 2016 *Opt. Mater.* **60** 86–93.

IOP Conf. Series: Materials Science and Engineering **496** (2019) 012019 doi:10.1088/1757-899X/496/1/012019

The effect of Ag-doping (x = 0 and 0.05) on structural and morphological $La_{0.8-x}Ag_xCa_{0.2}MnO_3$ material synthesized by sol-gel method

R Kamila and B Kurniawan

Department of Physics, Faculty of Mathematics and Natural Sciences (FMIPA) Universitas Indonesia, Depok 16424, Indonesia

Corresponding author: bkuru@fisika.ui.ac.id

Abstract. In this paper, we will report the Ag-doping effect on the structure and morphology of $La_{0.8}Ca_{0.2}MnO_3$ (LCMO) and $La_{0.75}Ag_{0.05}Ca_{0.2}MnO_3$ (LACMO) materials. LCMO and LACMO materials were synthesized via the sol-gel method and sintered at 900 °C for 24h. The structure of the materials was characterized using X-Ray Diffractometer (XRD) and found that both materials are in the orthorhombic structure (Pnma space group). Scanning Electron Microscopy (SEM) was used to observe the morphology of the material and obtained the homogeneity of the materials. With Ag doping, the grain size of the material will decrease. The results of Electron Dispersive Spectroscopy (EDS) confirm the composition of both materials.

1. Introduction

The perovskite manganite material which has the form $Rn_{1-x}A_xMnO_3$, with A as divalent cations from alkaline earth elements (Ca, Ba, Sr, or Pb) and Rn as trivalent cations of rare earth element (La, Nd, Cd, or Pr), has been widely studied because of its attractive properties such as physical, magnetic, or transport properties [1]. The attractive properties of this perovskite manganite provide many potential fields to be applied such as sensors, magnetic refrigeration, magnetic memory device, etc. [1]. The divalent alkaline earth cation element is doped to substitute the trivalent rare earth cation element will give variation to the valence of Mn, Mn^{3+} and Mn^{4+}. The existence of various Mn valences can produce the Double Exchange (DE) interaction phenomenon between $Mn^{3+}-Mn^{4+}$. In the DE phenomenon, for example, electrons can hop between Mn^{3+} and Mn^{4+} with oxygen as an intermediary. The strength of the DE interactions influences the magnetic and transport properties of $Rn_{1-x}A_xMnO_3$ material [2]. Doping other element to Rn, A, or Mn can cause lattice distortion and affect the DE phenomenon [3]. The inter-grain distance, such as the length of the Mn-O bond, plays a role in the transport properties of the material. Therefore, the structure and morphology of the material is important to be known.

In this paper, silver (Ag) will be doped to substitute Lanthanum (La). The structure and morphology will be observed as the results of Ag-doping. Ag doping is expected to reduce grain boundary. Ag was chosen because it can improve the electrical properties of LCMO materials [4]. Ag doping can also provide inter-granular conduction pathways [5]. It is expected that Ag doping to LCMO parents can enhance DE interactions, increase the Curie temperature, and provide a large magnetoresistance effect.

In previous studies, LCMO and LACMO materials were mostly synthesized by solid-state methods. In this paper, the materials were synthesized by the sol-gel method to obtain better homogeneity and only require low temperatures process. In addition, not many researchers conducted LCMO comparisons before and after La was substituted by Ag. Most studies doped Ag to change Ca concentrations.

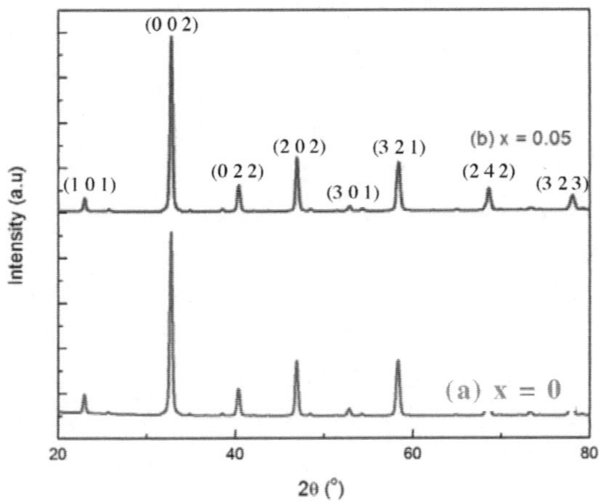

Figure 1. XRD pattern of $La_{0.8-x}Ag_xCa_{0.2}MnO_3$ (a) x = 0 and (b) x = 0.05.

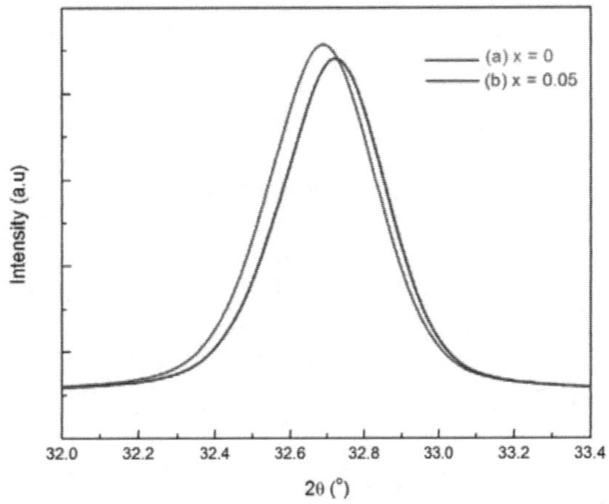

Figure 2. Local XRD patterns at the strongest peak (002) of $La_{0.8-x}Ag_xCa_{0.2}MnO_3$ (a) x = 0 and (b) x = 0.05.

2. Experimental

$La_{0.8-x}Ag_xCa_{0.2}MnO_3$ (x = 0 and 0.05) were prepared by the sol-gel method. The precursors used are nitrate based such as $AgNO_3$, $Ca(NO_3)_2.6H_2O$, and $Mn(NO_3)_2.4H_2O$. La_2O_3 powder is dissolved using HNO_3 solution to become a La nitrate solution. $C_6H_8O_7.H_2O$ was used as a catalyst of the solution. The PH of the solution was adjusted until it reaches 7 by an ammonia solution. The solution was mixed and stirred using a magnetic stirrer at a preserved temperature of 80-90 °C until it is formed into a highly viscous gel. Furthermore, the samples were dehydrated at 120 °C for 3 h, calcined at 500 °C for 6 h to decompose organic compounds, and pressed into pellets. Last, samples were sintered at 900 °C for 24h to grow crystal bonds.

The structural properties of the material were characterized using XRD (Rigaku-Smartlab, 3kV) and carried out at room temperature conditions of 18-25 °C. Scanning Electron Microscope (Hitachi SU-3500) was used to observe the morphology and homogeneity of LCMO and LACMO materials. The magnification of the SEM used is 20.000x using Secondary Electron (SE) and Backscattered Electron (BSE) modes. BSE mode indicates the material homogeneity and SE mode contains material topographic information. Material composition and its purity were confirmed by Energy Dispersive X-ray Spectroscopy (EDS) Horiba. The area chosen in the EDS characterization was carried out to ensure that Ag was successfully doped to substitute La, which supported the XRD results obtained.

3. Results and discussion

3.1. Crystal structure

LCMO and LACMO structures were characterized using XRD. The XRD results of the two materials are shown in figure 1. The XRD patterns show there no additional peak other than Ag after doping with Ag. The XRD patterns also show the occurrence of a small shift in position of the highest peak (figure 2). A shift to a slightly larger angle value indicates a decrease in lattice parameters with Ag doping [4].

The structural parameters of the refinement results for LCMO and LACMO materials are summarized in table 1. The structure of both LCMO and LACMO crystals are orthorhombic with the Pnma space group no 62. The structure of the two crystals remains the same with the Ag doping. This shows that Ag doping does not change the crystal structure. After doping by Ag elements, all crystal lattice parameter show a decrease in value. This occur because of the radius of the Ag^{+} ion is smaller than the radius of the La^{3+} ion, which results in a decrease in unit cell volume and other lattice parameters. The ion radii of the LCMO and LACMO materials are shown in table 2 [1,2]. Table 3 and

Table 1. Refined structural parameters of $La^{3+}_{0.8-x}Ag^{+}_{x}Ca^{2+}_{0.2}MnO_3$ (x = 0 and 0.05).

Structural parameters	x = 0	x = 0.05
Crystal structure	Orthohombic	Orthohombic
Space group	Pnma	Pnma
a (Å)	5.4696	5.4643
b (Å)	7.7361	7.7256
c (Å)	5.4969	5.4945
Volume (Å^3)	232.5991	231.9542
<Mn-O> (Å)	1.9636	1.9633
<Mn-O-Mn> (°)	162.185	161.149
χ^2	1.17	1.13

Table 2. Ionic radii of each ion in $La^{3+}_{0.8-x}Ag^{+}_{x}Ca^{2+}_{0.2}MnO_3$ (x = 0 and 0.05).

Ion	Radii (nm)
La^{3+}	1.36
Ag^{+}	1.28
Ca^{2+}	1.12
Mn^{3+}	0.65
Mn^{4+}	0.53
O^{2-}	1.4

Table 3. Complete crystallographic parameters of $La_{0.8}Ca_{0.2}MnO_3$ material

Atom	Atomic position			Wyckoff factor	Occupancy
	x	y	z		
La	0.4824	0.25	0.003	4c	0.8
Ag	-	-	-	-	-
Ca	0.4824	0.25	0.003	4c	0.2
Mn	0	0	0	4a	1
O1	0.2262	0.0327	0.2696	8d	1
O2	0.5024	0.25	0.5645	4c	1

Table 4. Complete crystallographic parameters of $La_{0.75}Ag_{0.05}Ca_{0.2}MnO_3$ material

Atom	Atomic position			Wyckoff factor	Occupancy
	x	y	z		
La	0.4824	0.25	0.003	4c	0.75
Ag	0.4824	0.25	0.003	4c	0.05
Ca	0.4824	0.25	0.003	4c	0.2
Mn	0	0	0	4a	1
O1	0.2262	0.0327	0.2696	8d	1
O2	0.5024	0.25	0.5645	4c	1

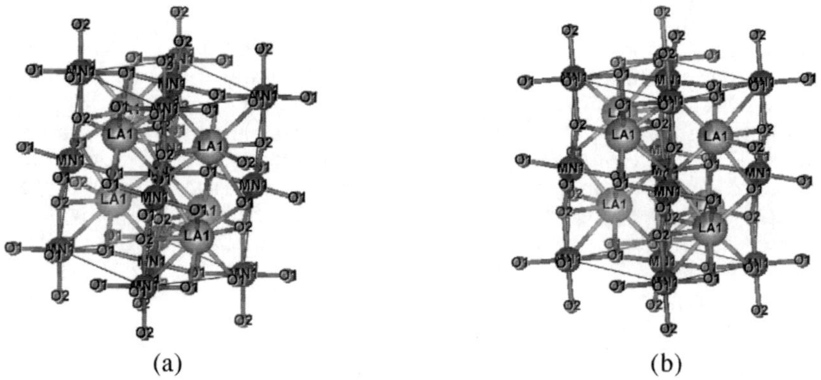

(a) (b)

Figure 3. Crystal structure of (a) $La_{0.8}Ca_{0.2}MnO_3$ and (b) $La_{0.75}Ag_{0.05}Ca_{0.2}MnO_3$ material

table 4 present complete crystallographic parameters, which consist of atomic position, Wyckoff factor, and occupancy of each atom that making up the material. The unchanging crystal structure can also be seen from the Wyckoff factor, which has not changed after being doped by Ag. The average crystallite size of LCMO and LACMO were calculated using Debye-Scherrer equation, each obtained 44.89 nm and 33.18 nm, respectively. However, changes in lattice parameters that occurred were not significant with Ag doping because of small-scale doping concentrations. This indicates a slight change in lattice parameters but does not change the crystal structure. The crystal structure of both materials is shown in figure 3.

IOP Conf. Series: Materials Science and Engineering **496** (2019) 012019 doi:10.1088/1757-899X/496/1/012019

(a) (b)

Figure 4. SEM images using (a) BSE and (b) SE mode on $La_{0.8}Ca_{0.2}MnO_3$ material

The length of the Mn-O bond and the Mn-O-Mn angle were obtained using refinement analysis of the XRD pattern (table 1). The results show a decrease in the length of Mn-O bond. Changes in the length of Mn-O bond could affect the nature of material transport [3]. Winarsih *et al.* [3] explained this by using a core-shell model, which uses distance of the grain as a half value of d, which is the distance of Mn and O ions [3]. Thus, if inter-grain distance gets wider, it could weaken the DE interaction and electron transport will be more difficult, which will result in increased material. Moreover, the DE interaction is also related to conduction bandwidth (W), where $W \sim \cos(\pi - \theta_{Mn\text{-}O\text{-}Mn}) / d_{Mn\text{-}O}$ with $\theta_{Mn\text{-}O\text{-}Mn}$ as Mn-O-Mn bond angle and $d_{Mn\text{-}O}$ as Mn-O bond distance [6]. It can be seen, if the distance of Mn-O ions becomes shorter, the bandwidth conduction value will be greater and strengthen the DE interaction, which will affect the material transport properties. By doping Ag^{+} ions to substitute La^{3+} ion, it can shorten the distance of Mn-O so that it is expected to increase the nature of material transport.

Furthermore, changes in the Mn-O bond distance can be obtained from the FT-IR measurements. Based on previous research by Daengsakul *et al.* [7] who studied the functional groups presence in $LaMnO_3$ material using FT-IR that there is a stretching mode on the absorption band at wavenumber 600 cm^{-1} because of internal motion that change the Mn-O bond distance [7]. Another study conducted by Shinde *et al.* [8] for $La_{0.8}Ca_{0.2}MnO_3$ material producing the FT-IR spectra, which show similar things that the main absorption band around wavenumber 600 cm^{-1} is associated with the presence of oxygen metal bonds in the perovskite structure that involves internal motion that changes the Mn-O bond distance [8]. Based on previous researches, doping Ag to substitute La^{3+} will produce a stretching mode with a change in the Mn-O bond distance from the XRD results. However, this cannot be explained more deeply because it is only based on XRD data.

3.2. Morphology

Figure 4 and Figure 5 show SEM images of LCMO and LACMO materials. From BSE mode, LCMO and LACMO materials have good homogeneity because the colour contrast is quite uniform. From SEM observations, it can be seen that the grain of both materials have a granular shape that are in accordance with previous studies [9,10]. Next, the grain size of LCMO and LACMO materials obtained grain sizes ranging from 200–698 nm with an average grain size of 420.8 nm for LCMO and with an average grain size of 347 nm for LACMO. The presence of various grain sizes occurs due to agglomeration so that grain size measurements are not appropriate. Observation of grain size from SEM characterization is greater than the crystallite size obtained from the Debye Scherrer equation using XRD results. The results also show that each grain from SEM results consists of several crystals.

3.3. Composition and purity

The material composition of LCMO and LACMO were confirmed by EDS measurements. The results of the observation are shown in table 5. The EDS results show that the atomic fraction values from measurement are close to the calculated atomic fraction values, which indicate that LCMO and LACMO materials are succeessfully synthesized according to the desired composition. The small

IOP Conf. Series: Materials Science and Engineering **496** (2019) 012019 doi:10.1088/1757-899X/496/1/012019

Table 5. Compositional results of $La_{0.8-x}Ag_xCa_{0.2}MnO_3$ (x = 0 and 0.05).

Doping concentrations	Element (% at.)									
	La		Ag		Ca		Mn		O	
	Calc	Obs	Calc	Obs	Calc	Obs	Calc	Obs	Calc	Obs
x = 0	16	18.06	0	0	4	3.44	20	17.69	60	60.82
x = 0.05	15	13.31	1	0.4	4	3.62	20	16.54	60	66.13

(a) (b)

Figure 5. SEM images using BSE and SE mode on $La_{0.75}Ag_{0.05}Ca_{0.2}MnO_3$ material

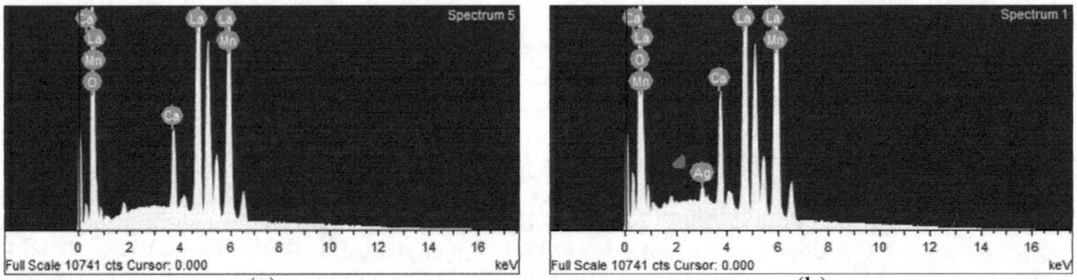

(a) (b)

Figure 6. Measurement results of EDS analysis $La_{0.8-x}Ag_xCa_{0.2}MnO_3$
(a) x = 0 and (b) x = 0.05.

difference between the atomic fraction from EDS measurement results and the results occurs because EDS provides a semi-quantitative analysis. Therefore, the results from the EDS are only close to the results of the calculation.

Furthermore, the measurement results of EDS in figure 6 shows the presence of Ag element in LACMO material. This indicates that the Ag element has successfully doped to the LCMO material and substituting La.

4. Conclusions

$La_{0.8-x}Ag_xCa_{0.2}MnO_3$ (x = 0 and 0.05) were successfully synthesized by the sol-gel method. The XRD results showed that Ag doping did not change the crystal structure of the materials and still remained the same as the orthorhombic. The lattice parameters and unit cell volume of the crystal decrease with Ag doping due to the smaller ionic radius of the Ag^+ ion than the ion radius of La^{3+}. Decreasing the length of the Mn-O bond due to Ag doping will make the DE phenomenon stronger and affect the material transport properties. Microstructure results from SEM showed that both of materials had a granular form grain with an average size of 257.17 nm for LCMO and 289.9 nm for LACMO. The

EDS results supported the XRD results and show that Ag is successfully doped into the material. The value of atomic fraction observed by EDS approaches the calculated value.

Acknowledgements
The research is funded entirely by PITTA Grant 2017 from Universitas Indonesia under contract number 633/UN2.R3.1/HKP.05.00/2017.

References

[1] Smari M, Hamouda R, Walha I, Dhahri E, Mompeán F J and Garcia-Hernández 2015 *J. Alloys Compd.* **644** 632–7
[2] Ye S L, Song W H, Dai J M, Wang K Y, Wang S G, Zhang C L, Du J J, Sun Y P and Fang J 2002 *J. Magn. Magn. Mater.* **248** 26–33
[3] Winarsih S, Kurniawan B, Manaf A, Saptari S A and Nanto D 2016 *J. Phys.: Conf. Ser.* **776** 012058
[4] Yin X P, Liu X, Yan Y Z and Chen Q M 2014 *J. Sol-Gel Sci. Technol.* **70** 361–5
[5] Huang Y H, Huang K F, Luo F, He L L, Wang Z M, Liao C S and Yang C H 2003 *J. Solid State Chem.* **174** 257–63
[6] Kurniawan B, Winarsih S, Ramadhan M R, Naomi A and Laksmi W 2017 *AIP Conf. Proc.* **1862** 030048
[7] Daengsakul S, Mongkolkachit C, Thomas C, Thomas I, Siri S, Amornkitbamrung V and Maensiri S 2009 *Optoelectron. Adv. Mater. Rapid Commun.* **3** 106–9
[8] Shinde K P, Oh S S, Baik S K, Kim H S, Sinha B B and Chung K C 2012 *J. Korean Phys. Soc.* **61** 2000–4
[9] Xi S, Lu W and Sun Y 2012 *J. Appl. Phys.* **111** 063922
[10] Kalyana L Y and Reddy P V 2010 *Solid State Sci.* **12** 1731–40

Synthesis and UV–Visible spectroscopy study of core–shell gold–palladium nanoparticles by iterative seeding of citrate

M B S Putra, E Saepudin and T A Ivandini

Department of Chemistry, Faculty of Mathematics and Natural Sciences (FMIPA), Universitas Indonesia, Depok 16424, Indonesia

Corresponding author: ivandini.tri@sci.ui.ac.id

Abstract. In this study, gold and palladium nanoparticles (Au NPs and Pd NPs) were successfully synthesized by the iterative seeding of citrate and ascorbate, respectively. UV–visible spectroscopy results revealed that Au NPs exhibit plasmonic resonance peaks at 540–560 nm, while Pd NPs exhibit peaks at 220 nm. Transmission electron microscopy images revealed particle size distributions of 19.3 ± 4.2 nm and 31.9 ± 6.1 nm for Au NPs and Pd NPs, respectively. Larger sizes were observed for palladium nanoparticles because of their aggregation caused by the deficiency of the capping agent, while gold particles exhibited size stability. Furthermore, gold–palladium (AuPd) nanoflower particles were synthesized by the iterative citrate seeding of Pd on the Au NP surface.

Keywords: Nanoparticle, Core–shell, Gold, Gold–palladium, Citrate seeding.

1. Introduction

Metal nanoparticles exhibit typically different properties from their corresponding bulk materials, which can be employed for many applications, including catalytic, optic as well as magnetic and sensing applications. These properties are affected by the shape of the nanoparticles. Therefore, control the shape and size is crucial factor in the synthesis of metal nanoparticles [1].

On the other hand, bimetallic heteronanostructures are known to provide synergic optic, electronic, and catalytic properties their monometallics. In some cases, a drastically improvement of properties was reported, such as at a gold and palladium bimetal [2].

Several studies have reported the synthesis of colloidal Au nanoparticles (Au NPs), including even some of the methods that start with the use of bulk metal. However, most of these methods begin with Au^{3+}, and using different reductants, particles of varying sizes are generated [3–5]. For example, reductants such as $NaBH_4$ or white phosphorus afford small Au particles with a diameter *d* of less than 10 nm [3]. Sodium citrate is the most widely used reductant. By the variation of the citrate: Au ratio, colloidal particles with d values ranging from 10 to 150 nm can be prepared [4]. On the other hand, the nucleation and growth of Pd nanoparticles (Pd NPs) have been successfully investigated using H_2PdCl_4 and *L*-ascorbic acid (AA) as the precursor and reductant, respectively [5]. This study focused on the synthesis of Au NPs, Pd NPs, and gold–palladium (AuPd) NPs (Au@Pd NPs). In addition, the comparison between their characteristics was discussed.

2. Experimental

2.1. Materials and instruments

Hydrogen tetrachloroaurate(III) ($HAuCl_4$), palladium chloride(II) ($PdCl_2$), and trisodium citrate ($Na_3C_6H_5O_7$) were purchased from Wako Pure Chemical Industries Ltd. Ultrapure deionized water with a resistivity of 18.2 MΩ·cm was obtained from a Millipore Direct® Q5-UV system.

UV–Visible spectra were recorded on a Thermo Scientific Multiskan GO system, and transmission electron microscopy (TEM) images were recorded on a Microscope Tecnai 200kV D2360 SuperTwin.

2.2. Preparation of Au-core/Pd-shell

2.2.1. Synthesis of Au NPs. Au NPs were synthesized using iterative citrate-seeded colloidal Au via the method reported by Brown and Natan with some modifications [3]. First, 100 mM of sodium citrate was added into a stirred solution of boiling 100 mL of 0.01% $HAuCl_4$. Second, this mixture was boiled until the solution reached the boiling point, followed by the dropwise addition of sodium citrate until the solution color changed from yellowish $HAuCl_4$ to dark red Au NPs. The solution was cooled with stirring for 10 min., and the solution color changed to red-pink, indicative of the complete reduction of $HAuCl_4$ into Au NPs.

2.2.2. Synthesis of Pd NPs. Pd NPs were synthesized using AA as the reductant. First, a solution of Pd(II) in 0.001 M of HCl was mixed with 0.1 M of fresh AA with vigorous stirring while simultaneously maintaining an extremely low temperature. Reaction completion was verified by the appearance of dark brown colloids of Pd NPs [6].

2.2.3. Synthesis of Au@Pd NPs. Au@Pd NPs were synthesized via the Hu *et al.* method reported previously [7]. Pd-coating procedures was performed by reduction reaction of Pd^{2+} solution by AA in the presence of Au NPs seeds. Briefly, an aliquot containing 30 mL of gold seeds was added into a flask containing 1.1 mL of 1.0 mM of H_2PdCl_4 and 9.9 mL of water. Second, the mixture was cooled down to ~4 °C in an ice bath with vigorous stirring. Then, 3.5 mL of 10 mM AA was slowly dropped in the mixture. After the addition of AA, the mixture was stirred for another 15 min. The mixture color gradually deepened from pinkish red for Au NPs to fully dark brown for coated Au@Pd NPs.

3. Results and discussion

3.1. Synthesis of Au, Pd, and Au@Pd NPs and its characterization

Au NPs were synthesized in one pot using 0.01% $HAuCl_4$ and 0.01 M sodium citrate as the precursor and reductant, respectively. Sodium citrate successfully reduced Au^{3+} into Au^0 nanoparticles at the boiling point of the solvent, leading to a color change from yellowish $HAuCl_4$ into pinkish-red Au^0. The reaction kinetics can be expressed by the following equation [8]:

$$Au^{3+} + Na_3C_6H_5O_7 \rightarrow Au^0 + 3Na^+ + citr \quad (1)$$

The UV–vis spectrum of colloidal Au NPs exhibiting a plasmon resonance peak at ~522 nm together with the TEM image of Au NP are shown in figure 1a and figure 1b, respectively. The correlation of the data revealed that the surface plasmon resonance of Au NPs is relative to the particle size as shown by the TEM image, which was ~19 nm with a standard deviation of 4.3 nm [9]. Kwolek and Wojnicki [10] have reported the size of this Au NPs is suitable as the core of a composite. Peak shifting was not observed (figure 1c), indicating that the Au NP size is stable with the increasing content of gold ions, i.e., double and triple the Au NP concentrations, which was probably related to the presence of citrate as the capping and reducing agent.

Meanwhile, Pd NPs (figure 2) has completely different spectrum compare with Au NPs. In the UV–vis spectrum shown in figure 2a for the obtained colloidal Pd NPs, two peaks were observed at ~220 nm and 246 nm, respectively, for the least amount of the reductant. The first peak corresponded to Pd(II) ions, whereas the second peak corresponded to AA. Both peaks were absent, indicative of the complete reduction of Pd, while the peak intensity increased at 300–400 nm, corresponding to Pd NPs (brownish dark). As shown in figure 2a, this Pd NPs peak is not observed, which was related to the low concentrations of the formed Pd NPs. TEM images (figure 2b, figure 2c and figure 2d) revealed a mean particle size of 31.9 nm with a standard deviation of ±6.1; this value is greater than that of the

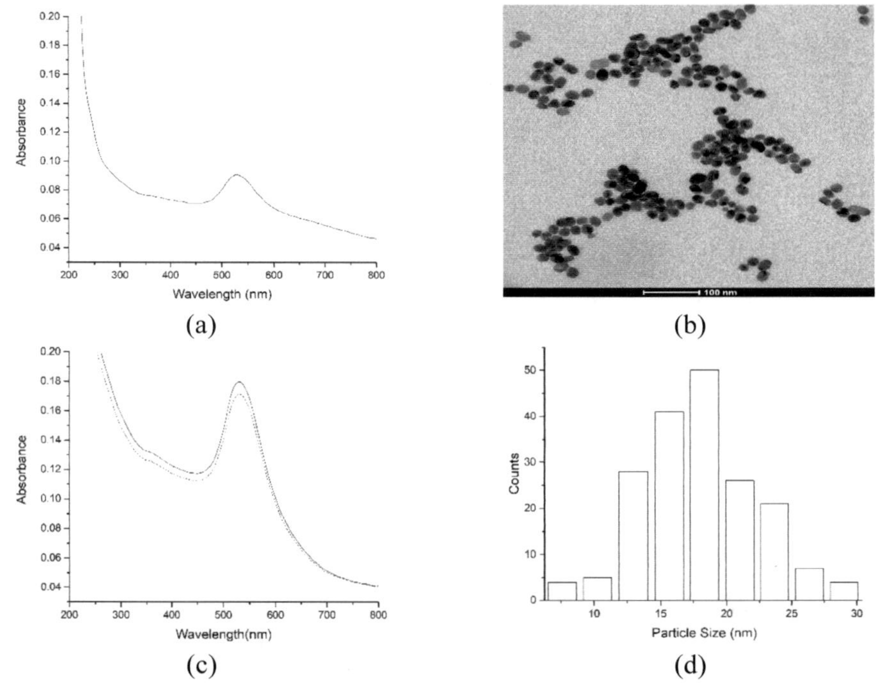

Figure 1. (a) UV–visible spectrum of Au NPs, (b) TEM image of Au NPs with 250× magnification, (c) UV–visible spectrum of Au NP at double (dash - - -) and triple concentrations (line —), and (d) particle size distribution of Au NPs.

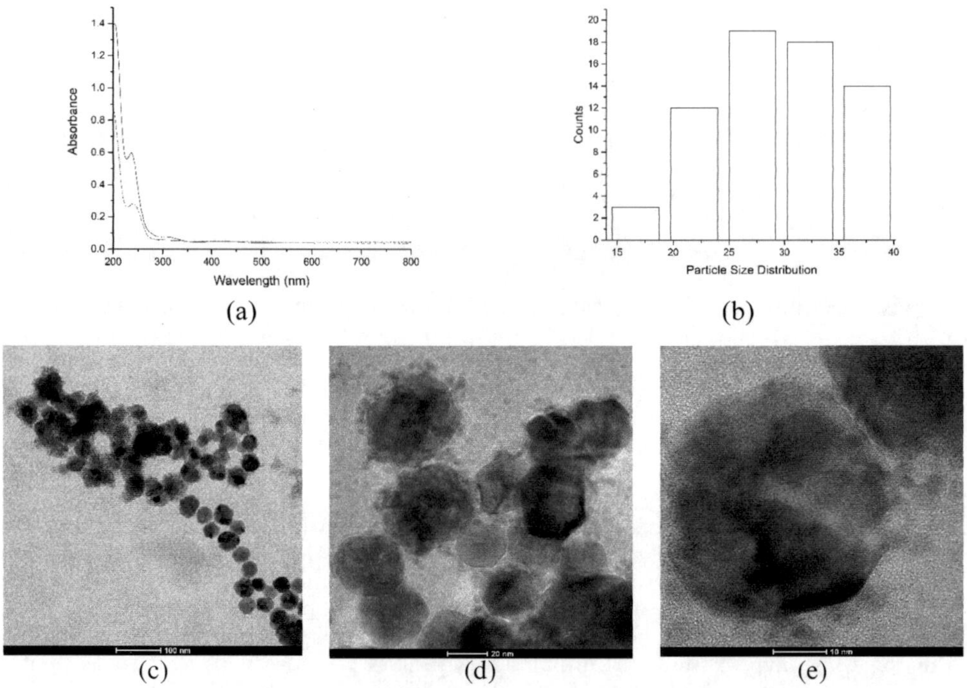

Figure 2. (a) UV–visible spectrum of H_2PdCl_4 (Line —) and Pd NPs (Dash - -), in addition to that of *L*-ascorbic acid, (b) particle size distribution of Pd NPs, (c) TEM images, and (d and e) HRTEM images of Pd NPs.

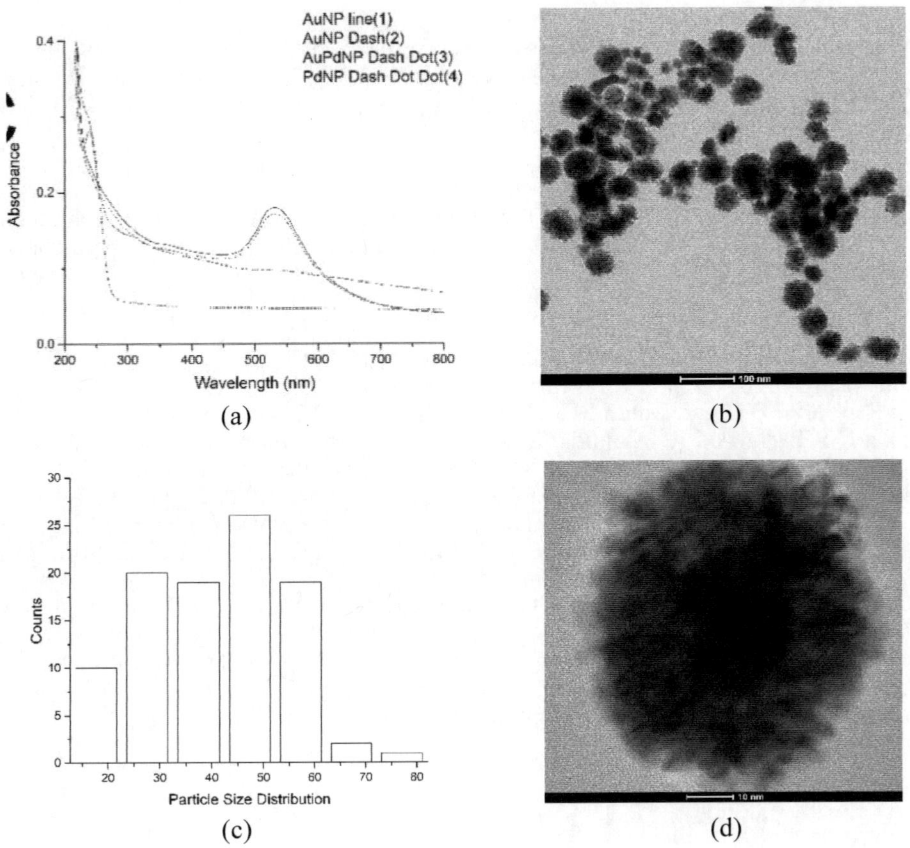

Figure 3. (a) Comparison of the UV–visible spectra of Au NPs, Pd NPs, and Au@Pd NPs, (b) TEM image of Au@Pd NPs and (c) its size distribution, and (d) HRTEM image of Au@Pd NPs.

Au NPs. The deficiency of the capping agent of Pd NPs tended to produce nanoparticle aggregates [5,6]. Despite their size, Pd NPs still exhibited catalytic property [2]. The reduction of Pd(II) as confirmed by the change in color confirmed the nucleation and growth of particles. Wojnicki *et al.* [5] have proposed two steps related to the reduction of Pd(II) to Pd(0) (nucleation) and auto-growth of Pd NPs:

$$\text{Pd(II)} + \text{AA} \rightarrow \text{Pd(0)} + \text{DHA} \tag{2}$$

$$\text{Pd (II)} + \text{AA} + \text{Pd(0)} \rightarrow 2\text{Pd(0)} + \text{DHA} \tag{3}$$

The comparison of the UV–visible spectra of the Au NPs, Pd NPs, and Au-core/Pd-shell NPs is shown in figure 3a. Wang [2] has reported that the optical properties of the Au@Pd core–shell nanostructures are strongly dependent on the thickness and morphology of the Pd shells. The figure clearly revealed the disappearance of the Au NP plasmon resonance peak at 522 nm for the Pd-shell-covered Au NPs. To examine the nanoparticle size, TEM and HRTEM images were recorded. The TEM images have shown in figure 3b and figure 3c revealed spherical particles with an average particle size of 45 ± 13.7 nm. The Pd-shell layer thickness determined by HRTEM images was calculated to be ~15 nm with a gold core diameter of ~23 nm [11,12], indicating that some Au NP aggregates are formed in case of Pd-shell-covered Au NPs. The HRTEM image also revealed the successful synthesis of the Au-core/Pd-shell nanoflower structure.

4. Conclusions

Au NPs, Pd NPs, and Au@Pd NPs were successfully synthesized by the iterative seeding of citrate and reduction using *L*-AA, leading to Au NPs, Pd NPs, and Au@Pd NPs with sizes of 19 ± 4.2 nm, 31 ± 6.1 nm, and 45 ± 13.7 nm, respectively. The Pd-shell covering the Au NPs successfully afforded a nanoflower structure for the Au@Pd NPs.

Acknowledgements

This research was funded by PITTA Grant 2017 from Universitas Indonesia with contract number 1822/UN.R3.1/PPM.00.01/2017. The authors would like to thank Einago (www.einago.com) for the English language review.

References

[1] Nikkoobakht B and El-Sayed M A 2003 *Chem. Mater.* **15** 1957–62
[2] Jing H and Wang H 2014 *Cryst. Eng. Comm.* **16** 9469–77
[3] Brown K R, Walter D G and Natan M J 2000 *Chem. Mater.* **12** 306–13
[4] Luty-Błocho M, Pacławski K, Wojnicki M and Fitzner K 2013 *Inorg. Chim. Acta* **395** 189–96
[5] Wojnicki M, Fitzner K and Luty-Błocho M 2016 *J. Colloid Interface Sci.* **465** 190–9
[6] Wojnicki M, Pacławski K, Rudnik E and Fitzner K 2011 *Hydrometallurgy* **110** 56–61
[7] Hu J W, Li J F, Ren B, Wu D Y, Sun SG and Tian Z Q 2007 *J. Phys. Chem. C* **111** 1105–12
[8] Pacławski K, Streszewski B, Jaworski W, Luty-Błocho M and Fitzner K 2012 *Colloids Surf. A Physicochem. Eng. Asp.* **413** 208–15
[9] Moraes Silva S, Tavallaie R, Sandiford L, Tilley R D and Gooding J J 2016 *Chem. Commun.* **52** 7528–40
[10] Kwolek P and Wojnicki M 2014 *J. Photochem. Photobiol. A Chem.* **286** 47–54
[11] Ma T, Liang F, Chen R, Liu S and Zhang H 2017 *Nanomaterials* **7** 239
[12] Harpeness R and Gedanken A 2004 *Langmuir* **20** 3431–4

Fabrication and characterization of thick films made of chitosan and nanofibrillar cellulose derived from pineapple leaf

B Amalia[1], C Imawan[2] and A Listyarini[1]

[1] Center for Chemical and Packaging (BBKK), Jalan Balai Kimia No.1 Pekayon, Jakarta 13710

[2] Department of Physics, Faculty of Mathematics and Natural Sciences (FMIPA) Universitas Indonesia, Depok 16424, Indonesia

Corresponding author: cuk.imawan@sci.ui.ac.id

Abstract. Research on the development of natural polymers is becoming increasingly important because of environmental reasons. In this study, the thick films manufactured using chitosan and reinforced with nanofibrillar cellulose that is derived from pineapple fibers have been successfully fabricated. Nanofibrillar cellulose is obtained by isolating the cellulose from pineapple fibers through chemical treatments and ultra-fine grinding. Thick chitosan-nanocellulose composite films are prepared by mixing the nanocellulose with a chitosan solution at concentrations of 1%, 3%, 5%, 7%, and 10%. The structures and properties of the obtained films are examined by X-ray diffraction (XRD), Ultraviolet and Visible (UV/Vis) spectroscopy, and Fourier transform infrared (FTIR) spectroscopy. Characterization of the optical properties of the materials by UV/Vis techniques exhibit that the highest transparency can be achieved using 3% nanocellulose. This result is also related to the XRD spectra, which depict that the addition of 1%–3% nanocellulose can effectively increase the degree of crystallinity of the composite films. These results prove that the addition of nanocellulose to a chitosan matrix depicts a good potential to improve the properties of composite films and that these films can be used as biodegradable packaging plastics.

Keywords: chitosan, nanofibrillar cellulose, pineapple leaf fiber

1. Introduction

Recently, research related to the development of biocomposites has attracted significant attention due to increasing environmental concerns. Biocomposites satisfy the requirements for a "green technology" in which the matrix and reinforcing agents are biodegradable materials that can be easily decomposed in the environment, thereby avoiding pollution [1]

Chitosan and nanocellulose are biodegradable materials that have been extensively developed due to their numerous advantages. Chitosan is one of the strongest natural polymers that are known; it is a hydrophilic cationic polysaccharide that is produced by the deacetylation of chitin, which is one of the most abundant materials on earth [2].

Cellulose is a natural ingredient that is used as a reinforcing agent in the polymer matrix. It is a fiber that can be isolated from certain plants, such as banana, pineapple, flax, cotton, jute, sisal and hemp, through chemical and mechanical processes [3]. Nanocellulose can be classified into three major sub-categories based on its dimensions, utilities, and preparation (depending on the source of cellulose and the processing conditions) [4]. The nomenclatures used are cellulose nanocrystals (CNC), cellulose nanofibers (CNF) and nanofibrillar cellulose (NFC). Cherian *et al.* [3] isolated cellulose from the pineapple-leaf fibers to ensure usage as a reinforcing agent in polyurethane. In this study, the isolated cellulose, which was obtained from the pineapple-leaf fibers, was used as a reinforcing agent in the chitosan matrix.

2. Materials and methods

2.1. Materials
The chitosan that was used in this study exhibited a deacetylation level that was greater than 80% and was purchased from Balai Penelitian Bioteknologi & Bioindustri, Bogor. The pineapple-leaf fiber was purchased from CV Hasanah Niaga, Bandung. The chemical reagents that were used included sodium hydroxide, sodium chlorite, hydrochloric acid, and acetic acid, which were ordered from PT. Merck Chemicals and Life Sciences.

2.2. Preparation of the composite film
The composite film was prepared by mixing the solution of a 2 % w/v acetated chitosan with NFC for several variation of w/w concentrations of 0%, 1%, 3%, 5%, 7%, and 10%. This solution was further heated for two h at 80 °C upon stirring and was ultrasonicated at 800 watt for 30 min. The chitosan/nanocellulose composite film was casted by pouring a 25 mL solution over a 20 × 20 cm mica, which was then dried at room temperature for 48 h.

2.3. Characterization

2.3.1. XRD analysis. The XRD characterizations are performed to obtain information about the crystal structures of metal and alloy materials, minerals, inorganic compounds, polymers, and organic materials. In this research, the studies were conducted using a Rigaku Smartlab XRD instrument at a voltage of 3 kV and a current of 30 mA. The source of X-ray radiation was Cu-Kα (α = 1,54 Å), and the samples were examined within the 2θ range of 5° to 80° at a scanning rate of 0.02°.

2.3.2. Optical properties. The absorbance of the plastic composite films was measured using a Thermo UV/Vis 10 S Genesys spectrophotometer at a resolution of 1 nm in the wavelength range of 200 to 800 nm.

2.3.3. FTIR analysis. An FTIR analysis was performed to study the effect of the addition of NFS to the chitosan film and to determine the IR band and its shift due to the reaction between nanocellulose and chitosan. The FTIR spectra were obtained for nanofibrillar cellulose, chitosan films, and chitosan composites containing nanofibrillar cellulose using an FTIR-ATR Thermo Scientific Nicolet iS5 device at 400–4000 cm^{-1}.

3. Results and discussion

3.1. XRD studies
Figure 1 illustrates the diffractograms of NFS, chitosan, and the composite film. It can be observed that NFS has a crystalline structure, exhibiting sharp peaks at 2θ = 15.51° and 22.84°. According to Cherian *et al.* [5], who also conducted a study on the isolation of nanocellulose from pineapple-leaf fibers, a sharp peak appeared at θ = 22.7°. Khan *et al.* [6] also reported the presence of sharp peaks at 2θ = 16.8° and 2θ = 20°–22.4° in the diffractogram of nanocellulose. These peaks correspond to the (1 1 0) and (2 2 0) planes of cellulose, respectively. Figure 1 depicts the peaks corresponding to chitosan at 2θ = 15.35 and 20.80. According to Celebi and Kurt [1], the wide peak of chitosan can be observed at 2θ = 20°, expressing the semi-amorphous properties of this compound. The addition of 3% nanocellulose to chitosan caused an increase in the intensities of the peaks that were observed at 2θ = 15.25° and 20.46°. This is likely to be caused due to the so-called transcrystallization effect [6], which is defined as the orientation of a semi-crystalline matrix perpendicular to the nanocrystalline cellulose [2].

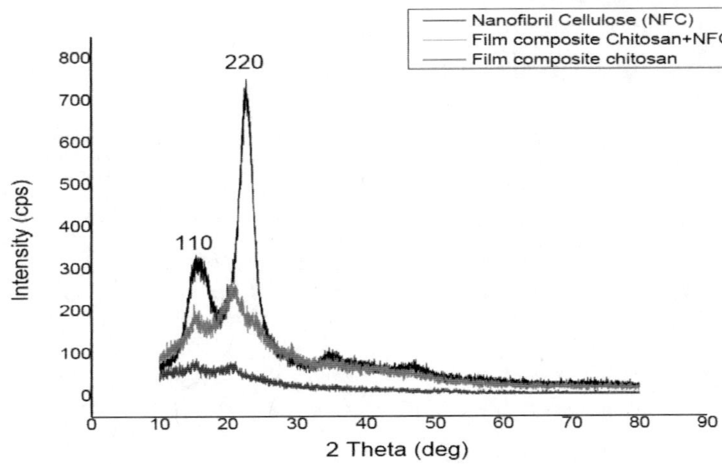

Figure 1. X-ray diffractograms of nanocellulose and film composites.

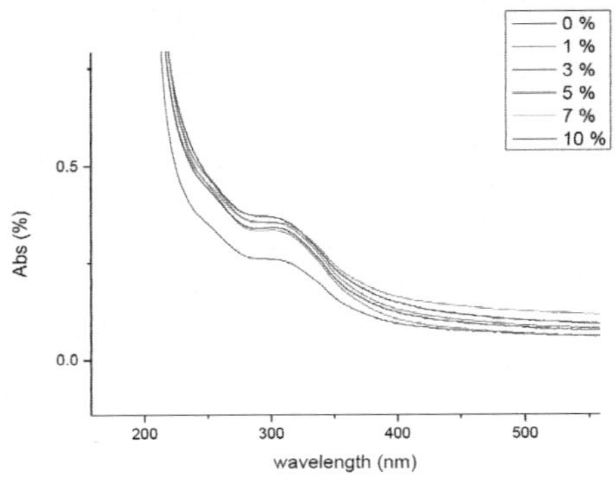

Figure 2. UV-Vis absorption spectra of composite film

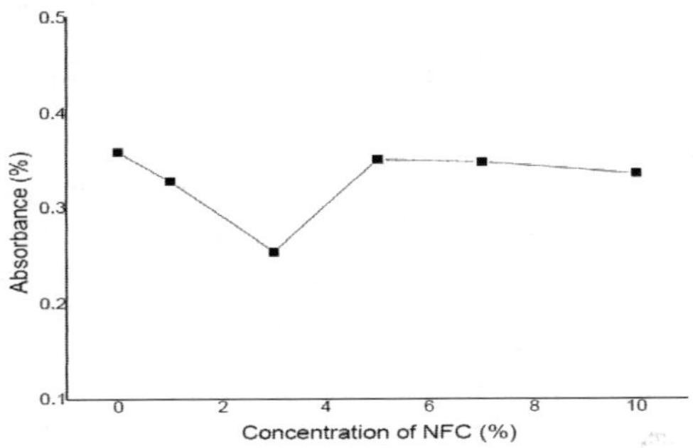

Figure 3. Absorption peaks of the composite films at 310 nm.

3.2. Analysis of the optical properties of the NFC-chitosan composite films

The optical properties of the NFC–chitosan composite films were evaluated between 200 and 800 nm. Figure 2 depicts the absorbance spectra of a pure chitosan film and chitosan–NFC composites with NFC concentrations of 1%, 3%, 5%, 7%, and 10%. It can be observed that the composite film that contains 3% NFC exhibited the lowest absorbance at a wavelength of 310 nm (the other concentrations caused similar absorbance values). This indicated that the chitosan–NFC (3%) composite was the most transparent one.

The absorbance-peak values that were measured for the composite films at a wavelength of 310 nm can be observed from figure. 3. Wu *et al.* [7] performed a study related to the optical properties of nanofibrillar cellulose and observed that the transmittance value of the composite film decreased with the increasing NFC concentrations. In this study, chitosan–NFC (3%) was the most transparent composite film, presumably because of its high crystallinity.

3.3. FTIR studies

The FTIR spectra of the pineapple-leaf fibers can be observed from figure 4. The chitosan absorption peaks were characterized by the intra- and intermolecular stretching of O–H and the vibration of CH_2OH at 3500–3250 cm^{-1}, which was overlapped by the stretching of $-NH_2$ at 3500–3280 cm^{-1}. There was also a C–H vibration at 2930–2880 cm^{-1}.

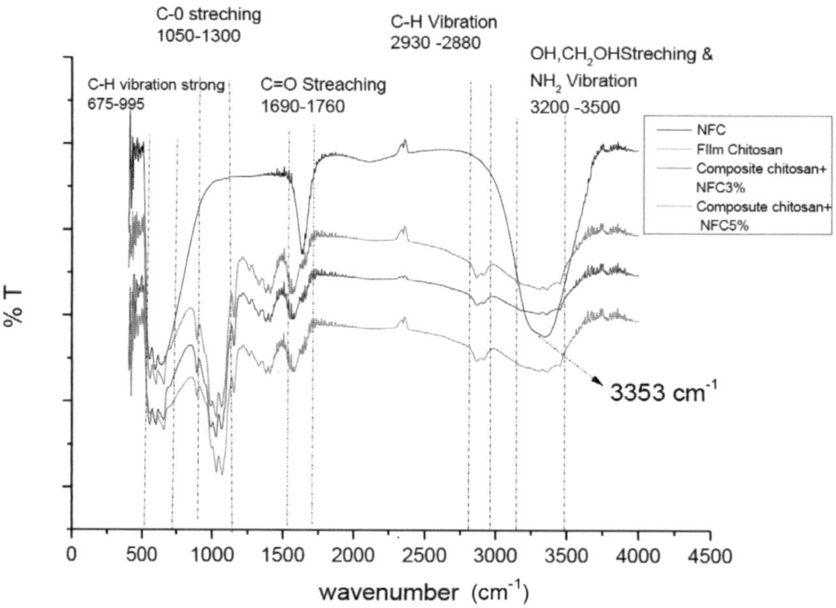

Figure 4. The FTIR spectra of NFC and the composite films.

In the nanocellulose spectrum, there was a wide transmission band at approximately 3400–3200 cm^{-1}, indicating the vibration of the O–H strain due to hydrogen bonding of the hydroxyl group of cellulose [8]. A sharp peak was also observed at approximately 3337 cm^{-1}, which corresponded to O–H vibration due to the occurrence of an intramolecular hydrogen bond [9]. Strong transmission bands are also observed at approximately 675–995 cm^{-1}, indicating the occurrence of C–H vibrations. At approximately 1160 cm^{-1}, there was a transmission band indicating a C=O strain.

The sharp peak that was observed in the FTIR spectrum of nanocellulose at approximately 3353 cm^{-1} cannot be observed in the spectra of pure chitosan films or composite chitosan–NFC films. Additionally, there was no signal corresponding to the C=O strain in the latter spectra. Hence, there were no significant differences between the FTIR spectrum of chitosan and the spectra of the composite films, which was due to the small concentration of NFC that was used.

4. Conclusions

In this study, we successfully prepared biocomposites that were derived from chitosan and nanocellulose obtained from the pineapple-leaf fibers. The composite film that contained 3% nanocellulose was the most transparent one among all the studied materials, as confirmed by the UV/Vis experiments. The results of the XRD analysis proved that the film crystallinity increases when the chitosan matrix is added with NFC. The low absorbance value of the composite chitosan–NFC film was most likely to be caused due to its high crystallinity. Our results demonstrated that biodegradable plastic with good transparency can be obtained from the composite film that has been successfully improved by the presence of NFC in the chitosan matrix.

Acknowledgements

The authors would like to thank Einago (www.einago.com) for the English language review.

References

[1] Celebi H and Kurt A 2015 *Carbohydr. Polym.* 133 284–93
[2] de Mesquita J P, Donnici C L and Pereira F V 2010 *Biomacromolecules* 11 473–80
[3] Cherian B M, Leão A L, de Souza S F, Costa L M M, de Olyveira G M, Kottaisamy M, Nagarajan E R and Thomas S 2011 *Carbohydr. Polym.* 86 1790–8
[4] Abdul-Khalil H P, Davoudpour Y, Islam M N, Mustapha A, Sudesh K, Dungani R and Jawaid M 2014 *Carbohydr. Polym.* 99 649–65

[5] Cherian B M, Leão A L, de Souza S F, Thomas S, Pothan L A and Kottaisamy M 2010 *Carbohydr. Polym.* **81** 720–5
[6] Khan A, Khan R A, Salmieri S, Le Tien C, Riedl B, Bouchard J, Chauve G, Tan V, Kamal M R, and Lacroix M 2012 *Carbohydr. Polym.* **90** 1601–8
[7] Wu T, Farnood R, O'Kelly K and Chen B 2014 *J. Mech. Behav. Biomed. Mater.* **32** 279–86
[8] Reddy J P and Rhim J W 2014 *Mater. Lett.* **129** 20–23
[9] Khan R A, Salmieri S, Le Tien C, Riedl B, Bouchard J, Chauve G, Tan V, Kamal M R and Lacroix M 2012 *Carbohydr. Polym.* **90** 1601–8

Synthesis and characterization of ZSM-5 zeolite with micro- and hierarchical pore structures from Bayat natural zeolite

S A Paramadini[1], R Ekananda[2] and Y K Krisnandi[1]

[1] Department of Chemistry, Faculty of Mathematics and Natural Sciences (FMIPA), Universitas Indonesia, Depok 16424, Indonesia

[2] Research and Development PT Pertamina (Persero), Jalan Raya Bekasi, Jakarta 13920, Indonesia

Corresponding author: yuni.krisnandi@sci.ui.ac.id

Abstract. Synthesis of ZSM-5 (Zeolite Socony Mobil–5) has been carried out using Bayat natural zeolite as silica and alumina source. Pretreatment on Bayat natural zeolite was performed through the process of activation, purification, dealumination, and fragmentation. The purpose of activation process is to remove the polar impurities, free oxide in the surface that cover up the pores and release the water that trapped in the pores of the crystals. The purification was conducted using Na-acetate buffer solution with ratio 1:3 (w/v). The dealumination process with 2M HCl was carried out to increase the Si/Al ratio. Fragmentation process then was carried out by using NaOH as much as 50% of NaOH was mixed with Bayat natural zeolite. Synthesis of ZSM-5 was performed using single templates method by adding TPAOH (Tetrapropylammonium hydroxide) as micropore structure directing agent. Furthermore, synthesis of hierarchical ZSM-5 using double templates method was also carried out in which PDDAM (Poly(acrylamide-co-diallyldimethylammonium chloride)) as mesoporous directing agent. The results showed XRD pattern of Microporous ZSM-5 and Hierarchical ZSM-5 had specific peaks at position 2 theta position of 7-10° and 22-25° which are the characteristics of ZSM-5 zeolite. The Si/Al molar ratio of microporous ZSM-5 and hierarchical ZSM-5 are 32.34 and 43.53.

Keywords: Bayat natural zeolite, ZSM-5 microporous, ZSM-5 hierarchical, TPAOH, PDDAM

1. Introduction

Indonesia is a country that has many volcanoes that potentially produce natural zeolite from volcanic dust. Zeolite is an aluminosilicate material with three-dimensional frameworks that consists of SiO_4 and AlO_4 tetrahedral arranged regularly, in which each oxygen atom is shared between two tetrahedral. The replacement of Si^{4+} with Al^{3+} in tetrahedral structure creates negative formal charge and counterbalanced by exchangeable cations [1]. But in the natural zeolite there are some impurities, therefore it is necessary to conduct pretreatment of natural zeolite to remove the impurities [2].

ZSM-5 (Zeolite Socony Mobil-5), which was developed in 1972, is usually synthesized using hydrothermal method from a liquid of alkaline silicate aluminate using tetrapropyl ammonium hydroxide (TPAOH) as a microporous template. ZSM-5 is one of the best working heterogeneous catalysts at industry especially in coal, oil and natural gas industries [3].

This research is focused on synthesis microporous and hierarchical ZSM-5 by using silica and alumina source from Bayat natural zeolite [4]. Before synthesis of zeolite, natural zeolite was

IOP Conf. Series: Materials Science and Engineering **496** (2019) 012022 doi:10.1088/1757-899X/496/1/012022

subjected to pretreatment, such as zeolite activation process, purification, dealumination and fragmentation. The synthesis of hierarchical ZSM-5 was carried out using double template method accompanied by addition to TPAOH as micropore directing agent and PDDAM (Poly(acrylamide-co-diallyldimethylammonium chloride)) as mesoporous directing agent [5]. As comparison microporous ZSM-5 was also synthesized, without addition of PPDA.

2. Experimental

2.1. Materials.

Bayat natural zeolite, deionized water, hydrochloric acid (HCl), solid sodium hydroxide (NaOH), glacial acetic acid (CH_3COOH), Na-citrate-bicarbonate solution, Na-acetate buffer ($C_2H_3NaO_2$), 30% H_2O_2 solution, Liquid Silica (Ludox 40%), solid sodium ditionate ($Na_2S_2O_4$), Tetrapropyl ammonium hydroxide (TPAOH) 1M and Poly(acrylamide-co-diallyldimethylammonium chloride (PDDAM).

2.2. Pretreatment of natural zeolite Bayat.

Pretreatment of natural zeolite bayat was performed through the process of activation, purification, dealumination, and fragmentation. Activation of zeolite was carried out by immersing the zeolite in distilled water with a ratio of 1:3 (w/v), and then purification was conducted using Na-acetate buffer solution with ratio 1:3 (w/v). In the dealumination process, the purified zeolite was added with 2M HCl and then stirred at 100 °C under reflux condition for 2 hours. Natural zeolite fragmentation was done through submolten salt technique system (SMS) [6] using a NaOH pellets with molar ratio of 2:1 (w/w) in deionized water. Characterizations of natural zeolite were conducted with Energy Dispersion X-Ray (EDX), X-Ray Diffraction (XRD) and Fourier-transform Infrared (FTIR) instruments.

2.3. Synthesis of ZSM-5 zeolites.

A mixture of fragmented Bayat natural zeolite, deionized water, TPAOH 1M with molar ratio of 64.34 SiO_2:1 Al_2O_3:10.08 $(TPA)_2O$:3571.67 H_2O, with addition of Ludox 40% (to reach the desired amount of Si) [2] was prepared under vigorous stirring at 100°C for 3 hours. Then, in order to control the pH of the mixture to be 11 ± 0.2, glacial acetic acid was added to the mixture. Stirring was carried out further for 3 hours at 100°C and at room temperature for 24 hours. Afterward, the mixture was transferred into an autoclave made from stainless steel with a teflon-lined inside for crystallization at 150 °C for 144 hours. After filtration, the white solid product was washed with the deionized water and dried at 60°C for 12 h and 105°C for 1 hour. Then, microporous ZSM-5 was calcined at 550°C for 8 hours and characterized using XRD, FTIR, EDX, BET and instruments. Similar procedure was employed in synthesis of hierarchical ZSM-5 zeolite, in which poly dimethyl dialyl ammonium chloride acrylamide (PDDAM) was added at room temperature, as a secondary template, continued with stirring for 24 hours, before transferred into the autoclave for further heated at 150 °C for 144 hours.

3. Results and discussion

3.1. Pretreatment of Bayat natural zeolite characterization

Before synthesis of ZSM-5 from Bayat natural zeolite, it was necessary that the natural zeolite to undergo pretreatment process, in order to remove impurities and to breakdown the structure of the zeolite framework, forming SiO_4^{4-} and AlO_4^{5-} active [1,4,6,7]. The result of FTIR (figure 1) showed typical spectra of aluminasilica compound, i.e. the peaks in the area of wave number 1250-1050 cm^{-1} and 1100-700 cm^{-1} assigned for the stretching vibration for SiO/AlO asymmetric and symmetric, respectively [8]. In the activation process, the zeolite can absorb a lot of water as indicated by the increased intensity of the H-OH bending vibration of water molecules at 1620 cm^{-1}.

There is a new peak assigned for stretching vibration of $Si\text{-}O^- Na^+$ $Al\text{-}O^- Na^+$ at 1461 cm^{-1} appeared in fragmented zeolite. Furthermore, in the region between 500-650 cm^{-1} and 1500-1700 cm^{-1}, peaks for finger print of zeolite framework and H-OH bending vibration disappeared in fragmented zeolite. It

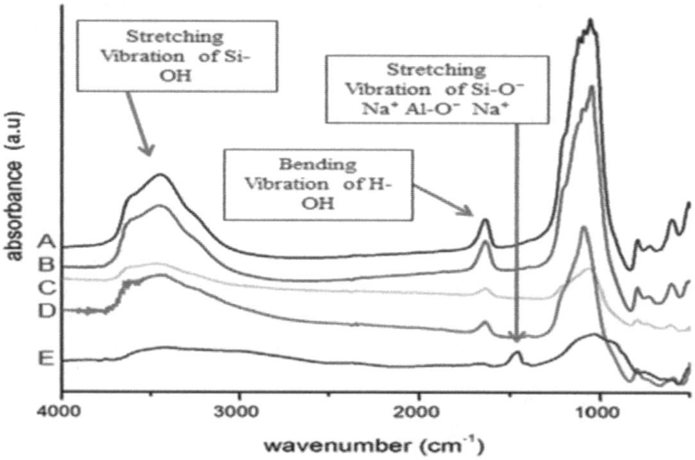

Figure 1. FTIR spectra of (a) raw, (b) activated, (c) purified, (d) dealuminated, (e) fragmented Bayat natural zeolite

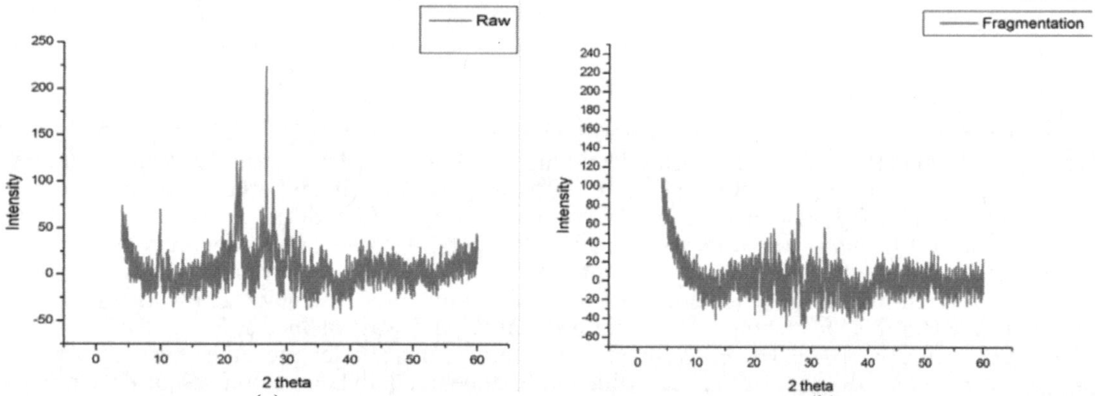

(a) (b)

Figure 2. Pattern of XRD (a) raw zeolite and (b) fragmented zeolite

indicates that the zeolite framework has been destructed [4]. This is supported by the XRD results in figure 2. Bayat raw zeolite diffraction pattern shown in figure 2a shows typical pattern of Mordenite structure with peaks at 2θ of 9.85°; 22.21°; and 25.61°. After fragmentation process, those peaks disappeared, indicative of the successful fragmentation process in which the zeolite structure was destroyed to become more amorphous material.

3.2. Zeolites ZSM-5 characterization

Si and Al active rearranged to form ZSM-5 zeolite framework by synthesis process. In the synthesis of ZSM-5, the purpose to use organic templates is to direct the shape and size of the pores in the structure [5]. In the synthesis of microporous ZSM-5 using single template, TPAOH was added to direct the development of micro-pores. Meanwhile, for the synthesis of hierarchical ZSM-5, using double template, PDDAM was also added to develop mesostructure. The existence of templates can be seen with the FTIR spectra (figure 3). From the results of both as-synthesis ZSM-5 zeolites, before calcination (figure 3a and figure 3b), the spectra contain peaks at 1470-1350 cm^{-1}, which is a absorption band of C-H bending vibration, and at 2960-2850 cm^{-1} which is an absorption band of C-H stretching vibration. After calcination, those peaks disappeared from the FTIR spectrum (figure 3c and

Table 1. EDX analysis of zeolite materials (% molar)

Elements	Raw	Fragmentation	Microporous ZSM-5	Hierarchical ZSM-5
Na	0.02	1.07	0.02	0.01
Al	0.24	0.10	0.06	0.04
Si	1.28	0.89	2.06	1.84
Ratio Si/Al	5.33	8.9	32.34	43.53

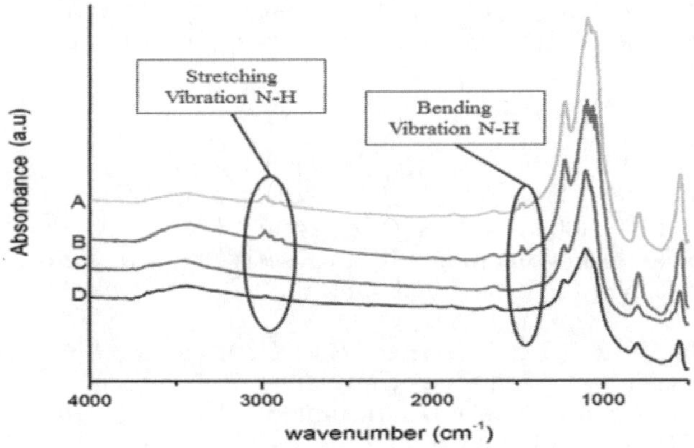

Figure 3. FTIR spectra of synthesized (a) microporous ZSM-5 and (b) hierarchical ZSM-5 before calcination; (c) microporous ZSM-5 and (d) hierarchical ZSM-5 after calcination.

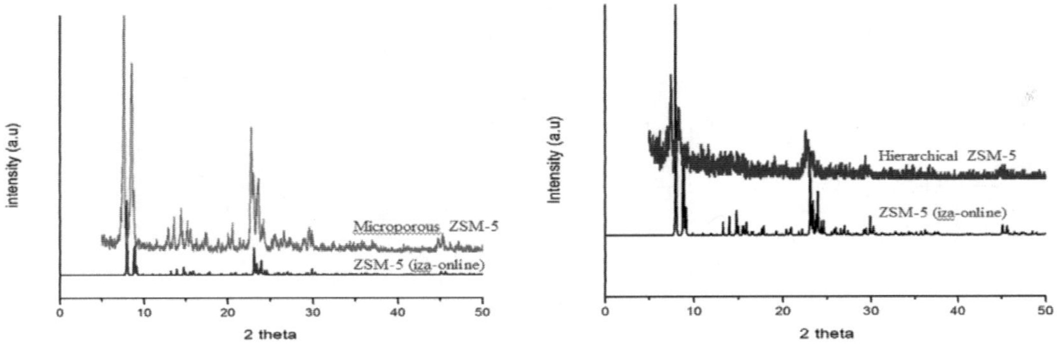

(a) (b)

Figure 4. Pattern of XRD for (a) microporous ZSM-5 and (b) hierarchical ZSM-5, both simulated from IZA online database and synthesized from Bayat natural zeolite

figure 3d). This indicates that the calcination process successfully removed organic templates, and zeolites become porous because there are empty space, which previously occupied by the template.

The result of EDX analysis (table 1) showed the raw Bayat natural zeolite has low Si/Al ratio of 5.33 and for the fragmentation ZAB of 8.9. The process of pretreatment is successful because there is an increase of Si/Al ratio in natural zeolite. After synthesis, ratio of Si/Al increase again, and it can be seen that the Si/Al molar ratio of hierarchical ZSM-5 (43.53) is higher than that of microporous ZSM-5 (32.4).

XRD result showed ZSM-5 sythesized has typical peaks at 2θ position of 7-10° and 22-25° [9]. Figure 4a and figure 4b showed the simulated pattern of ZSM-5 from IZA online database structure [10] and the patterns of ZSM-5 synthesized from Bayat natural zeolite. In general, they have similar specific diffraction patterns, suggesting that ZSM-5 has been successfully synthesized.

IOP Conf. Series: Materials Science and Engineering **496** (2019) 012022 doi:10.1088/1757-899X/496/1/012022

4. Conclusions

Synthesis of ZSM-5 zeolite with microporous and hierarchical structures using Bayat natural zeolite as silica and alumina source and addition of LUDOX is successful. This is evidenced by the XRD results addressing some typical zeolite ZSM-5 peaks. The Si/Al molar ratio of microporous ZSM-5 and hierarchical ZSM-5 are 32.34 and 43.53. Characterization by FTIR show that modification using ion-exchange method aims to produce zeolites that have high activity indicated has been successful.

Acknowledgements

The authors acknowledge Universitas Indonesia for PITTA Research Grant 2017 with contract number 702/UN2.R3.1/HKP.05.00/2017, Research and Development PT. Pertamina (Persero) for catalytic test and funding with contract number No. 004/E20400/2016-S0 and Einago (www.einago.com) for the English language review.

References

[1] Ming D W and Dixon J B 1987 *Clays Clay Miner.* **35** 463–8
[2] Barrer R M and Murphy E V T 1970 *J. Chem. Soc. A* 2506–14
[3] Kim D J and Chung H S 2003 *Appl. Clay Sci.* **24**, 69–77
[4] Rohayati, Krisnandi Y K and Sihombing R 2017 *AIP Conf. Proc.* **1862** 030094
[5] Wang L F, Zhang Z, Yin C, Shan Z and Xiao F 2010 *Micropor. Mesopor. Mater.* **131** 58–67 (2010).
[6] Yue Y, Liu H, Yuan P, Li T, Yu C, Bi H and Bao X 2014 *J. Catal.* **319** 200–210
[7] Mehra O P and Jackson M L 1953 *Clays Clay Miner.* **7** 317–327
[8] Moreno-Piraján J C, Garcia-Cuello V S and Giraldo L 2010 *J. Thermodyn. Catal.* **1** 1000101
[9] Treacy M M J and Higgins J B 1986 *Appl. Catal. B* **21** 388–9
[10] Baerlocher C 2017 *Database of Zeolite Structures: X-Ray Powder Pattern* (Zurich: Structure Commission of the International Zeolite Association, IZA-SC)

Preparation of Ag/Fe_3O_4/graphene composites for wastewater removal via the photo-Fenton reaction

A Taufik and R Saleh

Department of Physics, Faculty of Mathematics and Natural Sciences (FMIPA) Universitas Indonesia, Depok 16424, Indonesia

Integrated Laboratory of Energy and Environment, Faculty of Mathematics and Natural Sciences (FMIPA) Universitas Indonesia, 16424 Depok, Indonesia

Corresponding author: rosari.saleh@gmail.com

Abstract. In this study, Ag/Fe_3O_4/graphene composites were used as catalysts for the removal of wastewater via the photo-Fenton reaction. The Ag/Fe_3O_4/graphene composites were successfully synthesized by the hydrothermal method. The Ag content was varied from 15 and 25 wt%. The structural properties of the samples were investigated by X-ray diffraction and Raman spectroscopy, and their magnetic properties were examined using a vibrating sample magnetometer. Structural analysis revealed the presence of an inverse-cubic spinel structure corresponding to Fe_3O_4 and a cubic structure corresponding to Ag. The presence of graphene was confirmed by the presence of the D- and G-bands at 1300 and 1500 cm^{-1} in the Raman spectra, respectively, and the magnetic properties of the samples indicated that Ag/Fe_3O_4/graphene exhibits a ferromagnetic behavior at room temperature. The photo-Fenton reaction was examined in terms of the degradation of methylene blue from an aqueous solution. Control experiments were carried out to investigate the effects of the H_2O_2 content and solution pH. The maximum degradation of methylene blue was observed with 4 mL of H_2O_2 and at pH 13, reflecting that Ag/Fe_3O_4/graphene composites demonstrate promise for wastewater removal via the photo-Fenton process.

Keywords: surface phenomena, thermal analysis, composites, X-ray analysis

1. Introduction

Organic pollutants in wastewater are not only toxic but also recalcitrant [1]. Thus, these pollutants adversely affect the environment, mainly aquatic life and human health [1]. Therefore, it is imperative to develop an effective method to minimize and even remove organic pollutants from the environment. In this regard, advanced oxidation processes (AOPs) have been reported as technologies that appears to the efficiently eliminate organic pollutants, has high chemical stability and low biodegradability [2]. Among different AOPs, the photo-Fenton-like reaction has been emphasized because it can produce a high number of oxidative species such as hydroxyl radicals (OH) to accelerate the reaction [3].

Typically, iron oxides are used as Fenton-like catalysts because of their cost-effectiveness and relatively high activity [4]. Among iron oxides, magnetite (Fe_3O_4) has attracted the maximum attention [5]. Fe_3O_4 includes Fe^{2+} in its crystal structure (octahedral sites), which is crucial in Fenton reactions [6]. Moreover, Fe_3O_4 can be easily recovered from solution because of its magnetism, which is crucial for practical applications. However, Fe_3O_4 nanoparticles are limited because of rapid aggregation,

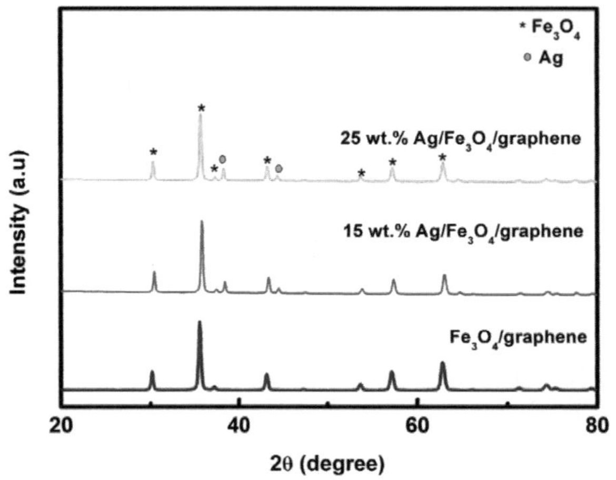

Figure 1. XRD spectra of Ag/Fe_3O_4/graphene composites with different weight percentages of Ag.

which hinders the efficiency of the photo-Fenton reaction. An effective strategy to improve the catalytic activity of Fe_3O_4 involves the formation of composites with metals [7], metal oxides [8], and carbon materials [9].

Therefore, in this study, Fe_3O_4 decorated with Ag and graphene materials was prepared to improve the catalytic activity of Fe_3O_4 materials, and the efficiency as a catalyst for the photo-Fenton reaction was investigated using model organic pollutant, methylene blue. In addition, the effects of the Ag and H_2O_2 contents were investigated.

2. Materials and methods

In the synthesis, $Fe(SO)_4{\cdot}7H_2O$ (iron sulfate heptahydrate), $AgNO_3$ (silver nitrate), NaOH (sodium hydroxide), and CH_3COOH (acetic acid) were purchased from Merck Chemicals and graphene was purchased from Angstron Materials. These materials were used without further purification.

Fe_3O_4 and Fe_3O_4/graphene composites were prepared according to a method reported previously by our group [10]. Ag nanoparticles were synthesized by the microwave-assisted method, and Ag/Fe_3O_4/graphene was synthesized by the hydrothermal method, as well as Fe_3O_4/graphene. First, graphene powder was dispersed into a solution, which is a mixture of alcohol and water. Second, graphene that has been dispersed is then sonicated using ultrasonic bath (frequency of 40 kHz) for 2 h. Third, the graphene dispersion added with nanoparticles Ag and Fe_3O_4 then stirred for 30 minutes. The mixture was heated for 30 min at a temperature of 120 °C, so that Ag and Fe_3O_4 can be deposited on graphene sheets.

All of the prepared samples were characterized by X-ray diffraction (Rigaku MiniFlex 600) operating at 30 kV and 15 mA with Cu-ka as the irradiation source. Thermogravimetric and differential thermal analysis was carried out on a Rigaku Thermo-Plus Evo TG8120, which was operated from room temperature to 100 °C under Ar. Vibrating Sample Magnetometer Oxford Type 1.2 T was used for measuring the magnetic hysteresis of the samples. UV–Vis measurement was carried out using a UV–Vis Hitachi UH5300 system.

The Fenton and photo-Fenton reactions were investigated by the degradation of methylene blue, serving as the model pollutant, from an aqueous solution. Typically, 100 mL of 20 mg/L of methylene blue was first stirred then kept for 30 min in the dark until equilibrium of adsorption–desorption was reached. Second, H_2O_2 was added into the solution, and the 40-W UV-C lamp was turned on. Degradation was monitored every 15 min and analyzed by UV–Vis spectroscopy. In this experiment, the effects of pH and H_2O_2 content were investigated.

3. Results and discussion

Figure 1 shows the X-Ray Difrraction (XRD) spectra of Ag/Fe_3O_4/graphene composites with different weight percentages of Ag. For comparison, the XRD pattern of Ag, Fe_3O_4, and graphene were also

Table 1. Lattice parameters and grain size of Ag/Fe_3O_4/graphene composites

Samples	Lattice parameters		Grain size	
	Ag	Fe_3O_4	Ag	Fe_3O_4
	a = b = c	a = b = c		
Fe_3O_4/graphene	-	8.332		39
15 wt% Ag/Fe_3O_4/graphene	4.0771	8.334	20	38
25 wt% Ag/Fe_3O_4/graphene	4.0777	8.346	22	38

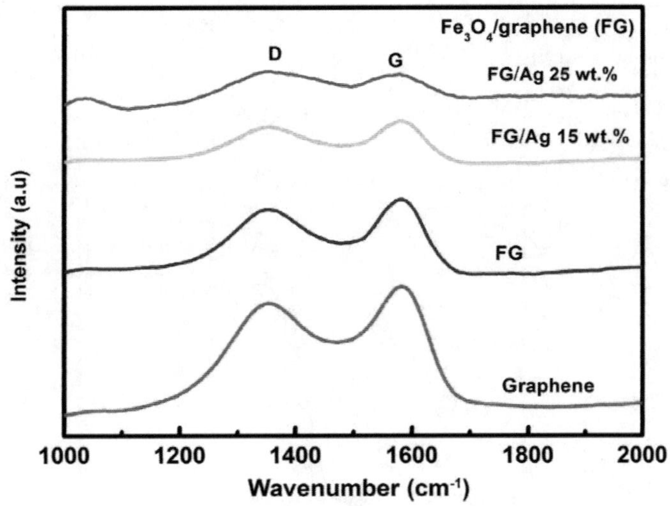

Figure 2. Raman spectra of the Ag/Fe_3O_4/graphene composites with different weight percentages of Ag.

recorded. The XRD pattern of Ag/Fe_3O_4/graphene composites revealed several diffraction peaks at $2\theta \approx 30.14°$, 35.49°, 43.28°, 53.76°, 57.20°, and 62.83°, corresponding to the (220), (311), (400), (442), (511), and (440) planes of the cubic spinel structures related to Fe_3O_4 nanoparticles. An additional peak was observed at $2\theta \approx 44.10°$, corresponding to the presence of Ag nanoparticles. Peaks corresponding to graphene were not detected, probably related to the low amount of graphene in the samples. Table 1 summarizes the results of the grain size of the samples and also lattice parameters obtained from the Rietveld refinement and Debye–Scherrer equation. The lattice parameters of Ag in the Ag/Fe_3O_4/graphene composites were not significantly different from those of pure Ag nanoparticles (table 1). Moreover, the lattice parameters of the Fe_3O_4 nanoparticles in Ag/Fe_3O_4/graphene composites were constant, indicating that the formation of Ag/Fe_3O_4/graphene composites does not change the structural properties of single-component materials. In addition, the calculated grain sizes for the samples revealed that the grain sizes of Ag and Fe_3O_4 in Ag/Fe_3O_4/graphene do not change significantly.

The presence of graphene in the sample was identified by using Raman spectroscopy. The measurement results are shown in figure 2. The Raman spectra of Ag/Fe_3O_4/graphene showed signals at 1300 and 1500 cm^{-1}, corresponding to the vibration modes of the D- and G-bands, respectively. The D-band involves the two zones of boundary phonons, which is a second-order double resonance between nonequivalent K points of the graphene Brillouin zone (it derived from transverse optical (TO) mode) for the D-band that are related to the out-of-plane breathing mode of sp^3 atoms. The G-band was the only Raman mode observed for graphene that is related to conventional first-order Raman scattering, and was associated with doubly degenerate phonon, in-plane and zone center modes (i.e., transverse optical and longitudinal optical modes, respectively) with $E_{2}g$ symmetry atoms [11]. Raman measurements confirmed the presence of graphene in the samples.

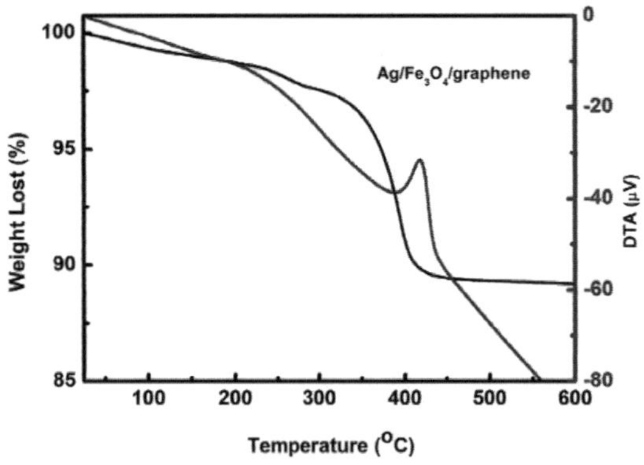

Figure 3. DTA/TGA analysis of Ag/Fe_3O_4/graphene composites.

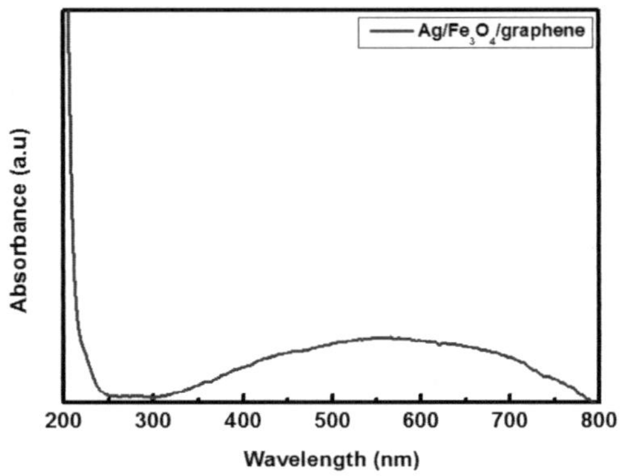

Figure 4. UV–Vis absorption spectrum of Ag/Fe_3O_4/graphene composites.

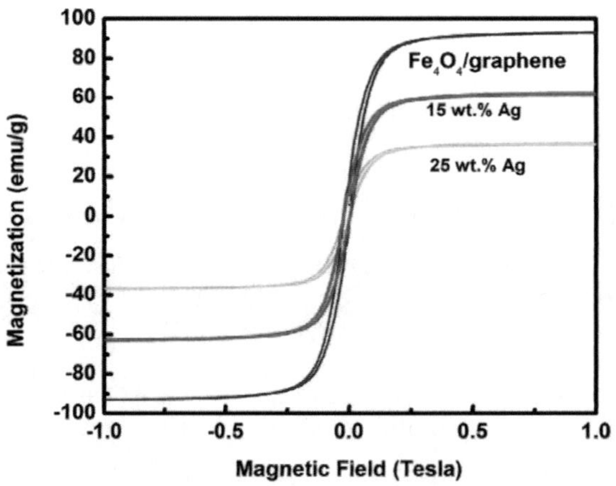

Figure 5. VSM measurement of Ag/Fe_3O_4/graphene composites.

Figure 3 plots the TGA curves of Ag/Fe_3O_4/graphene materials with two Ag concentrations. Ag/Fe_3O_4/graphene composites exhibited good thermal stability from room temperature until 400 °C. However, after 400 °C, a significant weight loss was observed, corresponding to the combustion of graphene materials. Notably, the combustion of graphene materials at 400 °C was an exothermic process. The UV–Vis absorption spectrum was recorded to identify the effect of silver nanoparticles on the optical characteristic of the samples. The UV–Vis spectrum of the Ag/Fe_3O_4/graphene composites is shown in figure 4. The UV–Vis spectrum of the Ag/Fe_3O_4/graphene composites exhibited broad absorption around the visible-light region (350–800 nm), corresponding to the presence of surface plasmon resonance (SPR) related to the silver nanoparticles [12]. Noble metal nanoparticles such as Au, Ag, and Cu show the phenomenon of SPR in the presence of electromagnetic radiation. This interesting optical property is related to the collective oscillations of conduction electrons that interact with the radiation.

Figure 5 shows the magnetic properties of the samples investigated using a VSM. All of the prepared samples exhibited ferromagnetic behavior at room temperature, with a coercive field of ~0.00136 T. With increasing Ag content, the magnetization strength gradually decreased, possibly related to the decrease in the net magnetic dipole moment of the samples in the presence of nonmagnetic materials. The high magnetization of the Ag/Fe_3O_4/graphene composites was suitable for the separation of Ag/Fe_3O_4/graphene composites from the aqueous solution by utilizing magnetic separation.

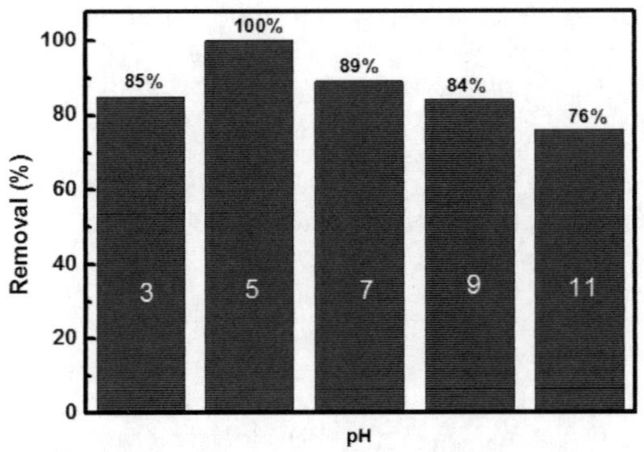

Figure 6. Effect of pH on the photo-Fenton process of Ag/Fe_3O_4/graphene.

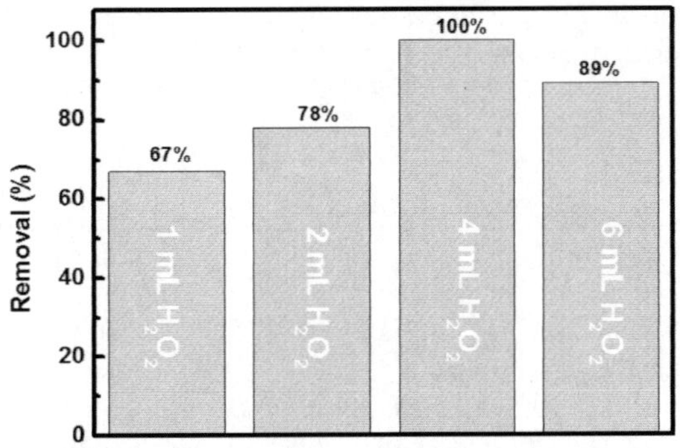

Figure 7. Effect of the H_2O_2 content on the photo-Fenton process of Ag/Fe_3O_4/graphene.

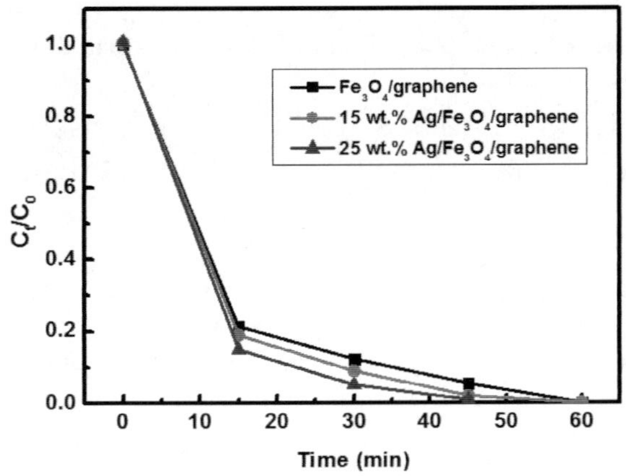

Figure 8. Effect of Ag loading on the photo-Fenton process of Ag/Fe_3O_4/graphene.

The photo-Fenton reaction was investigated at different solution pH values. The solution pH was varied from 3 to 13. Figure 6 shows the results obtained from the photo-Fenton reactions. The photo-Fenton reaction in the presence of Ag/Fe_3O_4/graphene composites was gradually enhanced with pH. From pH 3 to 5, the degradation percentages increased from 85% to 100% and then decreased thereafter. The decrease in the photo-Fenton reaction under alkaline conditions was related to the low degradation rate at high pH, which was caused by the precipitation of Fe^{3+} and the adsorption of iron hydroxides [13].

In addition, the effect of the H_2O_2 content was investigated to estimate the maximum H_2O_2 content for efficiency purposes. The H_2O_2 content was varied (1, 2, 4, and 6 mL). Figure 7 shows the results obtained for the effect of H_2O_2 on the photo-Fenton process. With the increase in the H_2O_2 content from 0 to 4 mL, the degradation ability increased.

The increased degradation ability with the increase in the H_2O_2 content from 0 to 4 mL was probably related to the increased generation of OH radicals with the increase in the H_2O_2 content. However, with the further increase in the H_2O_2 content up to 6 mL, the degradation ability decreased. The decreased degradation ability with the excessive H_2O_2 content was related to the excess H_2O_2, possibly related to the formation of undesirable hydroperoxyl radicals, but the decreased generation of hydroxyl radicals. Moreover, excess H_2O_2 also generated exuberant OH radicals, which is capable of dimerizing to form H_2O_2; which can inhibit the degradation rate of the dye due to the cumulative effects of H_2O_2.

In addition, the effect of the Ag content on the photo-Fenton reaction was investigated. Figure 8 shows the results. The addition of silver nanoparticles possibly improved the photo-Fenton reactions of Fe_3O_4/graphene, probably related to the presence of silver nanoparticles that can induce SPR, which can aid in the degradation of methylene blue from an aqueous solution. The photo-Fenton reaction mechanism of Ag/Fe_3O_4/graphene can be described as follows:

$$Ag/Fe_3O_4/graphene + h\nu \rightarrow Fe_3O_4\,(h^+) + graphene\,(e^-) + Ag\,(e^-)$$
$$O_2 + e^- \rightarrow O_2^-$$
$$H_2O_2 + O_2^- \rightarrow HO + OH^- + O_2$$
$$H_2O_2 + e^- \rightarrow HO + OH^- + O_2$$
$$Fe^{2+} + H_2O_2 \rightarrow Fe^{3+} + HO + OH^-$$
$$Fe^{3+} + H_2O_2 \rightarrow Fe^{2+} + HO_2 + H^+$$
$$HO + MB \rightarrow \text{Degraded product}$$

4. Conclusions

In this study, Ag/Fe_3O_4/graphene was successfully synthesized by the hydrothermal method. Structural properties revealed the presence of the cubic phase of Ag and the cubic spinel structure of Fe_3O_4. Raman spectra confirmed the presence of graphene, related to the detection of the D- and G-bands. The magnetic properties of the samples decreased with the incorporation of Ag nanoparticles, which was related to the decrease in the net magnetic moment of the samples. The optimum photo-Fenton process was observed under alkaline conditions with H_2O_2 content of 4 mL. The incorporation of Ag nanoparticles possibly enhanced the photo-Fenton reaction, related to the formation of surface plasmon resonance from Ag nanoparticles.

Acknowledgements

The authors would like to thank Einago (www.einago.com) for the English language review.

References

[1] Trapido M, Tenno T, Goi A, Dulova N, Kattel E, Kaluson D, Klein K, Tenno T and Viisimaa M 2017 *J. Water Process Eng.* **16** 277–82
[2] Dewil R, Mantzavinos D, Poulios I and Rodrigo M A 2017 *J. Environ. Manage.* **195** 93–9
[3] Xu T, Zhu R, Zhu G, Zhu J, Liang X, Zhu Y and He H 2017 *Appl. Catal. B* **212** 50–8
[4] Giannakis S, Liu S, Carratalà A, Rtimi S, Amiri M T, Bensimon M and Pulgarin C 2017 *J. Hazard. Mater.* **339** 223–31
[5] Sahar S, Zeb A, Liu Y, Ullah N and Xu A 2017 *Chin. J. Catal.* **38** 2110–9
[6] Yan S, He H, Wu D, Chen D, Ma Y, Li X, Zhu J and Yuan P 2009 *Ind. Eng. Chem. Res.* **48** 9915–21
[7] Huang B, Qi C, Yang Z, Guo Q, Chen W, Zeng G and Lei Chao 2017 *J. Catal.* **352** 337–50
[8] L. Xu, J. Wang, Environ. Sci. Technol. 46 (2012) 10145–10153
[9] Qiu B, Li Q, Shen B, Xing M and Zhang J 2016 *Appl. Catal. B* **183** 216–23
[10] Taufik A and Saleh R2016 Mater. Sci. Forum **864** 106–11
[11] Pastrana-Martínez L M, Morales-Torres S, Likodimos V, Falaras P, Figueiredo J L, Faria J L and Silva A M T *Appl. Catal. B* **158–159** 329–340
[12] Rather R A, Singh S and Pal B 2017 *Sol. Energy Mater. Sol. Cells* **160** 463–9
[13] Cao Z, Wen X, Chen P, Yang F, Ou X, Wang S and Zhong H 2018 Colloids Surf. A**549** 94–104

IOP Conf. Series: Materials Science and Engineering **496** (2019) 012024 doi:10.1088/1757-899X/496/1/012024

Magnetic and microwave absorbing properties of Zn-substituted Barium M-Hexaferrite in X-band frequency range

M C Aisiyah, M Zainuri and D Ristiani

Departement of Physics, Faculty of Science, Institute of Technology Sepuluh Nopember, Surabaya 60111, Indonesia

Corresponding author: zainuri@physics.its.ac.id

Abstract. In this study, Zn-doped Barium M-Heksaferrite (BaFe[illegible]Zn[illegible]O[illegible]) powders have been prepared by coprecipitation processing from iron sand, $BaCO_3$ and Zn powders, followed by calcined 1000°C for 5h. The powders were characterized by X-ray Diffractometry (XRD), Vibrating Sample Magnetometer (VSM), conductivity measurement and Vector Network Analyzer (VNA). The XRD patterns indicated that 2 phases, there were Barium M-Hexaferrite phase (dominant phase) and Hematite phase, with conductivity value on 6x10[illegible] S/cm and be include as semiconductor material. The hysterisis curve showed that was a soft magnetic with coercivity field (Hc) on 0.0506 kOe. The powders was mixed on paint 10% of total weight and coated on steel grade AH36. The maximum of reflection loss on x-band frequency was -29.1 dB on frequency of 9.34 GHz and have almost 87.70% electromagnetic energy absorbed by Zn-doped Barium M-Heksaferrite (BaFe[illegible]Zn[illegible]O[illegible]).

Keywords: barium m-hexaferite, zinc subtitution, reflection loss, radar absorbing material

1. Introduction

The widespread of technological devices that utilize electromagnetic waves (EM) was developed in wireless communiation system after the development of electromagnetic theory. Electromagnetic waves are emitted by resources of a certain wavelength used for different purposes, one of which is used as radio detection and ranging (RADAR) in military and commercial applications [1]. RADAR is an electromagnetic wave system that can be used as a detection tool, distance meter, and make a map of objects such as airplanes. By analyzing the radar reflected signal it can be determined the target location and through further analysis of the reflected signal can also be determined the target type. The radar detection on the target is based on Radar Cross Section (RCS). RCS is an area on the target that reflects signals coming by size, shape, or material of the target [2]. Along with the development of radar technology, a system was developed to avoid radar detection. This can be partially obtained by using materials that can absorb microwaves. This absorber uses magnetic and dielectric fillers to reduce the electric and magnetic field components from microwaves that reach the absorber [3].

In the process of absorption of electromagnetic radiation by absorbent materials, the characteristics of the complex permeability and permittivity of these materials play a very important role. Material that can be used a microwave absorber must be easily penetrated by microwaves and has a lossy filler that can convert microwave energy to another form of energy such as thermal energy. Magnetic material can be used as filler to covert microwave energy to thermal energy. In order to improve magnetic and microwave properties of magnetic fillers (i.e. hexaferrites in this case), the Fe^{3+} ion is usually substituted by divalent, divalent-tetravalent, trivalent cationic ion [4].

IOP Conf. Series: Materials Science and Engineering **496** (2019) 012024 doi:10.1088/1757-899X/496/1/012024

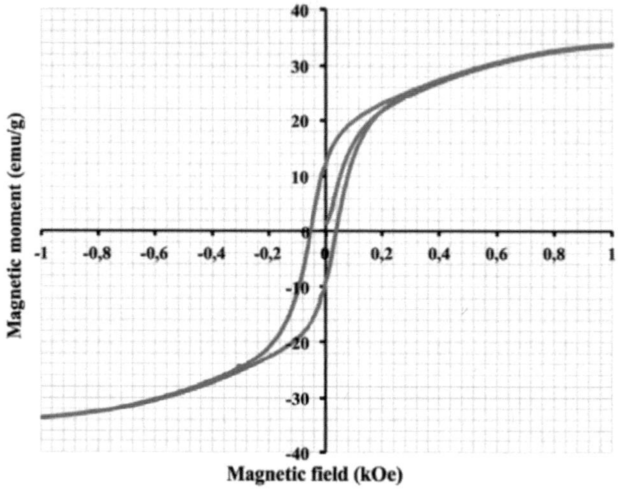

Figure 1. Results VSM analysis of Zn doped-Barium M-Hexaferite with x = 0.3

Recently, Barium M-Hexaferrite has been studied broadly because this material has excellent chemical stability and dielectric properties, high magnetocrystalline anisotropy and saturation magnetization and can be obtained at low cost and exhibits ferromagnetic resonance for microwave [5]. But the material has a high coercivity field, the field required for magnetization to be zero thus affecting the outer field is also high. This is a problem that should be avoided, because the high field coercivity causes the anisotropic properties are increasing so as to make the nature of the absorbs become increasingly weak. The problem can be solved by substitution of Fe^{3+} ions having radius of 0.69 Å with other metal cations having almost the same size [6]. In this study, the synthesis of barium m-hexaferrite used iron stone from Tanah Laut, south kalimantan, Indonesia as magnetic material with Fe^{3+} ion was replaced by Zn^{2+} ion to decrease magnetic properties. In the end was obtained Zn doped-Barium M-heksaferrite ($BaFe_{11.7}Zn_{0.3}O_{19}$).

2. Materials and methods

Following reagent chemicals were used to synthesize $BaFe_{11.7}Zn_{0.3}O_{19}$; iron stone (Fe_3O_4), $BaCO_3$, Zn, Hydrochloric acid and Ammonium hydroxide solution. This research began with the extraction process to produce iron powder from natural ironstones. Then, the powder is washed with alcohol (96%) using an ultrasonic cleaner to reduce and eliminate impurity. Hydrochloric acid solution is used as a solvent from a mixture of iron powder, $BaCO_3$ and Zn which is then stirred using a magnetic stirrer for 30 min at 70 °C. In the dissolution of the iron powder was filtered and evaporated at 100 °C, so that the resulting is $FeCl_2.FeCl_3$ solution. The third solution was mixed at 70 °C for 45 min and added ammonium hydroxide solution slowly until pH scale on 9 (precipitation). It was cleaned until pH scale on 7 and dried. The prekusor was calcined at 1000 °C for 5h to produce the crystalline phase. The Zn doped-Barium M-Hexaferrite was characterized by X-Rays Diffraction (Cu Kα radiation, λ = 1.54060 Å). The measurement of hysteresis of the samples was done using Vibrating Sample Magnetometer (OXFORD VSM 1.2H) and the sample conductivity value was obtained by two point probe methods.

The coating process requires Zn doped-Barium M-Hexaferrite, grade A type AH36 of steel plate, Agatha ship paint, dan thinner. The Zn doped-BaM and paint were mixed with ratio 1:10 and stirred for 10 min with velocity of ± 8000 rpm. The mixture is diluted with thinner and then used to coat the steel plate. Coating thickness is 4 mm. Vector Network Analyzer (VNA ADVANTEST type-3770) is used to measure microwave absorption on X-band frequency.

3. Results and discussion

3.1. *Analysis of magnetic properties*

The magnetism of Zn doped-Barium M-Hexaferrit material with x = 0.3 ($BaFe_{11.7}Zn_{0.3}O_{19}$) can be seen from the result of VSM measurement in figure 1. Based on the hysteresis curve $BaFe_{11.7}Zn_{0.3}O_{19}$ obtained the maximum magnetization value ~33.87 emu/g, coercivity field (Hc) ~0.0506 kOe, and magnetization remanesi (Mr) ~14.782 emu/g. If the magnitude is compared to the Barium M-Hexaferrit without doping material ($BaFe_{12}O_{19}$) in Rosyidah's study [7], it can be concluded that there is a decrease in the value of the coercivity field (Hc) and the increase of the maximum magnetization value and the magnetization of remanence (Mr) in the material.

The decrease in the coercivity field value of the Zn doped-Barium M-Hexaferrite ($BaFe_{11.7}$ $Zn_{0.3}O_{19}$) indicates a change of magnetic properties from hard magnetic ($BaFe_{12}O_{19}$) to soft magnetic ($BaFe_{11.7}Zn_{0.3}O_{19}$). The change can occur because of the Zn^{2+} ion as a doping that goes into the structure $BaFe_{12}O_{19}$ to replaces Fe^{3+} ion, so it is able to ruffle the magnetic domain arrangement and minimize the total magnetic moment. In hexagonal structures of Barium M-Hexaferrite consist of 8 Fe^{3+} ions with spin up condition and 4 Fe^{3+} ions with spin down condition [8]. In one unit of the molecule there are 6 ↑ Fe^{3+} ions occupying the octahedral position in the RS block, 1 ↑ Fe^{3+} ion occupies the bipiramidal trigonal position in the R block, 2 ↓ Fe^{3+} ions occupy the octahedral position on the block S, 2 ↓ Fe^{3+} ions occupies the tetrahedral position of the S block , and 1 ↑ Fe^{3+} ion occupy the octahedral position on the block S. Each Fe^{3+} ion contributes a magnetic moment of 5 μ_B, so that the total magnetic moment per unit of molecule is 20 μ_B ((6 + 1-2-2 + 1) * 5 μ_B = 20 μ_B). The substitution of Fe^{3+} ions by Zn^{2+} ions it can reduce the amount of magnet bohr from 5 μ_B (Fe^{3+}) and then be replaced to 0 μ_B (Zn^{2+}) in which substitution takes place simultaneously in spin-down and spin-up of sublattices, can decrease coercivity field [7].

Engineering of magnetism from hard magnetic to soft magnetic is useful in the absorption mechanism by the dissipation of electromagnetic wave energy of soft magnetic material whose magnetic domain is easily driven by a small magnetic field. The energy required in the magnetization process can be calculated from the hysteresis curve area where the soft magnetic material has a narrow curve so that the energy required in the magnetization process is relatively small. The properties of magnetic materials are also influenced by the magnetic material crystalline size where the size of the crystal can inhibit the movement of the domain wall. On the small crystal size, many the grain boundaries are formed so as to cause obstacles to the movement of the domain wall are more too, which is resulting in increased resistance to the demagnetization field. So it means the coercivity field (Hc) is also getting larger [9].

In determining the magnetic properties of a material not only seen from the field of coercivity but also can be seen from the parameters of maximum magnetization magnitude and magnetization remanence. The maximum magnetic value of Barium M-Heksaferrite increased from 16 emu/g to 34 emu/g and the magnetization remanence value of Barium M-Heksaferrite also increased from 8.334 emu/g to 14.782 emu/g after doping Zn^{2+}. This is because the substitution of Zn^{2+} ions is tend to occupy the tetrahedral position, and in the Barium M-Heksaferrite structure, the tetrahedral position has a structure opposite to the octahedral position. The octahedral position produces the total magnetic moment. Therefore, the reduction of negative magnetic moment can occur as a result of substitution of Zn^{2+} ions to Barium M-Heksaferrite structure. With the increase of the maximum magnetization value of Zn doped- Barium M-Heksaferrite, this means that the ability of particles to maintain the dominance of magnetic domains also increases. Physically, when the material is given an external magnetic field, the magnetic moment will be polarized in the direction of the given external magnetic field. The more of the orientation direction of the magnetic dipole moment, it will require more energy for the polarization process so that the maximum magnetization value is also higher.

3.2. Analysis of microwave absorption

The result of reflection loss measurements on AH36 steel before and after coating process is done by using VNA in the X-Band frequency range (8-12GHz) shown in figure 2. Based on result, the reflection loss value of the material is -29.1 dB at a frequency of 9.34 GHz with a reflection coefficient value of 0.0351 Γ so that 87.70% of the microwave energy was absorbed. The reflection loss is defines the energy value of the lost electromagnetic wave after the material (absorbed energy). While the reflection coefficient with symbol Γ is the ratio of the reflected wave to the coming wave. If the reflection coefficient Γ is zero, then it can be interpreted that the load impedance matched with channel impedance and no wave reflected by the material.

Magnetic material in this research is formed from two phases. By XRD results of the samples. Figure 3 shows characteristic diffraction peaks corresponding to the Barium M-Hexaferrite and Hematit structure. The dominant phase that is Barium M-Hexaferrite phase ($BaFe_{11.7}Zn_{0.3}O_{19}$) and

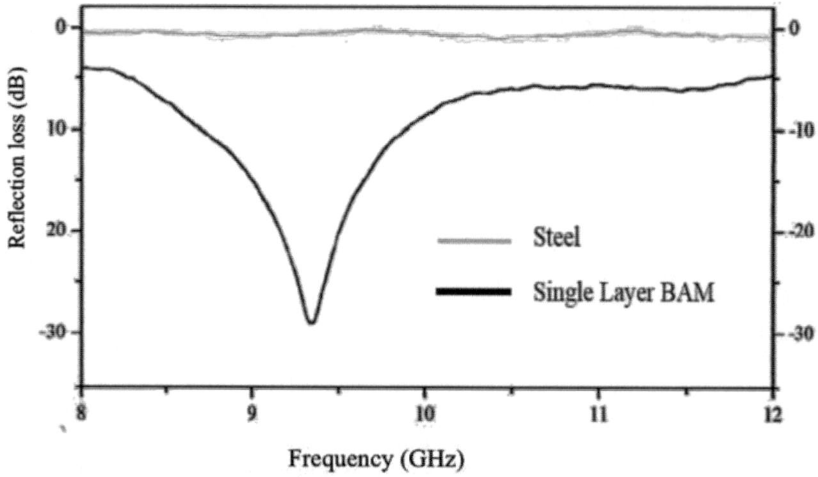

Figure 2. Graph of reflection loss at X-Band frequency

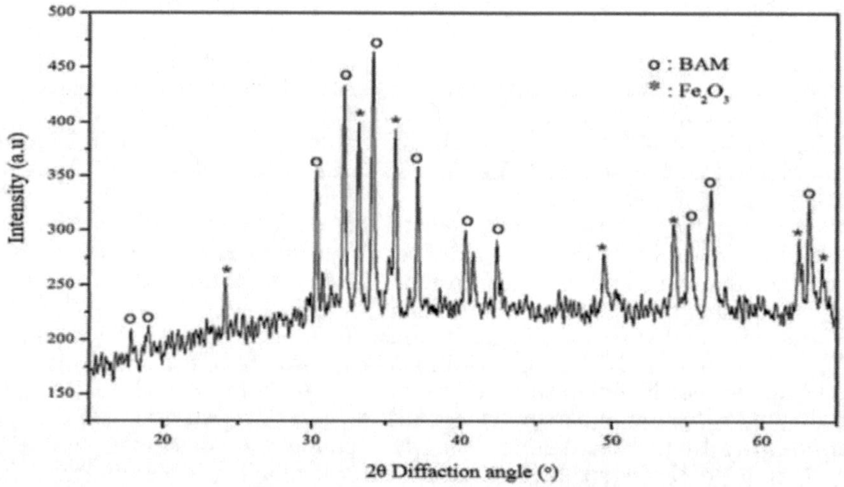

Figure 3. X-Ray Diffraction Pattern of Barium M-Hexaferite $BaFe_{11.5}Zn_{0.5}O_{19}$

secondary phase is Hematite phase (Fe_2O_3). For Barium M-Hexaferrite phase formed is 56.64% and Hematite phase formed is 43.36%. Physically, calcination process aims to getting formation of the desired phase. But, the hematite phase appears as a secondary phase after calcination at 1000 °C. This is because iron sand have Fe^{2+} and Fe^{3+} ions, only Fe^{3+} ions are required to form hexagonal structure. Furthermore, the remaining Fe^{2+} and Fe^{3+} ions in the solution bind to O^{2-} ions to form Fe_3O_4 compounds.

Heat treatment can form a new phase according the temperature, in this case the Barium M-Hexaferite precursor forms $BaFe_{11.5}Zn_{0.5}O_{19}$ and Fe_3O_4 transforms into the Hematite stable phase (Fe_2O_3). That is mean, in addition to eliminating impurities, this process can produce precursors into stable phase Barium M-Hexaferite have a crystal structures that bonds between particles are more compact/strong. When the particles gets heat energy, so that the atoms in the structure will have activation energy to vibrate and construct the structure in a more stable state. Atoms can diffuse interstitially and substitutionally. In this study, Fe^{3+} ions diffuses substitution by Zn^{2+} ions in the structure of Barium M-Hexaferite have occured. Referring to Amalia's study [10] that the Hematite particle also has the ability to absorb microwaves with a reflection loss of -28.69 dB at frequencies below X-Band (6-8 GHz). So that the secondary phase of hematite is expected not to reduce the ability in terms of microwave absorption, because basically ferrite materials have good magnetism properties.

Electromagnetic waves consist of two components, the electric field components and magnetic field components are perpendicular to each other. Electromagnetic wave energy is the resultant of electric field energy and magnetic field energy. To optimize the absorption of electromagnetic waves, the material must have an electrical loss component and a magnetic loss (magnetical loss) component capable of interacting with both electric and magnetic fields in electromagnetic waves. The absorption mechanism can be observed from distributed loading properties using electrical and magnetic materials that allow the absorption of microwave energy by the material. The electrical properties of materials are included in the semiconductor material range, but are still weak enough to affect the absorption mechanism of the electrical losses component. While the magnetism of material contribute in the mechanism of magnetic loss (magnetical losses). This material is a soft magnetic where the total magnetic moment is smaller. With a small total magnetic moment it means that the magnetic domain is easily moved by an external magnetic field small too, so that the electromagnetic energy can be dissipated as a magnetic loss. Based on the electrical conductivity value, the RAM (Radar Absorbing Material) is in the range of conductivity value of semiconductor material that is between $10^{[illegible]}$ - $10^{[illegible]}$ S/cm. Conductivity measurement of Zn doped-Barium M-Hexaferrite is relatively small (6x$10^{[illegible]}$ S/cm), this result indicates the ability of Barium M-Hexaferrite to deliver very weak electrical currents. Therefore this material needs to be combined with other materials with high conductivity values to improve microwave absorption properties. Basically, when microwaves arrive at microwaves absorbers, electric fields will form on the surface of the absorber. Then there will be a flow of current on the surface of the absorber. When the current flows on the absorber surface, the energy from the microwave will be converted thermal energy [11].

The results of absorption bandwidth in this study are able to cover the overall frequency of X-Band that is 8-12 GHz frequency, but there is a characteristic absorption peak at a certain frequency range. This indicates that the maximum material absorption capacity in that frequency. Single layer of coating materials tend to form absorption peaks at frequencies 8.6-10 GHz with 1.4 GHz absorption bandwidth. The bandwidth shows the effectiveness of the material to reduce the energy of electromagnetic waves, so that reflected electromagnetic waves are very weak.

4. Conclusions

Research on powder synthesis of Zn doped-Barium M-Hexaferrite by coprecipitation method has been conducted. Powder obtained by 56.64% of Barium M-Hexaferite phase and 43.36% of Hematite phase (Fe_2O_3), can absorbing 87.70% of microwave energy with RL 29.1 dB. However, it has a relatively low conductivity value ~6x$10^{[illegible]}$ S/cm. So it needs to be combined with high conductivity materials to improve the microwave absorbing properties. Single layer coating of Zn doped-Barium M-Hexaferrite have absorption bandwidth is 1.4 GHz on X-Band frequency range in 8.6-10 GHz bandwidth. So it is able to cover the overall frequency of X-Band is the frequency of 8-12 GHz.

References

[1] Gürer G 2010 *Design and Characterization of Electromagnetic Wave Absorbing Structural Composites* (Ankara: Middle East Technical University) Master Thesis
[2] Knott E F, Shaeffer J F and Tuley M T 2005 *Radar Cross Section* 2 ed (Raleigh: SciTech Publishing)
[3] Akman O, Durmus Z, Kavas H, Aktas B, Kurtan U, Baykal A and Sözeri H 2013 Phys. Status Solidi A **210** 395–402
[4] Tang X, Yang Y and Hu K 2009 *J. Alloys Compd.* **477** 488–92
[5] Qiu J, Gu M and Shen H 2005 *J. Magn. Magn. Mater.* **295** 263–8
[6] Sözeri H, Deligöz H, Kavas H and Baykal A 2014 *Ceram. Int.* 40 8645–57
[7] Rosyidah K C *Synthesis and Characterization of Structure, Magnetic Properties And Electrical Properties Composite Barium M-Hexaferite / Polianiline Core-Shell-Based Structure of Natural Iron Sand* (Surabaya: Institut Teknologi Sepuluh Nopember)
[8] Parween N 2014 *Study of Barium Hexaferrite ($BaFe_{12}O_{19}$) Synthesised by Sol Gel Auto-Combustion Technique* (Rourkela: National Institute of Technology Rourkela) Master Thesis
[9] Saidah I N and Zainuri M 2012 *Jurnal Sains dan Seni ITS* 1 B41–6
[10] Amalia L 2014 *Double Layer Coating with Dallenbach Layer Method Using Polyaniline and Barium M-Hexaferrite as Radar Absorbing Material (RAM)* (Surabaya: Institut Teknologi Sepuluh Nopember) Master Thesis
[11] Syamsir A and Astuti 2012 *Jurnal Fisika Unand* 1 48–52

Modification of alginate-modified natural zeolite for the degradation of 4-nitrophenol using H_2O_2 and microwaves

Y Yulizar, Foliatini, R Bakri, D O B Apriandanu and F Tiolita

Department of Chemistry, Faculty of Mathematics and Natural Sciences (FMIPA), Universitas Indonesia, Depok 16424, Indonesia

Corresponding author: yokiy@sci.ui.ac.id

Abstract. 4-Nitrophenol is used as a precursor or an intermediate in the explosive, pesticide, herbicide, and dye and pharmaceutical industries, and causes environmental pollution. One of the solutions for removing 4-nitrophenol is the degradation of dyes using H2O2, catalysts, and microwave assistance. In this study, alginate-modified zeolite was used as the catalyst. In addition, the effects of microwave power and degradation time were reported. Alginate-modified zeolite was characterized by UV–visible and Fourier transform infrared spectroscopy. The degradation of 4-nitrophenol (1.0×10^{-4} M) was observed in 120 s with 0.1 g of alginate-modified zeolite, 2.15×10^{-4} M of H_2O_2, and a microwave power of 800 W. The degradation of nitrophenol followed a first-order reaction with a k value of 8.95×10^{-2} min^{-1}.

Keywords: natural zeolite-alginate, degradation, 4-nitrofenol, microwave

1. Introduction

4-Nitrophenol is used as a precursor or an intermediate for manufacturing explosives, insecticides, pesticides, herbicides, fungicides, and dyes and pharmaceuticals [1–3]. It is obtained by the incomplete combustion of fossil fuels via nitration [4]. 4-Nitrophenol adversely affects the environment, causing headaches, nausea, kidney damage, and skin, eye, and liver irritation [1], similar to mutagenic and carcinogenic compounds [5] and phenol compounds and their derivatives. As a result, the US Environmental Protection Agency has designated mono-, di-, and poly-nitrophenol waste as toxic materials [6].

Several methods, including biological methods, adsorption, extraction, sonolysis, and reduction, have been reported for the removal of 4-nitrophenol [7–11]. However, these methods exhibit disadvantages, including incomplete waste removal limited only because of material absorption, high-energy requirements, and toxic compound production.

In recent years, advanced oxidation processes (AOPs), particularly catalytic wet hydrogen peroxide oxidation (CWPO), have been reported to be suitable for treating wastewater at room temperature and pressure. The CWPO process produces hydroxyl radicals (HO·), which can oxidize various pollutants into small nontoxic molecules such as CO_2 and H_2 [12]. AOPs using H_2O_2 as the oxidizing agent with microwave assistance are environmentally friendly [13]. H_2O_2 is a thermal initiator, which can evenly spread the converted heat energy formed by the microwaves, thereby increasing the reaction rate [14–15].

Several heterogeneous catalysts such as zeolites serve as support materials, where their structures, pores, and interlayers can be modified, and have been used in AOPs [16]. Zeolites are catalysts that contain uniform micropores and exhibit high hydrothermal stability [17]. However, zeolites have long, narrow pore channels; hence, they limit the diffusion of reactants and products during catalysis [18]. To modify natural zeolites, alginate can be used as an environmentally friendly material.

Alginate is a natural polymer that exhibits good water solubility. Furthermore, it can easily form gels by the addition of divalent cations such as Ca^{2+} [19]. Alginates, which belong to the polysaccharide family, are cell-wall constituents of brown seaweed, which produce carboxylate groups by extraction [20]. Alginate is composed of 1,4-linked β-D-mannuronic acid (M) connected to α-L-guluronic acid (G) residues [21]. Calcium alginate exhibits nontoxicity, biodegradability, nonimmunogenicity, water solubility, and thermal irreversibility as the immobilization material of catalyst material [22].

In this study, the degradation of 4-nitrophenol was examined using alginate-modified natural zeolites by utilizing AOPs and H_2O_2 as the oxidation agent with microwave assistance. In addition, the effects of microwave power and degradation time were examined, as well as degradation kinetics.

2. Experimental

2.1. Materials and methods

Natural zeolite was purchased from CV Transindo Citra Utama (Bandung, Indonesia). NaOH, HCl, NaCl, and 30% H_2O_2 were purchased from Merck (Germany). $AgNO_3$ was purchased from Sigma Aldrich, and 4-nitrophenol was purchased from PT Antam. Sodium alginate was purchased from Himedia Laboratories Pvt. Ltd. (Mumbai, India). Double-distilled water was purchased from PT Ikapharmindo Putramas.

2.2. Activation of natural zeolite

Natural zeolite was activated by washing in water at 70 °C and stirring for 1 h, and the result was precipitated overnight. The precipitate was dried at 150 °C for 4 h. In addition, natural zeolite was chemically activated by the addition of 0.05 M of NaOH and 0.05 M of HCl. The zeolite cation was uniformed by the addition of 1 M of NaCl. The mixture was stirred at 70 °C for 6 h and precipitated for 12 h. The precipitate was dried at 105 °C and then calcined at 300 °C for 2 h. The chloride ions of zeolite were removed by washing in aquabides. Finally, zeolite was treated with $AgNO_3$ to confirm chloride-free zeolite. Activated zeolite was characterized by Fourier transform infrared (FTIR) spectroscopy.

2.3. Modification of natural zeolites with alginates

Alginate was synthesized by dissolving solid sodium alginate with the desired mass into 100 mL of water and stirring for 12 h, yielding a viscous yellow solution [23]. Zeolite was modified using alginate by mixing zeolite with the desired mass and a sodium alginate solution for 2 h, followed by overnight precipitation. The precipitate was dried at 40 °C for ~12 h. The alginate-modified natural zeolite was characterized by FTIR spectroscopy.

2.4. Degradation of 4-nitrophenol using alginate-modified natural zeolite

First, 0.1 g of zeolite-alginate was mixed with 2.15×10^{-4} M of H_2O_2, followed by vortexing for 5 min. Next, 10^{-4} M of 4-nitrophenol was incorporated into the zeolite and H_2O_2 mixtures, followed by vortexing for 5 min. The mixture was subjected to a microwave power of 800 W for 120 s. The mixture was characterized by UV–visible spectroscopy.

2.5. Characterization

UV–visible spectra (Shimadzu 2600) were recorded to obtain the maximum wavelength of the alginate-modified natural zeolite and to examine the degradation activity of 4-nitrophenol. FTIR spectra were recorded on a Perkin-Elmer system for functional group identification.

3. Results and discussion

FTIR spectra were recorded to identify the functional groups in alginate-modified natural zeolites. After activation, significant changes were not observed for natural zeolite and Na-modified zeolite (figure 1a). FTIR spectra revealed an increased intensity for the peak at 421 cm^{-1}, indicative of the presence of the Na–O group in activated zeolite. The intensity increase indicated the presence of a considerable amount of Na–O in the zeolite pores. In other words, the activation of zeolite using NaCl

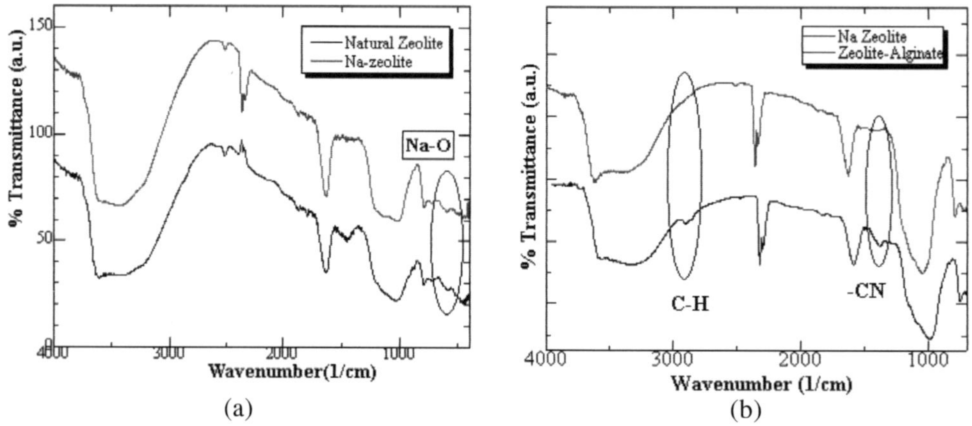

Figure 1. FTIR spectra of (a) natural zeolite and (b) alginate-modified zeolite

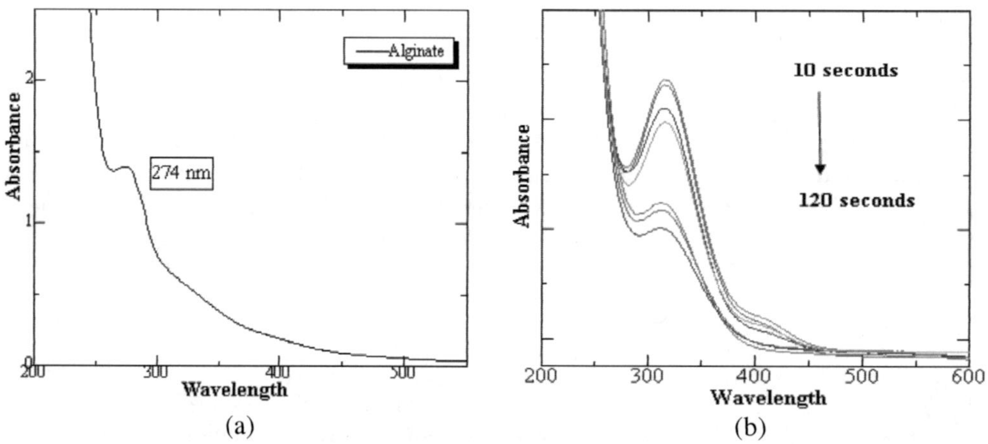

Figure 2. UV–visible absorption spectra of (a) alginate and (b) degradation of 4-nitrophenol in 120 s.

was successfully carried out. Chloride ions were removed; hence, NaCl is not formed again. The presence of chloride ions was confirmed by the addition of $AgNO_3$. The mixture did not form any AgCl precipitate. Hence, zeolites are free of chloride ions. The FTIR spectrum of alginate-modified natural zeolites is shown in figure 1b). Different peaks were observed for Na-modified zeolite and alginate-modified natural zeolite. C–H and C–N stretching peaks were observed at 2901 and 1387 cm^{-1}, respectively, with a predominance of the C–H groups in the alginates, indicative of the successful modification of zeolite by alginate.

The UV–visible absorption spectrum of alginate is shown in figure 2a. Alginate formation was confirmed at a maximum wavelength of 274 nm. UV–visible absorbance as a function of time for the degradation of 4-nitrophenol is shown in figure 2b. The decrease in the absorbance at maximum wavelengths of 318 and 405 nm can be used to estimate the percentage degradation of 4-nitrophenol. After degradation, the peaks almost disappeared at the maximum wavelengths of 318 and 405 nm, as confirmed by dissociation of the *p*-nitrophenoxide ion [24]. With increasing degradation time, the absorbance value decreased, and a considerable amount of 4-nitrophenol was degraded. From this study, 4-nitrophenol was first converted into smaller non-light-absorbing compounds, which were

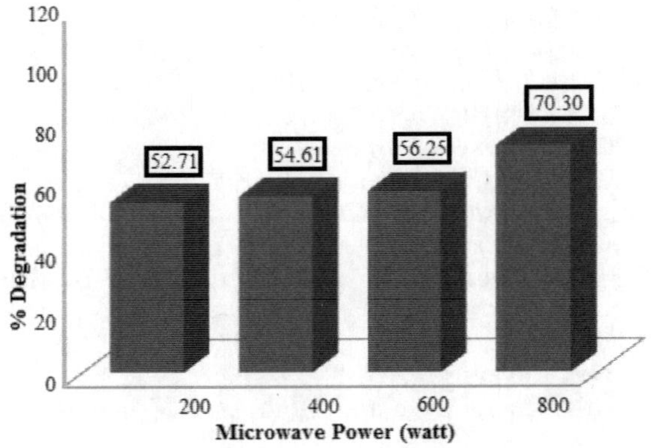

(a)

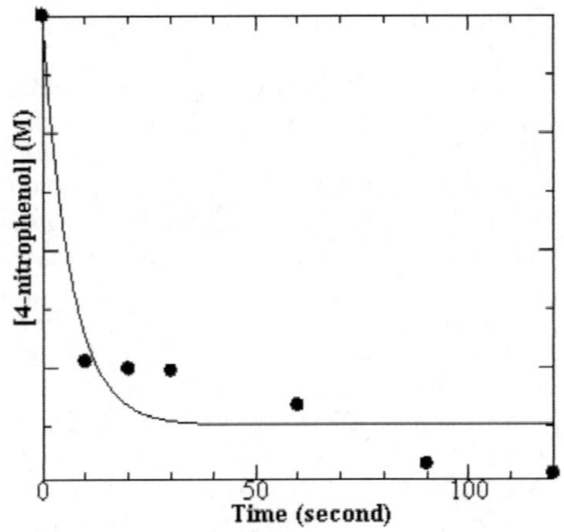

(b)

Figure 3. (a) Relationship between percentage degradation and microwave power. (b) Plot of 4-nitrophenol concentrations as a function of time.

subsequently converted into CO_2 and H_2O [24]. Na zeolite-alginate degraded 1.0 × 10^{-4} M of 4-nitrophenol in 120 s with 0.1 g of alginate-modified natural zeolite, an H_2O_2 concentration of 2.15 × 10^{-4} M, and a microwave power of 800 W.

The relationship between percentage degradation and microwave power is shown in figure 3a. A microwave power of 800 W gave the highest percentage degradation of 4-nitrophenol of 70.30%. Microwave energy causes the rotating molecules that trigger the H_2O_2 activity to produce OH radicals. It can form alginate radicals, which degrade 4-nitrophenol. The kinetics model for the degradation of 4-nitrophenol using alginate-modified natural zeolite was examined by plotting 4-nitrophenol concentrations as a function of time (figure 3b). The degradation of 4-nitrophenol followed a first-order reaction with a k value of 8.95 × 10^{-2} min^{-1}.

4. Conclusions

In this study, alginate-modified natural zeolites as the catalyst were used for the degradation of 4-nitrophenol by utilizing advanced oxidation processes and H_2O_2 as the oxidation agent with microwave assistance. Modification with alginate leads to the high activity of zeolites. A microwave power of 800 W led to the degradation of 70.30 % of 4-nitrophenol in 2 min. The results indicated that the degradation rate of 4-nitrophenol is affected by microwave power. The degradation of 4-nitrophenol followed a first-order reaction with a k value of 8.95 × 10^{-2} min^{-1}.

Acknowledgements

The authors would like to thank Universitas Indonesia for funding this research through PITTA Grant with contract number 700/UN2.R3.1/HKP.05.00/2017 and Einago (www.einago.com) for the English language review.

References

[1] Sun J H, Sun S P, Fan M H, Guo H Q, Lee Y F and Sun R X 2008 *J. Hazard Mater.* **153** 187–93
[2] Chen X, Murugananthan M and Zhang Y 2016 *Chem. Eng. J.* **283** 1357–65
[3] Parida K and Das D P 2004 *J. Photochem. Photobiol. A: Chem.* **163** 561–7
[4] Kulkarni M and Chaudhari A 2007 *J. Environ. Manage.* **85** 496–512
[5] Pang Y and Lei H 2016 *Chem. Eng. J.* **287** 585–92
[6] Tanaka K, Luesaiwong W and Hisanaga T 1997 *J. Mol Catal. A: Chem.* **122** 67–74
[7] Kulkarni M and Chaudhari A 2006 *Bioresour. Technol.* **97** 982–8

IOP Conf. Series: Materials Science and Engineering **496** (2019) 012025 doi:10.1088/1757-899X/496/1/012025

[8] Koubaissy B, Joly G, Batonneau-Gener I and Magnoux P 2011 *Ind. Eng. Chem. Res.* **50** 5705–13
[9] Gupta T, Pradhan N C and Adhikari B 2003 *J. Membr. Sci.* **217** 43–53
[10] Entezari M H, Mostafai M and Sarafraz-yazdi A 2006 *Ultrason. Sonochem.* **13** 37–41
[11] Jiang D, Xie J, Chen M, Li D, Zhu J and Qin H 2011 *J. Alloys Compd.* **509** 1975–9
[12] Garrido-Ramírez E G, Theng B K G and Mora M L 2010 *Appl. Clay. Sci.* **47** 182–92
[13] Bo L, Quan X, Chen S, Zhao H and Zhao Y 2006 *Water Res.* **40** 3061–8
[14] Menéndez J A, Inguanzo M and Pis J J 2002 *Water Res.* **36** 3261–4
[15] Lai T L, Lee C C, Wu K S, Shu Y Y and Wang C B 2006 *Appl. Catal. B: Environ.* **68** 147–53
[16] Caudo S, Centi G, Genovese C and Perathoner S 2006 *Top. Catal.* **40** 207–19
[17] Mielby J, Abildstrom J O, Wang F, Kasama T, Weidenthaler C and Kegnaes S 2014 *Angew. Chem. Int. Ed.* **53** 12513–6
[18] Hartmann M 2004 *Angew. Chem. Int. Ed.* **43** 5880–2
[19] Gombotz W R and Wee S F 1998 *Adv. Drug Deliv. Rev.* **31** 267–85
[20] McHugh D J, Hernández-Carmona G, Arvizu-Higuera D L and Rodríguez-Montesinos Y E 2001 *J. Appl. Phycol.* **13** 471–9
[21] Goh C H, Heng P W S and Chan L W 2012 *Carbohydr. Polym.* **88** 1–12
[22] Harikumar P S, Joseph L and Manjusha C M 2011 *Nano Science & Nano Technology: An Indian Journal* **5** 23–7
[23] Foliatini F, Yulizar Y and Hafizah M A E 2015 *J. Math. Fundam. Sci.* **47** 31–50
[24] Subbulekshmi N L and Subramanian E 2017 *J. Environ. Chem. Eng.* **5** 1360–71

High-temperature electrical properties of $LaFe_{1-x}Ti_xO_3$ (x = 0.4, 0.5, 0.6)

D Y Jufri, D Triyono and H Laysandra

Department of Physics, Faculty of Mathematics and Natural Sciences (FMIPA)
Universitas Indonesia, Depok 16424, Indonesia

Corresponding author: djoko.triyono@sci.ui.ac.id

Abstract. We report herein the high-temperature electrical properties of perovskite $LaFe_{1-x}Ti_xO_3$ with x = 0.4, 0.5, 0.6. All samples were synthesized by using the sol-gel method, and the electrical properties were investigated by using impedance spectroscopy in the frequency range of 100 Hz–1 MHz and at temperatures ranging from 348 to 548 K. Nyquist plots show a decrease in impedance with decreasing Ti content and the diameter of the Nyquist plot impedance semicircle decreases with increasing temperature. The dielectric constant increases with increasing temperature until reaches 10^5 at 548 K. In addition, the dielectric relaxation phenomenon appears at all temperatures. Finally, we used scanning electron microscopy to study morphology, grain size, and grain distribution of these materials. The average grain size is 360–370 nm and increases with increasing Ti content.

Keywords: perovskite, $LaFe_{1-x}Ti_xO_3$, electrical properties, sol-gel

1. Introduction

Perovskite oxide materials are composed of rare-earth element (A), transition metal (B) and oxygen, which form the ABO_3 structure that become the general structure of these materials. These materials are candidates for various applications in solid oxide fuel cells as a catalyst and as a sensor and magnetic material [1]. Lanthanum orthoferrite $LaFeO_3$ is an important perovskite oxide for oxygen sensor applications due to their attractive mixed ionic and electronic conductivity [1]. The substitution of a transition metal into the *A* and/or *B* sites and sample-preparation techniques have led to developments in the structural, electrochemical, and magnetic properties of $LaFeO_3$ [3]. Phokha *et al.* reported that, with increasing Ti content, the perovskites $LaFe_{1-x}Ti_xO_3$ with x = 0.1 and 0.2 develop a giant dielectric constant (up to 10^3) at low frequency (10^2 Hz) [2]. Previous studies reported the frequency-dependent dielectric properties of $LaFeO_3$, which exhibits a colossal dielectric constant at high temperature ($\varepsilon \geq 10^3$) [4,5]. Pb doping into the La site increases the dielectric constant even more ($\sim 10^5$) [4]. A material with colossal dielectric constant (CDC) (typically ($\varepsilon \geq 10^3$)) was recently investigated because of their ability to reduce the capacitive component [5]. Recently, the crystallite size was decreased from 47 ± 2 to 25 ± 2 nm and the magnetization was increased by decreasing the Ti content in calcined $LaFe_{1-x}Ti_xO_3$ (x = 0.1 and 0.2) samples. However, for ceramic sintered samples, an increase in Ti content increases the average grain size and dielectric constant [2]. However, to date, the high-temperature electrical properties, morphology, and elemental analysis of $LaFe_{1-x}Ti_xO_3$ with x = 0.4, 0.5, 0.6 have yet to be reported. Therefore, we investigated herein the electrical properties of $LaFe_{1-x}Ti_x$O3 with x = 0.4, 0.5 and 0.6.

2. Experimental

Lanthanum orthoferrite $LaFe_{1-x}Ti_xO_3$ with x = 0.4, 0.5 and 0.6 were prepared by using the sol-gel

IOP Conf. Series: Materials Science and Engineering **496** (2019) 012026 doi:10.1088/1757-899X/496/1/012026

Table 1. Average grain size of $LaFe_{1-x}Ti_xO_3$.

Ti content	Average grain size (nm)
$x = 0.4$	360
$x = 0.5$	364
$x = 0.6$	368

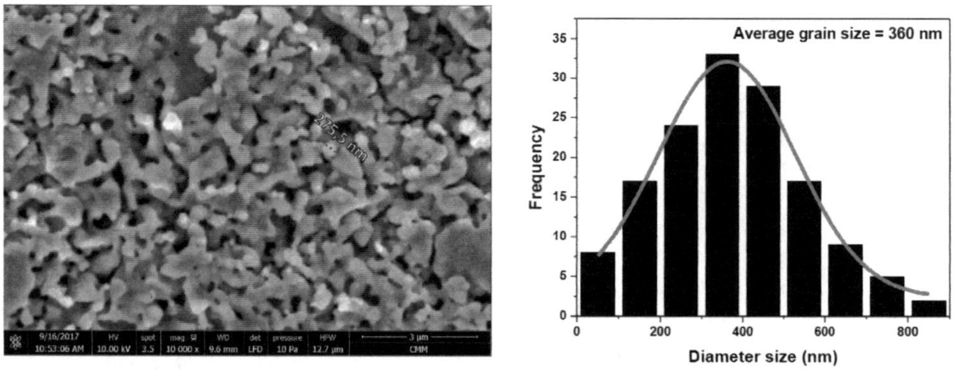

Figure 1. SEM micrograph and grain distribution of $LaFe_{1-x}Ti_xO_3$ with $x = 0.4$ ($x = 0.5$ and 0.6 not shown)

method and high-purity La_2O_3, $Fe(NO_3)_3{\cdot}9H_2O$, TiO_2, and $C_6H_8O_7{\cdot}H_2O$ as a precursors. The precursor solution was heated and stirred at 393 K until the gel phase formed. The gel was pre-calcined (dried) in a furnace at 473 K for 5 h. The precursor was then calcined at temperatures 1173 K for 6 h to form the perovskite phase of the compound. The calcined powder sample was subject to a pressure of 5 kN/m^2, then sintered for 3 h at 1573 K. Scanning electron microscopy (SEM) FEI QUANTA 650 was used to perform microstructural analysis. The bulk phase was coated with carbon by using a Quorum K975X Turbo Evaporator. The electrical properties of the bulk samples were characterized by using an RLC-Meter FLUKE-PM6306 over a wide frequency range (100 Hz to 1 MHz) at 348–548 K. Impedance data are represented on the Nyquist and Bode plots. The dielectric behavior was calculated by using the parallel-plate method.

3. Results and discussion

3.1. Microstructure analysis
Figure 1 shows that the grains are inhomogeneously distributed for $x = 0.4$. The samples gave similar results when $x = 0.5$ and 0.6 (data not shown). The average grain size was calculated, which reveals that the grain size increases with Ti content (see table 1). Similar results were reported by Phokha *et al.* [2]. These results indicate that increasing Ti content enhances the grain-boundary mobility [2].

3.2. Electrical properties
Figure 2 shows the temperature dependence of the complex impedance plot of $LaFe_{1-x}Ti_xO_3$ with $x =$ 0.4, 0.5 and 0.6. All samples show similar trends: the diameter of the semicircular arc decreases with increasing temperature, indicating an increasing conductivity [8] and decreasing resistance, which suggests the presence of semiconducting properties [13]. These results also imply that the impedance increases with increasing Ti content at all temperatures measured, which indicates a change in electrical properties.

Figure 3 shows the Bode plots of the real part Z' of the impedance for all samples. Z' decreases with increasing frequency and becomes constant at higher frequency. In addition, Z' decreases with increasing temperature, indicating an increase in ac conductivity [9]. The higher Z' at lower frequency indicates space-charge polarization in the samples [7]. Finally, the results for Z' at different temperatures converge at a high frequency, indicating that space charge is released as a result of the reduced barrier properties of the material [9,12].

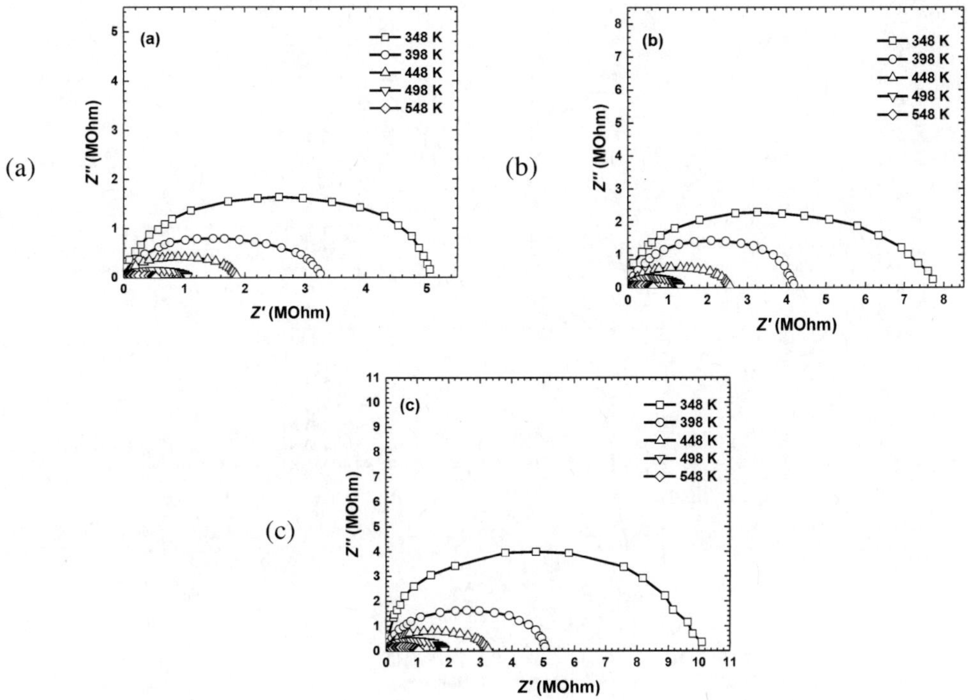

Figure 2. Nyquist plot of $LaFe_{1-x}Ti_xO_3$ with (a) $x = 0.4$, (b) $x = 0.5$. (c) $x = 0.6$ at 348–548 K.

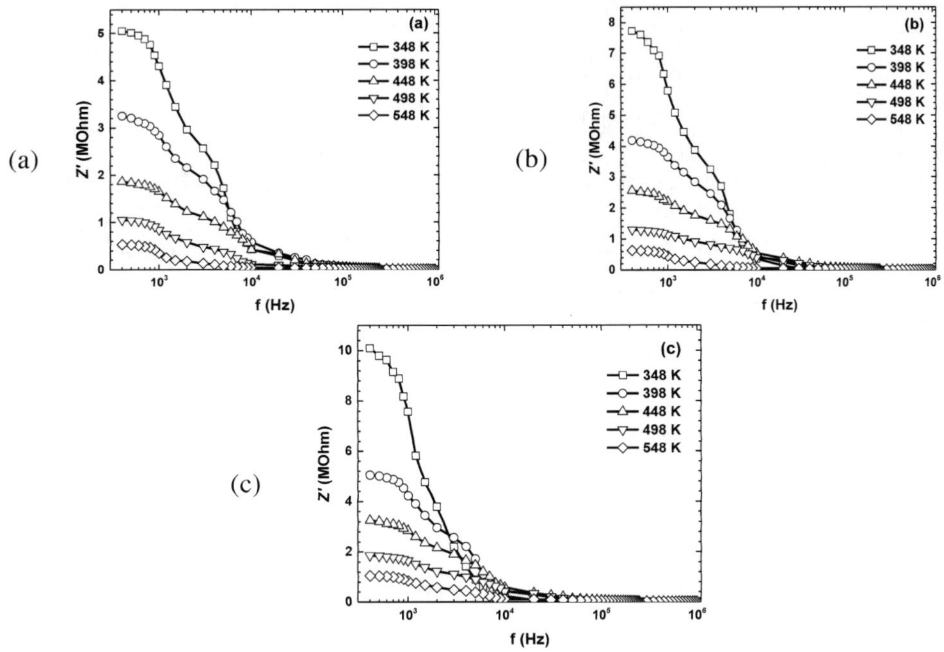

Figure 3. Bode plot of real part Z' of impedance of $LaFe_{1-x}Ti_xO_3$ with (a) $x = 0.4$, (b) $x = 0.5$, (c) $x = 0.6$ at 348–548 K.

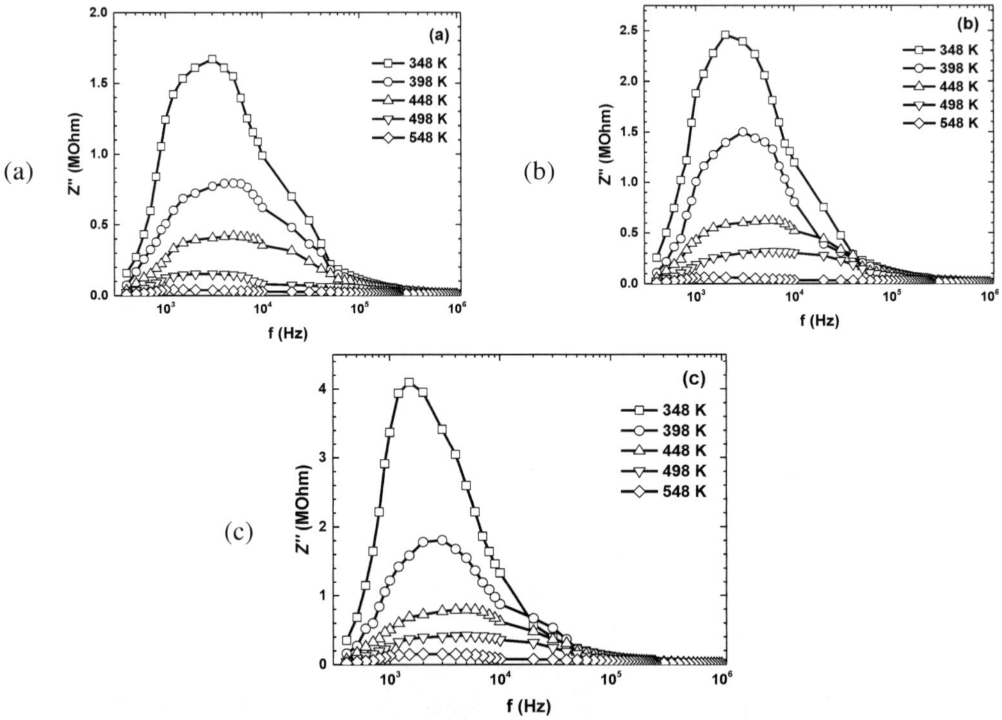

Figure 4. Bode plot of imaginary part Z'' of impedance for $LaFe_{1-x}Ti_xO_3$ with (a) $x = 0.4$, (b) $x = 0.5$, (c) $x = 0.6$ at 348–548 K

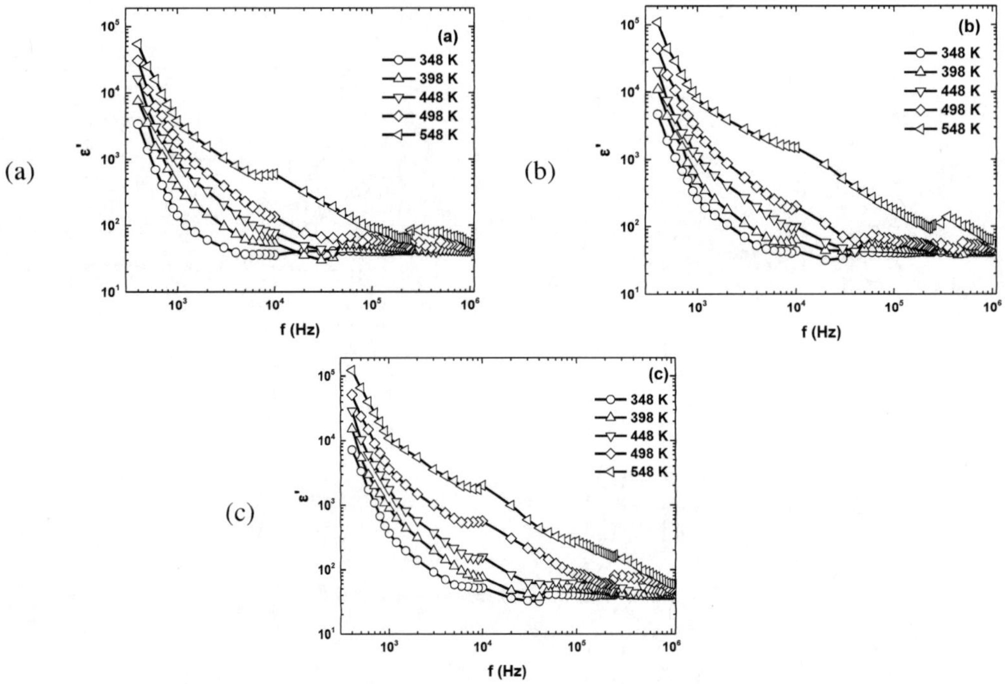

Figure 5. Real part of dielectric constant ε' for $LaFe_{1-x}Ti_xO_3$ for (a) $x = 0.6$, (b) $x = 0.5$, and (c) $x = 0.4$ at 348–548 K.

Figure 4 shows the Bode plots of the imaginary part Z'' for all samples. Z'' increases until it reaches a peak and then decreases monotonically, indicating a relaxation phenomenon [9]. Such a phenomenon suggests the presence of immobile charge localized at a low temperatures and the defects of the crystal at an elevated temperature [6,7]. The magnitude of the relaxation peak decreases with increasing temperature, which indicates the presence of a thermally activated relaxation mechanism [6,9]. The shifting of the relaxation peak to higher frequency indicates the accumulation of space charge in the samples [9,10].

Figure 5 shows the real part ε' of the dielectric constant as a function of frequency for all samples. The dielectric constant increases with increasing temperature and decreases with increasing frequency, which is the normal trend for a dielectric material [6]. In addition, the dielectric constant increases with increasing Ti content. For all samples, the colossal dielectric constant is attained at $\varepsilon' \geq 10^3$ [4,5]. A similar phenomenon was reported for perovskite compounds by Bhargav *et al.* [4] for $LaPb_{1-x}Fe_xO_3$, Idrees *et al.* [5] for $LaFeO_3$, and Phokha *et al.* [2] for $LaFe_{1-x}Ti_xO_3$ with $x = 0.1$ and 0.2. The colossal dielectric constant at low frequencies can be associated with the presence of all types of polarization in this frequency range [6], including electronic contribution, ionic, atomic, dipolar and interfacial and also from the electrode [5]. In ferrite compounds, electrons are considered to be the dominant charge carries, and the phenomenon of electron hopping occurs between Fe^{3+} and Fe^{2+} ions situated at different lattice sites [6]. The dispersion at low frequency may be explained by polarization of the interface and to the presence of dc conductivity [7,11]. Finally, the dielectric constant increases with increasing grain size, which indicates that ε' is dependent on the microstructure [5]. A similar phenomenon was reported for perovskite compounds by Phokha *et al.* [5] for $LaFe_{1-x}Ti_xO_3$ with $x = 0.1$ and 0.2.

4. Conclusions

We present the results of a study of the microstructure and dielectric properties of $LaFe_{1-x}Ti_xO_3$ with $x =$ 0.4, 0.5 and 0.6. The grains are inhomogeneously distributed with an increase in the average grain size with increasing Ti content. A colossal dielectric constant is observed ($\varepsilon' \geq 10^3$). Finally, the dielectric constant increases with increasing Ti content.

Acknowledgements

The authors would like to appreciate PITTA Grant from Universitas Indonesia for the provided financial support under contract No. 2257/UN2.R3.1/HKP05.00/2018 and Einago (www.einago.com) for the English language review.

References

[1] Fujii T, Matsusue I, Nakatsuka D, Nakanishi M and Takada J 2011 *Mater. Chem. Phys.* **129** 805–9
[2] Phokha S, Hunpratup S, Pinitsoontorn S, Putasaeng B, Rujirawat S and Maensiri S 2015 *Mater. Res. Bull.* **67** 118–25
[3] Phokha S, Hunpratub S, Usher B, Chenlek N, Bootchanont A, Maensiri S and Chindaprasirt P 2017 *J. Alloys Compd.* **708** 605–11
[4] Bhargav K K, Ram S and Majumder S B 2015 *J. Alloys Compd.* **638** 334–43
[5] Idrees M, Nadeem M, Atif M, Shiddique M, Mehmood M and Hassan M M 2011 *Acta Mater.* **59** 1338–45
[6] Nath S, Barik S K, Choudhary R N P and Barik S K 2017 *Phys. Lett. A* **381** 2174–80
[7] Dhak P, Dhak D, Das M, Pramanik K and Pramanik P 2009 *Mater. Sci. Eng. B* **164** 165–71
[8] Mahato D K, Saha S and Sinha T P 2016 *J. Phys. Chem. Solids* **92** 45–52
[9] Benali A, Bejar M, Dhahri E, Graça M F P and Costa L C 2015 *J. Alloys Compd.* **653** 506–12
[10] Prasad B V, Rao B V, Narsaiah K, Rao G N, Chen J W and Babu D S 2015 IOP Conf. Ser.: *Mater. Sci. Eng.* **73** 012061
[11] Jacob R, Nair H G and Isac J 2015 *Processing and Application of Ceramics* **9** 73-79.
[12] Verma K C, Ram M, Singh J and Kotnala R K 2011 *J. Alloys Compd.* **509** 4967–71
[13] Shanker J, Prasad B V, Suresh M B, Kumar R V and Babu D S 2017 *Mater. Res. Bull.* **94** 385–98

IOP Conf. Series: Materials Science and Engineering **496** (2019) 012027 doi:10.1088/1757-899X/496/1/012027

Dynamic susceptibility spectra analysis of ferromagnetic spheres via micromagnetic simulations

D Djuhana[1], C Kurniawan[2], and A T Widodo[1]

[1]Department of Physics, Faculty of Mathematics and Natural Sciences (FMIPA) Universitas Indonesia, Depok 16424, Indonesia

[2]Center for Physics Research, The Indonesian Institute of Sciences (LIPI), Kawasan Puspitek No. 441-442 Serpong, Banten 15314, Indonesia

Corresponding author: dede.djuhana@sci.ui.ac.id

Abstract. Herein, we report the dynamic susceptibility spectra of ferromagnetic spheres with diameters varying from 50 to 100 nm determined via micromagnetic simulations. The dynamic susceptibility spectra were calculated based on magnetization and the external field response in the frequency domain using Fourier transforms. The peak of the imaginary part of the dynamic susceptibility spectra, which indicates the resonance frequency of the ferromagnetic material, is interestingly near Kittel's resonance frequency.

1. Introduction

Understanding the high-frequency response of magnetic systems is important for both fundamental and applied research [1,2]. A previous study explained the relation between high-frequency response and magnetization is characterized using the dynamic susceptibility tensor [3]. Many studies have focused on the microscale dynamic susceptibility spectra in the shaped magnetic systems, such as nanopillars [4,5], rectangular structures [6], nanodots [7], and stripes [8]. However, there has been insufficient research on the microscale spheres of ferromagnetic materials.

This study reports the simulated results of the dynamic susceptibility spectra of ferromagnetic spheres modeled in a public-domain micromagnetics software package. The dynamic susceptibility spectra are calculated based on magnetization and the external field response in the frequency domain. The results show that the peak of imaginary part of the dynamic susceptibility spectra is close to Kittel's resonance frequency. It means that the micromagnetic dynamic susceptibility of ferromagnetic nanosphere materials can be predicted both numerically and analytically.

2. Micromagnetic simulation

Herein, we used the public-domain micromagnetic simulator OOMMF [9] to numerically solve the Landau–Lifshitz–Gilbert (LLG) equation [10]. The diameter of the ferromagnetic spheres modeled herein was varied from 50 to 100 nm. A dynamic exponential magnetic field, $H(t) = 1000\exp(-10^9 t)$, $t \geq 0$ ($H(t)$ in A/m and t in s), was applied along the x-direction of the sphere and perpendicular to the spin configuration (y-direction) [4], as shown in figure 1. The material and parameters are listed in table 1 [11]. The damping factor was α = 0.05, and the cell size was $2.5\times 2.5\times 2.5\ \text{nm}^3$. The dynamic susceptibility spectra were calculated based on magnetization and the external field behavior in the frequency domain using Fourier transform.

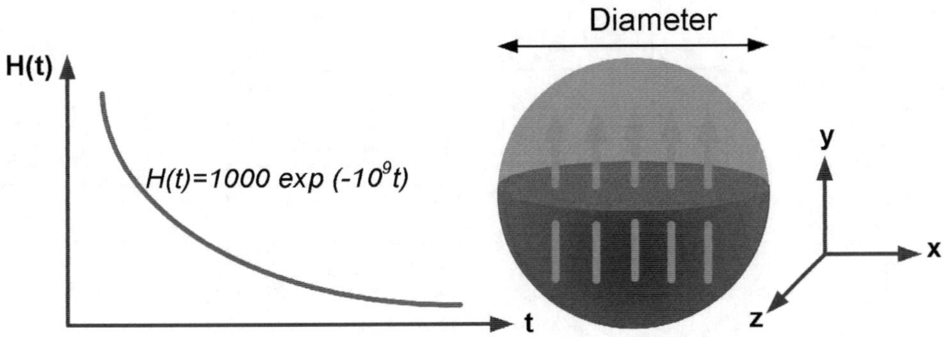

Figure 1. An exponential magnetic field is applied to a ferromagnetic sphere model along the x-direction and perpendicular to the spin configuration (y-direction) in simulations. The sphere diameter is varied from 50 to 100 nm.

Table 1. Material and parameters are used in micromagnetic simulation [11].

Materials	Magnetization saturation M_s (Am^{-1})	Exchange stiffness A (Jm^{-1})	Anisotropy constant K (Jm^{-3})
Co	1400×10^3	14×10^{-12}	53×10^4
Fe	1700×10^3	21×10^{-12}	48×10^3
Ni	490×10^3	9×10^{-12}	-5.7×10^3
Py	860×10^3	13×10^{-12}	0

3. Results and discussion

According to the resonance principle, the dynamic susceptibility of a ferromagnetic sphere model can be calculated from the relation $\chi(\omega) = M(\omega)/H(\omega) = \chi(\omega)' - j\chi(\omega)''$, where $M(\omega)$ and $H(\omega)$ represent the magnetization and the external field, respectively, in the frequency domain. We used fast Fourier transforms (FFTs) to convert the time series of the magnetization response and external field into the frequency domain. The imaginary part of the dynamic susceptibility spectra $\chi(\omega)''$ is related to resonance frequency behavior [3,4]. The real and imaginary parts of the dynamic susceptibility spectra are plotted in figure 2. In this figure, the dynamic susceptibility spectra are around the GHz range, as expected at the nanometer scale.

For comparison, we also calculated the resonance frequency of a ferromagnetic sphere model using the Kittel's formula $\omega_r = \gamma\sqrt{[H_0 + H_k + (N_y - N_x)M_s][H_0 + H_k + (N_z - N_x)M_s]}/(2\pi)$ [12], where γ is the gyromagnetic ratio $2.21 \times 10^5\ mA^{-1}s^{-1}$ [13], H_0 is the static magnetic field, and $H_k = 2K/(\mu_0 M_s)$ is the anisotropy field. In the sphere model, the demagnetizing factors $N_x = N_y = N_z$ and $N_x + N_y + N_z = 1$, so Kittel's resonance frequency is $\omega_r = \gamma(H_0 + H_k)/(2\pi)$ [4]. Figure 3 plots the resonance frequency predicted from Kittel's formula and the frequency peak of the imaginary part of the simulated dynamic susceptibility spectra for Co, Fe, Ni, and Py. Interestingly, the frequency peaks of the imaginary dynamic susceptibility spectra are close to Kittel's resonance frequency: Co was 42 GHz, Fe was 29.5 GHz, Ni was 28.2 GHz, and Py was 27.9 GHz. The resonance frequency varied with the diameter of the nanospheres, as predicted by Kittel's formula, which is expected from the geometry of the demagnetizing factors. The small differences between the micromagnetic simulations and Kittel's formula originated from the discretization of the cubic model in micromagnetic calculations [14]. Further, the resonance frequency originated from dipolar interactions [4,15].

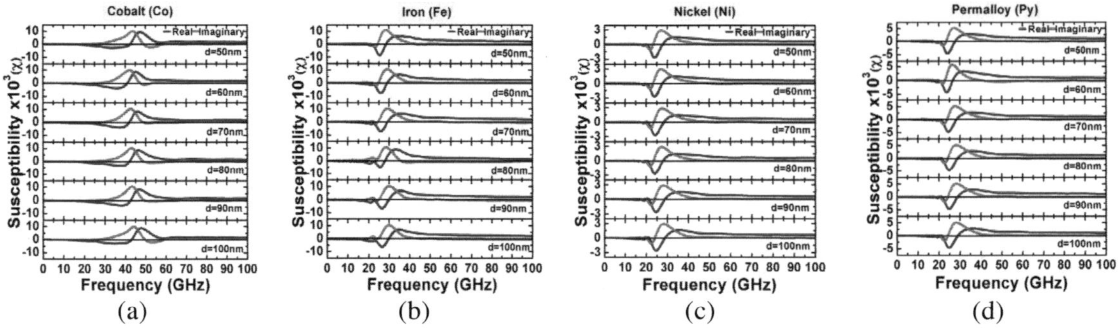

Figure 2. The dynamic susceptibility of ferromagnetic sphere model profiles for (a) Co, (b) Fe, (c) Ni, and (d) Py for diameter ranging from 50 to 100 nm. Blue lines indicate the real part, and red lines indicate the imaginary part.

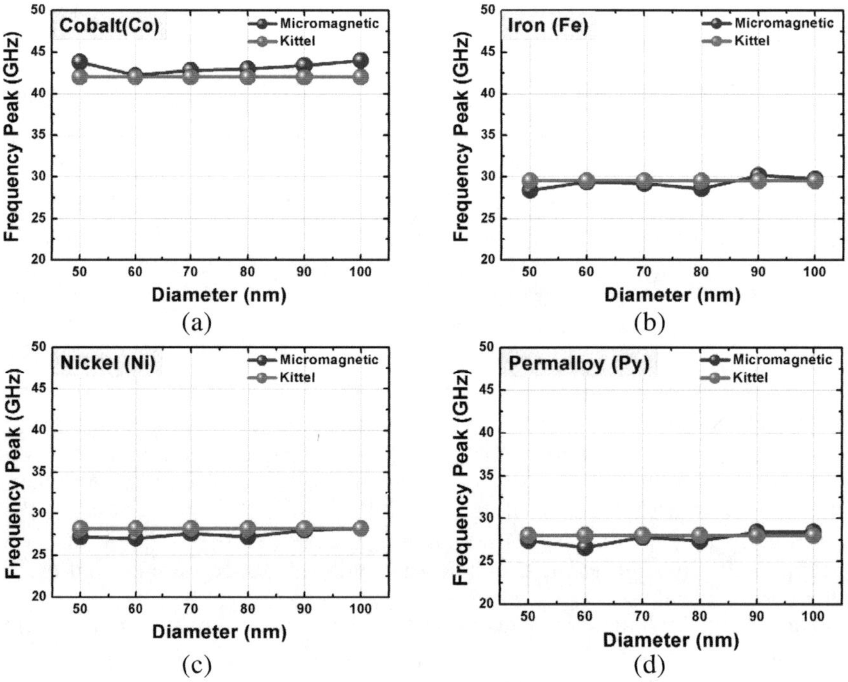

Figure 3. Frequency peak in the imaginary part of the dynamic susceptibility spectra (blue line) and Kittel's equation predictions (red line) for (a) Co, (b) Fe, (c) Ni, and (d) Py spheres with diameters from 50 to 100 nm.

4. Conclusions

Our findings help in understanding the resonance frequency behavior of nanoscale ferromagnetic spheres via micromagnetic simulations, which yielded the dynamic susceptibility spectra. We found that the imaginary part of the frequency peak in the dynamic susceptibility agrees well with analytical solutions for the resonance frequency of the ferromagnetic nanospheres. Engineers can therefore use Kittel's equation or magnetic simulation to effectively predict resonance responses when designing magnetic devices with microscale ferromagnetic spheres.

Acknowledgements

This work was supported by PUPT Grant 2017 (Contract No. 2700/UN2.R3.1/HKP05.00/2017) funded by Ministry of Research, Technology and Higher Education Republic of Indonesia. The author would like to thank Einago (www.einago.com) for the English language review.

References

[1] Russek S E, Kaka S and Donahue M J 2000 *J. Appl. Phys.* **87** 7070–2
[2] Neudecker I, Woltersdorf G, Heinrich B, Okuno T, Gubbiotti G and Back C H 2006 *J. Magn. Magn. Mater.* **307** 148–56
[3] Vukadinovic N 2002 *IEEE Trans. Magn.* **38** 2508–13
[4] Dao N, Donahue M J, Dumitru I, Spinu L, Whiitenburg S L and Lodder J C 2004 *Nanotechnology* **15** S634–38
[5] Bing C W, Gui H M, Hao Z, Yu O and Jiang D L 2010 *Chin. Phys.* B **19** 087502-1–7
[6] Gérardin O, Le Gall H, Donahue M J and Vukadinovic N 2001 *J. Appl. Phys.* **89** 7012–14
[7] Vukadinovic N and Boust F 2007 *Phys. Rev. B* **75** 014420.1–8
[8] Gérardin O, Youssef J B, Le Gall H, Vukadinovic N, Jacquart P M and Donahue M J 2000 *J. Appl. Phys.* **88** 5899–903
[9] Donahue M J and Porter D G 2008 *OOMMF User's Guide, Version 1.0* (Gaithersburg: National Institute of Standards and Technology)
[10] Gilbert T L 2004 *IEEE Trans. Magn.* **40** 3443–9
[11] López-Urias F, Torres-Heredia J J and Muñoz-Sandoval E 2005 *J. Magn. Magn. Mater.* **294** e7–12
[12] Kittel C 1948 *Phys. Rev.* **73** 155–61
[13] Guimarães A P 2009 *Principle of Nanomagnetism* (Berlin: Spinger-Verlag)
[14] Donahue M J and McMichael R D 2007 *IEEE Trans. Magn.* **43** 2878–80
[15] Djuhana D, Rohman L and Kim D H 2017 *J. Magn.* **22** 364–68

IOP Conf. Series: Materials Science and Engineering **496** (2019) 012028 doi:10.1088/1757-899X/496/1/012028

Effects of heat treatment temperature and atmosphere on the morphology and structure of calcium phosphate

D R Tresnasari, A L Juwono, D S Soejoko and N N Mulyaningsih

Department of Physics, Faculty of Mathematics and Natural Sciences (FMIPA) Universitas Indonesia, Depok 16424, Indonesia

Corresponding author: ariadne.laksmidevi@ui.ac.id

Abstract. Calcium phosphate can exist as several phases including hydroxyapatite (HA) and β-tricalcium phosphate (βTCP). Both HA and βTCP are chemically similar to calcified tissue, resulting in high biocompatibility. In this study, we investigated the effects of heat treatment temperature (800 °C and 900 °C) and atmosphere (ambient air and argon) on the morphology, crystalline structure, and functional groups of calcium phosphate powder. Calcium phosphate without heat treatment was evaluated for comparison. The powders were characterized by Fourier transform infrared spectroscopy, X-ray diffraction, and transmission electron microscopy. Under both atmospheric conditions, the effect of heat treatment at 800 °C led to the formation of both HA and βTCP, which was caused by the reduction of CO_3^{2-} bond in calcium phosphate. In contrast, the heat treatment of calcium phosphate at 900 °C generated C–H and C=O groups, resulting in βTCP alone.

Keywords: β-tricalcium phosphate, functional group bonds, calcium phosphate, crystalline structure, hydroxyapatite, morphology

1. Introduction

Nanotechnology involves the process of manipulating, controlling, and integrating of atoms and molecules in the formation of nanoscale materials, structures, components, devices, and systems [1]. Nanosized calcium particles have become additives to food products since they can help absorb calcium in the body, especially its absorption by bone, have no effect on the appearance and taste of food.

The calcium phosphate phases hydroxyapatite (HA) and β-tricalcium phosphate (βTCP) have similar chemical structures as the inorganic components of teeth and bones, resulting in good biocompatibility [2]. HA and βTCP play significant roles in the formation of biphasic calcium phosphate (BCP), which has the ability to form strong bonds with the host bones that form a fibrous interface, resulting in bioinertness in bones [3].

Previous studies have evaluated how temperature and other conditions affect the transformation of calcium phosphate powder into different phases. This study will focused on the morphology, crystallinity, and functional groups of calcium phosphate after the heat treatment. The treatment was done at temperature of 800 °C and 900 °C under air and argon atmospheres. Therefore, this study can provide the proper method in obtaining each phase of hydroxyapatite, β-tricalcium phosphate or BCP.

2. Experimental

2.1. *Preparation of calcium phosphate*

Commercial calcium phosphate powder was purchased from CV Surchana Agung. Five calcium phosphate samples were evaluated in this study: calcium phosphate without heat treatment (WH),

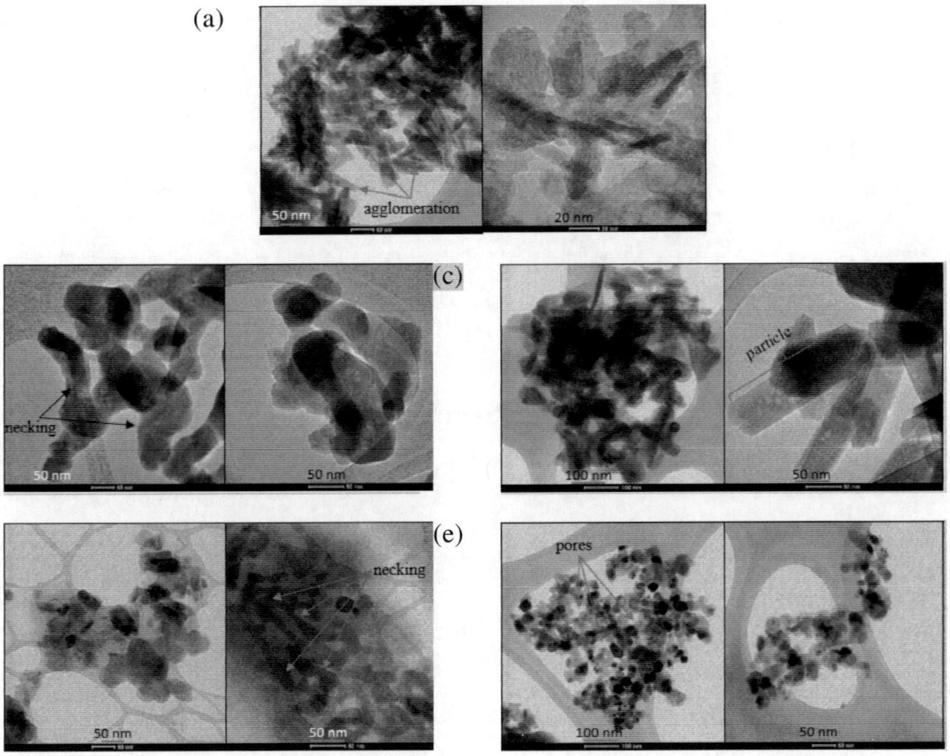

Figure 1. TEM images of calcium phosphate samples: (a) WH, (b) 800°C, (c) 800°C-Ar, (d) 900°C, and (e) 900°C-Ar.

calcium phosphate heated at 800 °C in ambient air (800 °C), calcium phosphate heated at 800 °C in argon atmosphere (800 °C–Ar), calcium phosphate heated at 900 °C in ambient air (900 °C), and calcium phosphate heated at 900 °C in argon atmosphere (900 °C–Ar). The masses of calcium phosphate heated in ambient air were about 5 g and meanwhile the masses of calcium phosphate heated in argon atmospheres were about 3 g.

2.2. Heat treatment of calcium phosphate

Heat treatment was performed at either 800 °C or 900 °C [3,4] in a Nobertherm P330 furnace for 4 h at a heating rate of 30 °C/min under ambient air or argon atmosphere. For heat treatment in argon atmosphere, the furnace was flushed with argon eight times followed by vacuuming for 10 min.

2.3. Characterization of calcium phosphate

The calcium phosphate samples were examined before heat treatment and thereafter by using transmission electron microscopy (TEM; Tecnai 200 kV D2360 SuperTwin) with an emitted electron current of 2.94 μA. X-ray diffraction (XRD) was carried out using a Rigaku Miniflex 600 with operating power at 40 kV and 1 kV/step control mode with CuKα radiation at a wavelength of 1.54 Å. Diffraction lines were recorded in the 2θ range of 5°–90°. The percentages of HA and βTCP phases and full width at half maximum (FWHM) were determined from the XRD data using X'Pert HighScore Plus. Fourier transform infrared (FTIR) spectroscopy was carried out according to ASTM E168 standards in the wavenumber range 450–4000 cm^{-1}.

3. Results and discussion

The TEM image of the WH sample (figure 1a) shows agglomerated calcium phosphate nanoparticles with sizes of ~20 nm. In the TEM images of the 800 °C and 800 °C–Ar samples (figures 1b and figure 1c), respectively, the nanoparticles exhibited necking resulting from the heat treatment, and the

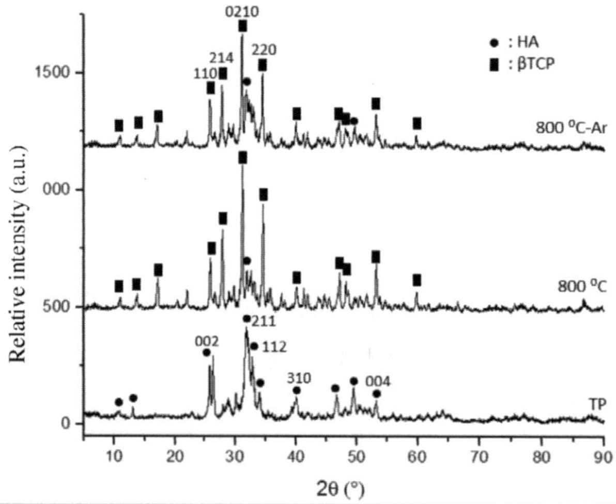

Figure 2. XRD patterns of the WH, 800 °C, and 800 °C–Ar samples.

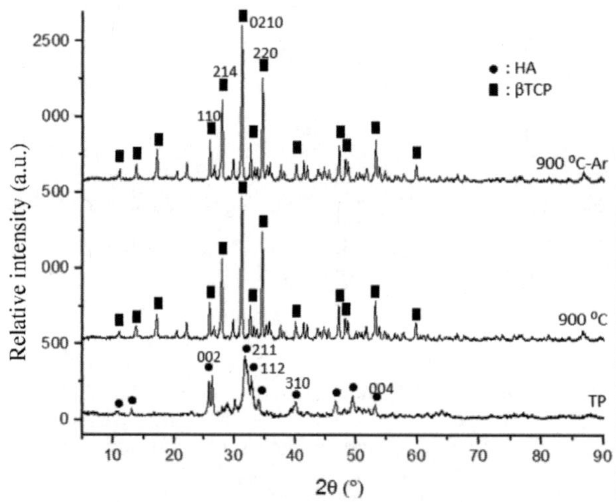

Figure 3. XRD patterns of the WH, 900 °C, and 900 °C–Ar samples.

particle size was ~50–100 nm. This necking phenomenon was attributed to agglomeration and densification in volume or grain boundary of calcium phosphate particles, in agreement with the findings of Bernache-Assollant *et al.* [5]. Otherwise, heat treatment under argon atmosphere resulted in increased particle size and prevented necking. Figures 1d and figure 1e show TEM images of the 900 °C and 900 °C–Ar samples, respectively. Under ambient air, the particles exhibited necking and were ~60–110 nm in size. The increase in particle size can be attributed to the high temperature, as reported by Zhou [4]. The particle microstructure exhibited densification, due to heat treatment of 900 °C in ambient air. The particles in the 900 °C–Ar sample were spherical with sizes of ~20 nm, and grains were clearly formed. The grains indicated a microstructure with poor densification that was marked by the pores. Thus, high crystallization was caused by argon atmosphere during heating.

Figure 2 shows the XRD patterns of the WH, 800 °C, and 800 °C–Ar samples. Peaks corresponding to the (002), (211), (112), (310), and (004) planes of HA were observed in the spectrum of WH, with the (211) peak having the highest intensity. Heat treatment at 800 °C led the formation of HA and βTCP phases, as noted by Victoria [3]. The refinement results from X'Pert High Score Plus software denoted the percentage of relative intensity of calcium phosphate, which heated in ambient air, comprised HA phases as much as 23.1% and βTCP phases as much as 76.9% respectively. Meanwhile, the heat treatment of calcium phosphate using argon atmosphere resulted in the changes in the relative intensity of HA into 42.7% and βTCP into 57.3%.

Table 1. XRD refinement data of all samples for (220) peak

	2θ (°)	FWHM	Crystallite size (Å)
800°C	34.49	0.24	383.22
800°C-Ar	34.43	0.25	180.48
900°C	34.52	0.20	496.65
900°C-Ar	34.51	0.19	537.50

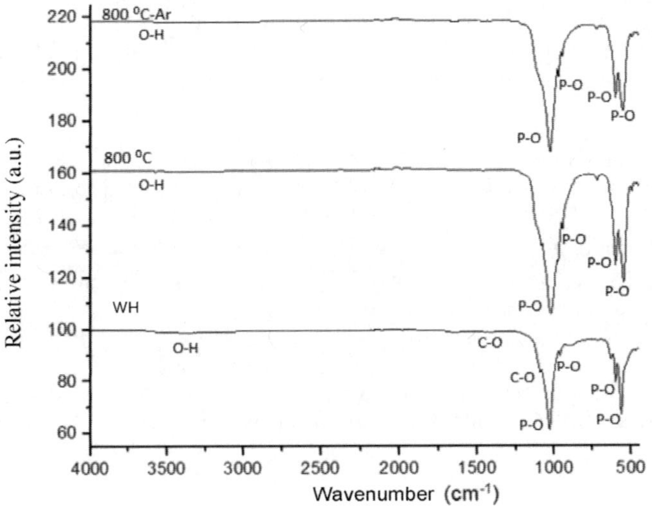

Figure 4. FTIR spectra of the WH, 800 °C, and 800 °C–Ar samples

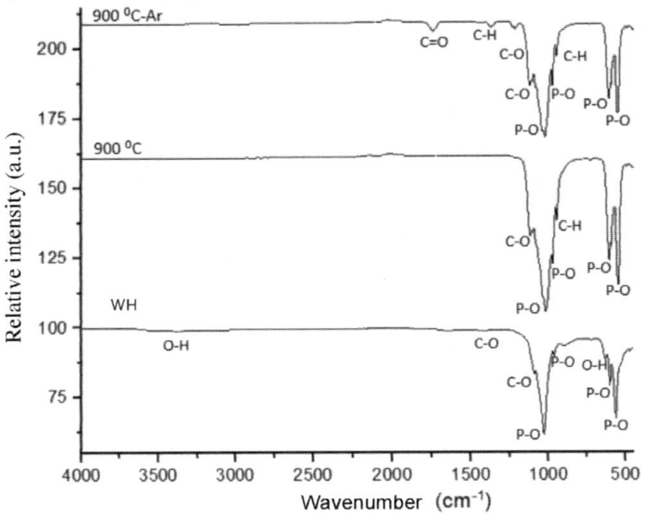

Figure 5. FTIR spectra of the WH, 900 °C, and 900 °C–Ar samples.

Heat treatment at 900 °C led to the formation of βTCP alone (figure 3). The low FWHM values of the 900 °C and 900 °C–Ar samples indicate that the high temperature led to good crystallinity [6], in agreement with the TEM image in figure 1e.

Table 1 shows the FWHM values of all samples for the (220) peak, which corresponds to βTCP [6]. The (220) peak was not observed in the XRD spectrum of the WH sample, which contained only HA.

Figure 4 shows the FTIR spectra of the WH, 800 °C, and 800 °C–Ar samples. Peaks corresponding to the bonds of various functional groups including OH^-, PO_4^{3-}, and CO_3^{2-} were identified in the spectra [3,7]. OH^- bond peaks appeared at 3386, 3572 and 3571 cm^{-1} in the WH, 800 °C, and 800 °C–Ar samples, respectively. Peaks corresponding to CO_3^{2-} bonds were observed in the spectrum of the WH sample at 1029, 1023, and 1025 cm^{-1}. Meanwhile, peaks of CO_3^{2-} disappeared in both 800 °C and 800 °C-Ar samples due to heat treatment temperature. We concluded that the condensation CO_3^{2-} occurred and BCP that comprised HA and βTCP phases were formed [3].

Figure 5 shows the FTIR spectra of the WH, 900 °C, and 900 °C–Ar samples. Peaks of OH^-, PO_4^{3-}, and CO_3^{2-} bonds were identified [4]. The peak associated with OH^- stretching at 3386 cm^{-1} was found only in the spectrum of the WH sample. The high temperature of 900 °C reduced –OH bond at the wave number of 3386 and 630 cm^{-1} in both 900 °C and 900 °C–Ar samples, corresponded to Victoria [3]. The C–H groups of alkenes were indicated by peaks at 945 and 946 cm^{-1} in the spectra of the 900 °C and 900 °C–Ar samples, respectively. CO_3^{2-} bonds were indicated by a peak at 1116 cm^{-1} in the spectrum of the 900 °C sample and by peaks at 1118 and 1217 cm^{-1} in the spectrum of the 900 °C–Ar sample. Heating under argon atmosphere resulted in the formation of new bonds, as indicated by peaks corresponding to the C–H and C=O groups of alkenes and esters at 1366 and 1739 cm^{-1}, respectively.

4. Conclusions

Before heat treatment, the agglomerated HA particles were ~20 nm in size and exhibited poor crystallinity. Heating in air atmosphere caused the HA particles to undergo necking, resulting in increased particle size. Heating under argon atmosphere might reduce the necking of the calcium phosphate particles compared with ambient air. Both HA and βTCP phases formed after heat treatment at 800 °C. Heating at 900 °C resulted in the formation of only βTCP. Heating at this temperature under argon atmosphere resulted in the formation of alkene and ester functional groups.

Acknowledgements

This research was funded by PITTA Grant 2017 from Universitas Indonesia with contract number 624/UN2.R3.1/HKP.05.00/2017. The authors would like to thank Einago (www.einago.com) for the English language review.

References

[1] Filipponi L and Sutherland D 2013 *Nanotechnologies: Principles, Applications, Implications and Hands-on Activities* (Luxembourg: European Commission Directorate-General for Research and Innovation Industrial Technology)

[2] Vallet-Regí M and González-Calbet J M 2004 *Prog. Solid State Chem.* **32** 1–31

[3] Victoria E C and Gnanam F D 2002 *Trends Biomater. Artif. Organs* **16** 12–14

[4] Zhou W Y 2007 *J. Mater. Sci. Mater. Med.* **19** 103–10

[5] Bernache-Assollant D, Ababou A, Champion E and Heughebaert M 2005 *J. Eur. Ceram. Soc.* **23** 229–41

[6] Pankaew P, Hoonnivathana E, Limsuwan P and Naemchanthara K 2010 *Journal of Applied Sciences* **10** 3337–42

[7] Elfalaky A, Hashem H and Mohamed T 2014 *Nanoscience and Nanoengineering* **2** 10–16

Effects of magnetic field and temperature on the density of states of the poly(dA)-poly(dT) DNA molecule

A M Risqi and E Yudiarsah

Department of Physics, Faculty of Mathematics and Natural Sciences (FMIPA)
Universitas Indonesia, Depok 16424, Indonesia

Corresponding author: e.yudiarsah@sci.ui.ac.id

Abstract. Charge-transport behavior in poly(dA)-poly(dT) DNA molecules is indicated by the distribution of density of states (DOS) at different electron energies. The DOS of DNA molecules in the presence of magnetic field and temperature is calculated using the tight-binding Hamiltonian and Green's function. The molecule in question consists of 32 base pairs of adenine (A) and thymine (T) in the interior, and a sugar phosphate backbone as the exterior. Electron hopping along the backbone sites is possible with this structure. The DNA molecule is connected to metallic electrodes at both ends. In the model, an external magnetic field modifies the electron-hopping constant through the Peierls transition. Temperature affects the molecule by generating a twisting motion among the base pairs. The results show that temperature and magnetic field limit the DOS of the DNA molecule for a given electron energy. Electrons can occupy all states in the energy spectrum, which are localized states, at temperatures considerably higher than zero.

Keywords: density of states, poly(dA)-poly(dT) DNA, temperature, magnetic field

1. Introduction

DNA (deoxyribonucleic acid) is the molecule responsible for storing genetic information in all known forms of life. The primary structure of DNA consists of two chains of nucleotide polymers, the base pairs and backbone of the molecule. This structure was discovered by Watson and Crick [1]. As an important biopolymer, DNA has great potential for use in molecular-scale electronics [2] due to its unique self-assembling and self-recognizing properties. This potential for easy fabrication of specific DNA molecules has attracted research attention to the details of the charge-transport mechanisms that may be exploited in the design of electronic circuits [3,4]. Experimental and theoretical studies of charge transport along the DNA molecule suggest a variety of possible electronic behaviors. DNA molecules in different configurations, e.g., sequences and contact to electrodes, can act as insulators [5–7], semiconductors [8–10], or conductors [11,12]. This variety of functions is possible because the charge-transport mechanisms in the DNA molecule are very sensitive to intrinsic influences like the base-pair sequence, backbone structure, and hydrogen bonds between base pairs as well as extrinsic influences like the metal-molecule contacts and temperature [2].

To study a realistic model of charge transport through DNA molecule, we model the poly(dA)-poly(dT) DNA extended-ladder molecule [13], which is schematically diagrammed in figure 1. In this model, the electron-hopping amplitudes between sites up to the next-nearest neighbour are calculated under the influence of an external magnetic field and varying temperatures. In this paper, we calculate the density of states (DOS) over a range of electron energies to study the band-gap structure of the DOS spectra [13]. Furthermore, we calculate the DOS at zero temperature for several values of magnetic field as well as the DOS at zero magnetic fields for several temperatures. The band-gap structure indicates the electronic properties of the DNA molecule, and is affected by external magnetic field and temperature. In the next section, the Hamiltonian equation of DNA poly(dA)-poly(dT) and the method for calculating the DOS as a function of external magnetic field and temperature will be discussed. The results of our calculations will be discussed in the third section, and a final section summarizes the conclusions of this research.

IOP Conf. Series: Materials Science and Engineering **496** (2019) 012029 doi:10.1088/1757-899X/496/1/012029

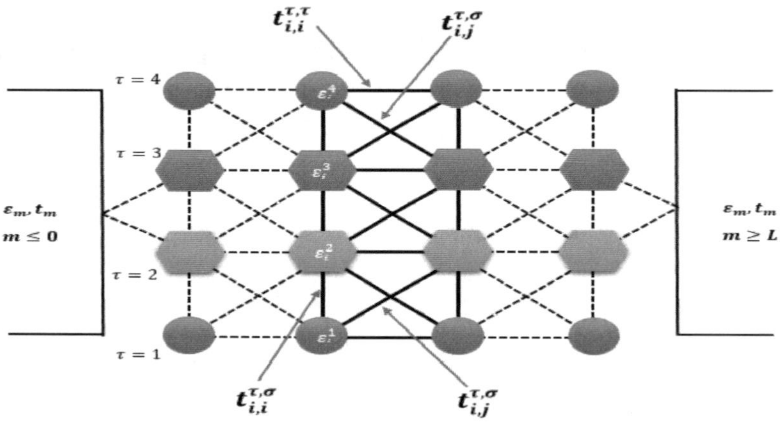

Figure 1. DNA poly(dA)-poly(dT) extended ladder model. Hexagons represent thymine and adenine bases, circles represent backbone sugars. Both ends of the molecule DNA are in contact with electrodes with onsite energy ε_m and hopping contant t_m.

2. Theoretical model

A DNA molecule with 32 base pairs of adenine and thymine is studied in this paper. Every base in this DNA structure is connected to the phosphate-sugar backbone, and each end of the molecule is connected to an electrode. The system is diagrammed schematically in figure 1. To form a realistic model of the charge-transport mechanism in DNA, electrons in the model are allowed to propagate as far as two sites away from their present site in each step of the calculation. The Hamiltonian of the poly(dA)-poly(dT) DNA molecule that used the tight binding approximation is written in equation (1).

$$H_{DNA} = \sum_{i=1}^{L}\sum_{\tau=1}^{4}\left\{\varepsilon_i^{\tau} + \phi_i^{\tau}\,|i,\tau\rangle\langle i,\tau| + \sum_{\sigma=1}^{4}\left(t_{i,i}^{\tau,\sigma}\,|i,\tau\rangle\langle i,\sigma| + t_{i,i+1}^{\tau,\sigma}\,|i,\tau\rangle\langle i+1,\sigma| + h.c\right)\right\}$$

with $\phi_i^{\tau} = b_{i,i-1}^{\tau}(\cos\Delta_{i,i-1}^{\tau} - \cos\varphi_{eq}) + b_{i,i+1}^{\tau}(\cos\Delta_{i,i+1}^{\tau} - \cos\varphi_{eq})$

and, $\Delta_{i,i-1}^{\tau} = \varphi_{eq} - (\varphi_{i-1}^{\tau} - \varphi_i^{\tau})$ also, $\Delta_{i,i+1}^{\tau} = \varphi_{eq} - (\varphi_i^{\tau} - \varphi_{i+1}^{\tau})$ (1)

The length of the DNA molecule, L is set to 32 base pairs. The sites of bases and backbone pairs are identified with the index *i*. The indexes τ and σ represent the strand number at a site, which are 2 and 3 for the two DNA bases and 1 and 4 for the two backbone molecules. The probability of electron hopping between bases and from a base to a backbone molecule at a site is represented with parameter $t_{i,i}^{\tau,\sigma}$. The parameter $t_{i,i+1}^{\tau,\sigma}$ is calculated to represent steps of the simulation. The value of each hopping constant, which depends on the angle of twist in the structure, is calculated by expansion of the Slater-Koster formula [14,15]. The symbol ε_i^{τ} is the onsite energy of base and backbone molecules at site i and in strand τ. In this study, the energy for backbone molecule was set at 8.36 eV, and the onsite energy used for thymine was 9.06 eV and for adenine was 8.22 eV [15].

In the Hamiltonian, the effect of twisting vibrations on the base onsite energy is represented by parameter ϕ_i^{τ}.. The coupling of charge and vibration is represented by the parameters $b_{i,i-1}^{\tau}$ and $b_{i,i+1}^{\tau}$. Twisting vibration changes the twist angle between base pairs to the instantaneous twist angle φ_i. The twist angle formed by two neighboring base pairs at equilibrium is chosen to be 36° [14]. The instantaneous twist angle is set to follow a Gaussian distribution with average of zero and standard deviation $\sqrt{T/f}$. f is proportional to the square of the twisting vibrational frequency and T is the temperature (in Kelvin) [16]. In this study, the twisting vibrational frequency is set to 0.51 meV [15].

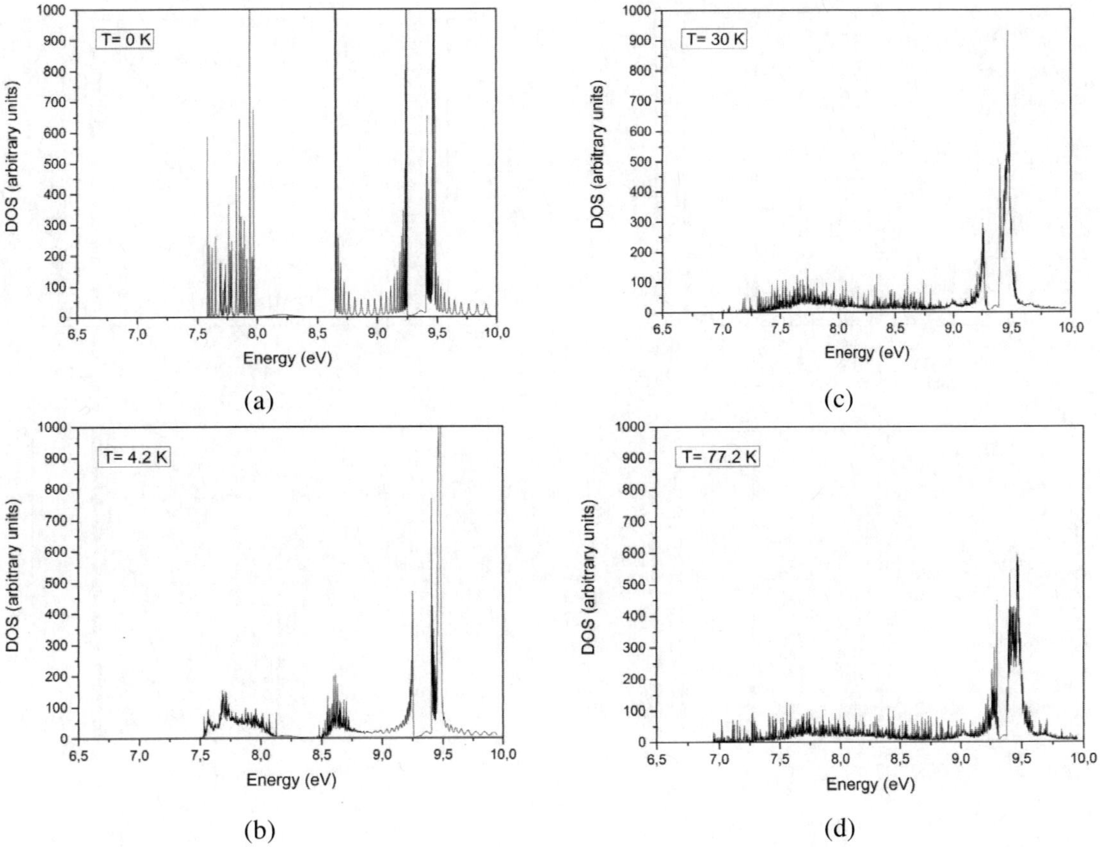

Figure 2. Density of states vs energy for poly(dA)-poly(dT) DNA with twisting-motion frequency 0.51 meV and magnetic field 0 T at several temperatures: (a) 0 K, (b) 4.2 K, (c) 30 K, and (d) 77.2 K.

The external magnetic field is applied alongside to the symmetrical axis of the DNA molecule. In order to simplify the problem, the Zeeman effect is ignored in the calculation. The presence of an external magnetic field affects the electron-hopping amplitudes through Peirl's phase transition [16], from $t_{r,r+\Delta}$ to $t_{r,r+\Delta}\exp(ie\theta_{r,r+\Delta}/\hbar)$. The symbol $\theta_{r,r+\Delta}$ is the line integral of the vector potential $\boldsymbol{A}$ from r to $r+\Delta$. For an external magnetic field $\boldsymbol{B}$ parallel to the DNA molecule's symmetrical axis, the line integral can be written as

$$\int_r^{r+\Delta} \boldsymbol{A}\, d\boldsymbol{r} = -\frac{eB}{2\hbar}(x_j - x_i)(y_j + y_i) \tag{2}$$

where x and y represent the positions of base and backbone molecules in Cartesian coordinates.

The electronic properties of the poly(dA)-poly(dT) DNA molecule are indicated by the DOS calculated at each electronic energy using Green's function [17].

$$G(\vec{k},z) = \frac{1}{(z-\mathrm{Im}\,\eta)[I] - H(\vec{k}) - \sum(\vec{k})} \tag{3}$$

In equation (3), the variable z represents energy of an electron that travels across the DNA molecule. Symbol Im and η respectively represent the imaginary part and an infinitesimal number. The identity matrix is represented by $[I]$ and the Hamiltonian in matrix form is represented by $H(\vec{k})$. The variable $\sum\vec{k}$ represents the correction on electron Green's function in the poly(dA)-poly(dT) DNA molecule due to the left and right electrodes.

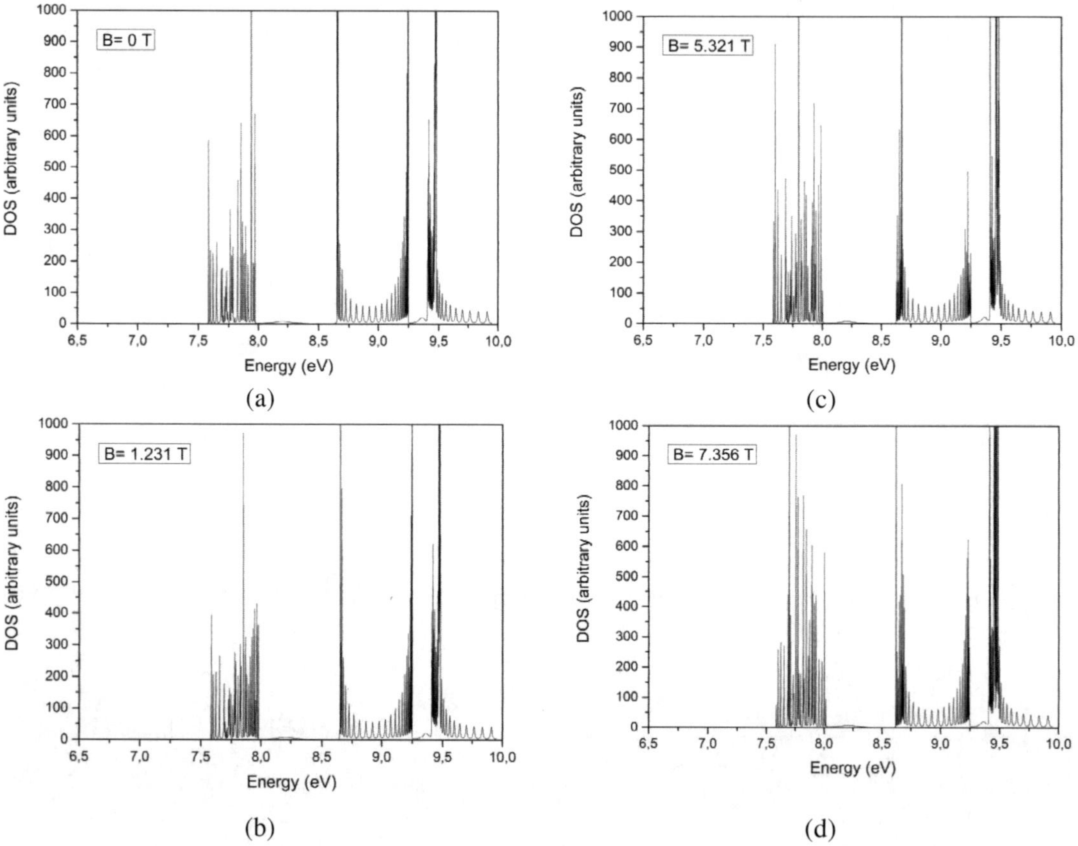

Figure 3. Density of states vs energy for poly(dA)-poly(dT) DNA with twisting-motion frequency 0.51 meV at zero temperature and several magnetic field strengths (a) 0 T, (b) 1.231 T, (c) 5.321 T, and (d) 7.356 T.

3. Results and discussion

Figure 2 shows DOS spectra with zero external magnetic field, twisting-motion frequency 0.51 meV, and at a selection of temperatures. At 0 K (figure 2a), the DOS spectrum shows three ranges of electron energies at which DOS value is zero: 7.0–7.6 eV, 7.9–8.3 eV and 9.25–9.3 eV. This means that no electrons can occupy these energy levels. As temperature increases, the overall DOS values decrease but the band gaps narrow (figure 2b, figure 2c and figure 2d)). At 0 K, the maximum DOS value is over 1000. When the temperature is 77.2 K the maximum DOS value is around 600. Nonzero temperatures cause phonons to appear in the DNA structure, which affect the charge-transport mechanisms. These phonons widen the distribution of twisting angles between A and T bases, introducing disorder to the DNA structure. Therefore, as the temperature increases, the DOS tends to decrease, but the gaps in the DOS spectra narrow. Next, at temperatures far from absolute zero, similar to conductor materials, electrons can occupy every energy level in poly(dA)-poly(dT) DNA molecules. However, at these temperatures, these states are localized [15]; thus, the molecule behaves as an insulator.

DOS spectra at zero temperature, twisting-motion frequency 0.51 meV, and a selection of magnetic field strengths are shown in figure 3. The width of the band gaps in the allowed energy levels remains constant as magnetic field strength increases in figure 3b, figure 3c and figure 3d. These gaps form because of the electron structure of the three nucleotides that make up the poly(dA)-poly(dT) DNA molecule, which are adenine, thymine, and backbone. The presence of a magnetic field increases the peak DOS values, and a higher value means that electrons are more likely to occupy the energy level. This characteristic is caused by the modification of complex hopping amplitudes by the Peierl transition, which changes the value of inter-base hopping amplitude along the molecule with increasing external magnetic field. This disorder in the electron-hopping amplitude along the molecule structure reduces the freedom of electrons with certain energies to move among the sites on the DNA molecule.

4. Conclusions

The results of the calculation of DOS versus energy have been succeeded in showing the influence of effects of external magnetic field and temperature on the charge-transport properties of poly(dA)-poly(dT) DNA. An external magnetic field modifies the electron-hopping integral of the DNA molecule and can be represented with the Peirels phase factor. This modification affects charge transport as the strength of the magnetic field increases, so the peak values of DOS increase as magnetic field increases and electrons tend to occupy certain sites on the DNA molecule. The gaps in the DOS spectra become narrower as the temperature increases at the expense of lower DOS value. Since all available energy states are localized states at temperatures much above absolute zero, the molecule behaves as an insulator.

Acknowledgements

This research is supported by Universitas Indonesia through Hibah Publikasi Internasional Terindeks untuk Tugas Akhir (PITTA Grant) 2017 under contract number 648/UN2.R3.1/HKP.05.00/2017. The authors would like to thanks to CISCO laboratory, Department of Physics, FMIPA Universitas Indonesia for the facility and Einago (www.einago.com) for the English language review.

References

[1] Watson J D and Crick F H C 1953 *Nature* **171** 737–8
[2] Cuniberti G, Maciá E, Rodríguez A and Römer R A 2007 *Charge Migration in DNA: Perspectives from Physics, Chemistry and Biology*, ed T Chakraborty (Berlin: Springer) pp 1–20.
[3] Keren K, Berman R S, Buschtab E, Sivan U and Braun E 2003 *Science* **302** 1380–2
[4] Reif J H, LaBean T H and Seeman N C 2001 *Lecture Notes in Computer Science* vol 2054, ed A Condon and G Rozenberg (Berlin: Springer-Verlag) pp. 173–98
[5] de Pablo P J, Moreno-Herrero F, Colchero J, Herrero J G, Herrero P, Baro A M, Ordejon P, Soler J M and Artacho E 2000 *Phys. Rev. Lett.* **85** 4992–5
[6] Storm A J, van Noort J, de Vries S and Dekker C 2001 *Appl. Phys. Lett.* **79** 3881–3
[7] Zhang Y, Austin R H, Kraeft J, Cox E C, and Ong N P 2002 *Phys. Rev. Lett.* **89** 198102–5
[8] Cohen H, Nogues C, Naaman R and Porath D 2005 *Proc. Natl. Acad. Sci. U S A* **102** 11589–93
[9] Shapir E, Cohen H, Calzolari A, Cavazzoni C, Ryndyk D A, Cuniberti G, Kotlyar A, Di Felice R and Porath D 2008 *Nat. Mater.* **7** 68–74
[10] Roy S, Vedala H, Roy A D, Kim D H, Doud M, Mathee K, Shin H K, Shimamoto N, Prasad V, and Choi W 2008 *Nano Lett.* **8** 26–30
[11] Kasumov A Y, Kociak M, Gueron S, Reulet B, Volkov V T, Klinov D V and Bouchiat H 2001 *Science* **291** 280–2
[12] Xu B, Zhang P, Li X and Tao N J 2004 *Nano Lett.* **4** 1105–8
[13] Malakooti S, Hedin E R and Joe Y S 2013 *J. Appl. Phys.* **114** 014701-1–8
[14] Endres R G , Cox D L and Singh R R P 2004 *Rev. Mod. Phys.* **76** 195–214
[15] Suhendro D K, Yudiarsah E and Saleh R 2010 *Phys. B: Condens. Matter* **405** 4806–11
[16] Peirels R 1933 *Z. Phys.* **80** 763–91
[17] Kang D W, Sun M L, Zuo Z W, Wang H X, Lv S J, Li X Z and Li L B 2016 *Phys. Lett. A* **380** 977–82

Electrochemical detection of As^{3+} and As^{5+} by anodic stripping voltammetry at a gold electrode

D T Lestarini and T A Ivandini

Department of Chemistry, Faculty of Mathematics and Natural Sciences (FMIPA), Universitas Indonesia, Depok 16424, Indonesia

Corresponding author: ivandini.tri@sci.ui.ac.id

Abstract. In this article, electrochemical detection of arsenite (As^{3+}) and arsenate (As^{5+}) by stripping voltammetry at a gold electrode has been investigated. As^{3+} and As^{5+} are reduced to arsine (AsH_3) in 0.1 M sodium borohydride before deposition on the electrode surface. 0.1 M HCl was used as the supporting electrolyte, where Cl^- ions act as an ionic bridge, facilitating reduction at the working electrode. As^{3+} and As^{5+} can be deposited as arsine by simple electrochemical reduction at –0.5 V (versus Ag/AgCl). Anodic stripping voltammograms of arsine at a gold electrode showed similar peak potentials for As^0 oxidation (≈0.31 V versus Ag/AgCl) under the optimum conditions (deposition potential, deposition time, and scan rate of –0.5 V, 60 s, and 200 mV/s, respectively). The stripping voltammograms for As^{3+} and As^{5+} presented similar peak shapes and current intensities. A reaction mechanism involving reduction to As^0 at a mild potential in a strong acid, followed by oxidation of As^0 to As^{3+} at the gold electrode, was proposed. The method developed offers a wide linear concentration range for As^{3+} (0–5 ppm) and As^{5+} (0–20 ppm) and very low limits of detection for As^{3+} (0.85 ppm) and As^{5+} (6.07 ppm).

Keywords: anodic stripping voltammetry, arsenic, arsenate, sodium borohydride, gold electrode.

1. Introduction

Arsenic is one of the most toxic elements encountered in the environment. It is commonly found in marine environments and is highly resistant to chemical degradation. Arsenic exists in four oxidation states (i.e., As^{3-}, As^0, As^{3+}, and As^{5+}), with As^{3+} and As^{5+} being the most widespread forms in nature [1]. Arsenic is mostly to be found in its inorganic form in the natural water, namely oxyanions of trivalent arsenite (As^{3+}) or pentavalent arsenate (As^{5+}). Arsenic can also exist in organic forms, such as monomethyl arsenic acid (MMA) and dimethylarsinic acid (DMA), which are less toxic, and in volatile organoarsenic species such as methylarsine (h (AsH_3) [2].

These different arsenic species have very different toxicities. Their acute toxicity decreases in the order arsine (AsH_3) >> arsenite (As^{3+}) > arsenate (As^{5+}) >> dimethylarsinate > monomethylarsonate >> arsenobetaine and arsenocholine, both of which are organoarsenic compounds that are found in marine organisms and, therefore, seafood [3]. Arsine gas is formed when arsenic or arsenic-containing materials react with acids. It is often applied as dopant in the industry such as semiconductor and crystals manufacture for fiber optics and computer chips, and in lesser usage are for galvanizing, soldering, etching, burnishing, and lead plating. Arsine is highly toxic, and its threshold limit value, is only 10 μg/m^3 [4].

Determination of arsenic is conventionally based on the Gutzeit method or inductively coupled plasma spectrometry [5]. However, electroanalytical methods of anodic stripping voltammetry (ASV) and differential pulse anodic stripping voltammetry exhibit an excellent sensitivity and allow the detection of arsenic in distinct oxidation states using low-cost portable equipment, making them ideal

IOP Conf. Series: Materials Science and Engineering **496** (2019) 012030 doi:10.1088/1757-899X/496/1/012030

for field analysis [6]. The determination of arsenic using ASV at a gold electrode was first investigated by Forsberg *et al.* [7] in 1975. They reported that reducing As^{5+} to As^{3+}, which they achieved by heating with sodium sulfite (Na_2SO_3) in a concentrated acid solution, is necessary because As^{5+} is electroinactive [7].

ASV comprises two steps: (1) the analyte preconcentration or deposition by reduction process on an electrode surface; (2) the stripping process of preconcentrated analyte on the electrode surface by oxidation. Gold is the most suitable material for arsenic determination, due to the stable intermetallic compound of Au–As can be formed during the deposition stage [8,9]. Yamada *et al.* [8] have reported that inorganic As^{3+} and As^{5+} in mixed solutions can be detected at gold-modified boron-doped diamond electrodes by ASV at detection limits of 5 and 100 ppb, respectively.

In the present study, we developed and assessed an electrochemical sensor that is based on the reduction of inorganic arsenic using sodium borohydride to form arsine. An HCl solution is used as the electrolyte to provide the very low pH conditions under which arsine can be reduced to As^0 at mild potentials [10]. A linear current response is observed within the range of concentration from 2–10 ppm. Moreover, the current response's stability is demonstrated, as well as a feasible reaction mechanism.

2. Experimental

2.1. Chemicals and materials

Sodium arsenite ($NaAsO_2$), sodium arsenate ($Na_2HAsO_4{\cdot}7H_2O$), hydrogen chloride (HCl), sodium borohydride ($NaBH_4$), and all other compounds (ACS-certified reagent grade) were obtained from Wako Inc. and applied without further purification. Aqueous solutions were made in deionized water (Merck Millipore). Fresh solution of 0.1 M $NaBH_4$ was prepared daily.

2.2. Electrochemical measurements

Electrochemical measurements were performed in a room temperature with an ASV technique using an electrochemical cell completed with a magnetic stirrer. Gold plate (area of around 0.25 cm^2) was used as the working electrode, while platinum spiral and Ag/AgCl were respectively employed as the counter and the reference electrodes. HCl solution (0.1 M) was used as the electrolyte and ED410 eDAQ e-corder was used as the potentiostat. All measurements were conducted with stirring at room temperature.

3. Results and discussion

3.1. Reduction of arsenite and arsenate to arsine

It was previously stated that the ASV method comprises the analyte preconcentration by reduction process on the electrode surface and the preconcentrated analyte is stripped on the electrode surface by oxidation [8]. Initially, As^{3+} or As^{5+} is reduced to As^0 with an adequately negative potential, and on the gold electrode surface, a film will be formed [6], via the following reactions:

$$As^{3+} + 3e^{-} \rightarrow As^{0}$$

$$As^{5+} + 5e^{-} \rightarrow As^{0}$$

In this study, ASV detection of arsenic was conducted on the basis of the reduction of arsenite and arsenate to arsine using sodium borohydride in acidic conditions; that is,

$$As^{5+} + 2BH_4 + 6H_2O \rightarrow As^{3+} + 2B(OH)_3 + 7H_2$$

$$As^{3+} + 3BH_4 + 9H_2O \rightarrow AsH_3 + 3B(OH)_3 + 9H_2$$

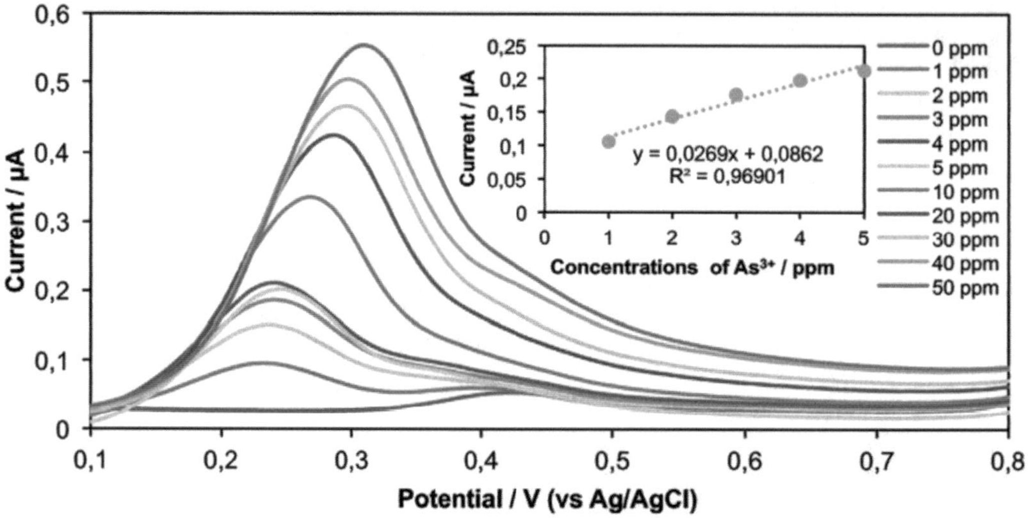

Figure 1: Background-subtracted voltammograms of 0.1 M HCl solutions containing of 0.1 M $NaBH_4$ and various concentrations of As^{3+} (0–50 ppm). Stripping voltammetry parameters were E_{dep} = –0.5 V, t_{dep} = 60 s, and scan rate = 200 mV/s. The inset shows the linear calibration curve extracted from the related stripping voltammograms.

At the stripping step, arsine is oxidized to As^{3+}. Two mechanisms are proposed for this oxidation reaction. The first mechanism involves a two-step oxidation process where AsH_3 is oxidized to As^0, further to As^{3+} through an electrochemical oxidation at the gold electrode, as showed by the below equations:

$$AsH_3 \rightarrow As^0 + 3H^+ + 3e^-$$

$$As^0 \rightarrow As^{3+} + 3e^-$$

The second proposed mechanism involves the direct oxidation of AsH_3 to As^{3+}. Preliminary ASV experiments showed that the optimum peak oxidation currents for As^{3+} and As^{5+} are achieved by maintaining the potential at –0.5 V (versus Ag/AgCl) (figure 1 and figure 2).

3.2. ASV conditions for As^{3+}

In order to achieve the optimal conditions for As^{3+}, we investigated the deposition potential (E_{dep}), deposition time (t_{dep}), and scan rate. A deposition of –0.5 V (versus Ag/AgCl) was applied for As^{3+} as this potential has been previously reported to be optimal [11]. A 60 s deposition time resulted the highest signal-to-background (S/B) ratio for ASV conducted in 0.1 M HCl. The deposition time is the time required to reduce As^{3+} to As^0 at the electrode surface. At low concentrations, longer deposition is required for the detection of As^{3+}, while at high concentration of As^{3+}, short deposition times is to be applied in order to prevent saturation of the electrode surface [12]. HCl was used as the electrolyte, because Cl^- ion serves as an ionic bridge between the working electrodes to facilitate their reduction. The polarograms for As^{3+} in acidic media present two waves [6].

The scan rate parameters were also optimized. The correct scan rate must be applied to completely strip all the deposited As^0. At low scan rate, the deposited As^0 is also stripped, however will result in a poor current response. Conversely, a high scan rate results in a higher current, however should the scan rate be too high, the stripping process will not be completely occurred on the deposited As^0 leads to a distortion of the peak shape [8]. The optimum ASV conditions for As^{3+} were found to be E_{dep} = –0.5 V, t_{dep} = 60 s, and scan rate = 200 mV/s. The background-subtracted stripping voltammograms and the related linear calibration curve (r = 0.97) are shown in figure 1. Limit of detection (LOD) of 0.85 ppm was estimated.

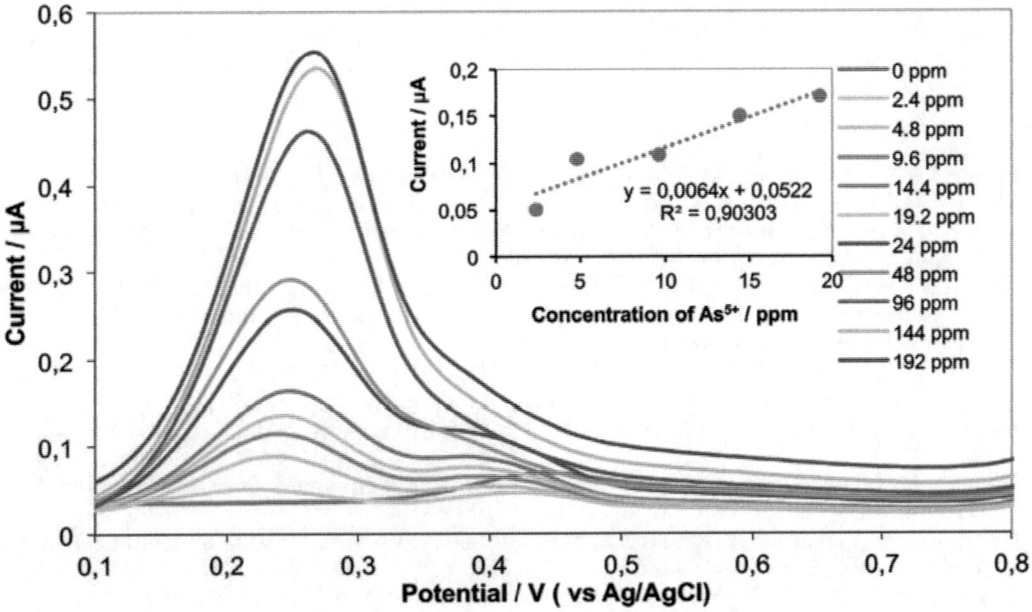

Figure 2. Background-subtracted stripping voltammetry for various concentrations of As^{5+} (0–20 ppm) in 0.1 M HCl solutions using 0.1 M $NaBH_4$. Stripping voltammetry parameters were E_{dep} = –0.5 V, t_{dep} = 60 s, and scan rate = 200 mV/s. The inset shows the linear calibration curve extracted from the related stripping voltammograms.

3.3. *ASV conditions for* As^{5+}

As^{5+} deposition requires a very low negative potential in order to resolve the high activation energy of As^{5+} reduction. Nevertheless, at high negative potentials, evolution of hydrogen can occur, and the formed hydrogen could be adsorbed on the electrode surface, inhibit the As^0 deposition [8]. Conversely, the application of very low potentials can also lead to the generation of hydrogen gas by water reduction. In a previous study, the adsorption of hydrogen gas was minimized by controlling the deposition of As^{5+} in two steps, including the reduction step to be As^{3+} followed by the reduction of As^{3+} to be As^0 [8]. In that work, the direct deposition of As^{5+} was predict to occur at a potential of –1.5 V (versus Ag/AgCl).

In the present study, ASV detection of As^{5+} was performed on the basis of the reduction of arsenate to arsine with sodium borohydride. Followed by its deposition on a gold electrode at a potential of –0.5 V (versus Ag/AgCl). Reduction of arsenate to arsine was performed in order to avoid the need for a very low negative potential. The preconcentration of As^{5+} was proposed to comprise two steps, including the reduction of As^{5+} to arsine and deposition to be As^0 on the electrode surface, whereas the stripping step was based on the oxidation of As^0 to As^{5+}. The background-subtracted stripping voltammograms and linear calibration curve (r = 0.90) with estimated LOD of 6.07 ppm as shown in figure 2. The optimal stripping voltammetry conditions for As^{5+} were found to be E_{dep} = –0.5 V, t_{dep} = 60 s, and scan rate = 200 mV/s.

4. Conclusions

Detection of As^{3+} and As^{5+} was successfully achieved using ASV. As^{3+} can be deposited by the reduction of As^{3+} to arsine and subsequently detected via two-step oxidation (arsine to As^0, followed by As^0 to As^{3+}). For As^{5+}, the deposition step was based on the reduction of As^{5+} to As^{3+}, followed by the reduction of As^{3+} to arsine, whereas the stripping step was based on the oxidation of arsine to As^0, followed by oxidation of As^0 to As^{5+}. The optimum deposition potential, deposition time, and scan rate were found to be –0.5 V, 60 s, and 200 mV/s, respectively. Good linear responses were observed for standard solutions of As^{3+} and As^{5+}. Using our method, detection limits of 0.85 and 6.07 ppm can be achieved for As^{3+} and As^{5+}, respectively.

Acknowledgements
This work was funded by PITTA Grant Universitas Indonesia with contract number 1822/UN2.R3.1/PPM.00.01/2017. The authors would like to thank Einago (www.einago.com) for the English language review.

References
[1] Luong J H T, Majid E and Male K B 2007 *Open Anal. Chem. J.* **1** 7–14
[2] Mestrot A, Planer-Friedrich B and Feldmann J 2013 *Environ. Sci. Proc. Imp.* **15** 1639–51
[3] Hata N, Kasahara I, Taguchi and Goto K 1993 *Encyclopedia of Food Sciences and Nutrition*, ed Caballero B et al. (London: Elsevier) pp.304–311
[4] Czerczak S and Fishbein L [5] C. International and C. Assessment,2003 A*rsine: Human Health Aspects*, Concise International Chemical Assessment Documents 47 (Geneva: World Health Organization (WHO))
[5] Yan X P, Kerrich R and Hendry M J 1998 *Anal. Chem.* **70** 4736–42
[6] Cavicchioli A, La-Scalea M A and Gutz I G R 2004 *Electroanalysis* **16** 697–711
[7] Forsberg G, O'Laughlin J W, Megargle R G and Koirtyihann S R 1975 *Anal. Chem.* **47** 1586–92
[8] Yamada D, Ivandini T A, Komatsu M, Fujishima A and Einaga Y 2008 *J. Electroanal. Chem.* **615** 145–153
[9] Ivandini T A, Yamada D, Watanabe T, Matsuura H, Nakano N, Fujishima A and Einaga Y 2010 *J. Electroanal. Chem.* **645** 58–63
[10] Nagaoka Y, Ivandini TA, Yamada D, Fujita S, Yamanuki M, Einaga Y 2010 *Chem. Lett.* **39**, 1055-57
[11] Ivandini T A, Wijaya L, Gunlazuardi J and Einaga Y 2012 *Makara J. Sci.* **16** 9–14
[12] Dai X, Nekrassova O, Hyde M E and Compton R G 2004 *Anal. Chem.* **76** 5924–9

IOP Conf. Series: Materials Science and Engineering **496** (2019) 012031 doi:10.1088/1757-899X/496/1/012031

Synthesis of one-dimensional nanotubes from an organic semiconducting polymer (PDPPBTT) by the centrifuge method

N I Pratiwi, V Fauzia and C Imawan

Department of Physics, Faculty of Mathematics and Natural Sciences (FMIPA) Universitas Indonesia, Depok 16424, Indonesia

Corresponding author: vivi@sci.ui.ac.id

Abstract. The synthesis of the nanostructures of low-band-gap organic semiconducting materials has attracted significant attention as a technique to attain high surface areas. This study aims to synthesize PDPPBTT nanotubes using an anodic aluminum oxide (AAO) template and centrifugation method at three different speeds (3000, 3500, and 4000 rpm). The results show the effect of the rotational speed on the structural, morphological, and optical properties of PDPPBTT nanostructures. The results showed that the use of high rotational speeds resulted in shorter nanotubes with lower absorption intensities but enhanced photoluminescence in comparison with those formed at a lower rotational speed. These results demonstrate an important advancement toward materials that better facilitate light absorption, charge separation, and charge diffusion, and are expected to contribute toward the future development of organic solar cells, light-emitting diodes, and sensors on flexible substrates.

Keywords: PDPPBTT, low band gap, semiconductor, nanotubes, anodic aluminum oxide template, centrifuge method

1. Introduction

Low-band-gap organic semiconducting materials have attracted significant attention owing to their unique properties, such as controllable electrical and optical properties, low cost, flexibility, and ease of synthesis. These polymer-based materials have the potential to be applied in organic optoelectronic devices, such as organic solar cells, light-emitting diodes, and sensors on flexible substrates [1–3]. One such material is PDPPBTT (poly {2,2 '- (2,5-bis (2-octyldodecyl) -3,6-dioxo-2,3,5,6-tetrahydropyrrolo [3,4-c] pyrrole-1,4-diyl) dithieno [3,2-b] thiophene-5,5'-diyl-alt-thiophen-2,5-diyl}).

Generally, organic semiconductor materials are synthesized as thin films, which have very limited surface areas. Therefore, some researchers have attempted to fabricate the nanostructures organic materials, such as nanotubes, nanoflower and nanorods [4–6]. Such nanostructures are expected to increase the surface area to achieve better light absorption, facilitate charge separation, and provide a shorter distance for charge diffusion. Nanostructures are generally created by either bottom-up or top-down methods. Top-down method is a relatively simple and inexpensive method. It uses a soft or hard template, such as polycarbonate, a silica membrane, or a nanoporous anodic aluminum oxide (AAO) template. The infiltration of the polymer material into the AAO template can be facilitated by spin coating, dip coating, electrodeposition, or centrifugation [6–9].

In this study, PDPPBTT nanotubes were synthesized using an AAO template by centrifugation to facilitate the infiltration method. The objective is to observe the effects of the centrifugation speed on the structural, morphological, and optical properties of the PDPPBTT nanostructures. Till date, no data has been reported on PDPPBTT nanostructures so this study has the potential to provide new insight that can be applied to the future development of organic optoelectronic devices.

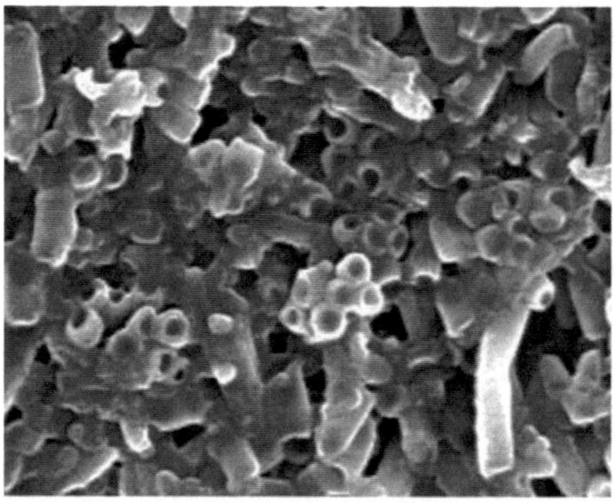

Figure 1. FESEM image of PDPPBTT nanotubes

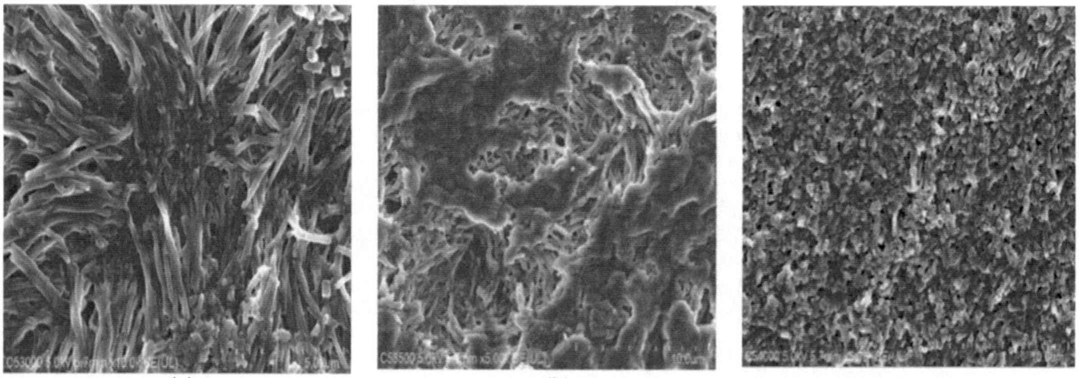

(a) (b) (c)

Figure 2. The FESEM images of PDPPBTT nanotubes formed with rotational speeds of (a) 3000 (b) 3500, and (c) 4000 rpm

2. Methods

The commercial semiconducting polymer, PDPPBTT, was obtained from Luminescence Technology Corp. and used without further purification. The nanoporous AAO template with a pore size of 20–200 nm and a thickness of 60 μm was purchased from Whatman Anodisc (Sigma-Aldrich, St. Louis, USA). Prior to the infiltration process, the template was cleaned by sonication in acetone for 15 minutes followed by rinsing in deionized water and dried completely at room temperature. The cleaned AAO template was inserted into a modified centrifuge tube as described previously [6]. PDPPBTT (5 mg) was dissolved in 1 mL of chloroform and 100 μL of this solution was dropped onto the template. The template was then rotated in a centrifuge (Kubota 2420) at either 3000 (1008 g), 3500 (1370 g), or 4000 rpm (1790 g). The PDPPBTT nanostructures were then detached from the AAO templates on copper tape, immersed in 3 M natrium hydroxide for 24 hours, and subsequently rinsed with deionized water.

The structural and optical properties were characterized by Raman spectroscopy (Renishaw RE 04), UV-vis spectroscopy (Shimazu UV-3101 PC), and photoluminescence spectroscopy (Renishaw RE 04). Morphological characterization was conducted using a field-emission scanning electron microscope (FESEM, Hitachi SU-8030).

3. Results and discussion

Figure 1 shows a cross-sectional FESEM image of the PDPPBTT nanostructure formed with a rotational speed of 3000 rpm. It can be seen that the nanotube structure was successfully formed,

Table 1. Vibration modes from the Raman spectra of PDPPBTT nanotubes synthesized with different rotational speeds

Peak	Rotational speeds (rpm)			Vibration modes
	3000	3500	4000	
V1	712	712	710	C–H bending
V2	1150	1149	1147	C–H bending and C=C stretching
V3	1329	1328	1327	C–H bending of BTT
V4	1374	1373	1371	C=C stretching of DPP
V5	1498	1497	1496	C=C stretching/thiophene ring stretching
V6	1516	1516	1515	C–C stretching

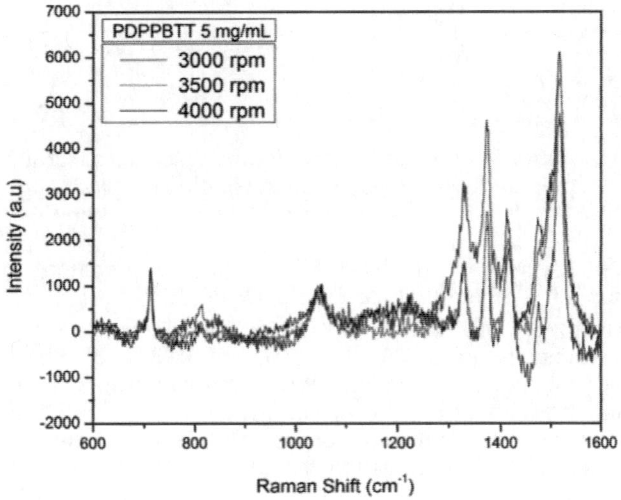

Figure 3. Raman spectra of PDPPBTT nanotubes prepared with three different rotational speeds

indicating that the polymer solution can infiltrate into the pores and tends to wet the pore wall rather than bind with other PDPPDTT molecules. This implies that the adhesion force between the PDPPBTT molecules and the pore wall is greater than the cohesion force between PDPPBTT molecules. The use of a readily evaporated chloroform solvent promoted the formation of nanotubes instead of nanorods [5,6].

The FESEM images of PDPPBTT nanotubes prepared at three different rotational speeds are shown in figure 2. The nanotubes prepared at 3000 rpm were longer than those prepared at 3500 rpm and 4000 rpm. This can be attributed to the effect of the greater centrifugal force that takes place during centrifugation at a higher rotational speed. For the centrifuge used here [6], a rotational speed of 3000 rpm yields a centrifugal force of 1000.2 mN. This force can easily push a low-concentration polymer solution deep into the pores of the template. However, the use of a higher rotational speed results in a higher centrifugal force of 2000.2 mN, which may be too high and cause the nanotubes to break, resulting in shorter nanotubes [6].

Figure 3 shows the Raman spectra for the three different samples. Six peaks were observed in these spectra. The peaks and the bonds they correspond to are summarized in Table 1. The vibration modes indicate C–H bending, C=C bending, C–H bending in BTT, C=C stretching in DPP, C=C stretching in thiophene, and C–C stretching [10].

These results show that an increase in the rotational speed causes a peak shift toward shorter wave number (downward), indicating that the rotational speed affects the bonds between the molecules of PDPPBTT. Moreover, the greater rotational speeds resulted in higher peak intensities at 1329 and 1374 cm^{-1} and lower peak intensities at 1498 and 1516 cm^{-1}. This may be related to the FESEM observation that higher rotational speeds yield shorter nanotubes.

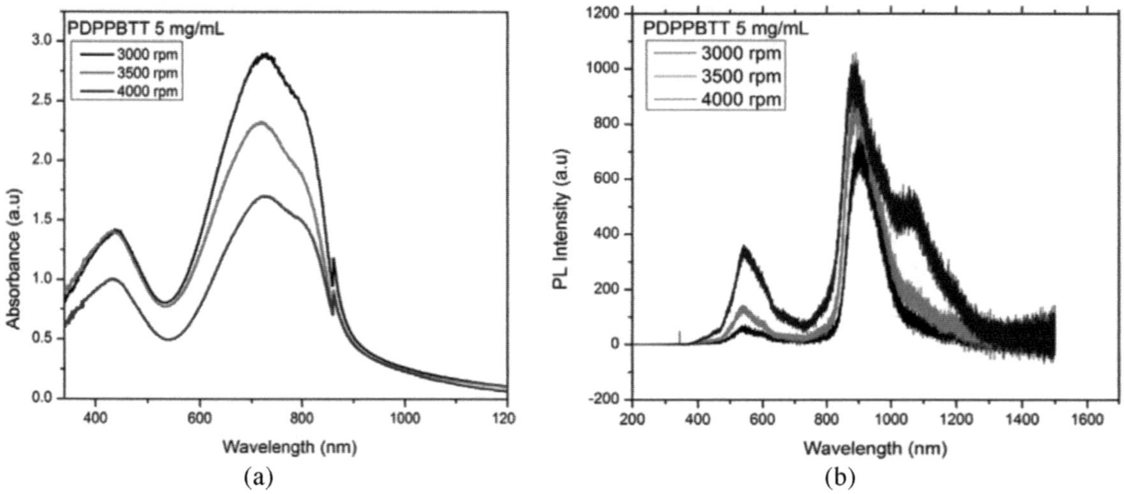

Figure 4. (a) UV-Vis Absorption spectra and (b) photoluminescence spectra of PDPPBTT nanotubes prepared with three different rotational speeds

The UV-vis absorption spectra of the PDPPBTT nanotubes prepared with different rotational speeds are shown in figure 4a. The main absorption peak of PDPPBTT was located at 419 nm; this corresponds to the absorbance of DPP due to excitation from the π band to the π^* band [4,11] and another peak at 729 nm corresponding to the absorbance of BTT [12]. The results indicate that the samples prepared at higher rotational speeds have lower absorption intensities. This may be because longer nanotube structure can absorb light more effectively. Alternatively, the high level of coplanarity and more effective intra-molecular charge transfer from the donor unit, BTT, to the acceptor unit, DPP, may increase the conjugation length and, thus, increase the absorption intensity [13].

Figure 4b shows the photoluminescence spectra of the PDPPBTT nanotubes prepared at different rotational speeds. The spectra exhibit two emission peaks, i.e., one at a wavelength of 540 nm and another in the infrared region at 890 nm. These peaks result from the radiative transitions when the material absorbs light at wavelengths of 419 and 729 nm, respectively, as shown in absorption spectra in figure 4a. Moreover, greater rotational speeds enhance the photoluminescence intensity in contrast to the absorption intensity, which was decreased at the greater rotational speeds. The mechanism leading to this effect is not clearly understood based on the information available. However, we hypothesize that the observation that high absorption intensity is not always accompanied by high radiative transitions indicates that there are non-radiative transitions occurring in the PDPPBTT nanotubes prepared at higher rotational speeds.

4. Conclusions

In this study, low-band-gap semiconducting PDPPBTT nanotubes were successfully synthesized with a nanoporous AAO template via the centrifuge method at three different rotational speeds (3000, 3500, and 4000 rpm). The results show that an increase in the rotational speed results in longer nanotubes with Raman peaks shifted toward shorter wave numbers, reduced absorption intensity, and enhanced photoluminescence intensity. All samples exhibited the two typical absorption peaks at 419 and 729 nm and emission peaks at 540 and 890 nm. However, the reason for the opposing effects of the rotational speed on the absorption and photoluminescence intensities is not clearly understood yet. We hypothesize that the longer PDPPBTT nanotubes absorb the light more effectively but that this absorption is not accompanied by high radiative transitions. This phenomenon should be further investigated in future studies.

Acknowledgements

The authors would like to thank Einago (www.einago.com) for the English language review.

IOP Conf. Series: Materials Science and Engineering **496** (2019) 012031 doi:10.1088/1757-899X/496/1/012031

References

[1] Bubb D M, Corgan J, Yi S, Khan M, Hughes L, Gurudas U, Papantonakis M and Mcgill R A 2010 *Appl. Phys. A Mater. Sci. Process.* **100** 523–31
[2] Houng M, Lu W, Yang T and Lee K 2014 *Journal of Nanomaterials* **2014** 130716
[3] Halizan M Z M, Makinudin A H A and Supangat A 2016 *RSC Adv.* **6** 37574–82
[4] Fakir M S, Supangat A and Sulaiman K 2014 *Nanoscale Res. Lett.* **9** 225
[5] Bakar N A, Supangat A and Sulaiman K 2014 *Nanoscale Res. Lett.* **9** 600
[6] Doris M, Aziz F, Alhummiany H, Bawazeer T, Alsenany N, Mahmoud A, Zakaria R, Sulaiman K and Supangat A 2017 *Nanoscale Res. Lett.* **12** **6**7
[7] Pasquali M, Liang J and Shivkumar S 2010 *Mater. Lett.* **64** 1943–6
[8] Xu J, Xia J, Wang J, Shinar J and Lin Z 2006 *Appl. Phys. Lett.* **89** 133110.
[9] Shingubara S 2003 *J. Nanopart. Res.* **5** 17–30.
[10] Francis C, Fazzi D, Grimm S B, Paulus F, Beck S, Hillebrandt S, Pucci A and Zaumseil J 2017 *J. Mater. Chem. C* **5** 6176–84
[11] Zhang K and Tieke B 2008 *Macromolecules* **41** 7287–95.
[12] Kim M J, Choi J Y, An G, Kim H, Kang Y, Kim J K, Son H J, Lee J H, Cho J H and Kim B 2016 *Dyes Pigm.* **126** 138–46
[13] Le Borgne M, Quinn J, Martín J, Stingelin N, Wantz G and Li Y 2016 *Dyes Pigm.* **131** 160–7

IOP Conf. Series: Materials Science and Engineering **496** (2019) 012032 doi:10.1088/1757-899X/496/1/012032

Partial oxidation of methane to methanol over M_xO_y/ZSM-5 (Mn, Fe, Co, and Ni) hierarchical transition metal oxide catalysts

B G Nurgita, D A Nurani and Y K Krisnandi

Department of Chemistry, Faculty of Mathematics and Natural Sciences (FMIPA), Universitas Indonesia, Depok 16424, Indonesia

Corresponding author: yuni.krisnandi@sci.ui.ac.id

Abstract. The selective oxidation of methane to methanol is a key challenge in catalysis. The catalytic activities of nickel, cobalt, iron, and manganese oxides supported by hierarchical ZSM-5 zeolites for the partial oxidation of methane to methanol were studied. Hierarchical ZSM-5 was synthesized using a double-template method, in which tetrapropylammonium hydroxide was used as an MFI structure-directing agent and PDDA-M was used as a mesopore-directing agent. The synthesized hierarchical ZSM-5 was characterized by X-ray diffraction, Fourier-transform infrared spectroscopy, and scanning electron microscopy with energy dispersive spectroscopy. M_xO_y/ZSM-5 catalysts were prepared using a wet impregnation method and were also extensively characterized. The catalytic test was performed at 423 K in a batch reactor with a CH_4 to N_2 ratio of 0.5:2 and 0.5 g of M_xO_y/ZSM-5 catalyst for 2 h. The hierarchical Fe_xO_y/ZSM-5 catalyst showed the highest yield of methanol, whereas Mn_xO_y/ZSM-5 tended to produce formic acid.

Keywords: partial oxidation, hierarchical ZSM-5, metal oxide catalyst, methane, methanol

1. Introduction

Natural gas has been known as a clean and effective energy source since its combustion generates fewer greenhouse gases than coal or petroleum fuels. However, its storage and transport is not easy, making its derivatives uncompetitive with those of other fossil fuels [1]. Conventionally, an indirect route for conversion of natural gas to methanol was through the production of syngas. However, the process is an energy intensive process, (temperature reaction of 800 °C–1000 °C) and more than 25 % of the feed (natural gas) has to be burned to provide the heat of reaction [2]. Hence, such a sustainable direct conversion of methane to methanol is more desirable compared with the current technology [3].

Research into the direct partial oxidation of methane has progressed rapidly, and several developments have been made, including the use of oxidants such as N_2O, O_2, and H_2O_2. The catalysts used are metal oxides such as Co_3O_4 and MoO_3 in conjunction with zeolites and some transition metals as active sites, such as Fe, Co, and Cu. Nadzeya *et al.* [4] carried out partial oxidation of methane using a Co-impregnated agglomerate of ZSM-5 zeolite as a catalyst and obtained methanol and formaldehyde as the major products. Furthermore, Krisnandi *et al.* [5] used a hierarchical CoO_x/ZSM-5 catalyst, and the reaction showed a high selectivity toward methanol conversion [5]. In this work, we compared the activities of four hierarchical ZSM-5 supported catalysts based on metal oxides Mn_xO_y, Fe_xO_y, Ni_xO_y, and Co_xO_y for partial oxidation of methane.

2. Experimental

2.1. Materials

All materials in this work were used without further purification. Tetraethyl orthosilicate (TEOS, 98%, Sigma Aldrich), sodium aluminate (Sigma Aldrich), tetrapropylammonium hydroxide (TPAOH, 1 M, Sigma Aldrich), poly(diallyldimethylammonium chloride) (PDDA, 35%, Aldrich), and deionized water were used for hierarchical ZSM-5 synthesis. While cobalt(II) nitrate ($Co(NO_3)_2{\cdot}6H_2O$, Sigma Aldrich), iron(III) nitrate ($Fe(NO_3)_3{\cdot}9H_2O$, Sigma Aldrich), manganese (II) nitrate ($Mn(NO_3)_2{\cdot}4H_2O$, Sigma Aldrich), nickel(II) nitrate ($Ni(NO_3)_2{\cdot}6H_2O$, Sigma Aldrich) were used to prepare M_xO_y/ZSM-5. Finally, methane gas (99.99% BOC), N_2 gas (99.99% UHP) were used in catalytic test.

2.2. ZSM-5 synthesis and M_xO_y/ZSM-5 preparation
The synthesis was carried out following method reported by Wang *et al.* [6] with some modifications. Hierarchical ZSM-5 was synthesized by mixing the solution of sodium aluminate, TEOS, TPAOH and deionized water in a molar ratio of 1 [Al_2O_3] : 64 [SiO_2] : 10 [$(TPA)_2O$] : 3571 [H_2O] at 373 K for 3 h. Afterward, this mixture was added dropwise by 1 g PDDA stirred for 24 h, then hydrothermal processed in a Teflon-lined stainless steel autoclave at 423 K for 144 h. The then obtained zeolite (white powder) was dried in room temperature before calcined at 820 K. Metal oxide insertion into the zeolite was carried out using a wet impregnation method, where 1 g of ZSM-5 was added to 0.2495 M metal solution prepared from metal nitrate hydrate and stirred for 24 h at ambient condition, followed by calcination at 823 K. Respectively, the hierarchical ZSM-5 and M_xO_y-ZSM-5 counterparts were characterized by X-ray diffraction (XRD), Fourier-transform infrared spectroscopy (FTIR, Shimadzu Prestige 21), scanning electron microscopy with energy dispersive spectroscopy (SEM-EDX, SEM FEI quanta FEG 450), and atomic absorption spectroscopy (AAS, Shimadzu 13000).

2.3. Catalytic test: partial oxidation of methane
The methane partial oxidations were performed using each M_xO_y/ZSM-5 catalysts (c.a. 0.5 g) in a 200 mL stainless steel batch reactor. Prior to use, the catalyst (0.5 g) was activated at 773 K and flushed with flowing nitrogen gas. The catalytic reaction was carried out at 423 K for methane and nitrogen pressures of 0.75 and 2 bar, respectively. After the reactor was left to cool down to room temperature, reaction product was extracted from the zeolite using 3 mL of ethanol and analyzed using gas chromatography (GC Shimadzu 2010) with a Carbowax column and flame ionization detector (Shimadzu).

3. Results and discussion

3.1. Characterization of ZSM-5

3.1.1. XRD Pattern. The ZSM-5 zeolite was synthesized using TPAOH as an MFI structure-directing agent and PDDA polycation as a mesopore template. During the formation of zeolites with Si/Al ratios ranging from 10.5 to 36.5, competition between ZSM-5 and mordenite occurred during the hydrothermal crystallization process [5]. However, the positively charged PDDA interacts with the negatively charged ZSM-5 crystal-to-be-structure, so that the ZSM-5 framework could be established around the PDDA polication [6]. The XRD pattern of as-synthesized hierarchical ZSM-5 (figure 1) shows certain peaks in the 2θ range from 7° to 9° and from 22° to 25°, which is consistent with the standard ZSM-5 pattern [7], indicating that the ZSM-5 structure was successfully synthesized [6].

3.1.2. FTIR Spectroscopy. Figure 2 shows the FTIR spectra of ZSM-5 before (a) and after (b) calcination. The spectrum of ZSM-5 before calcination shows several extra peaks at 2960–2850 cm^{-1} (C–H stretching) and at 1470–1350 cm^{-1} (C–H bending) compared with that of ZSM-5 after calcination, which can be attributed to the templates used in the synthesis of zeolite, i.e., TPAOH and PDDA. Calcination was performed to remove the organic templates used in the synthesis of zeolites. It can be seen from figure 2b that the template was removed as indicated by the loss of the C–H stretching and bending vibration bands. Moreover, the pores and channels in the zeolite became unoccupied after calcination.

3.1.3 SEM-EDX. To determine the crystal structure and the Si/Al ratio of the synthesized zeolite, EDX analysis was performed, from which the mass% of the elements in the zeolite can be determined, in which the Si/Al ratio was 37.4. Table 1 shows the mass% of the elements present in a ZSM-5 crystal.

IOP Conf. Series: Materials Science and Engineering **496** (2019) 012032 doi:10.1088/1757-899X/496/1/012032

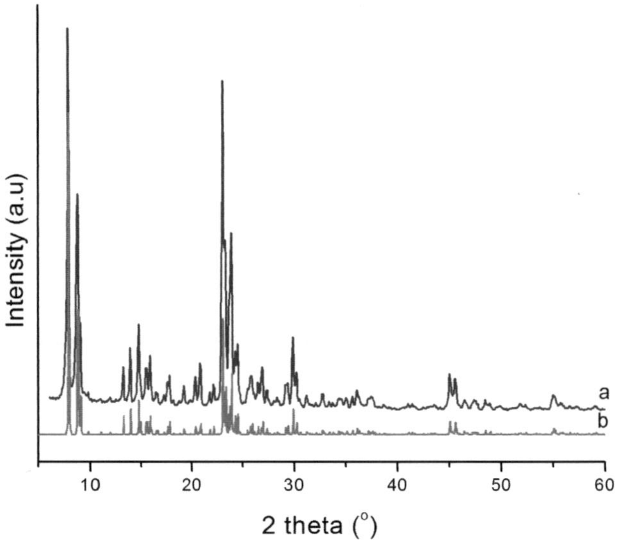

Figure 1. XRD patterns of (a) synthesized ZSM-5 and (b) standard ZSM-5.

Table 1. SEM-EDX analysis

Element	Composition
C	1.17
O	50.04
Na	0.41
Si	46.42
Al	1.24

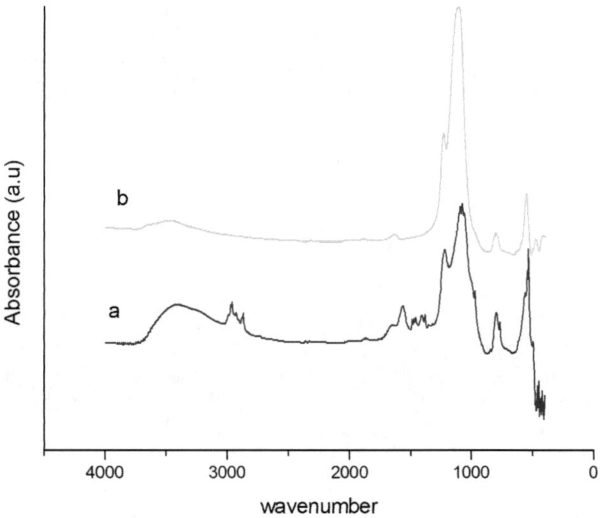

Figure 2. FTIR Spectra ZSM-5 (a) before and (b) after calcination at 823 K.

3.2. Preparation of M_xO_y/ZSM-5

M_xO_y/ZSM-5 heterogeneous catalysts were prepared using a wet impregnation method. The elemental analysis with AAD indicates that the loading of each metal on ZSM-5 zeolites was about 2.1%–2.5%.

Figure 3 shows the XRD patterns of modified ZSM-5 catalysts. The peaks at 2θ = 31°, 47°, and 58° in the diffraction pattern of Ni_xO_y/ZSM-5 are attributed to Ni_2O_3 [8]. The peaks at 2θ = 31°, 44°, and 45° in the diffraction pattern of Fe_xO_y/ZSM-5 originates from Fe_2O_3 [9]. Moreover, the peaks at 2θ = 31°, 45°, 55°, and 60° in the diffraction pattern of Co_xO_y/ZSM-5 are attributed to Co_3O_4 [10], and the peaks at 2θ = 35° and 40° are attributed to CoO [11]. The peaks at 2θ = 19° and 25° in the diffraction pattern of Mn_xO_y/ZSM-5 originate from Mn_2O_3, and the peaks at 2θ = 16°, 31°, and 65° originate from Mn_3O_4 [12].

IOP Conf. Series: Materials Science and Engineering **496** (2019) 012032 doi:10.1088/1757-899X/496/1/012032

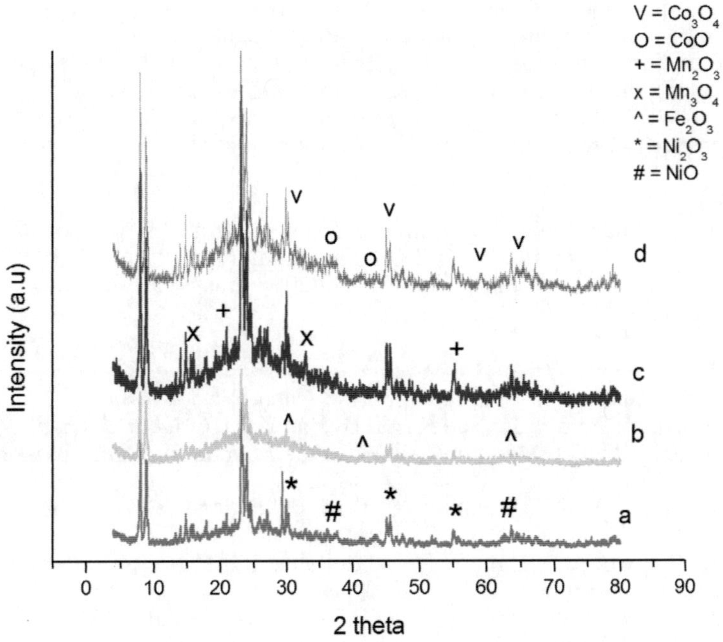

Figure 3. XRD patterns of (a) Ni_xO_y/ZSM-5, (b) Fe_xO_y/ZSM-5, (c) Mn_xO_y/ZSM-5, and (d) Co_xO_y/ZSM-5.

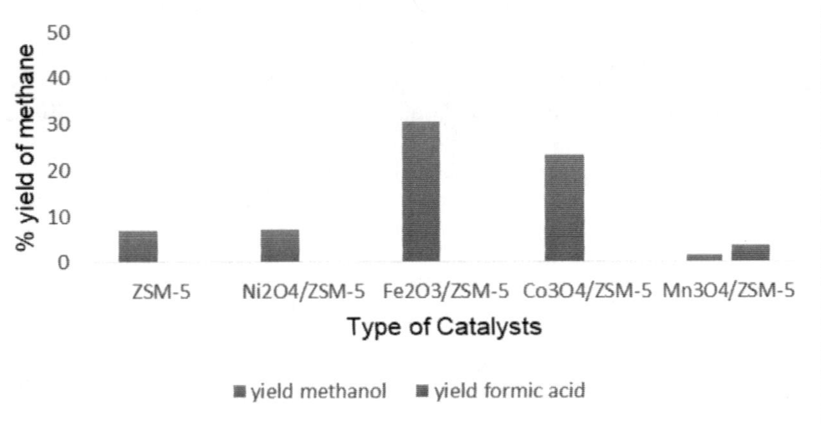

Figure 4. Results of partial oxidation of methane using unmodified ZSM-5 and metal oxide-modified ZSM-5.

3.3. Methane partial oxidation catalytic test

The results of methane partial oxidation using as synthesized ZSM-5 catalysts are shown in figure 4. ZSM-5 without metal oxides produced methanol in 6.8% yield. The impregnation of most metal oxides in ZSM-5 increased the % yield of methanol, which indicates that the metal oxide is capable of being the active site of the catalyst, thereby increasing the catalytic activity for oxidizing methane to methanol. However, when using Mn_3O_4-modified ZSM-5, formic acid was formed with higher yield than methanol, indicating Mn_3O_4/ZSM-5 has selectivity toward the formation of formic acid.

4. Conclusions

Catalysts based on ZSM-5 modified with four metal oxides have been successfully synthesized, as confirmed by XRD data. The presence of metal oxides increased the number of catalyst active sites, resulting in high yield of methanol. It was found that the type of metal oxide species influenced the yield of methanol. Among the four types of metal oxide-modified ZSM-5 catalysts, Fe_2O_3/ZSM-5 is the most promising for partial oxidation of methane to methanol with the highest yield of 30.5%, whereas Mn_3O_4/ZSM-5 has selectivity toward the formation of formic acid.

Acknowledgements

Authors wishing to acknowledge financial funding and support from Ministry of Research Technology and Higher Education Republic of Indonesia (Kemenristekdikti RI) through PUPT BOPTN research grant with contract number 2719/UN2.R3.1/HKP05.00/2017 and also Einago (www.einago.com) for the English language review.

References

[1] York A P E, Xiao T and Green M L H 2003 *Top. Catal.* **22** 345–58
[2] Han B, Yang Y, Xu Y, Etim U J, Qiao K, Xu B, Yan Z 2016. *Chin. J. Catal.* **37** 1206–15
[3] Arutyunov VS 2014 *Direct Methane to Methanol: Foundations and Prospect of the Process.* (Amsterdam: Elsevier)
[4] Beznis N V, Van Laak A N C, Weckhuysen B M and Bitter J H 2011 *Micro. Meso. Mater.* **138** 176–83
[5] Krisnandi Y K, Putra B A P, Bahtiar M, Abdullah I and Howe R F 2015 *Procedia Chem.* **14** 508–15
[6] Wang L, Zhang Z, Yin C, Shang Z and Xiao F S 2010 *Micro. Meso. Mater.* **131** 58–67
[7] Treacy M M J and Higgins J B 2001 *Collection of Simulated XRD Powders Patterns for Zeolites* (London: Elsevier)
[8] Shaban M, Abukhadra M R, Hamd A, Amin R R, A Abdel Khalek A 2017 *J. Environ. Manag.* **204** 189–99
[9] Ganachari S V, Joshi V K, Bhat R, Deshpande R, Salimath B and Rao N V 2012 Large scale synthesis and characterization of Fe_2O_3 nanoparticlesd by self-propagating low temperature combustion method. *Int. J. Sci. Res.* **1** 77–79
[10] Yue W and Zhou W 2007 *J. Mater. Chem.* **17** 4947–52
[11] Ghosh M, Sampathkumaran E V and Rao C N R 2005 *Chem. Mater.* **17** 2348–52
[12] Dhaouadi H, Ghodbane O, Hosni F and Touati F 2012 *ISRN Spectroscopy* **2012** 1–8

IOP Conf. Series: Materials Science and Engineering **496** (2019) 012033 doi:10.1088/1757-899X/496/1/012033

Study of $LaFe_{1-x}Mn_xO_3$ (x = 0.05, 0.1, 0.15, 0.2) perovskite materials by impedance spectroscopy

A W Anugrah, D Triyono and H Laysandra

Department of Physics, Faculty of Mathematics and Natural Sciences (FMIPA)
Universitas Indonesia, Depok 16424, Indonesia

Corresponding author: djoko.triyono@sci.ui.ac.id

Abstract. We use room-temperature impedance spectroscopy to investigate $LaFe_{1-x}Mn_xO_3$ (x = 0.05, 0.1, 0.15 and 0.2) perovskite-type oxides. The samples were synthesized using the sol-gel method followed by a sintering process to form bulk samples. The microstructural and chemical composition were analyzed using scanning electron microscopy and x-ray fluorescence, with the results indicating a grain size of 373–305 nm and an atomic composition compatible with the chemical formula $LaFe_{1-x}Mn_xO_3$. Applying impedance spectroscopy to $LaFe_{1-x}Mn_xO_3$ at room temperature reveals a decreasing impedance and relaxation time with increasing Mn content. The dielectric spectra show a increasing dielectric constant with increasing of Mn content.

Keywords: sol-gel method, impedance spectroscopy, $LaFe_{1-x}Mn_xO_3$

1. Introduction

Perovskites have an ABX_3 structure where A and B are cations and X is usually an anion such as $LaFeO_3$, $LaMnO_3$, etc. Lanthanum orthoferrite ($LaFeO_3$), an orthoferrite perovskite compound, is the focus of the present work because of its structural stability and suitability in gas-sensor applications [1]. Previous research has reported substitution into the La and/or Fe sites of $LaFeO_3$, which enhances its electronic properties by producing an increased dielectric constant, decreased dielectric loss, and high thermal stability [2,3]. Cao *et al.* [2] reported that Na substitution into the La site of $La_{1-x}Na_xFeO_3$ ($x = 0, 0.1, 0.2$) results in a higher dielectric constant, with the maximum being 10^5 for $x = 0.2$. Phokha *et al.* [3] reported that $LaFe_{1-x}Ti_xO_3$ ($x = 0, 0.1, 0.2$) has a higher dielectric constant, with the maximum at $x = 0.2$ being to the order of 10^4. In addition, Fe substitution into the Mn site of $LaMn_{1-x}Fe_xO_3$ (x = 0.15, 0.3, 0.5, 0.7, 1) was investigated by Karmakar *et al.* [4], who reported an increasing dielectric constant with increasing Fe content. Zeng *et al.* [5] reported that $LaMn_{0.5}Fe_{0.5}O_3$ nanocrystalline materials have crystallite sizes and a dielectric constant of ~25 nm and ~4300 at room temperature, respectively.

Building on previous research, we investigate how Mn doping into the Fe site of $LaFe_{1-x}Mn_xO_3$ with x = 0.05, 0.1, 0.15, 0.2 affects the microstructural and dielectric properties at room temperature. The samples were synthesized by using the sol-gel method followed by sintering to form bulk samples. The chemical composition and microstructural properties were examined by using x-ray fluorescence and scanning electron microscopy (SEM), respectively. RLC-Meter FLUKE-PM 6303 with frequency range of 100 Hz to 1 MHz is used to characterized the electrical properties of the samples. The data were analyzed by using impedance spectroscopy and the parallel-plate method.

2. Experimental

By using the sol-gel method, $LaFe_{1-x}Mn_xO_3$ perovskite compounds were prepared with x = 0.05, 0.10, 0.15, and 0.20 with lanthanum (III) oxide La_2O_3, manganese (II) chloride tetrahydrate $MnCl_2.4H_2O$, and

Table 1. Results of x-ray fluorescence analysis of $LaFe_{1-x}Mn_xO_3$ with $x = 0.05, 0.10, 0.15$, and 0.20 after calcination at 1173 K.

$LaFe_{1-x}Mn_xO_3$	(%) weight experimental			(%) weight calculated		
	La	Fe	Mn	La	Fe	Mn
$x = 0.05$	65.48	25.50	0.99	71.34	27.40	1.40
$x = 0.10$	61.48	22.70	1.97	71.35	25.81	2.82
$x = 0.15$	63.72	22.16	3.40	71.37	24.39	4.23
$x = 0.20$	65.22	21.25	4.40	71.39	22.96	5.64

(a)

Figure 1. SEM images of $LaFe_{1-x}Mn_xO_3$ with x = (a) 0.05, (b) 0.1, (c) 0.15 and (d) 0.2.

iron (III) nitrate nanohydrate $Fe(NO_3)_3 9H_2O$ (analytical grade) as raw materials. The materials were mixed stoichiometrically with aquades and citric acid monohydrate as a solvent and stirred at 393 K until a gel formed. The gel was dried at 473 K for 5 h then calcined at 1173 K to obtain the nanoparticle form. To obtain nanocrystalline form, the material was pressed into bulk samples by applying 3 kN/m^2 and sintered at 1273 K for 12 h. The chemical composition was checked by x-ray fluorescence. The morphology was analyzed by using SEM (FEI QUANTA 650). To apply the parallel-plate capacitor model, the materials were coated with carbon by using a Quorum K975x Turbo Evaporator. RLC-Meter FLUKE-PM 6303 with frequency of 100 Hz to 1 MHz is used to measured the electrical properties of the samples at room temperature.

3. Results and discussion

X-ray fluorescence analysis confirms an atomic composition consistent with the chemical formula $LaFe_{1-x}Mn_xO_3$. Table 1 shows the composition of $LaFe_{1-x}Mn_xO_3$ with x = 0.05, 0.10, 0.15, 0.20 after calcination at 1173 K. The results of these measurements differ slightly from the calculation, which may be attributed to preparation conditions and insufficient measurement accuracy.

Figure 1 shows typical SEM images of Mn-doped $LaFeO_3$ sintered at 1273 K for 12 h. The average grain sizes are in the range of 373–305 nm. The SEM images show a homogenous surface morphology with spherical grains [6]. Zeng *et al.* [5] reported a similar morphology for $LaFe_{0.5}Mn_{0.5}O_3$.

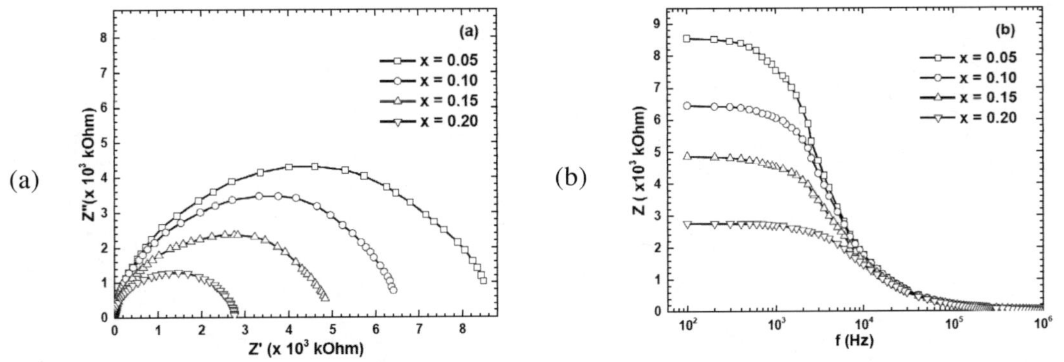

Figure 2. (a) Nyquist plot and (b) impedance Z as a function of frequency for $LaFe_{1-x}Mn_xO_3$ with $x = 0.05, 0.1, 0.15$, and 0.2 at room temperature.

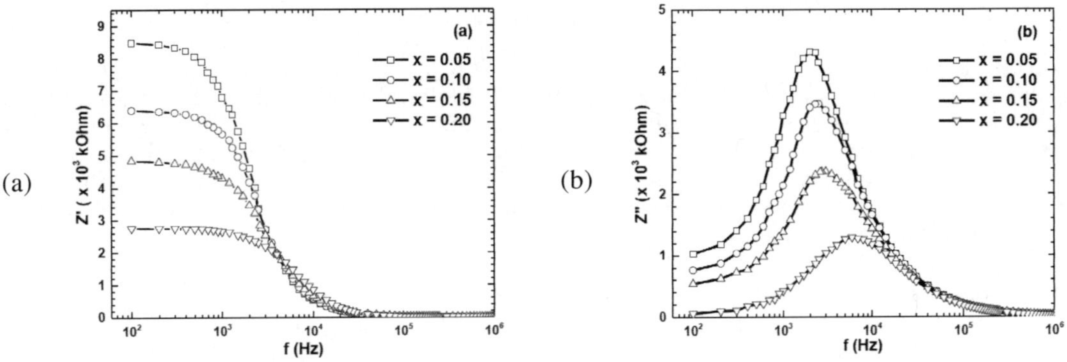

Figure 3. (a) Real and (b) imaginary part of impedance as a function of frequency (Bode plot) for $LaFe_{1-x}Mn_xO_3$ with $x = 0.05, 0.1, 0.15$, and 0.2 at room temperature.

Figure 2a shows a Nyquist plot of $LaFe_{1-x}Mn_xO_3$ with x = 0.05, 0.10, 0.15 and 0.20 at room temperature. The diameter of the semicircle changes slightly as Mn content increases. Figure 2b shows a Bode plot of the impedance $Z(f)$ for $LaFe_{1-x}Mn_xO_3$. The impedance magnitude $|Z|$ decreases with increasing Mn content at low frequency, and decreases at high frequency for all materials.

Impedances of $LaFe_{1-x}Mn_xO_3$ at room temperature are shown in figure 3. The real part $Z'(f)$ of impedance decreases with increasing frequency for all materials (figure 3a). Figure 3b shows the imaginary part $Z''(f)$ of $LaFe_{1-x}Mn_xO_3$, which increases to a maximum value and then starts to decrease at high frequencies, indicating that a relaxation phenomenon is active in the system. The peak of Z'' gives the relaxation time for the dielectric-relaxation process.

Figure 4a shows the dielectric constant $\varepsilon'(f)$ for all materials, which increases with increasing frequency and Mn content. This phenomenon is similar to that in $BaTi_{1-x}Mn_xO_3$ (with x = 0.00, 0.01, 0.02, 0,03, 0.04) [7], where the dielectric constant also increases with increasing Mn content. Figure 4b shows the dielectric loss tan $\delta(f)$ of $LaFe_{1-x}Mn_xO_3$ at room temperature, which shows that the dielectric loss depends on Mn content. In addition, all materials exhibit the dielectric-relaxation phenomenon. The varying dielectric constant and dielectric loss may be due to the number of grain boundaries and to the Mn doping. $LaFe_{0.80}Mn_{0.20}O_3$ has a higher dielectric constant than the others, which indicates that it has a smaller grain size and fewer grain boundaries. Riaz *et al.* [8] reported variations in dielectric constant and dielectric loss due to several factors, including grain size and grain boundaries. The decrease in grain size reportedly changes the dielectric constant in $BiFe_{1-x}Mn_xO_3$ [8]. In addition, Islam *et al.* [7] reported a decrease in the dielectric constant of $BaMn_xTi_{1-x}O_3$ with increasing Mn content and increasing grain size, and Verma *et al.* [9] reported a small dielectric constant resulting from a large number of grain boundaries. Riaz *et al.* [8] also reported increased dielectric parameter values due to a large number of grain boundaries and Mn doping.

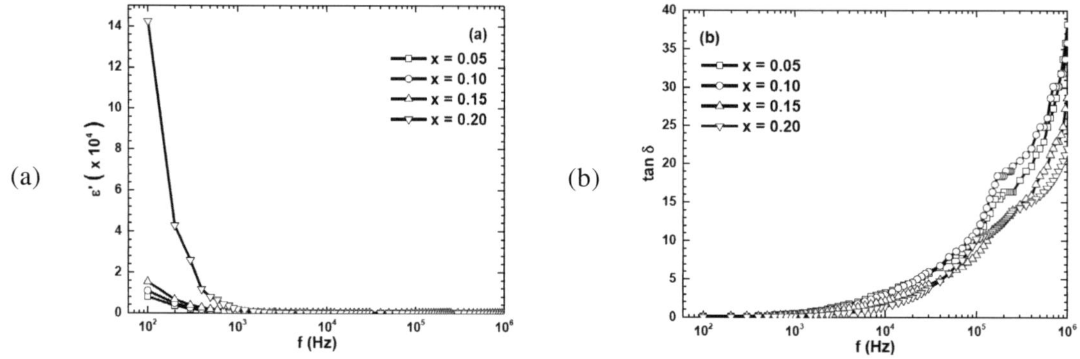

Figure 4. Frequency dependence of the (a) real part of dielectric constant and (b) dielectric loss of $LaFe_{1-x}Mn_xO_3$ with $x = 0.05, 0.1, 0.15$, and 0.2 at room temperature.

4. Conclusions

Perovskites $LaFe_{1-x}Mn_xO_3$ with four variations of Mn (x = 0.05, 0.1, 0.15, and 0.2) were successfully synthesized by using the sol-gel technique followed by sintering. The results of x-ray fluorescence spectroscopy indicate an atomic composition consistent with the chemical formula $LaFe_{1-x}Mn_xO_3$. SEM images show a morphology consisting of homogeneously distributed spherical grains. The average grain size decreases with increasing Mn content, and Mn doping decreases the impedance and increases the dielectric constant.

Acknowledgements

The authors would like to appreciate PITTA Grant from Universitas Indonesia for financial support that was provided under contract number 2257/UN2.R3.1/HKP05.00/2018 and Einago (www.einago.com) for the English language review.

References

[1] Cyza A, Kopia A, Cieniek L and Kusinski J 2016 *Mater. Today Proc.* **3** 2707–12
[2] Cao E, Qin Y, Cui T, Sun L, Hao W and Zhang Y 2017 *Ceram. Int.* **43** 7922–8
[3] Phokha S, Hunpratup S, Pinitsoontorn S, Putasaeng B, Rujirawat S and Maensiri S 2015 *Mater. Res. Bull.* **67** 118–25
[4] Karmakar A, Majumdar S, Singh A K and Giri S 2009 *J. Phys. D: Appl. Phys.* **42** 092004
[5] Zeng C, He Y, Li C and Xu Y 2013 *Ceram. Int.* **39** 5765–9
[6] Sultan K, Ikram M and Asokan K 2015 *RSC Adv.* **5** 93867–76
[7] Islam R, Choudhury S, Rahman S N and Rahman M J 2012 *J. Ceram. Process Res.* **13** 248–51
[8] Riaz S, Shah S M H, Akbar A, Atiq S and Naseem S 2015 *J. Solgel Sci. Technol.* **74** 329–39
[9] Vermaa K C, Ram M, Singh J and Kotnala R K 2011 *J. Alloys Compd.* **509** 4967–71

IOP Conf. Series: Materials Science and Engineering **496** (2019) 012034 doi:10.1088/1757-899X/496/1/012034

Synthesis of zircon powder by polyvinyl alcohol addition

Musyarofah, A D C Putri, G Elsandika and S Pratapa

Department of Physics, Faculty of Sciences, Institute of Technology Sepuluh Nopember, Surabaya 60111, Indonesia

Corresponding author: seminar_pratapa@physics.its.ac.id

Abstract. The effect of polyvinyl alcohol (PVA) addition on the formation of zircon ($ZrSiO_4$) powders was studied. The powders were prepared with a stoichiometric Zr/Si (1:1 mol%) mixture of zirconium oxychloride aqueous solution ($ZrOCl_2.H_2O$) and tetraethoxysilane (TEOS) ($Si(OC_2H_5)_4$) by a sol-gel method after PVA addition. The amount of added PVA varied between 1, 2, 3, and 4% by weight. The addition of PVA was aimed to be a catalyst which can accelerate the formation of zircon. The sol-gel process was followed by a mechanical activation via ball-milling and by calcination at 1000-1200 °C for 10 h. The phase composition of the resulted powders was determined through Rietveld analysis of X-ray diffraction (XRD) data using *Rietica* software. Results showed that amorphous zirconia transformed to a transient tetragonal zirconia phase at a calcination temperature of 1100 °C for all added PVA compositions, then reacted with silica to form zircon at 1200 °C. The 1, 2, and 4 wt% PVA-added samples crystallized to form zircon with the secondary phases of *t*-zirconia, *m*-zirconia, and cristobalite. The single phase of zircon was obtained in the 3 wt% PVA added-powder calcined at 1200 °C.

1. Introduction

The thermo-mechanical properties of zircon ($ZrSiO_4$) are excellent i.e. its heat conductivity coefficient at room and high temperature is very low, its mechanical strength is good and its thermal expansion coefficient is low [1–3], which do not even alter at high temperature above 1400 °C [4,5]. Zircon has the potential to be used in refractory applications because of its durability, stability, and resistance to thermal shock. Ferry *et al.* [6] reported the Gibbs energy ΔG_r of formation of zircon from the oxides at a value of -18.44 kJ/mol which was derived from a phase-equilibrium study. The high yield of pure zircon synthesis is quite difficult to obtain because the formation energy is relatively small [7], the diffusion kinetics are slow and the kinetic activation energy required for the combination reaction between the oxide components of zirconium and silica [8]. In addition, large changes in the crystal structure of zirconium between the structure of zirconia (ZrO_2) and zircon ($ZrSiO_4$) and the very low solubility of the solids of SiO_2 in ZrO_2 increase the difficulty of obtaining such high yield synthesis [8].

Curtis and Sowman [9] found that zircon formation occurs minimum at temperature around 1320 °C. They obtained these results based on their research for zircon formation from a mixture of zirconia and tridimite, zirconia and quartz and zirconia and cristobalite. Furthermore, in our previous work [10,11], we have synthesized the stoichiometric SiO_2-ZrO_2 polymorphic combination powders by mechanical activation using ball mill followed by a calcination at the varied temperatures, as the four reactions, i.e. *a*-SiO_2 + *a*-ZrO_2, *a*-SiO_2 + *t*-ZrO_2, *c*-SiO_2 + *a*-ZrO_2, *c*-SiO_2 + *t*-ZrO_2 (with *a* amorphous, *t* tetragonal, and *c* cristobalite). The results revealed that the mixture of amorphous oxides of silica and zirconia (SiO_2-ZrO_2) would be crystallized to form zircon at a lower calcination temperature if the fraction of one component in the composition was greater than the other components, c.a. 82.9 mol%. However, the method based on the solid-state reaction of a mixture of amorphous powder of oxide zirconium and silica (ZrO_2-SiO_2) reveals that the crystallization that occurs in the powder is incomplete with the heat treatment.

Several studies to obtain pure zircon powder have been attempted through micro-emulsion process, reverse micelle process, chemical reaction like aerosols, sol-gel routes and hydrothermal. Seeding of the oxide precursor powders with zircon powder to initiate zircon phase formation from the reaction between precursors has been reported [4,12,13]. Very high results have been achieved when using such seeding, which is carried out through the synthesis of sol-gel and hydrothermal methods. On the other hand, efforts to produce zircon previously involving only the sintering process of zirconia and silica powder require very long calcination time and relatively low yield (<50%) [8,13].

The addition of polyvinyl alcohol (PVA) was used in several works to synthesize some oxides by a sol-gel method such as $LaFeO_3$ [14], $BiFeO_3$ [15], and $LiMn_2O_4$ [16] etc. The function of PVA is as a binder in solid-state reactions of magnesium and silica [17,18]. In our previous works [10,11], we also used the addition of PVA with a fixed amount in the SiO_2-ZrO_2 mixture with solid-state reaction method. There is no literature that reports the work in examining the effect of the amount of PVA on the zircon formation. In the present paper, the synthesis of zircon powder with sol-gel method followed by the addition of PVA through mechanical activation using ball mills is reported. The effect of calcination temperature and PVA content on zircon formation were investigated.

2. Experimental

Zircon powders were synthesized from zirconium oxychloride aqueous solution ($ZrOCl_2.H_2O$) and tetraethoxysilane (TEOS) ($Si(OC_2H_5)_4$) in a Zr/Si stoichiometric ratio. The precursors were mixed under stirring at a temperature of 100 °C for 24 hours. Then an amount of 3 M ammonia solution was added into the mixture until pH 11 and form gel, then followed by washing and drying to obtain the powder. A variation content of polyvinyl alcohol (PVA), i.e., 1, 2, 3, and 4 wt%, was added to the powders followed by the mechanical activation using ball milling. The resulting powder is then calcined for three temperature variations of 1100, 1200, and 1300 ° C for 10 hours. The Diffractometer of Philips XPert Powder (CuKα λ = 1.54060 Å and step size 0.0170 °) was used to obtain phases from X-ray diffraction data (XRD) for all samples. Analysis of the XRD patterns is done by Match! 2 for the phases identification and the accurate powder phase composition can be obtained by using Rietica [19] and Rietveld refinement.

3. Results and discussion

Figure 1 presents the X-ray diffraction patterns of the PVA-added powders prepared by the sol-gel method followed by mechanical activation and calcination at 1100, 1200, and 1300 °C. All calcined powder shows phases which are a combination of zircon (PDF No. 9-582), cristobalite (SiO_2) (COD No. 1-010-938), tetragonal-zirconia (COD No. 1-525-706) and monoclinic-zirconia (PDF No. 89-426). There is no zircon phase identified for all PVA-added powders calcined at 1100 °C, only the tetragonal zirconia is identified. The zircon phase is identified for all PVA-added powders calcined at 1200 °C but accompanied by the presence of the other phases from oxide components, i.e., SiO_2 (cristobalite) and ZrO_2 (tetragonal and monoclinic), except for 3 wt. % PVA-added powders. The addition of 3 wt.% PVA presents only the zircon phase without the presence of the other phases. Itoh [20] has explained the mechanism of amorphous zirconia and amorphous silica powder in forming zircon. Zircon formation is occurs as a result of the reaction between tetragonal zirconia and amorphous silica. At the calcination temperature of 1100 °C, the thermal energy is not sufficient for zircon formation but rather used for the crystallization process of tetragonal zirconia. The onset of zircon formation is detected at 1200 °C. The remaining amorphous silica will change to cristobalite silica. During the cooling process, unreacted zirconia will transform from tetragonal to monoclinic [20].

In a previous study, Du *et al.* [21] worked on the synthesis of zircon powder by yttria and zircon seed addition without the addition of PVA. The almost complete reaction was found when the powder was calcined at 1500 °C. In our work, the pure zircon can be achieved at a lower calcination temperature of 1200 °C by adding 3 wt% of PVA. Therefore, the addition of PVA is effective to lower the temperature of zircon formation. However, zircon can dissociate into silica and zirconia at temperatures more than 1500 °C (determined by the condition of the system) [3,9,22,23] by following the reaction:

$$ZrSiO_4 \rightarrow ZrO_2 + SiO_2 \quad (1)$$

IOP Conf. Series: Materials Science and Engineering **496** (2019) 012034 doi:10.1088/1757-899X/496/1/012034

Table 1. Results of quantitative analysis of X-ray diffraction patterns of PVA-added powders calcined at varied temperatures using *Rietica* [19].

Samples		Zircon	*t*-Zirconia	*m*-Zirconia	Cristobalite
1 wt%	1100	-	100	-	-
	1200	16.3 ± 1	1.5 ± 0	77.3 ± 2	4.9 ± 1
	1300	95.3 ± 1	0.1 ± 0	3.0 ± 4	1.6 ± 5
2 wt%	1100	-	100	-	-
	1200	11.1 ± 2	2.7 ± 1	71.6 ± 3	14.5 ± 3
	1300	86.3 ± 8	0.1 ± 0	4.5 ± 3	9.1 ± 5
3 wt%	1100	-	100	-	-
	1200	100	-	-	-
	1300	87.6 ± 2	0.1 ± 0	2.5 ± 7	9.7 ± 9
4 wt%	1100	-	100	-	-
	1200	9.3 ± 2	2.2 ± 1	80.7 ± 3	7.8 ± 2
	1300	91.1 ± 9	0.1 ± 0	0.3 ± 4	8.6 ± 5

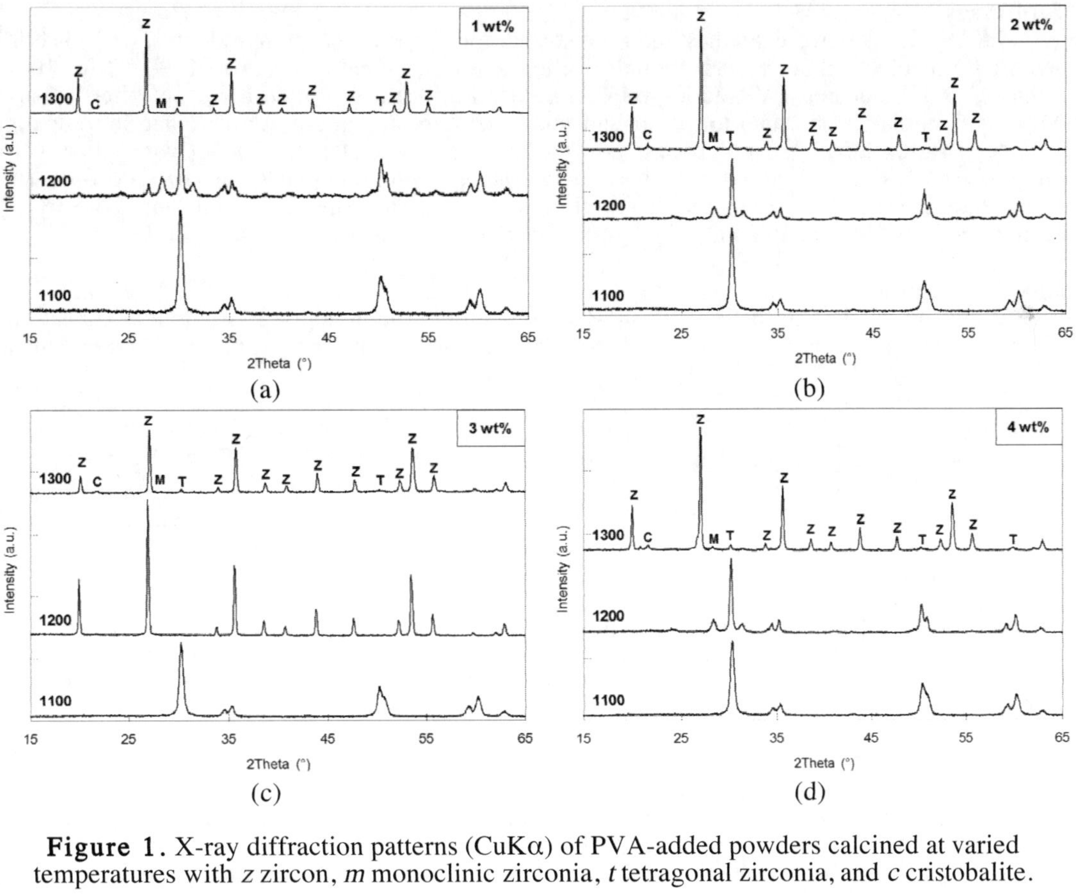

Figure 1. X-ray diffraction patterns (CuKα) of PVA-added powders calcined at varied temperatures with *z* zircon, *m* monoclinic zirconia, *t* tetragonal zirconia, and *c* cristobalite.

The silica can be present in a glassy phase that is not detected by XRD. The glassy phase of silica can be crystallized into quartz or cristobalite phases, both of which can be observed with XRD [2]. Zirconia, can also be identified by XRD. At room temperature, zirconia is generally stable in monoclinic phase (*m*-ZrO_2). However, if the size of the particles is small enough, zirconia may present in tetragonal phase (*t*-ZrO_2) that has higher energy state, at room temperature [2]. In this work, with the 3 wt.% PVA-added powder that calcined at 1300 °C, there is a dissociation of zircon which produces silica in a cristobalite phase and zirconia in tetragonal and monoclinic phases, in addition to zircon crystal formation. The zircon dissociation temperature is obtained lower than 1500 °C with the addition of PVA.

Table 1 presents the relative mole fraction of each phase as a result of quantitative analysis of X-ray diffraction patterns based on the Rietveld method for all samples. For each addition of PVA, zircon content shows an increase with increasing temperature during the calcination process, i.e., up to 95.3 mol%, except for the 3 wt.% PVA addition. For this PVA addition, the reaction straightly completed to yield a single-phase of zircon at 1200 °C, but calcination at 1300 °C makes zircon dissociate into silica and zirconia so that the final zircon content is only 84.6 wt.% mol.

The main function of PVA is to provide polymer networks that can inhibit cation mobility so that local stoichiometry allows to be maintained and it also helps minimizing precipitation of unexpected phases [24]. PVA may also act as a binder [25] so that the wet ability between the powders can be increased during sintering [10–11]. It is therefore argued that these roles of PVA which have enhanced the reaction between raw powders and hence produces a higher zircon content. The PVA content that can optimize the reaction is 3 wt.%. However, the higher PVA content produces smaller zircon content. It is estimated that increasing the PVA addition cause the raw material particles to become entrapped in entangled network from organic polymers so that it will hinder the reaction.

4. Summary

Zircon ($ZrSiO_4$) powders were synthesized by a sol-gel method with polyvinyl alcohol (PVA) addition followed by a mechanical activation via ball-milling and by calcination at 1100-1300 °C for 10 h. In the 1100 °C of all calcined PVA-added powders, only *t*-zirconia was identified, but zircon was not. Amorphous zirconia transformed to a transient tetragonal zirconia phase at a calcination temperature of 1100 °C for all added PVA compositions. Tetragonal zirconia started to react with silica to form zircon at 1200 °C. The 1, 2, and 4 wt% PVA-added samples in particular crystallized to form zircon with the secondary phases, i.e. *t*-zirconia, *m*-zirconia, and cristobalite. The addition of 3 wt% PVA was able to optimize the reaction until the single phase of zircon was formed at 1200 °C.

Acknowledgements
The authors would like to thank Ministry of Research, Technology and Higher Education of The Republic of Indonesia and LPPM ITS to support finance through program PMDSU Financial Year 2017 No. 491/PKS/ ITS/2017.

References

[1] Rendtorff N M, Grasso S, Hu C, Suarez G, Aglietti E F and Sakka Y 2012 *Ceram. Int.* **38** 1793–99
[2] Suárez G, Acevedo S, Rendtorff N M, Garrido L B and Aglietti E F 2015 *Ceram. Int.* **41** 1015–21
[3] Rendtorff N M, Grasso S, Hu C, Suarez G, Aglietti E F and Sakka Y 2012 *J. Eur. Ceram. Soc.* **32** 787–93
[4] Shi Y, Huang X X and Yan D S 1994 *J. Eur. Ceram. Soc.* **13** 113–9
[5] Mori T, Hoshino H, Yamamura H, Kobayashi H and Mitamura T 1990 *J. Ceram. Soc. Jpn.* **98** 1017–22
[6] Ferry J M, Newton R C and Manning C E 2002 *Am. Miner.* **87** 1342–50
[7] Ellison A J G and Navrotsky A 1992 *J. Am. Ceram. Soc.* **75** 1430–3
[8] Kanno Y 1989 *J. Mater. Sci.* **24** 2415–20
[9] Curtis C E and Sowman H G 1953 *J. Am. Ceram. Soc.* **36** 190–8
[10] Musyarofah, Nurlaila R, Muawwaqor N F, Saukani M, Kuswoyo A, Triwikantoro and Pratapa S 2017 *J. Phys.: Conf. Ser.* **817** 012033
[11] Nurlaila R, Musyarofah, Muawwaqor N F, Triwikantoro, Kuswoyo A and Pratapa S 2017 *AIP Conf. Proc.* **1788** 030122
[12] Shi Y, Huang X X and Yan D S 1994 *Mater. Lett.* **21** 79–83
[13] Mori T, Yamamura H, Kobayashi H and Mitamura 1993 *J. Mater. Sci.* **28** 4970–3
[14] Feng J, Liu T, Xu Y, Zhao J and He Y 2011 *Ceram. Int.* **37** 1203–7

[15] Liu T, Xu Y and Zhao J 2010 *J. Am. Cerm. Soc.* **93** 3637–41
[16] Lu C H and Saha S K 2001 *J. Solgel Sci. Technol.* **20** 27–34
[17] Dewa E, Musyarofah, Nurbaiti U, Triwikantoro, Firdaus S and Pratapa S 2015 *Adv. Mater. Res.* **1112** 262–5
[18] Aini F N, Musyarofah, Triwikantoro, Mashuri, Firdaus S and Pratapa S 2015 *Adv. Mater. Res.* **1112** 385–8
[19] Hunter B A 2000 *Rietica - a visual Rietveld program* (Menai: Australian Nuclear Science and Technology Organization)
[20] Itoh T 1992 *J. Cryst. Growth* **125** 223–8
[21] Du C, Yuan Q and Yang Z 1999 *J. Mater. Sci. Lett.* **18** 965–6
[22] Kaiser A, Lobert M and Telle R 2008 *J. Eur. Ceram. Soc.* **28** 2199–211
[23] Rendtorff N M, Garrido L B and Aglietti E F 2010 *Ceram. Int.* **36** 781–8
[24] Gulgun M A, Nguyen M H and Kriven W M 1999 *J. Am. Ceram. Soc.* **82** 556–60
[25] Pratapa S, Handoko W D, Nurbaiti U and Mashuri 2017 *Ceram. Int.* **43** 7172–6

Effects of Ag doping on the microstructural, morphological, and magnetic properties of a perovskite material, $(La_{1-x}Ag_x)_{0.8}Ca_{0.2}MnO_3$

S D Rosanti and B Kurniawan

Department of Physics, Faculty of Mathematics and Natural Sciences (FMIPA) Universitas Indonesia, Depok 16424, Indonesia

Corresponding author: bkuru@fisika.ui.ac.id

Abstract. In this study, perovskite $(La_{1-x}Ag_x)_{0.8}Ca_{0.2}MnO_3$ material was synthesized using the sol-gel method with different Ag doping levels (x = 0.03 and 0.1). The precursors, La_2O_3, $AgNO_3$, $Ca(NO_3)_2.4H_2O$, and $Mn(NO_3)_2.4H_2O$ were combined stoichiometrically and sintered at 900°C for 24 h. Energy dispersive X-ray spectroscopy was used to confirm the purity of each sample. Rietveld refinement of the X-ray diffraction pattern showed that the samples had a single phase and orthorhombic structure with the Pnma space group. Scanning electron microscopy revealed density, porosity, and granular morphology that affected the characteristics of the sample. The magnetic hysteresis curves indicate that the material was paramagnetic and material with x = 0.1 has a higher magnetic susceptibility than with x = 0.03.

Keywords: perovskite, Ag-doping, sintering, sol-gel

1. Introduction

$LnMnO_3$ is one of fascinating material to study. This material is an antiferromagnetic (AFM) insulator with Neel temperature (T_N, the temperature above which the material becomes paramagnetic) around 140 K. The Ln site can be substituted with alkali element or alkali earth element (*A*) to form $Ln_{1-x}A_xMnO_3$ (where Ln is a trivalent rare earth element and A is alkaline elements or alkaline earth elements which shows unique physical properties around the phase transition temperatures (T_C) of paramagnetic (PM)–ferromagnetic (FM), including colossal magnetoresistance and magnetocaloric effects [1–4].

LCMO perovskite materials can be applied in the fields of, magnetic cooling, sensors, medicine, biology, medicine, catalysts and memory devices [5]. Changes its chemical composition and applying several treatments to the manganite perovskite material can modify the microstructure so that the nature of its physical characteristics of this material can be improved [6-7].

In this study, perovskite $(La_{1-x}Ag_x)_{0.8}Ca_{0.2}MnO_3$ material were synthesized by the sol-gel processes. Ag doping effects on structural, morphological, and magnetic properties of materials will be studied.

2. Experimental

$(La_{1-x}Ag_x)_{0.8}Ca_{0.2}MnO_3$ materials were prepared by the sol-gel method for x = 0.03 and 0.1. Precursors, La_2O_3, $AgNO_3$, $Ca(NO_3)_2\cdot 4H_2O$, and $Mn(NO_3)_2\cdot 4H_2O$ dissolved with nitric acid. Citric acid ($C_6H_8O_7\cdot H_2O$) was added in a molar ratio of 1:1.2 for citric acid to metal ions in total. Adjustment of the pH solution until pH 7 is achieved by adding ammonia solution. Then the magnetic stirrer with a speed of 400 rpm was used to homogenize the mixture which was carried out at temperature of 80–90 °C.

IOP Conf. Series: Materials Science and Engineering **496** (2019) 012035 doi:10.1088/1757-899X/496/1/012035

Table 1. Lattice parameters and unit cell volumes of $(La_{1-x}Ag_x)_{0.8}Ca_{0.2}MnO_3$ (x = 0.03 and 0.1)

Samples	a (Å)	b (Å)	c (Å)	Volume $(Å)^3$
$(La_{0.77}Ag_{0.03})_{0.8}Ca_{0.2}MnO_3$	5.47081	7.73883	5.50146	232.919
$(La_{0.7}Ag_{0.1})_{0.8}Ca_{0.2}MnO_3$	5.50063	7.74575	5.47364	233.2125

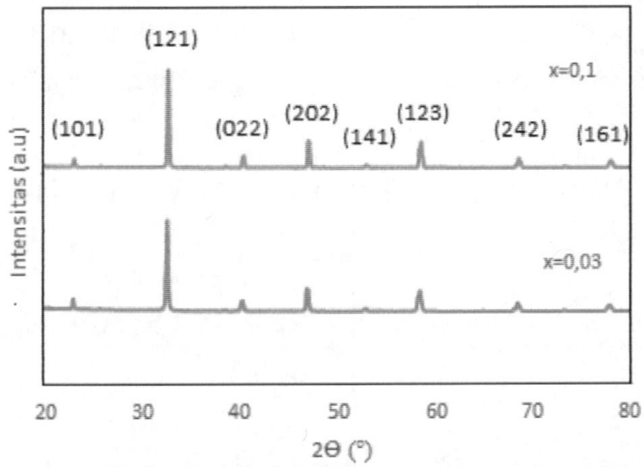

Figure 1. XRD pattern of $(La_{1-x}Ag_x)_{0.8}Ca_{0.2}MnO_3$ (x = 0.03 and 0.1)

(a) 

(b) 

Figure 2. SEM image of $(La_{1-x}Ag_x)_{0.8}Ca_{0.2}MnO_3$ for (a) x = 0.03 and (b) x= 0.1

The resulting gel is then dried by heating it for 3 h at a temperature of 120 °C so that the gel becomes high viscosity. The decomposition of organic precursor was carried out on the high viscosity gas by calcining it for 5 h at 500 °C. Then proceed with sintering process for 24 h at 900 °C. Characterisation of the sample with X-ray diffraction (XRD) was done to identify phases and analyse sample structures. Scanning electron microscopy (SEM) was carried out to observe the morphology of the sample. Determination of the elemental composition contained in the sample was carried out by Energy Dispersive X-ray (EDX) spectroscopy. Finally, vibrating sample magnetometer (VSM) was used to obtain the M-H hysteresis curve of the samples.

3. Results and discussion

Refined results from the XRD pattern of $(La_{0.77}Ag_{0.03})_{0.8}Ca_{0.2}MnO_3$ and $(La_{0.7}Ag_{0.1})_{0.8}Ca_{0.2}MnO_3$ material that indicate characteristic peaks of the sample compound, are shown in figure 1. At room temperature, the system shows the perovskite structure belonging to the Pnma space group in the orthorhombic crystal system.

IOP Conf. Series: Materials Science and Engineering **496** (2019) 012035 doi:10.1088/1757-899X/496/1/012035

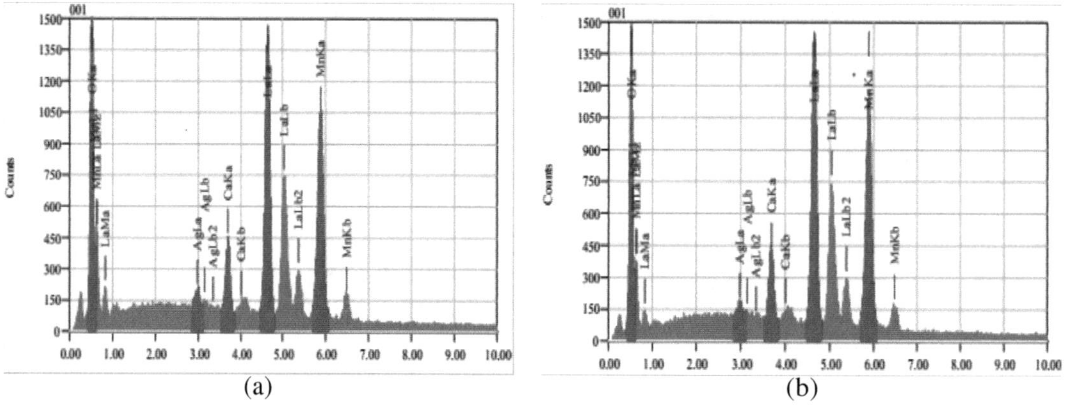

Figure 3. EDX image of $(La_{1-x}Ag_x)_{0.8}Ca_{0.2}MnO_3$ for (a) x = 0.03 and (b) x= 0.1

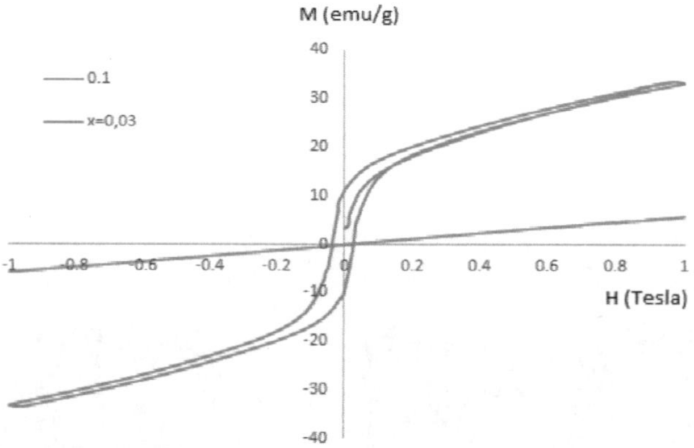

Figure 4. M-H loop showing the hysteresis of $(La_{1-x}Ag_x)_{0.8}Ca_{0.2}MnO_3$ for x = 0.03 and x=0.1

The unit cell volume and the lattice parameters of $(La_{0.77}Ag_{0.03})_{0.8}Ca_{0.2}MnO_3$ material are shown in table 1. The values are smaller than $(La_{0.7}Ag_{0.1})_{0.8}Ca_{0.2}MnO_3$ material. This is because the radius of the Ag^+ ion (1.28 Å) as a dopant is greater than La^{3+} ion (1.22 Å).

Figure 2 shows the SEM patterns of $(La_{1-x}Ag_x)_{0.8}Ca_{0.2}MnO_3$ when x = 0.03 and 0.1. Grain size increases with higher Ag doping levels. The $(La_{0.77}Ag_{0.03})_{0.8}Ca_{0.2}MnO_3$ sample was found to have agglomerated particles. This is because of the increase in magnetic moments due to suppression of antiferromagnetic ordering in material [5].

Figure 3 shows the chemical composition and semi-quantitative EDX measurements results. There were no impurities present in both samples (x = 0.03 and 0.1). EDX also shows that there was no quantifiable loss of any elements.

The VSM characterization results of $(La_{1-x}Ag_x)_{0.8}Ca_{0.2}MnO_3$ materials are shown in figure 4. The results show that the sample with x = 0.03 is paramagnetic because there was no Hc value. However, sample with x = 0.1 was found as soft magnetic with a magnetization value of 9 emu/g. The susceptibility of $(La_{1-x}Ag_x)_{0.8}Ca_{0.2}MnO_3$ when x = 0.1 is greater than x = 0.03 at room temperature. The hysteresis curves indicate that both samples are paramagnetic and that Ag doping with x = 0.1 produces a higher magnetic susceptibility than twith x = 0.03.

4. Summary

In this study, $(La_{1-x}Ag_x)_{0.8}Ca_{0.2}MnO_3$ materials were synthesized by a sol-gel process with two different Ag doping levels (x = 0.03 and 0.1). The microstructure of the obtained perovskite materials has more favorable characteristics on doping with x=0.1. The results of SEM and EDX indicate that the characteristics of the sample were likely derived from density, porosity, and granular morphology. Moreover, grain sizes increases with higher Ag-doping levels.

Acknowledgements
The author would like to greatly appreciate for financial funding from Universitas Indonesia through PITTA (Publikasi Internasional Terindeks untuk Tugas Akhir Mahasiswa Universitas Indonesia) Grant 2017 with contract number of 634/UN2.R3.1/HKP.05.00/2017 and Einago (www.einago.com) for the English language review.

References
[1] Thanh T D, Linh D C, Manh T V, Phan T L and Yu S C 2016 *IEEE Trans. Magn.* **52** 2501004
[2] Riahi K, Messaoui I, Cheikhrouhou-Koubaa W, Mercone S, Leridon B, Koubaa M and Cheikhrouhou A 2016 *J. Alloys Compd.* **688** 1028–38
[3] Ye S L, Song W H, Dai J M, Wang K Y, Wang S G, Zhang C L, Du J J, Sun Y P, Fang J 2002 *J. Magn. Magn. Mater.* **248** 26–33
[4] Debnath J C and Strydom A M 2015 *J. Alloys Compd.* **621** 7–11
[5] Xia W, Li L, Wu H, Xue P and Zhu X 2017 *Ceram. Int.* **43** 3274–83
[6] Pękała M and Drozd V 2008 *J. Non-Cryst. Solids* **354** 5308–14
[7] Pekała M, Drozd V, Fagnard J F and Vanderbemden P 2010 *J. Alloys Compd.* **507** 350–5

A UV dosimeter label made from films of polyvinyl alcohol and *Swietenia mahagoni* bark extract

Y Priasetyono, P Widodo, M Harahap, A Listiarini, D Djuhana and C Imawan

Department of Physics, Faculty of Mathematics and Natural Sciences (FMIPA) Universitas Indonesia, Depok 16424, Indonesia

Corresponding author: cuk.imawan@sci.ui.ac.id

Abstract. In daily life, we are all exposed to ultraviolet (UV) radiation, which, in excessive doses, is very harmful to skin health. To measure UV intensity in everyday life, UV labels must be developed to determine the UV dose received in a simple and practical way. This paper reports on the development of a UV dosimeter label that is fabricated from polyvinyl alcohol (PVA) film and natural dye extract. The film labels were prepared by mixing extracts of *Swietenia mahagoni* and PVA with dye to 30% concentration. The film was characterized by using Fourier transform infrared spectrometry, UV-visible spectrometry, and colorimetry. The response of film labels to UV exposure was studied by exposing them to UV radiation at intensities of 0, 2, 3, 4, and 5 kJ/m^2. The effects of time and storage conditions on the color stability of the film labels were evaluated for seven days. After exposure to UV radiation, the film labels change color from brown before exposure to increasingly dark brown with increasing UV intensity. The results of this experiment show that film labels made of PVA and *Swietenia mahagoni* extract provide a sensitive measure of the intensity of UV radiation.

Keywords: ultraviolet, label, *Swietenia mahagoni*, polyvinyl alcohol

1. Introduction

Ultraviolet (UV) radiation is nonionizing electromagnetic radiation in the wavelength range from 200 to 400 nm [1]. Depending on the dose, UV radiation may have negative or positive effects on health. Exposure to UV at specific doses is very beneficial for the synthesis of vitamin D and for healing rickets, eczema [2], and psoriaris [3]. However, excessive exposure is extremely dangerous, especially for the eyes and skin, and can result in sunburn, tanning, photokeratitis, photoconjunctivitus, photoaging, clouding of the lens, cataracts, and/or skin cancer [2,4]. Thus, it is important to known the UV dose received by the body, which is the motivation for developing a personal UV dosimeter.

Personal dosimeters play an important role in determining UV dose and typically take the form of small film labels. Several material dosimeters have been reported for detecting UV dose; these are divided into two groups; namely, dosimeters with and without coloring agents. Dosimeters without coloring agents include phenothiazine film [5], biological dosimeters (uracil molecules, bacteria, DNA) [6], and polyphenylene oxide [7]. Several other researchers have reported the use of coloring agents to indicate UV radiation [3,4,8]. Coloring agents can be obtained from synthetic and natural dyes, and natural coloring agents are environmentally friendly because they are much more biodegradable than synthetic coloring agents [9]. However, few studies have focused on the use of coloring agents from natural dyes. In one such work, Bhatti *et al.* [10] reported the use of Eucalyptus bark as a UV-sensitive dye in a dosimetry label. In their study, one of the coloring agents was immobilized on the polymer polyvinyl alcohol (PVA).

PVA is often used as a dosimeter label matrix [11] and is an environmentally friendly synthetic polymer. To prepare film labels, the properties of PVA must be considered: its solubility in water, mechanical strength, biodegradability, and thermal stability [12].

This paper reports the development of a UV dosimeter label made from PVA and *Swietenia mahagoni* (SM) bark extract. SM is an Indonesian tree species whose extract is used to color batik [13]. The sensitivity of the proposed UV dosimeter label to UV radiation is tested by using colorimetry.

2. Experimental

2.1. Materials
The SM bark used as coloring agent in this research was obtained from a local market. The absolute ethanol solution used to extract the SM bark was purchased from J. T. Baker. The PVA used as matrix was supplied by Merck.

2.2. Extraction of Swietenia mahagoni bark
The extraction of SM bark was done in several steps: First, to reduce the water content, the SM bark was heated to 40 °C for three days in an oven. After drying, the bark was cut into small pieces and weighed at 40 g, mixed with 150 mL of ethanol, and left for one day at room temperature. The extract solution thus produced had pH of 5.9. To remove residue, the solution was filtered twice by using Whatman no. 1 paper. Finally, the solution was stored at 5 °C in the dark.

2.3. Preparation of labels
The film labels were made from a coloring agent extract and the PVA matrix. The PVA matrix was prepared by diluting 20 g of PVA powder into 1000 mL of Aquades 2% (v/v) that was then stirred at 300 rpm at 55 °C until a homogeneous solution was reached. The PVA solution was then mixed with 30% extract of SM bark, stirred for 10 minutes, and then 50 mL of the solution was poured into a mold on a 15 cm × 15 cm acrylic board to create the film. The PVA-SM films were then dried at 20 °C for 48 hours. After drying, the labels were stored in the dark at ambient temperature.

2.4. Test of stability of label colour
The stability of the label color was tested under dark and bright conditions. The label stored in dark conditions was enveloped in UV-resistant material, and the label stored in bright conditions was illuminated by a lamp with very low UV emission. Both samples were placed in a transparent room temperature chamber with a relative humidity of 53%.

2.5. UV irradiation
All samples were irradiated by using an ATLAS UV2000 (Chicago, USA). The UV chamber contained four UV-A lamps, each emitting 40 W at 340 nm. The UV-A intensity on the label was 0.77 W/m^2. The sample was then cut into 2 cm × 3 cm labels, placed in a special holder, and then put into the UV chamber. The labels were exposed to UV radiation at doses of 2, 3, 4, and 5 kJ/m^2 at 30 °C.

2.6. Characterization of labels
The pH of the PVA, SM extract, and PVA-SM was recorded by using a pH meter (Horiba Sciencetific, Japan). The label color before and after UV exposure was measured by using a Nix Pro colorimeter (Nix sensor Ltd, Ontario, Canada). The label absorbance was measured by using a Spectral UV-vis (Thermo Genesys 10S) over the wavelength range 200–800 nm with 1 nm resolution. The functional groups of the labels were obtained by using Fourier transform infrared spectroscopy (FTIR) with a Nicolet iS5 spectra (Thermo Scientific) at wavenumbers 500–4000 cm^{-1}.

3. Results and discussion

3.1. Effects of UV radiation on label colour
Extraction of SM bark yields a brownish-red solution, which is consistent with the experience of Fauziyah *et al.* [14]. The bark extract mixed with the PVA was light brown in color and produced brown film labels with a total RGB of 165.44, as shown in figure 1a.

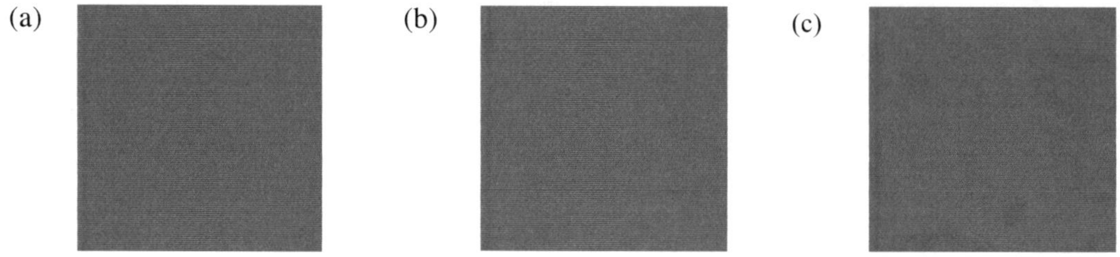

(a)	(b)	(c)
RGB Total= 165.44	RGB Total = 156.05	RGB Total = 153.59
R = 213.2	R = 208.2	R = 205.2
G = 157.8	G = 147.2	G = 145.4
B = 79.6	B = 65.4	B = 60.4

Figure 1. Label colors (a) before UV exposure, (b) after UV 3 kJ/m^2 exposure, and (c) after 5 kJ/m^2 exposure.

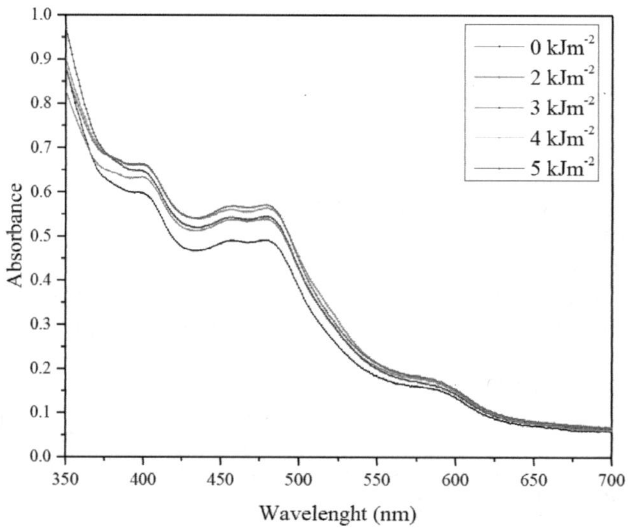

Figure 2. Absorbance spectra of UV dosimeter film before and after UV exposure.

Upon UV exposure, the PVA-SM label changed color as shown in figure 1b and figure 1c. The label changed from brown to darker brown upon exposure to 5 kJ/m^2 of UV radiation. The optical properties of the label were measured by using the NIX pro UV-Vis colorimeter. The decreased total RGB reveals a color change to a darker shade of brown. The decrease in RGB occurs in the blue (B) and green (G), which decrease significantly with increasing UV radiation. The largest decrease in RGB occurs in B (19.2 points), with G decreasing by 12.4 points, and red (R) decreasing by 8 points, as observed from RGB control.

3.2. UV-Vis spectra of label

The absorbance of the labels before and after UV exposure was measured by using the UV-vis spectrometer over the wavelength range 200–800 nm. Figure 2 shows the increase in the label's absorbance after UV exposure. The maximum absorbance before and after radiation exposure has four peaks: 399, 455, 480, and 579 nm. Although the magnitude of the first peak is the largest, it is in the UV, so it may be neglected. The second and third peaks are close in wavelength; it is these peaks that provide the dominant contributions to the color of the label. At 455 and 480 nm, the label reflects the complimentary colors of yellow and orange, and the mix of these colors give the complimentary color seen on the label, which is brown. The biggest change in absorbance (0.005) occurs between the

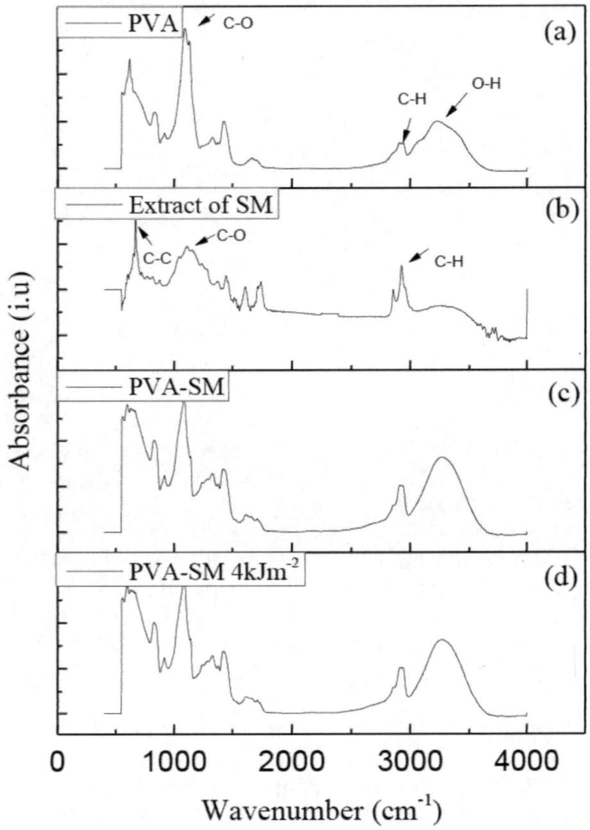

Figure 3. FTIR spectra of (a) PVA film (b) SM extract (c) PVA-SM film before UV exposure (d) PVA-SM film after 4 kJ/m2 UV exposure.

second and third peaks. Finally, the fourth peak has the lowest absorbance. The wavelength λ_{max} of maximum absorbance at each peak stays constant despite the increase in UV dose. The absorbance at 480 nm is greater than that at 455 and 579 nm. Thus, the 480 nm absorbance peak is suitable for calibrating the label.

3.3. FTIR spectra of labels

Figure 3 shows the FTIR spectra of the PVA film, SM-bark extract, and PVA-SM film labels before UV exposure and of PVA_SM film labels after UV exposure to 4 kJ/m^2. The optical properties of PVA labels were acquired at 1096, 2908, and 3226 cm^{-1}. C–O, C–H, and O–H stretching vibrations appear at 1096, 2908, and 3226 cm^{-1}, respectively [15]. The SM extract has absorption bands at 665–730, 1087–1124, and 2840–3000 cm^{-1}, which correspond to C=C, C–O, and C–H stretching vibrations, respectively. The PVA-SM labels have absorption bands at 665–730, 1085, 2915, and 3285 cm-1, which correspond to C=C, C–O, C–H, and O–H stretching vibrations, respectively. The FTIR absorbance spectrum of PVA-SM film is similar to that of PVA film. The absorbance bands of the SM extract are not very apparent because the extract was not sufficiently concentrated. However, the label is brown in color, which differentiates it from the PVA film. The absorbances of the PVA-SM label before and after UV exposure has similar FTIR spectra, which indicates that UV exposure does not change the chemical structure of the labels.

3.4. Effects of UV exposure on stability of colour of film label

The stability of the color of the film label before UV exposure was tested in both dark and bright conditions (see figure 4). The color intensity of the labels was calculated at seven-day intervals by

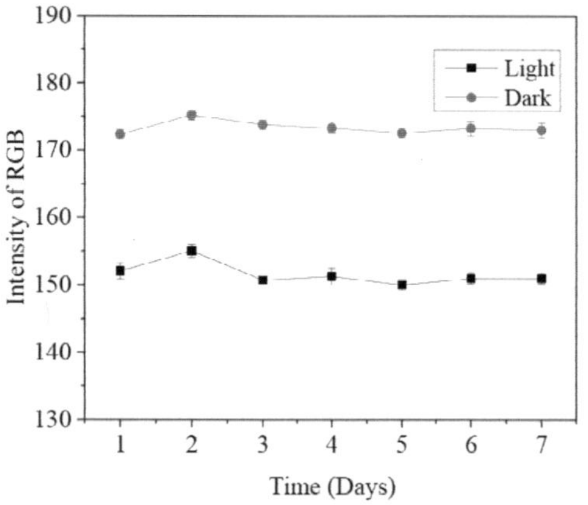

Figure 4. Stability of film labels during storage in dark and bright conditions at room temperature with 53% relative humidity.

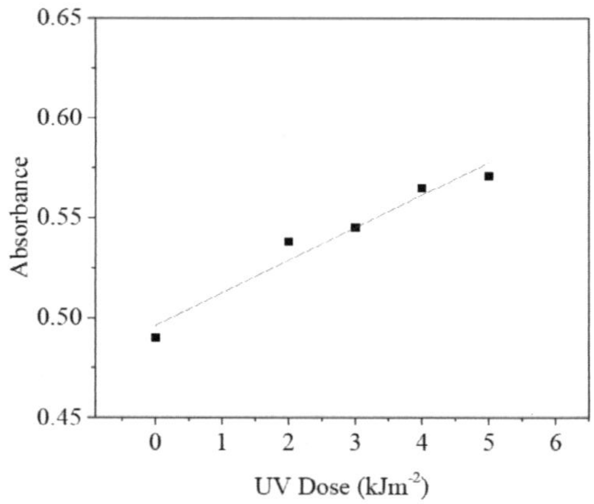

Figure 5. Absorbance at λ_{max} =480 nm of PVA-SM film label as a function of UV dose

using the colorimeter. The color of the labels remains relatively stable when stored in both dark and bright conditions. Although differences in intensities appeared because the labels had different homogeneities, storage in dark or bright clearly did not change the color of labels. In bright conditions, the labels were stable because the lights in the laboratory emit very little UV.

3.5. Calibration curve

Figure 5 shows the absorbance at λ_{max} = 480 nm as a function of UV dose for the PVA-SM film labels before and after UV exposure. This curve identifies the absorbance increase of the labels corresponding to UV exposures of 2, 3, 4, and 5 kJ/m^2. The minimum (maximum) absorbance is 0.49 (0.57) at 5 kJ/m^2. This increase can be seen by eye on the labels, which change from brown to dark brown in color. This curve is essentially linear, so it can be used to calibrate these UV dosimeter labels.

4. Conclusions

We fabricated UV dosimeter labels from a PVA matrix with *Swietenia mahagoni* bark as coloring agent. The dosimeter is sensitive to exposure to UV radiation up to 5 kJ/m^2. The color of the dosimeter label changes continuously from brown to dark brown with increasing UV radiation dose. Color gradation is demonstrated by using UV-Vis spectroscopy and a NIX Pro colorimeter to monitor the increase in absorbance and decrease in color intensity. The maximum absorbance is 480 nm and increases linearly with UV exposure up to 5 kJ/m^2. These UV dosimeter labels are also stable in both dark and bright conditions when not exposed to UV irradiation.

Acknowledgements

The authors would like to thank Einago (www.einago.com) for the English language review.

References

[1] Koutchma T N, Forney L J and Moraru C I 2009 *Ultraviolet Light in Food Technology Principles and Applications.* (New York: CRC Press)
[2] World Health Organization, World Meteorological Organization, United Nations Environment Programme & International Commission on Non-Ionizing Radiation Protection 2002 *Global Solar UV Index: A Practical Guilde* (Geneva: World Health Organization)
[3] Mills A, Grosshans P and McFarlane M 2009 *J. Photochem. Photobiol. A* **201** 136–41
[4] Kozicki M and Sasiadek E 2011 J. Radiat. Meas. **46** 510–26
[5] Parisi A V, Kimlin M G, Turnbull D J and Macaranas J 2005 *Photochem. Photobiol. Sci.* **4** 907–10
[6] Heydenreich J and Wulf H C 2005 *Photochem. Photobiol.* **81** 1138–44
[7] Turnbull D J and Schouten P W 2008 *Atmos. Chem. Phys.* **8** 2759-62

[8] Khankaew S, Mills A, Yusufu D, Wells N, Hodgen S, Boonsupthip W and Suppakul P 2017 Sens. Actuators B Chem. **238** 76–82
[9] Haque M A, Khan G M A, Razzaque S M A, Khatun K, Chakraborty A K and Alam M S 2013 *Indian J. Fibre Text.* **38** 280–4
[10] Bhatti I A, Adeel S, Rani S, Gulzar T, Naz S and Mansour H F 2012 *Research Journal of Textile and Apparel* **16** 62–7.
[11] Miranda T M R, Gonçalves A R and Amorim M T P 2001 *Polym. Int.* **50** 1068–72
[12] Gaaz T S, Sulong A B, Akhtar M N, Kadhum A A H, Mohamad A B and Al-Amiery A A 2015 *Molecules* **20** 22833–47
[13] Febriana I D, Kusuma H S, Gala S and Mahfud M 2016 *Asean Journal of Chemical Engineering* **16** 45–50.
[14] Fauziyah N and Hakim L 2015 *Journal of Indonesian Tourism and Development Studies* **3** 41–44
[15] Gohil J M, Bhattacharya A and Ray P 2006 *J. Polym. Res.* **13** 161–9

IOP Conf. Series: Materials Science and Engineering **496** (2019) 012037 doi:10.1088/1757-899X/496/1/012037

Ag modified $LaMnO_3$ nanoparticles for methylene blue degradation via photosonocatalytic activities

Y D Susanti, N Afifah and R Saleh

Department of Physics, Faculty of Mathematics and Natural Sciences (FMIPA) Universitas Indonesia, Depok 16424, Indonesia

Integrated Laboratory of Energy and Environment, Faculty of Mathematics and Natural Sciences (FMIPA) Universitas Indonesia, 16424 Depok, Indonesia

Corresponding author: rosari.saleh@gmail.com

Abstract. Catalyst incorporating Ag nanoparticles with $LaMnO_3$ nanoparticles have been successfully synthesized using the co-precipitation method. The characteristics of the materials are characterized using x-ray diffraction (XRD), UV-Visible spectrometer and thermogravimetric analysis (TGA). The results show that the Ag/$LaMnO_3$ composite show orthorhombic structure of the $LaMnO_3$ nanoparticle and the cubic structure of Ag nanoparticle. The Ag/$LaMnO_3$ composite shows better photosonocatalytic activities of methylene blue (MB) than $LaMnO_3$ nanoparticle under the combination of ultrasonic irradiation and visible light irradiation. Ag can be used as an electron capture in the photosonocatalytic process, because Ag can capture the electron transport along $LaMnO_3$ nanoparticles leading to a high recombination rate of electrons and holes.

1. Introduction

Perovskite nanomaterial with a general structure (ABO_3) has attracted attention from many researchers because of its excellent performance as catalyst materials under visible light irradiation. Unfortunately, similar to other catalysts electron and hole recombination still easily occurs in pure perovskite nanomaterial during the catalysis process. Many methods have been explored to determine an effective way to inhibit the recombination of electrons and holes further increasing the degradation of organic pollutants such as methylene blue (MB). Coupling it with semiconductor or modifying it with noble metal are some methods that have been explored by researchers to increase the degradation of organic pollutants [1–2]. Furthermore, noble metal such as silver (Ag) have been widely studied to modify some catalyst nanomaterials to hinder the electron and hole recombination [3–4].

Hu *et al.* [5] reported that Ag could be used as an agent to capture electrons so holes cannot be recombined with electrons, which can improve the photocatalytic process of $LaMnO_3$/graphene composite. Furthermore, Ag nanoparticles could be used as emerging visible-light driven photocatalysts that has been attracted many researchers.

This work, we reports the synthesis of Ag modified $LaMnO_3$ nanoparticles via the co-precipitation method. In particular, the photosonocatalytic of Ag modified $LaMnO_3$ nanoparticles was evaluated by the photosonodegradation of aqueous methylene blue as an organic pollutant. In addition, the effects of various scavengers were also evaluated in this research.

2. Experimental

2.1. Materials

All the chemical reagents were analytical grades so further purification is not carried out in their use.

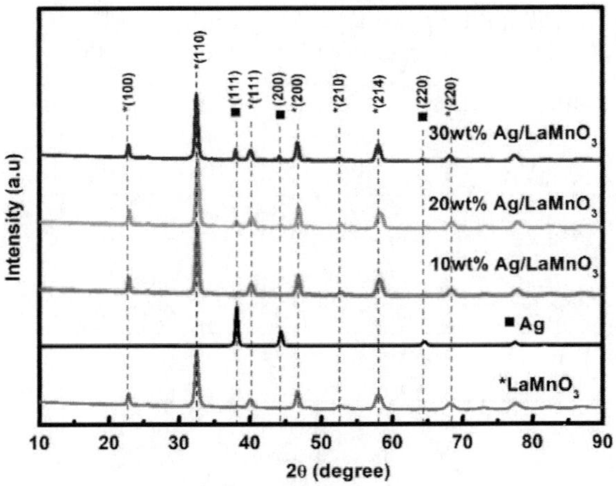

Figure 1. XRD pattern of $LaMnO_3$ and $Ag/LaMnO_3$ composite.

Lanthanum (III) chloride heptahydrate ($LaCl_3.7H_2O$, Merck, 99%), manganese (II) chloride tetrahydrate ($MnCl_2.4H_2O$, Merck, 99%), sodium hydroxide (NaOH, Merck, 99%)), silver nitrate ($AgNO_3$), sodium dodecyl sulfate (SDS), sodium carbonate (Na_2CO_3), and ethylene glycol (EG, Merck).

2.2. Preparation of $Ag/LaMnO_3$
$LaMnO_3$ nanoparticle were synthesized using the co-precipitation method as described previously [6] but without vacuum treatment in the calcination process. In contrast, Ag nanoparticles were synthesized using microwave-assisted methods [7]. Then, the $Ag/LaMnO_3$ composite was synthesized by the same method as the $LaMnO_3/Fe_3O_4$ nanocomposite [8].

2.3. Characterizations
Instruments used to analyse the samples are X-ray diffraction (XRD), thermogravimetric analysis (TGA) and UV-Visible spectrometer. XRD measurements were done by using Rigaku Miniflex 600 (Cu-Kα (λ = 1.54060) with 20 mA and 40 kV) to get the phase and crystal structure of the sample. Thermal stability of the samples is obtained from the results of TGA measurement carried out from room temperature to temperature of 1000 °C. UV-Visible spectrometry was used to investigate the surface plasmon resonance of the sample and was performed using Hitachi 5300.

2.4. Photosonocatalytic experiment
Methylene blue is an organic pollutant used in this study as a model. The photosonocatalytic of $Ag/LaMnO_3$ was investigated by observing the degradation of MB for concentrations of 20 mg/L and pH 13 at room temperature with a combination of two irradiations by visible light and ultrasonic waves simultaneously. The catalyst was dissolved in 100 mL of a MB solution and was continuously stirred using a magnetic stirrer in the dark to ensure that the adsorption-desorption process from MB molecule to the catalyst surface reaches its equilibrium. Then, the solution was irradiated under visible and ultrasonic irradiation and several solution samples were collected every 15 minutes through centrifugation. Then, it was measured using Hitachi 5300 UV-vis spectrophotometer. The solution was irradiated using a 40 W Xe lamp light as visible light irradiation and using ultrasonic bath that was operated at a fixed frequency of 40kHz and 150 W as ultrasonic irradiation.

3. Results and discussion
Figure 1 shows the XRD pattern of $LaMnO_3$ nanoparticle and Ag modified $LaMnO_3$ nanoparticle with different Ag nanoparticle weight percentages. The XRD pattern of Ag modified $LaMnO_3$ nanoparticles confirm the orthorhombic structure of $LaMnO_3$ nanoparticle and cubic phase of silver nanoparticles. Figure 1 shows that when the Ag content is less than 20wt.% the peak of Ag cannot be seen or only

IOP Conf. Series: Materials Science and Engineering **496** (2019) 012037 doi:10.1088/1757-899X/496/1/012037

Table 1. The lattice parameter and grain size <D> of $LaMnO_3$ nanoparticle and Ag/$LaMnO_3$ composite.

Samples	Lattice Parameter $LaMnO_3$			Lattice Parameter Ag	<D> $LaMnO_3$	<D> Ag
	a (Å)	b (Å)	c (Å)	a=b=c=(Å)	(nm)	(nm)
$LaMnO_3$	5.46323	5.5341	7.7499	-	41	-
10wt.% Ag/$LaMnO_3$	5.47462	5.5358	7.7527	1.4877	28	10
20wt.% Ag/$LaMnO_3$	5.51255	5.4447	7.7646	4.0827	22	13
30wt.% Ag/$LaMnO_3$	5.53014	5.4749	7.7839	4.0996	17	15

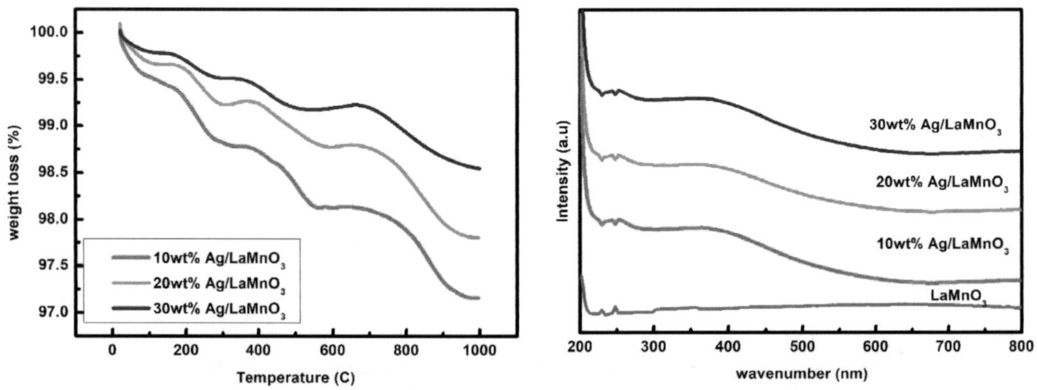

(a) (b)

Figure 2. (a) TGA and (b) SPR spectra of Ag/$LaMnO_3$ composite.

the $LaMnO_3$ phase was observed. The peaks of Ag appear sharp and intense at 38° and 44° in 20wt.% of Ag on $LaMnO_3$. When the Ag content was increase from 20 wt.% up to 30 wt.%, the peak of Ag increased. This indicates that increasing Ag content on $LaMnO_3$ nanoparticles could improve the degree of Ag crystallization. Furthermore, the crystallite size of $LaMnO_3$ nanoparticle could be decreased as increasing the Ag content on $LaMnO_3$ nanoparticles is increased. The lattice parameter and grain size obtained from the Rietveld refinement and Debye Scherrer's formula are presented in table 1. The value of lattice parameter and grain size from all nanocomposite do not significantly derivative from pure $LaMnO_3$ nanoparticle.

The thermogravimetric analyses of the sample Ag/$LaMnO_3$ composite with different wt. % of Ag are depicted in figure 2a. The overall weight loss of the sample was associated to decomposition of the organic matter. The first stage weight loss of the sample from 30 ℃ to 300 ℃ was attributed to the removal of water [9]. The second stage from 300 ℃ to 550 ℃ was attributed to the decomposition of organic compounds. Furthermore, the weight loss from 820 ℃ to 1000 ℃ was attributed to the oxidation of lanthanum manganite in the perovskite structure. Moreover, the addition of Ag on $LaMnO_3$ increased the thermal stability of the sample from room temperature to 1000 ℃ [10]. Figure 2b shows the surface plasmon resonance of the sample $LaMnO_3$ nanoparticles and Ag/$LaMnO_3$ composite. The Ag/$LaMnO_3$ show the characteristic an expected strong broad absorption in visible region around 400–500 nm after the addition of Ag nanoparticle on $LaMnO_3$ nanoparticle. Furthermore, as the weight percentage of Ag increased, the absorption of Ag/$LaMnO_3$ composite increased in the region of 400–500 nm.

The photosonocatalytic activities of $LaMnO_3$ nanoparticles and Ag modified $LaMnO_3$ nanoparticles with different Ag weight percentages are shown in figure 3a. Figure 3a shows that Ag modified $LaMnO_3$ nanoparticles have superior photosonocatalytic activity than pure $LaMnO_3$ nanoparticles. In addition, the degradation of MB was increased by adding Ag modified $LaMnO_3$ nanoparticles until 20 wt.% of Ag was obtained. This effect can be due to the following reasons. First, Ag can extend the

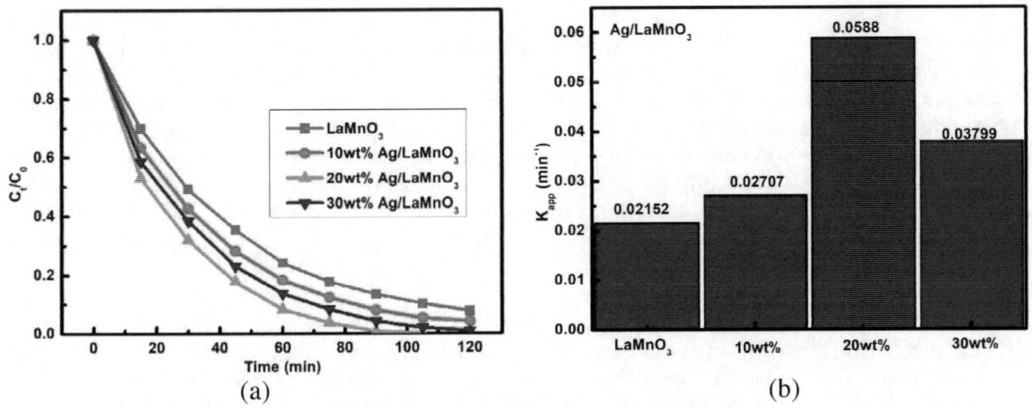

Figure 3. (a) The photosonocatalytic activity and (b) rate constant of Ag modified $LaMnO_3$ nanoparticles.

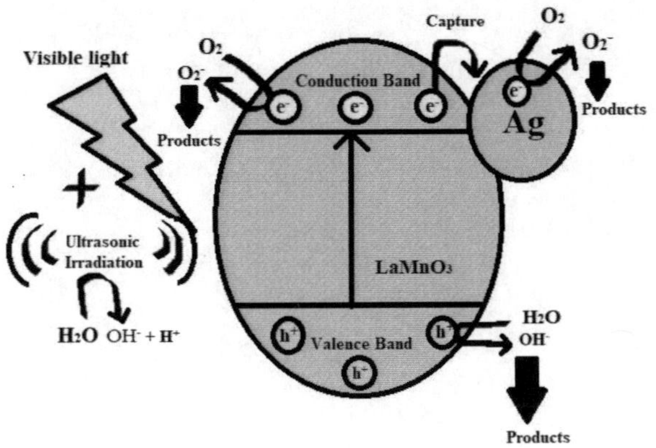

Figure 4. Proposed photosonocatalytic mechanism of MB Using $Ag/LaMnO_3$ composite.

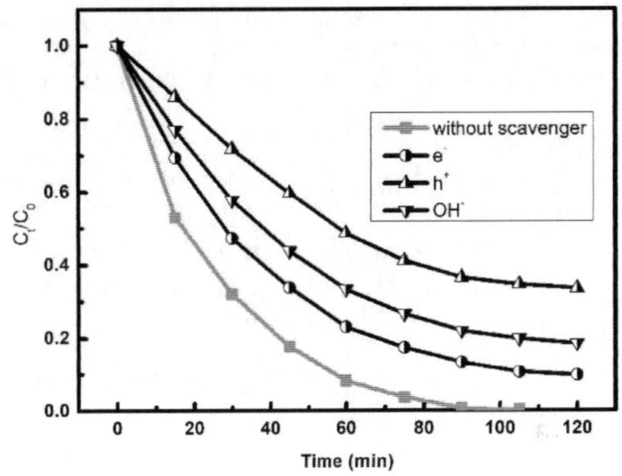

Figure 5. The photosonocatalytic of $Ag/LaMnO_3$ by the addition of several scavenger.

utilization of the light adsorption spectrum, and Ag can capture electrons that are transported along $LaMnO_3$ nanoparticles and inhibit the recombination of electron and hole pairs [5]. Moreover, the degradation of MB using Ag modified $LaMnO_3$ nanoparticles decrease by adding more Ag to $LaMnO_3$ nanoparticles. The degradation of MB was decreased at 30 wt.% of Ag, which is caused by the decreased electron generation [11]. Furthermore, the excess of Ag could hinder the incoming of visible light so it does not irradiate the $Ag/LaMnO_3$ composite. Figure 3b shows a constant rate of $LaMnO_3$ nanoparticles and $Ag/LaMnO_3$ composite. The figure also shows that the $Ag/LaMnO_3$ composite with 20 wt.% of Ag has the highest constant rate.

The proposed photosonocatalytic mechanism of MB is shown in figure 4. The $Ag/LaMnO_3$ composite was irradiated using visible light and ultrasonic irradiation. The visible light irradiation can make the electrons from valence band excite to the conduction band and leave holes in the valence band. Then, the electrons from the valence band of $LaMnO_3$ nanoparticles can react with oxygen (O_2) and form a superoxide radical (O_2). In addition, the holes in the valence band of the $LaMnO_3$ nanoparticles can react with a water molecule and form a hydroxyl radical (OH-). Both O_2- and OH- directly oxidize organic pollutants such as MB. Furthermore, ultrasonic irradiation can form microbubbles because of the cavitation effect. However, microbubbles can suddenly collapse and create hotspots and sonoluminescence. Hotspots with high temperature can make water (H_2O) split and form H· and OH. For sonoluminescence, ultrasonic irradiation could act as a light irradiation and have a wide wavelength because of the cavitation effect. Furthermore, the process of sonoluminescence follows the same role as visible light irradiation and could oxidize organic pollutants MB. Moreover, the Schottky barrier that is formed between Ag and $LaMnO_3$ nanoparticles can inhibit the electrons and holes recombination. The Ag could act as an electron capture, leading to the high recombination of electrons and holes and having superior photosonocatalytic than pure $LaMnO_3$ nanoparticles.

Figure 5 shows the photosonocatalytic of MB by adding several scavengers, electrons (e-), holes (h+) and hydroxyl radicals (OH-). Figure 5 shows that the addition of several scavenger can inhibit the degradation of MB. Furthermore, the figure shows that the hole is the most active species in the photosonocatalytic activities of the Ag/$LaMnO_3$ composite.

4. Conclusions

Ag/$LaMnO_3$ with various weight percentages of Ag has been successfully synthesized using the co-precipitation method. The XRD measurement of Ag/$LaMnO_3$ composite confirms the cubic phase of Ag and the orthorhombic phase of $LaMnO_3$ nanoparticles. The photosonocatalytic activity of MB was investigated under the combination of visible light and ultrasound irradiation. Furthermore, Tthe addition of Ag on $LaMnO_3$ could increase the degradation of MB until 20 wt.%, and it willthe degradation of MB decreases on at 30 wt.% of Ag. By adding some scavengers on in the photosonocatalytic process, it can be found that the results show that holes as are the most active species to degrade MB using Ag/$LaMnO_3$.

Acknowledgements

The authors would like to thank Einago (www.einago.com) for the English language review.

References

[1] Zhu Y Z, Sun Y Q, Niu X Y, Yuan F L and Fu H G 2010 *Catal. Lett.* **135** 152–8
[2] Cao J F, Ji Y X, Tian C B and Yi Z G 2014 *J. Alloys Compd.* **615** 243–8
[3] Zhang S, Zhang S, Song L, Wu X and Fang S 2014 *Chem. Eng. J.* **243** 24–30
[4] Lu J, Wang H, Dong S, Wang F and Dong Y 2014 *J. Alloy. Compd.* **617** 869–76.
[5] Hu J, Liu Y, Men J, Zhang L and Huang H 2016 *Solid State Sci.* **61** 239–45
[6] Afifah N and Saleh R 2016 *Mat. Sci. Forum* **864** 99–105
[7] Kristianto Y, Taufik A and Saleh R 2017 *IOP Conf. Ser.: Mater. Sci. Eng.* **202** 012061
[8] Susanti Y D, Afifah N and Saleh R 2017 *AIP Conf. Proc.* **1862** 030039
[9] Lonkar S P, Pillai V and Abdala A 2019 *Appl. Surf. Sci.* **465** 1107–1113
[10] Durham J L, Pelliccione C J, Zhang W, Poyraz A S, Lin Z, Tong X, Wang F, Takeuchi E S, Marschilok A C and Takeuchi K J 2018 *Appl. Mater. Today* **10** 142–52
[11] Li S, Zhao Y, Wang C, Li D and Gao K 2016 *Mater. Lett.* **170** 122–5

Identification of metal porphyrins in Duri crude oil

S T Kumolo[1], Y Yulizar[1], H Haerudin[2], I Kurniawaty [1,2] and D O B Apriandanu[1]

[1] Department of Chemistry, Faculty of Mathematics and Natural Sciences (FMIPA), Universitas Indonesia, Depok 16424, Indonesia

[2] Research and Development PT Pertamina (Persero), Jalan Raya Bekasi, Jakarta 13920, Indonesia

Corresponding author: yokiy@sci.ui.ac.id

Abstract. Crude oil comprises several complex hydrocarbon mixtures because of their slow natural formation process, including the toluene-soluble asphaltene component. The comparison result of Soxhlet extraction with and without silica was 55.15% and 86.75%, respectively. The Duri crude oil residue is mixed with silica gel, and Soxhlet extraction was performed with toluene. Further Soxhlet extractions and purifications obtain free porphyrin and two metal porphyrins. The Fourier Transform Infrared (FTIR) spectrum of n-heptane column chromatographic separation indicated the presence of a pyrrole ring at wavenumber 800 cm^{-1} in the fingerprint region as a porphyrin-formed group. UV-Vis absorption spectra of the n-heptane separation showed free porphyrin spectra at a maximum wavelength of 440 nm and those of metal porphyrins ranging 500–700 nm. The results of qualitative analysis of toluene fraction and n-heptane fraction showed the presence of C, Si, S, Ni, V, Fe , and Al. With this combination of spectroscopic techniques (FTIR, UV-Vis), Liquid Chromatography–Mass Spectrometry (LC-MS), and Energy Dispersive X-Ray Spectroscopy (EDX), nickel porphyrin ($C_{32}H_{36}N_4Ni$, 534 m/z) and etioporphyrinato vanadyl (IV) ($C_{32}H_{36}N_4VO$, 541 m/z) are isolated and identified from the silica gel chromatography fraction of the n-heptane Soxhlet extract of the toluene Soxhlet extract obtained from the Duri crude oil residue.

Keywords: silica gel, crude oil Duri, metalporphyrin, toluene

1. Introduction

Almost 50% of the crude oil heavy fraction can be used as renewable oil [1]. The renewable oil is obtained from heavy fraction and remediation of soil pollution that are interesting to be investigated due to the presence of asphaltenes and resins. For example, the percentage of asphaltenes and resins in oil samples reaches from Mexico and China are 46% to 80% and 33% to 53%, respectively [2].

Crude oil is produced from the natural processing of microorganisms, animals, and plants, and this occurs slowly over a period of thousands of years. The main constituents of crude oil are hydrocarbons and organic compounds that contain sulfur, nitrogen, oxygen, and organometallics. The composition of crude oil depends on the location, age, and depth of the mine. Hydrocarbons are classified into four main groups based on their polarity and solubility. This classification is known as SARA, which includes saturated (alkanes and cyclopranin), aromatics (hydrocarbons, monoatomics, diatomics, and polyatomics), resins (polar molecules with N, O, and S), and asphaltenes (large molecular weight resin with a polyatomic nucleus) [3].

It is difficult for metal complexes to be integrated into the asphaltenes fraction [4]. The amount of metal in crude oil varies, and as much as 1000 ppm of metal has been observed. Commonly found

Table 1. The percentage of toluene fraction in crude oil residue.

Toluene Fraction	Percentage
Crude oil residue with silica	55.15%
Crude oil residue without silica	86.75%

metals include lithium, sodium, calcium, potassium, strontium, silver, copper, manganese, vanadium, zinc, lead, tin, cobalt, gold, nickel, chromium and titanium [5].

The metal complexes that cause problems in the organic fraction of fossil fuels are vanadium and nickel [6]. Vanadium and nickel usually form naphthenic acids in the form of organometallic complexes such as metalloporphyrins. Some of these destructive porphyrins can be isolated from oil. In addition to their destructive effects, these porphyrins also act as important geochemical markers [7].

In this study, metal porphyrins were isolated from crude oil, and their structure and content in the oil were determined. The Duri crude oil residue was mixed with silica gel, and Soxhlet was extracted with toluene. A series of further Soxhlet extractions and purifications resulted in the isolation of the target compounds. The resulting compounds were examined by Fourier Transform Infrared (FTIR) spectroscopy, Ultraviolet–Visible (UV-Vis) spectroscopy, Liquid Chromatography–Mass Spectrometry (LC-MS), and Energy Dispersive X-Ray Spectroscopy (EDX).

2. Experimental

2.1. Materials and methods
Crude oil was obtained from Duri, Riau, Indonesia. Methanol, n-heptane, and silica gel 60 (70–230 mesh) were purchased from Merck (Germany). Toluene was obtained from Mallinckrodt, and double distilled water was used as a solvent.

2.2. Distillation of Duri crude oil
The crude oil was distilled at 370 °C. The residue with a boiling point that was greater than 370 °C was filtered as the crude oil residue, and the metal content was determined.

2.3. Soxhlet extraction of the crude oil residue using toluene
The crude oil residue was mixed with pure silica (1:1) to form a dispersion. The mixture was Soxhlet extracted using toluene, filtered, and concentrated at 111 °C under reduced pressure. Soxhlet extraction of the crude oil residue without silica was also conducted under identical conditions.

2.4. Soxhlet extraction of the toluene fraction using n-heptane
The toluene fraction was mixed with pure silica to form dispersion. The mixture was Soxhlet extracted using n-heptane, filtered, and concentrated at 98.4 °C.

2.5. Soxhlet extraction of the n-heptane residue using methanol
The n-heptane fraction was mixed with pure silica (1:1) to form a dispersion. The mixture was Soxhlet extracted using methanol, filtered, and concentrated at 64.7 °C.

2.6. Silica gel column chromatography (Flash chromatography)
The n-heptane fraction was mixed with pure silica to form dispersion. The mixture was injected onto a silica-gel-filled column to perform chromatography. The eluent composition was n-heptane and ethyl acetate (8.5:1.5). The obtained fractions were characterized by UV–Vis spectroscopy to determine the compound-containing fractions, FTIR spectroscopy to identify the functional groups present in each fraction, EDX to determine the elemental content and composition of the fractions, LC–MS to investigate the determination of the molecular formula, and structural information through fragmentation pattern analysis.

3. Results and discussion
The distillation process was conducted to separate the oil fractions from the Duri crude oil. The crude oil residue was Soxhlet extracted using toluene with and without the added silica gel. The mass percentages of the crude oil residue Soxhlet extracted using toluene are presented in table 1. The

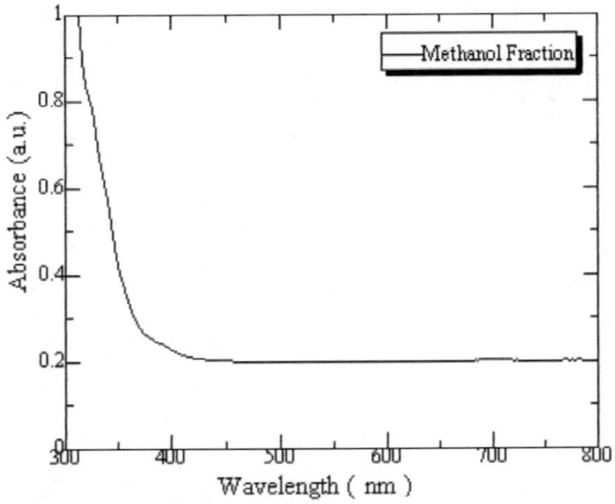

Figure 1. UV–Vis absorption spectrum of the methanol-extracted oil residue fraction.

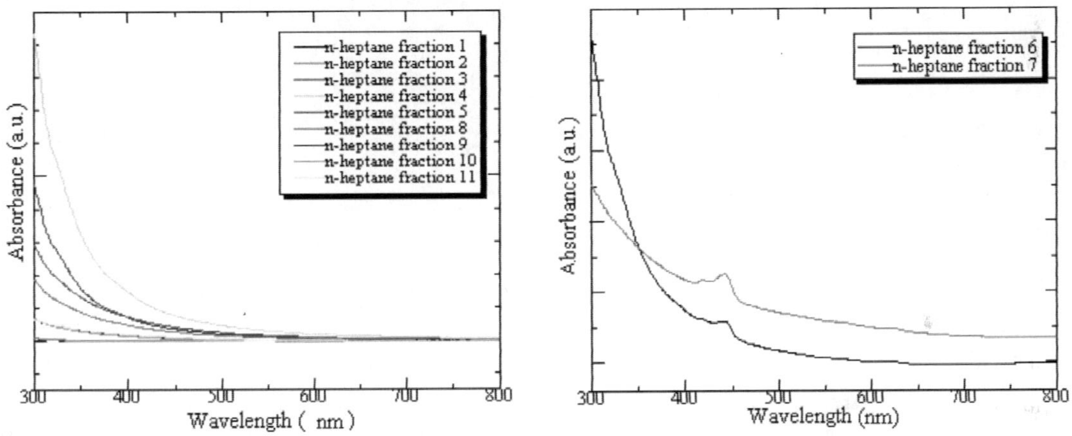

Figure 2. UV–Vis absorption spectra of the column chromatography fractions.

smallest percentage value was observed for the crude oil residue that was mixed with silica. This indicated that silica adsorbed the polar organic compounds from the crude oil and prevented toluene from extracting these compounds. The addition of silica can enlarge the contact surface area of metal porphyrin and asphaltenes with solvents during separation.

The methanol Soxhlet extraction of the n-heptane residue on silica was investigated (figure 1). No UV–Vis absorption peaks were observed in the visible range. This indicated that there were no porphyrins in the methanol fraction. The presence of free porphyrins have been reported between 380 and 450 nm, whereas free porphyrins were absent in the methanol fraction [8].

The n-heptane fraction was further separated by column chromatography, and 11 fractions were collected. Each fraction was characterized by UV–Vis spectroscopy (figure 2). Figure 2a depicted no peak in the visible range, whereas figure 2b depicted an absorption peak at 440 nm. The presence of free porphyrins is indicated by the peak at 440 nm [8]. Figure 3 depicts that the concentrated fractions (6 and 7) display an absorption maximum at 575 and 627 nm, respectively; this indicated the presence of a metal porphyrin [8].

Table 2. Elemental composition of the crude oil and residues.

Sample	% wt								
	C	Si	O	S	V	Fe	Ni	Al	Total
Crude oil	97.28	-	-	0.23	-	0.27	-	2.22	100
Crude oil residue	97.07	0.19	-	0.20	-	0.49	-	2.04	100
Toluene Soxhlet extract of oil residue without silica gel	96.96	-	-	0.21	0.08	0.46	1.35	2.94	100
Toluene Soxhlet extract of of oil residue with silica gel	15.5	26.5	57.9	-	-	-	-	-	100
n-Heptane Soxhlet extract	77.38	0,33	-	0.20	0.09	0.54	0.44	20.04	100

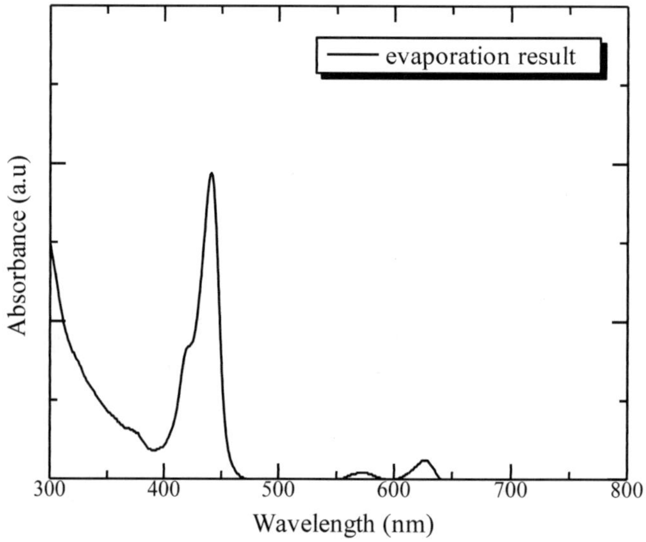

Figure 3. UV–Vis absorption spectra of column chromatography fractions 6 and 7 after the eluent was removed by evaporation.

FTIR characterization was conducted to identify the functional groups in each fraction. The FTIR spectra comparison of the concentrated crude n-heptane Soxhlet extract and concentrated fractions 6 and 7 after column chromatography are presented in figure 4. The presence of C–H groups was indicated by the peaks at 2850 and 2922 cm^{-1}. The presence of C–C groups was indicated by the peak at 1490 cm^{-1}. The presence of a pyrrole C–N bond was indicated by the peak at 775 cm^{-1}. Therefore, the presence of a porphyrin in these fractions was clearly indicated [9].

EDX characterization was conducted to determine the elemental composition of the sample. EDX data depicted the presence of elements that are commonly observed in petroleum, as presented in table 2.

LC–MS characterization was conducted to determine the molecular formula and structures of the isolated products. Figure 5 depicts the peak of the pseudo molecular ion [M+H+ACN]+ at 576 m/z. The true compound exhibits a molecular weight of 534. According to Liu *et al.* [10], the molecular weight peak at 534 m/z can be assigned to $C_{32}H_{36}N_4Ni$, a nickel-bound porphyrin compound [10]. Using the method described by Liu *et al.* [10], we observed a porphyrin compound with a molecular weight of 541 m/z. The molecular weight indicated that the nickel ions were replaced with vanadium oxide. In figure 5, the pseudo molecular ion peak [M+H+ACN]+ can be observed at 583 m/z. This indicated that the compound was etioporphyrinato vanadyl (IV), with a molecular formula of $C_{32}H_{36}N_4VO$, as depicted in figure 6. These results were supported by the UV–Vis characterization in figure 3, which indicated the presence of a metal porphyrin.

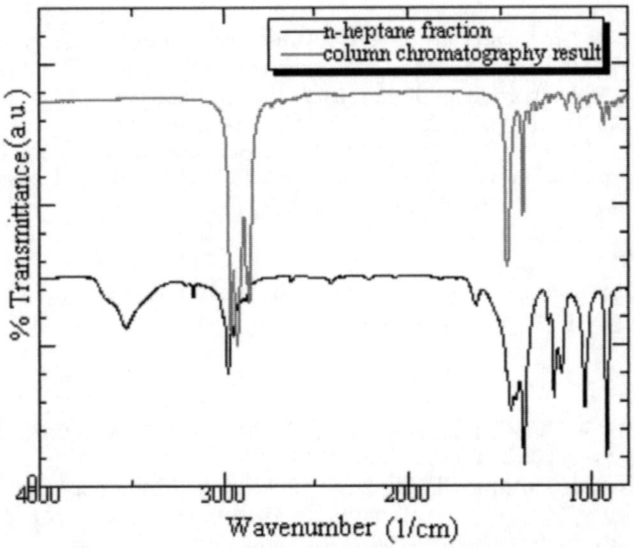

Figure 4. FTIR spectra of the concentrated crude n-heptane soxhlet extract and concentrated fractions 6 and 7 after performing column chromatography.

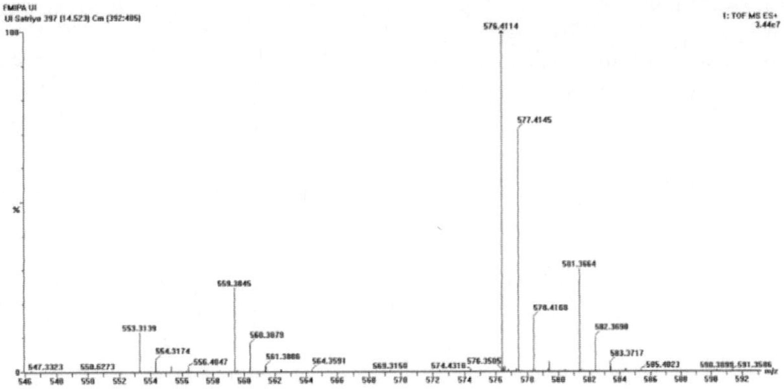

Figure 5. Mass spectrogram of fraction 7 from the heptane soxhlet extract that was subjected to column chromatography.

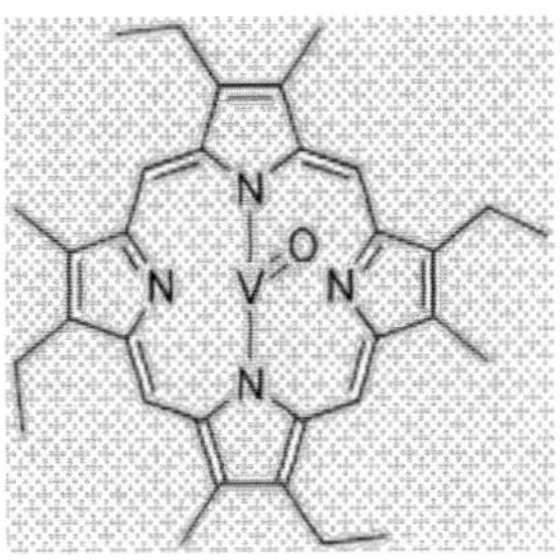

Figure 6. Molecular structure of etioporphyrinato vanadyl (IV).

IOP Conf. Series: Materials Science and Engineering **496** (2019) 012038 doi:10.1088/1757-899X/496/1/012038

4. Conclusions

The toluene Soxhlet extracts of the Duri crude oil residue, both aided and unaided by silica gel, accounted for 55.15% and 86.75% of the crude residue, respectively. The addition of silica caused the polar organic compounds to be adsorbed and increased the contact surface area of porphyrin and asphaltenes with the solvents during separation. UV–Vis spectroscopy of the n-heptane Soxhlet fraction indicated the presence of a free porphyrin with absorption maximum at 440 nm and metal porphyrins at 575 and 627 nm. The mass spectrum result of LC–MS indicated the presence of Ni(II) porphyrin ($C_{32}H_{36}N_4VO$, 534 m/z) and etioporphyrinato vanadyl (IV) ($C_{32}H_{36}N_4VO$, 541 m/z).

Acknowledgements

This work was funded by Pertamina, as a joint research between and Research & Development PT Pertamina (Persero) and Department of Chemistry, Faculty of Mathematics and Natural Sciences, Universitas Indonesia in 2017.

References

[1] Ghannam M T, Hasan S W, Abu-Jdayil and Esmail N 2012 *J. Pet. Sci. Eng.* **81** 122–8

[2] Yang Z, Gu X and Zhang L 1987 *Acta Pet. Sin.* **8** 37–47

[3] Moy F, Mangani F and Hofhine T R 2013 A Novel Petrochemistry Solution: SARA Fractionation of Crude Oil Using an Adaptable Gilson MPLC System (Middleton: Gilson, Inc. World Headquarters)

[4] Yen T F 1975 *The Role of Trace Metals in Petroleum* (Ann Arbor: Ann Arbor Science Publishers)

[5] Ali M F and Abbas S 2006 *Fuel Process. Technol.* **87** 573–84

[6] Dechaine G P and Gray M R 2010 *Energy Fuels* **24** 2795–808

[7] Branthaver J F 1990 *Fuel Science and Technology Handbook* ed J G Speight (New York: Marcel Dekker Inc.)

[8] Rio Y, Rodrıguez–Morgade M S and Torres T 2008 Org. Biomol. Chem. **6** 1877–94

[9] Mishra S, Kaur S, Tripathi S K, Mahajan C G and Saini G S S 2006 *J. Chem. Sci.* **118** 361–9.

[10] Liu H, Mu J, Wang Z, Ji S, Shi Q, Guo A, Chen K and Lu J 2015 *Energy Fuels* **29** 4803–13

Analysis of morphology and absorption of calcium and magnesium for calcium phosphate $Ca_3(PO_4)_2$ in rat's spine

M N Ramahwati, A L Juwono, D S Soejoko and N N Mulyaningsih

Department of Physics, Faculty of Mathematics and Natural Sciences (FMIPA) Universitas Indonesia, Depok 16424, Indonesia

Corresponding author: ariadne.laksmidevi@ui.ac.id

Abstract. The decline in bone mass is a sign of metabolic diseases called osteoporosis. Osteoporosis can be treated by adding the calcium level in the form of nano sized $Ca_3(PO_4)_2$ One type of calcium which can build and strengthen the bones is $Ca_3(PO_4)_2$ This study aimed to determine the effectively of nano sized $Ca_3(PO_4)_2$ addition which can be absorbed by rat bones and to identify the crystalline structure that was formed in rat bones with a specific purified diet. Nano sized $Ca_3(PO_4)_2$ was varied in a composition 0.5X, 1.0X and 1.5X compared to the normal needs in the purified diet. The rats were harvested in every two weeks, then their spines were taken to be investigated. The rat's spines were examined by using X-Ray Diffraction (XRD) and Atomic Absorption Spectroscopy (AAS). The effective amount of calcium for the rats was at the value of 0.5X. The amount of Ca as a bone mineral increased as the increasing amount of nano sized $Ca_3(PO_4)_2$, while the amount of Mg tended to fluctuate. The crystalline structure of hexagonal type was found in the rat bones with. Lattice parameters of a = 9.50 A and c = 6.83 A.

Keywords: $Ca_3(PO_4)_2$, purified diet, atomic absorption spectroscopy, X-Ray Diffraction, hydroxyapatite.

1. Introduction

Osteoporosis is a disease that affects bone in elderly people, which the initial symptom is not visible whether it can be perceived. International Osteoporosis Foundation (IFO) said that one of four women in Indonesia with an age range of 50-80 years was infected by osteoporosis. According to WHO, in 2050 the number of osteoporosis sufferers may be increased three times. The risk of being an osteoporosis sufferer in Indonesian women is four times higher than men, it is due to osteoporosis that may afflicts the women in menopause period. The loss of oestrogen after menopause increases the risk of osteoporosis. There are several causes of osteoporosis, such as unhealthy lifestyle, lacking exercise, eating junk food, smoking and drinking alcoholic beverages.

Researches on osteoporosis treatment have been conducted. For example Rollo *et al.* [1] analyzed microarstektur the trabecullar on the osteoporosis. Several similar studies have been made, such as research conducted by Darmadi Ts [2] and Elkomy and Elsaid [3] about the treatment of osteoporosis using a tungak and herbal plants. Other research had been conducted by Aulyani *et al.* [4] which is about absorption $Ca_3(PO_4)_2$ in rats with unapplied ovariectomy.

Research on the treatment of osteoporosis and the absorption of calcium on bone has been conducted. From the previous studies, research on the absorption of calcium phosphate in ovariectomized rats has not been done. Therefore, this study aims to observe the absorption capacity and the size of the crystalline by the addition of nano- $Ca_3(PO_4)_2$ due to osteoporosis bones.

Table 1. Mineral composition of rat's bone sample by AAS (in ppm)

Sample Code	Variation	Time	Ca	Mg
			ppm	
TO		Non ovariectomized	207	68.56
O2M	Control	Ovariectomized (2 weeks)	192	54.8
O4M		Ovariectomized (4 weeks)	168.5	59.98
2M0.5X		2 weeks	5390	66,8
4M0.5X	0,5	4 weeks	16070	254.5
6M0.5X		6 weeks	16220	224
2M1.0X		2 weeks	6108	65.4
4M1.0X	1,0	4 weeks	16345	207
6M1.0X		6 weeks	16770	222.5
2M1.5X		2 weeks	9140	106.5
4M1.5X	1,5	4 weeks	17500	173.5
6M1.5X		6 weeks	17955	206.5

2. Materials and methods

2.1. Preparation of sample

The sample were spines of *Rattus Norvegicus* rat with Wistar strain on the age of seven weeks. The rats fed with nano sized $Ca_3(PO_4)_2$ in their purified diet. The compositions of the nano size $Ca_3(PO_4)_2$ were 0.5X, 1.0X, and 1.5X of the normal needs. So that the samples were identified of four groups, namely rat control, rat with nano sized $Ca_3(PO_4)_2$ 0.5X, rat with nano sized $Ca_3(PO_4)_2$ 1.0X, and rat with nano sized $Ca_3(PO_4)_2$ 1.5X. The rat control consisted of three distinct types of rats, such as non-ovariectomized rats, ovariectomized rats with the age of two weeks, and ovariectomized rats with the age of four weeks for all groups. Table 1 shows the value of Ca and Mg, which belongs to non-ovariectomized control sample and ovariectomized sample after two weeks and four weeks.

2.2. Characterization of sample

The samples were pulverized to facilitate the process characterization. Atomic Absorption Spectroscopy (AAS) and X-Ray Diffraction (XRD) were used to characterize the samples. XRD was performed by using Rigaku Miniflex 600 1 kV with CuKα radiation and wavelength of 1.54 Å. The phases of the samples were determined by the Highscore Plus software to analyze the diffraction lines in a range of 5°–90°. The mineral concentration such as Calcium (Ca) and Magnesium (Mg) were examined by AAS.

3. Results and discussion

Table 1 presents the results of the calcium and magnesium compositions in the spin rat's samples. The value of calcium (Ca) in the sample after being ovariectomized decreases than sample without ovariectomy. The composition of Ca increased because of the addition of varied nano sized $Ca_3(PO_4)_2$ in feed might be well absorbed, so the bones contained high calcium level. However, the value of Ca after the treatment is quite high, so that the Ca level in blood is much as well. It is due to the content of Ca in the crystals are constant; less addition of Ca value can increase the absorption level of Ca immediately. High level of calcium that found in bone was the deposition of high calcium in blood plasma. Mg value is fluctuating caused by its role as inhibitor in bone that can detain repair for bone [5].

Figure 1 exhibit the diffraction pattern of bone samples with the variation of composition of feed and ages showing the counts versus angle 2θ (°) From these difractograms, the peak of (002) which refers to HAp can be observed. The size of $Ca_3(PO_4)_2$ particle was in a range of ±75 nm and formed hexagonal crystalline structure of hydroxyapatite. The lattice parameters can be observed in table 2. In

Table 2. Lattice parameters of the rat's spinal

Sample Code	Variation	Time	a (Å)	c (Å)
TO		Non ovariectomized	9.42	6.88
O2M	Control	Ovariectomized (2 weeks)	9.39	6.75
O4M		Ovariectomized (4 weeks)	9.48	6.73
2M 0.5X		2 weeks	9.45	6.79
4M 0.5X	0.5X	4 weeks	9.57	6.83
6M 0.5X		6 weeks	9.45	6.80
2M 1.0X		2 weeks	9.59	6.85
4M 1.0X	1.0X	4 weeks	9.46	6.79
6M 1.0X		6 weeks	9.72	7.11
2M 1.5X		2 weeks	9.59	6.78
4M 1.5X	1.5X	4 weeks	9.42	6.78
6M 1.5X		6 weeks	9.45	6.81

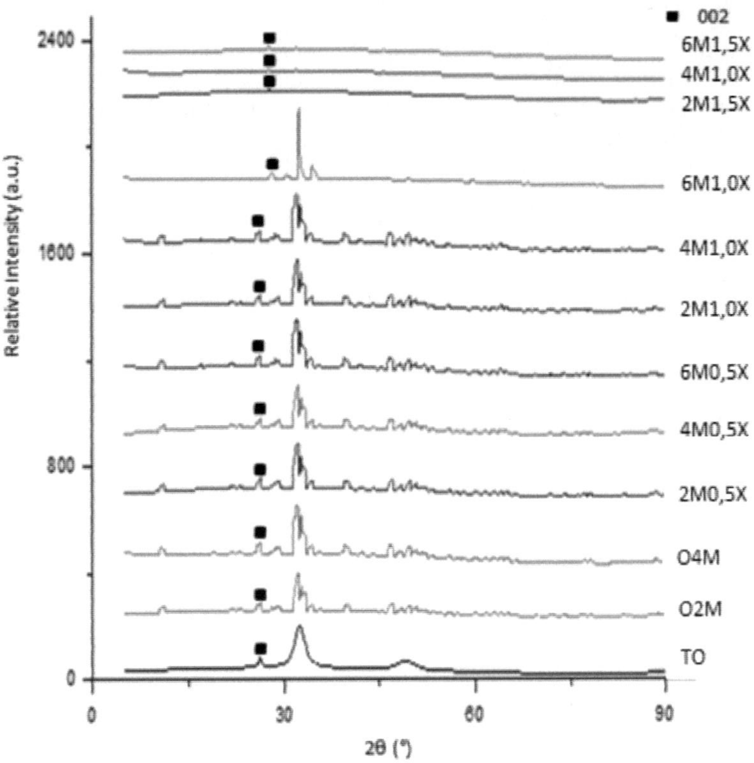

Figure 1. XRD patterns for the all samples

addition to the presence of crystal structure, the amorphous phase can be found in the bone as well because the relatively small peaks were formed. So, this phenomenom denotes the presences of crystalline and amorphous phase. In ovariectomized sample, the amorphous phase decreased and the incresing value of Ca caused increasing value of amorphous phase as well until the amorphous level would become similar as non-ovariectomized. This phenomenon can be known, because the small

Table 3. Crystalline size of the rat's spinal

Sample Code	Variation	Time	A (amorphous peak)	β (FWHM)	2θ (°C)	D (Crystalline size)
TO		Non ovariectomized	52	1.49	26	62.52
O2M	Control	Ovariectomized (2 weeks)	48	2.25	26.6	43.92
O4M		Ovariectomized (4 weeks)	45	2.38	25.5	31.74
2M0,5X		2 weeks	50	1.43	26.5	69.96
4M0,5X	0,5X	4 weeks	48	1.91	26.7	50.19
6M0,5X		6 weeks	52	1.64	26.3	67.11
2M1,0X		2 weeks	53	1.99	26.8	45.57
4M1,0X	1,0X	4 weeks	50	1.53	26.4	63.56
6M1,0X		6 weeks	51	1.33	26.1	74.21
2M1,5X		2 weeks	53	2.58	28	39.77
4M1,5X	1,5X	4 weeks	54	1.57	28.2	44.06
6M1,5X		6 weeks	48	1.55	28.4	53.98

peaks that are formed indicate the existence of crystalline and amorphous phase. The amorphous in the control has decreased because on the osteoporosis, crystals, which is experiencing a reduction in, andleaving huge crystal that led to the top of the amorphous. The value that determined crystalline size of bone can be observed in table 3.

4. Conclusions

The Ca value on the rat bones increased as the increasing amount of nano sized $Ca_3(PO_4)_2$ in purified diet. The content of Mg was fluctuating for the varied composition. The effective result of nano sized $Ca_3(PO_4)_2$ that was absorbed by the rats was shown by the purified diet composition at 0.5X. The bone samples before and after the treatment applied had the same crystal structure with the form of hexagonal, which referred to hydroxyapatite.

Acknowledgements

The authors acknowledge Universitas Indonesia for the financial support, this research was funded by PITTA UI Grant with contract number 624/UN2.R3.1/HKP.05.00/2017.

References

[1] Rollo J M D A, Boffa R S, Cesar R, Schwab D C and Leivas T P 2015 *Procedia Eng.* **110** 8–14
[2] Darmadi Ts D, Nurdiana N and Norahmawti E 2011 *Jurnal Kedokteran Brawijaya* **26** 151–5
[3] Elkomy M M and Elsaid F G 2015 *J. Basic Appl. Zool.* **72** 81–8
[4] Aulyani T L, Astuti D A and Hermana W 2016 *Pemberian Kalsium Nano Ca3(PO4)2 terhadap Efektivitas Penyerapan Kalsium Tulang Hewan Model Tikus Putih Rattus novergicus* (Bogor: Institut Pertanian Bogor) Undergraduate Thesis
[5] Mulyaningsih N N, Tresnasari D R, Ramahwati M R, Juwono A L, Soejoko D S and Astuti D A 2017 *AIP Conf. Proc.* **1862** 030066

IOP Conf. Series: Materials Science and Engineering **496** (2019) 012040 doi:10.1088/1757-899X/496/1/012040

Modification of boron-doped diamond electrodes with platinum-iridium for carbon dioxide electroreduction

S R Atriardi, R K P Dewandaru, J Gunlazuardi and T A Ivandini

Department of Chemistry, Faculty of Mathematics and Natural Sciences (FMIPA), Universitas Indonesia, Depok 16424, Indonesia

Corresponding author: ivandini.tri@sci.ui.ac.id

Abstract. Boron-doped diamond (BDD) is reportedly a good electrode material for the electroreduction of CO_2 to produce formaldehyde with high Faradaic efficiency. In this study, BDD electrodes were modified with a platinum-iridium composite (Pt-Ir). Deposition of Pt-Ir on the surface of BDD was performed by two different methods. One method involved the modification of the BDD surface, wherein terminal-H atoms were substituted with N atoms. The N-terminal BDD was then immersed in a Pt–Ir nanoparticle solution to form Pt–Ir–BDD (1). The Pt–Ir nanoparticles were synthesized from a 3:4 molar ratio mixture of H_2PtCl_6 and K_3IrCl_6 by a hydrothermal method. Characterization of Pt–Ir nanoparticles by using UV-Vis spectrophotometry and Transmission Electron Microscopy (TEM) image showed an absorbance peak at 420 nm with an average particle size of 2.6±0.6 nm. In the second method, Pt-Ir was deposited onto BDD using chemically assisted electrodeposition by applying a reduction potential of −0.5 V (vs. Ag/AgCl) to form Pt–Ir–BDD (2). Scanning Electron Microscope (SEM) images showed that the second method provides better distribution of Pt–Ir on the BDD surface. Furthermore, cyclic voltammetry of CO_2, dissolved in 0.1 M NaCl showed a reduction potentials at −1.7, −1.3, and −1.2 V (vs. Ag/AgCl) at BDD, Pt–Ir–BDD (1) and Pt–Ir–BDD (2), respectively, indicating that modification with Pt–Ir allows the conversion of CO_2 with higher Faradaic efficiency.

Keywords: Boron-doped diamond, iridium, platinum, surface modification, CO_2 electroreduction

1. Introduction

Global warming is caused by industrial emission of greenhouse gases (GHGs), such as COx, NOx, SOx, CFCs, and CH_4, and has many negative environmental effects. Carbon dioxide (CO_2) is the most abundant GHG in the atmosphere, accounting for ≈65% of total atmospheric GHGs [1]. Therefore, reducing atmospheric CO_2 levels is imperative if the impact of global warming is to be limited.

Electrochemical conversion is a promising technique for the removal of atmospheric CO_2. Furthermore, it can be used to produce a range of possible products. However, electrochemical reduction of CO_2 requires the development of suitable electrode materials. One such material is boron-doped diamond (BDD), which has many advantages over other electrode materials, including its extremely wide potential window and very low capacitance as well as its high chemically inertness and physical durability [2]. The use of BDD for CO_2 electroreduction in methanol solution has been reported to afford formaldehyde with a high faradic efficiency (74%) [3].

Platinum is known to be an excellent catalyst for a range of processes, largely owing to its large surface area. However, the catalytic effectiveness of platinum can be increased by compositing it with other metals to form platinum-based multimetallic composites (Pt-X, X = Cu, Au, Ag, etc.) [4]. These composites exhibit different optical, electronic, thermal, and catalytic characteristics to their monometallic counterparts For example, platinum–iridium (Pt–Ir) provides significantly higher current density in comparison with that of Pt metal alone [5] and has been successfully applied to nitrate electroreduction, in which it exhibits higher activity than pure Pt or Ir metal [6].

Accordingly, in the current study, BDD electrodes were modified with Pt–Ir nanoparticles (NPs) as a means to improve their ability to convert CO_2 to bulk products. Similar strategies involving the modification of BDD electrodes with Pt and IrO_2 have proven to be effective previously [7,8].

2. Experimental

2.1. Materials and instruments

BDD electrodes with a B:C ratio of 1:1000 (i.e., 0.1% BDD) were fabricated using microwave plasma-assisted chemical vapor deposition at the Faculty of Science and Technology, Keio University, Japan. Hydrogen hexachloroplatinate (IV) (H_2PtCl_6), potassium hexachloroiridate (IV) (K_2IrCl_6), and sodium borohydride ($NaBH_4$) were purchased from Wako Pure Chemical Industries Ltd. Sodium hydroxide (NaOH), sodium chloride (NaCl), 2-propanol, and formaldehyde were purchased from Merck. Sulfuric acid (JT Baker), allylamine (Sigma Aldrich), CO_2, N_2, and ultra-pure deionized water from a Millipore Direct® Q5-UV with a resistivity of 18.2 MΩcm were also used.

Electrochemical measurements were performed using a three-electrode cell (eDAQ e-corder 410). A platinum wire and an Ag/AgCl system in a saturated KCl solution were employed as the counter and the reference electrodes, respectively. Others techniques used in this research were UV-Vis spectrophotometry (Thermo Scientific Multiscan Go), a transmission electron microscopy (TEM) (Tecnai 200kV D2360 SuperTwin), scanning electron microscopy (SEM) (JEOL/EO JCM-6000Plus, Japan), and energy dispersive X-ray analysis (EDX).

2.2. Preparation of Pt–Ir-BDD electrodes

2.2.1. Synthesis of platinum-iridium NPs and preparation of Pt–Ir–BDD (1). Synthesis of platinum-iridium nanoparticles (Pt–Ir NPs) was conducted using a hydrothermal method as previously described [5] with some modifications. Briefly, 1 mM H_2PtCl_6 (15 mL) and 1 mM K_2IrCl_6 (20 mL) were mixed and boiled with stirring until the volume of the mixture was reduced to 10 mL. Then, the mixture was transferred to an autoclave containing 5% formaldehyde (0.5 mL) and heated at 180 °C for 10 h. Separately, a BDD electrode was converted into an N-terminated BDD electrode by the addition of allylamine to the BDD surface under illumination with a UV lamp (λ = 254 nm) [9,10]. Pt–Ir–BDD (1) was prepared by the immersion of the N-terminated BDD into a Pt–Ir NP solution for 12 h.

2.2.2. Preparation of Pt–Ir–BDD (2). The deposition of Pt–Ir NPs onto the BDD surface involved two steps: (1) wet-chemical deposition of uniformly distributed Pt–Ir seeds and (2) electrochemical overgrowth of the Pt–Ir seeds [11]. First, a solution of 1.0 M $NaBH_4$ in 0.1 M NaOH (10 μL) was dropped at the BDD surface to allow the adsorption of BH_4 ions The formation of Pt–Ir seeds was performed by then adding a drop (40 μL) of 1 mM H_2PtCl_6:K_2IrCl_6 (3:4 molar ratio) solution onto the adsorbed $NaBH_4$. The electrode was then rinsed with water and dried under flowing N_2. Repeating this process three times resulted in the generation of bigger Pt–Ir seeds and higher surface coverage. Electrochemical overgrowth of these Pt–Ir seeds was conducted at a constant applied potential of −0.5 V (vs. Ag/AgCl) for 10 min in 1 mM H_2PtCl_6:K_2IrCl_6 solution (3:4 molar ratio).

2.3. CO_2 electroreduction

Electrochemical reduction of CO_2 was performed in a three-electrode cell. A solution of 0.1 M NaCl was used as the electrolyte. BDD, Pt–Ir–BDD (1), or Pt–Ir–BDD (2) was used as the working electrode, while Ag/AgCl and Pt spiral wire was used as the reference and the counter electrode, respectively. Then, the electrolyte solution was aerated with N_2 for 15 min to remove other gases, followed by CO_2 aeration. Then, the electrochemical reduction of CO_2 was assessed using cyclic voltammetry (CV) analysis over the potential range 1.0–2.0 V (vs. Ag/AgCl).

IOP Conf. Series: Materials Science and Engineering **496** (2019) 012040 doi:10.1088/1757-899X/496/1/012040

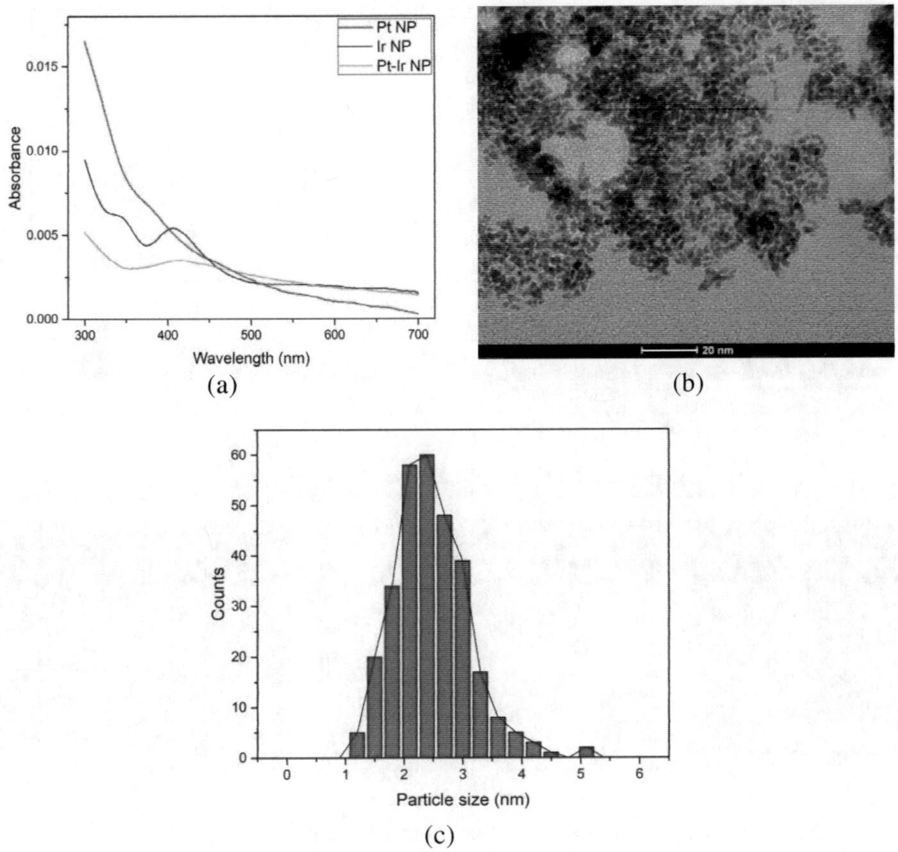

Figure 1. (a) UV-Vis spectra of Pt NPs, Ir NPs, and Pt–Ir NPs, (b) TEM image of Pt–Ir NPs at 250,000× magnification, and (c) Size distribution of Pt–Ir NPs.

3. Results and discussion

3.1. Synthesis of Pt–Ir NPs

Pt–Ir NPs were successfully synthesized from 1 mM H_2PtCl_6:K_2IrCl_6 (3:4) precursor solutions using a hydrothermal method. The addition of formaldehyde was employed to reduce Pt(IV) to Pt(0) and Ir(VI) to Ir(0). The process is expressed by the following reactions [5]:

$$H_2PtCl_6 + HCHO + H_2O \rightarrow Pt\downarrow + CO_2 + 6H^+ + 6Cl^-$$

$$K_2IrCl_6 + HCHO + H_2O \rightarrow Ir\downarrow + CO_2 + 4H^+ + 2K^+ + 6Cl^-$$

Figure 1a shows the results of UV-Vis characterization for different NPs, with the absorbance peak for the Pt–Ir NPs visible at 420 nm. However, Pt NPs do not present peaks and Ir NPs present a peak at 405 nm. The optical properties of these NPs are highly dependent on their diameters [12]. The TEM image of the Pt–Ir NPs shows that they are well distributed and sometimes agglomerated (figure 1b). The average particle size of the Pt–Ir NPs is 2.6 ± 0.6 nm (figure 1c).

The EDX spectrum of the Pt–Ir NPs is given in figure 2, showing peaks corresponding to the investigated sample (Pt,Ir), clearly demonstrating that Pt–Ir NPs were successfully synthesized, as well as signals from the carbon-coated cooper TEM grid (C, Cu) and impurities from the nanoparticle synthesis (O).

3.2. Modification of BDD with Pt-Ir and its characterization

Pt–Ir–BDD (1) was prepared by immersing the N-terminated BDD surface into a Pt–Ir NP solution. Conversely, Pt–Ir–BDD (2) was prepared using a seeding technique with $NaBH_4$ as a reducing agent

IOP Conf. Series: Materials Science and Engineering **496** (2019) 012040 doi:10.1088/1757-899X/496/1/012040

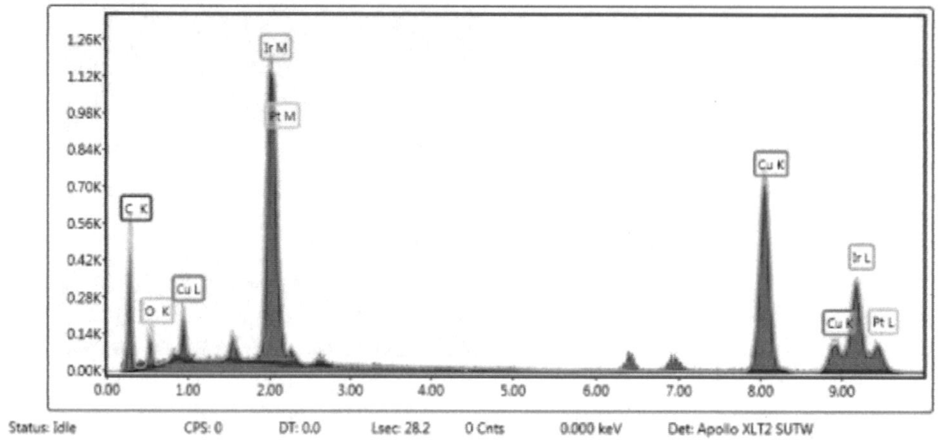

Figure 2. EDX spectrum of Pt–Ir NPs.

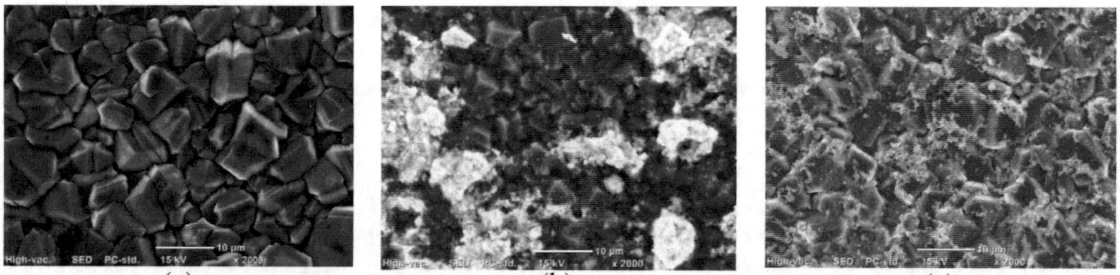

(a) (b) (c)

Figure 3. SEM Images (2000× magnification) of (a) BDD, (b) Pt–Ir–BDD (1), and (c) Pt–Ir–BDD (2).

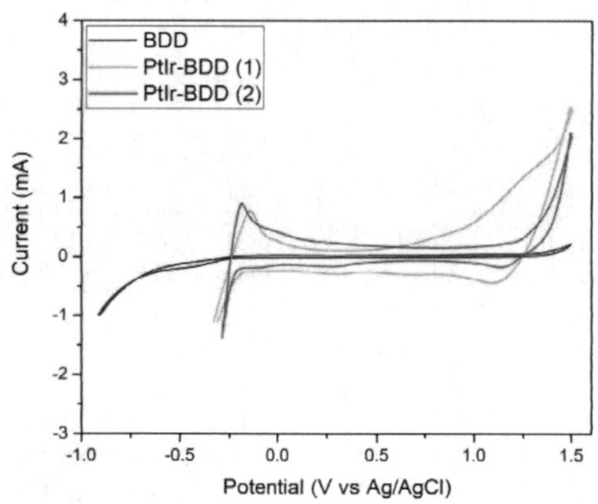

Figure 4. CV characterization of BDD, Pt–Ir–BDD (1), and Pt–Ir–BDD (2) in 0.5 M H_2SO_4.

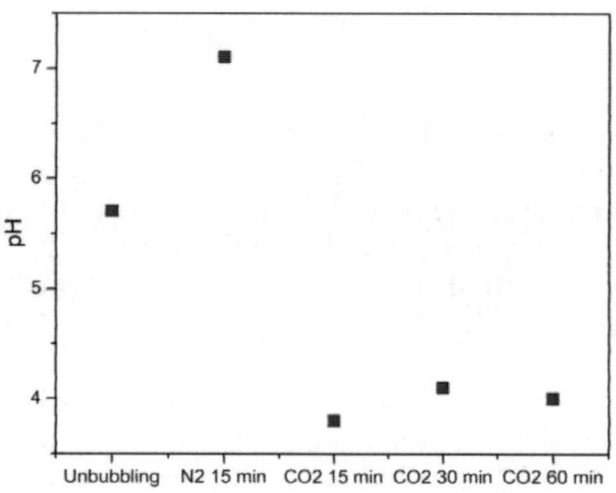

Figure 5. The influence of the aeration time of CO_2 into 0.1 M NaCl solutions on the pH values.

[11]. The Pt–Ir NP seeds deposited by this process were then enlarged using electrodeposition by applying a potential of –0.5 V. Figure 3 shows the surface morphologies of the BDD, Pt–Ir–BDD (1), and Pt–Ir–BDD (2) electrodes. Pt–Ir particles on BDD surface are observed with the white point on BDD surface. It can be seen that the Pt–Ir NPs prepared by the seeding method are more evenly distributed than those deposited by the immersion method.

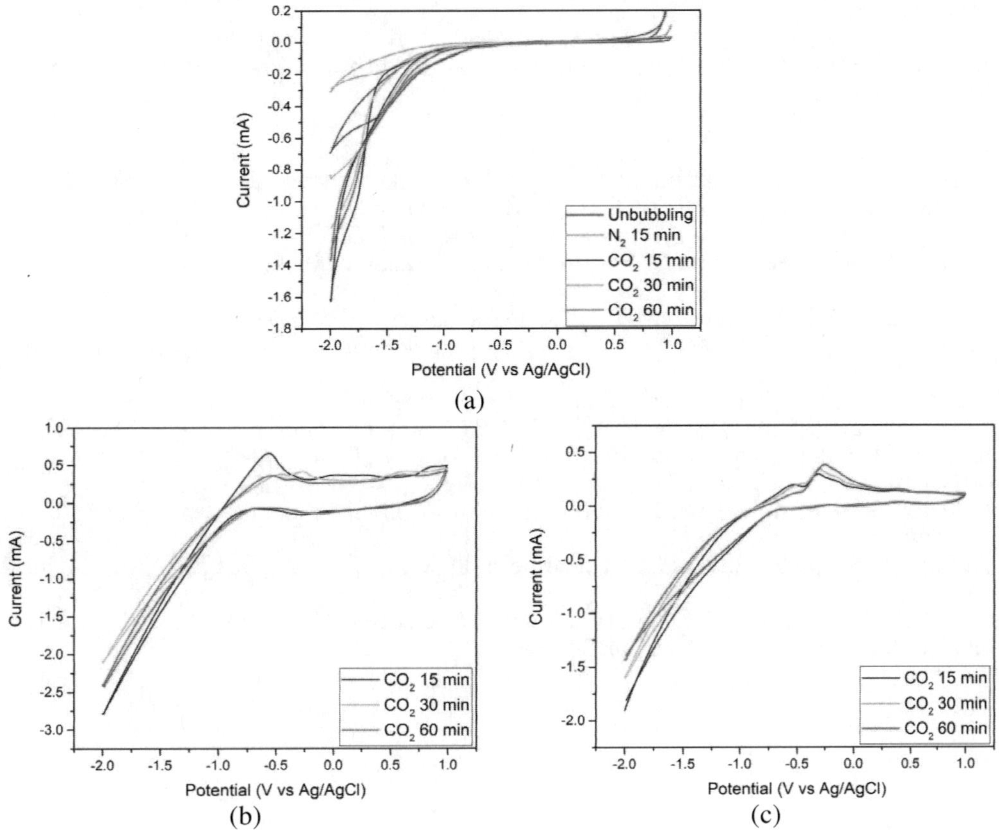

Figure 6. Preliminary CV results for CO_2 dissolved in 0.1 M NaCl at (a) BDD, (b) Pt–Ir–BDD (1), and (c) Pt–Ir–BDD (2) in several bubbling times of CO_2

CV analysis of the BDD electrodes in 0.5 M H_2SO_4 in figure 4 shows that they were successfully modified with Pt–Ir. A scan rate of 100 mV/s was applied throughout this study. The CV characteristics of Pt–Ir–BDD (1) and Pt–Ir–BDD (2) are similar, exhibiting hydrogen evolution at –0.3 V (vs. Ag/AgCl) and oxygen evolution at +1.5 V (vs. Ag/AgCl), as well as reduction-oxidation peaks for Pt and Ir (figure 4).

3.3. Preliminary study of CO_2 electroreduction

To reduce CO_2, it was dissolved in 0.1 M NaCl solution. The bubbling of CO_2 was conducted for 15, 30, and 60 min. to identify the optimum CO_2 aeration time. The optimum aeration time of CO_2 was found to be 15 min. providing a solution pH of 3.8 (figure 5).

Furthermore, CV analysis of the CO_2 solutions at the different electrodes revealed reduction peaks at approximately –1.7 V (vs. Ag/AgCl) for BDD, –1.3 V (vs. Ag/AgCl) for Pt–Ir–BDD (1), and –1.2 V (vs. Ag/AgCl) for Pt–Ir–BDD (2) as shown in figure 6. The results indicated that the direct electrochemical reduction of CO_2 in 0.1 M NaCl solution is promising to be conducted at similar potentials using a chronoamperometric technique.

4. Conclusions

Pt-Ir NPs with an average size of 2.6 ± 0.6 nm were successfully synthesized using a hydrothermal method. Pt-Ir-BDD electrodes were successfully prepared by immersion in this Pt-Ir NP solution (Pt-Ir-BDD (1)) or by chemical seeding followed by electrodeposition (Pt-Ir-BDD (2)). However, the seeding method was found to provide a better distribution of nanoparticles than the immersion method. The potentials observed for CO_2 electroreduction were –1.7 V (vs. Ag/AgCl) for BDD, –1.3 V (vs. Ag/AgCl) for Pt-Ir-BDD (1), and –1.2 V (vs. Ag/AgCl) for Pt-Ir-BDD (2).

Acknowledgements
This work was funded by PITTA Grant Universitas Indonesia with contract No.1821/UN2.R3.1/PPM.00.01/2017. The authors would like to thank Einago (www.einago.com) for the English language review.

References
[1] Intergovernmental panel on climate change 2014 Climate change 2014: Mitigation of climate change, 12th session of IPPC WG III (Berlin, Germany)
[2] Ishii Y, Ivandini T A, Murata K and Einaga Y 2014 *Anal. Chem.* **85** 4284–8
[3] Nakata K, Ozaki T, Terashima C, Fujishima A and Einaga Y 2014 *Angew. Chemie - Int. Ed.* **53** 871–4
[4] Karthikeyan B and Murugavelu M 2012 *Sens. Actuators B Chem.* **163** 216–23
[5] Yi Q, Chen A, Huang W, Zhang J, Liu X, Xu G and Zhou Z 2007 *Electrochem. Commun.* **9** 1513–8
[6] Ureta-Zañartu S and Yáñez C 1997 *Electrochim. Acta* **42** 1725–31
[7] Jasril, Gunlazuardi J and Ivandini T A 2017 *IOP Conf. Ser.: Mater. Sci. Eng.* **188** 012037
[8] Ichzan A M, Gunlazuardi J and Ivandini T A 2017 *IOP Conf. Ser.: Mater. Sci. Eng.* **188** 012030
[9] Dewandaru R K P, Gunlazuardi J and Ivandini T A 2018 *AIP Conference Proceedings* **2023** 020098
[10] Ivandini TA, Saepudin E, Wardah H, Harmesa, Dewangga N and Einaga Y 2012 *Anal. Chem.* **84** 9825–32
[11] Gao F, Yang N, Smirnov W, Obloh H and Nebel C E 2013 *Electrochim. Acta* **90** 445–51
[12] Bilankohi S M 2015 *Orient. J. Chem.* **31** 293–7

A simple gamma dosimeter using a film label made of polyvinyl alcohol and bark of *Peltophorum ferrugineum* extract

M Harahap, P Widodo, Y Priasetyono, A Listyarini, D Djuhana and C Imawan

Department of Physics, Faculty of Mathematics and Natural Sciences (FMIPA) Universitas Indonesia, Depok 16424, Indonesia

Corresponding author: cuk.imawan@sci.ui.ac.id

Abstract. Gamma radiation is considerably dangerous, but at certain doses, it can be beneficial to humans, for example, in food sterilization applications. Therefore, the development of simple and practical dosimeter devices is essential. Herein, we developed a simple dosimeter label made of *Peltophorum ferrugineum* bark extract separated from its fat content and mixed in a polyvinyl alcohol (PVA) matrix (prepared using an ammonium peroxide crosslinker). The film obtained from this mixture had a yellowish–brown color that changed to dark brown after gamma irradiation. The film label structure was investigated via Fourier transform infrared spectroscopy, and the absorbance properties of the film labels were measured using Ultraviolet-Visible (UV-Vis) spectrophotometer. The intensity of the color change was quantified using a colorimeter. The response of the film label to gamma radiation was measured via irradiation at 0–66 kGy. We found that gamma radiation caused a decrease in the intensity of the R and G color components of the film. Additionally, swelling tests demonstrated that the films exhibited good plastic properties after gamma irradiation. These results suggest that film labels made of PVA-extracted *P. ferrugineum* can be used as a simple dosimeter label for sterilization purposes.

Keywords: Dosimeter, Bark of *Pelthophorum ferruginum* , PVA

1. Introduction

The enormous energy of gamma radiation can be used for sterilization purposes in the food and medical equipment industries. The use of radiation for food sterilization must adhere to international standards [1], and the absorbed dose of radiation is a crucial factor to observe. Dosimeters, which are used to determine the exposure to gamma rays and other types of ionizing radiation, play an important role in providing this information. There are two types of dosimeters: *active dosimeters*, which can read data immediately, and *passive dosimeters*, which need further data processing.

Radiochromic devices are a type of passive dosimeter that is often used to measure the exposure to ionizing radiation. Such dosimeters have been developed in the form of gels, liquids, and films and can be applied in a varying dose range. Gel-type devices irradiated by low-dose (1–15 Gy) ^{60}Co gamma rays have been prepared using ferrous xylenol orange gel for radiotherapy measurements [2]. Radiochromic dosimeters for medium absorption doses (10–1000 Gy) have also been developed using synthetic nitro blue tetrazolium chloride (NBT). Such devices have been used for food radiation, pest control, and other applications, as reported by in a previous study [3]. Dosimeters for high absorption doses (1–125 kGy), obtained from diethyl fumarate (DEF) in a dimethylformamide (DMF) solution, have also been studied for routine dosimetry purposes [4].

Development and innovation in the fabrication of dosimeters are crucial factors for monitoring radiation exposure. Materials for dosimeters comprise a matrix and a coloring agent. In previous studies, plastic film matrixes made of polyvinyl alcohol (PVA) [5], polyvinyl chloride [6], polyvinyl butyral [7], and other polymers have been used for this purpose. PVA has been chosen for this research because it is a water-soluble and non-toxic biodegradable synthetic polymer [8], and these safety parameters are important for radiation on food and medical equipment.

The coloring agents for radiochromic dosimeters used in previous studies were synthesized from synthetic matter. Natural coloring agents that have already been used in previous research include turmeric roots, walnut bark, and henna leaves in gel form [9]. Natural coloring agents of local origin are mostly used for textiles and batik processes. One of them is synthesized from the bark of *Peltophorum ferrugineum*, which is taxonomically classified into the *Fabaceae* family that includes the *Caesalpiniodeae* sub-family containing flavonoids and tannins [10,11].

The purpose of this research is to produce radiochromic film labels from a PVA matrix using the natural coloring agent synthesized from the bark of *P. ferrugineum*. This should result in color changes in the film label after irradiation by high-dose gamma rays. The film label can thus function as a simple dosimeter, which can be used to sterilize food and medical equipment at permitted dose levels.

2. Experimental

2.1. Preparation of film labels using P. ferrugineum extract

The *P. ferrugineum* bark used as the coloring agent was obtained from local production. The bark was first dried at 40°C for 48 h to remove water and then milled using a dry milling process to obtain a powder. Then, 40 g of this powder was mixed with 150-ml ethanol (J.T Baker; 100%) and stirred at 300 rpm and 55°C for 4 h using a magnetic stirrer. Next, the solution was filtered using Whatman No. 1 paper. The resulting solution of the *P. ferrugineum* extract had a yellowish-brown color.

Film labels were then made using this coloring extract and PVA. Fully hydrolyzed PVA, which was used as the matrix, was obtained from Merck, Germany. The PVA powder (2 g) was diluted in 100 ml (w/v) of distilled water, and the resulting solution was stirred at 300 rpm and room temperature until it reached homogeneity. The solution was then mixed with ammonium peroxodisulfate [APS; $(NH_4)_2S_2O_8$] at 4% of the PVA weight. APS, obtained from Emsure, Germany, served as a crosslinker for PVA. The PVA–APS solution was then stirred at 200 rpm and room temperature for 4 h. Next, 50 ml of the solution was mixed with 10 ml of brown *P. ferrugineum*, after which the mixture was stirred for 15 min. This final solution was poured into an acrylic mold (size: 15 cm × 15 cm) and dried at room temperature for 48 h to synthesize the film labels.

2.2. Characterization of the dosimeter film

A Thermo Genesys 105 UV/Vis spectrophotometer was used to measure the maximum absorbances of a PVA film label and a film label containing *P. ferrugineum* at wavelengths between 300 and 800 nm. The optical spectra were measured before and after irradiation, and the Origin 9.1 software package was used to calculate the maximum absorbances and optical spectra for both film labels.

The RGB intensity of film labels containing *P. ferrugineum* was measured using a colorimeter purchased from Nix Pro, Canada. The transparent film labels were placed on a board with a white background, and the intensity measurements were performed five times to obtain the standard deviation. The total RGB intensity was calculated using the luminance formula (Y) and the color variance recommended by the International Telecommunication Union ITU-R BT.601-7:

$$Y = 0.299(\mathrm{R}) + 0.587(\mathrm{G}) + 0.114(\mathrm{B}). \quad (1)$$

The molecular structures of the bonds in the film labels were determined via Fourier transform infrared (FTIR) spectroscopy (Thermo Scientific, Nicolet iS5). The tested samples included film labels made of PVA, PVA–APS, PVA–APS–*P. ferrugineum* before being irradiated at 0 kGy and PVA–APS–*P. ferrugineum* after being irradiated at 66 kGy. The spectra were recorded at 400–4000 cm^{-1}. Absorbance spectra of all the four film labels were then compared. Origin 9.1 was used to illustrate the film labels' spectra.

Swelling tests were performed on the *P. ferrugineum* extract film labels before and after gamma irradiation. Both film labels were weighed separately using an analytical balance and then washed with distilled water for 30 min. Next, they were kept aside and reweighed. These steps were repeated for washing periods of 60, 90, and 120 min.

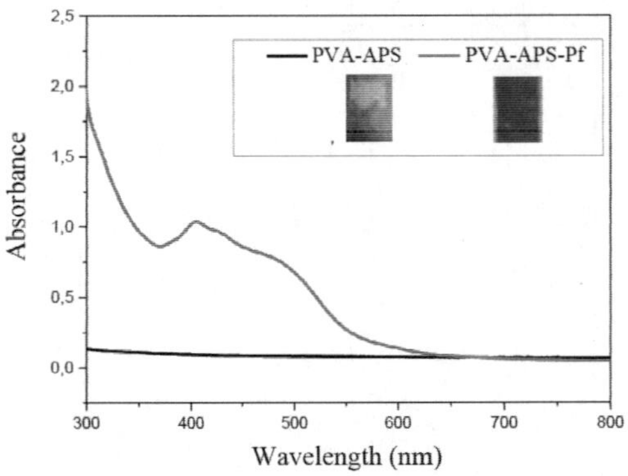

Figure 1. Absorbance spectra of polyvinyl alcohol (PVA) film labels and film labels containing *Peltophorum ferrugineum*.

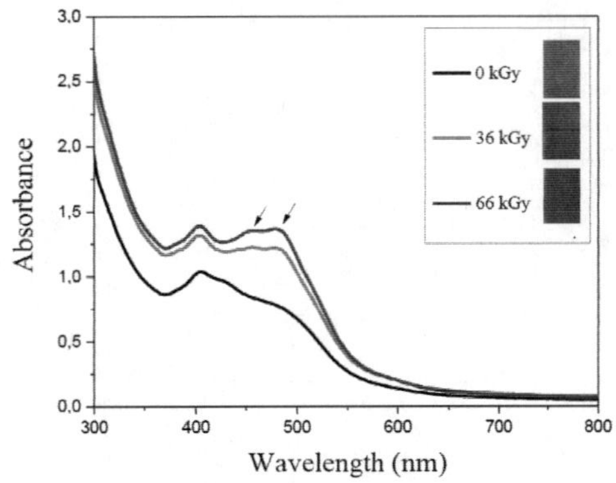

Figure 2. Absorbance spectra of film labels containing *P. ferrugineum* before and after irradiation at varying doses.

The film labels containing *P. ferrugineum* were cut into small pieces (2 cm × 3 cm) and irradiated using a Gamma Cell ^{60}Co device at the Irradiation Facility owned by BATAN, Jakarta. The radiation dose of the gamma cell was set to 6 kGy/h with a measurement range of 36–66 kGy. The irradiated *P. ferrugineum* film labels were then analyzed for color changes using the UV/Vis spectrophotometer, colorimeter, and FTIR equipment.

3. Results and discussion

3.1. Absorbance spectra of the film labels

The absorbance spectra of the PVA film labels and the film labels containing *P. ferrugineum* prior to gamma irradiation are shown in figure 1. The absorbance spectra of the transparent PVA film labels show a straight line without any peaks in the wavenumber range 300–800 nm. On the contrary, the film labels containing *P. ferrugineum* show an absorbance peak at 403 nm. This wavelength at 403 nm shows a complimentary color of yellowish green. These results show that the color of the PVA film labels, which are identified as the matrix, does not affect the film labels containing *P. ferrugineum*, which exhibit a yellowish-brown color. The colors of the film labels are dominated by the dark color of the bark extract.

The absorbance spectra of film labels containing *P. ferrugineum*, irradiated at 36 and 66 kGy, can be are shown in figure 2. These film labels undergo a visually perceptible color change: their yellowish-brown color changes to solid brown after irradiation. This color change phenomenon in films containing synthetic colors, such as methyl viologen [12], bromothymol blue (BTB) [13], and methyl orange [14], when irradiated with a high dose of gamma rays has been reported before. The film label spectra reported herein show an increase in absorbance parallel to an increase in the gamma-ray dose. Gamma radiation ionizes the film labels and creates hydrogen radicals from PVA [14]. Additionally, the optical spectra of the irradiated film labels show a different maximum absorbance in comparison with the spectra taken before irradiation, and this spectral difference is marked by the appearance of new peaks at 455 and 482 nm, which show a color change from yellowish brown to solid brown. However, the FTIR results for all four-film labels do not show different peaks after irradiation.

3.2. RGB Intensity of film labels containing P. ferrugineum

The film labels containing *P. ferrugineum* were evaluated to determine the relative RGB intensities before and after irradiation, as illustrated in figure 3. After irradiation, the R, G, and total RGB intensities of the film labels decreased depending on the given radiation dose, while the intensity of B remained considerably low. It can be observed that the irradiation dose is inversely proportional to the intensity of R and G, with the relative intensity of G undergoing a more significant decrease compared to the other values. This result is consistent with the color changes in the film labels (from yellowish

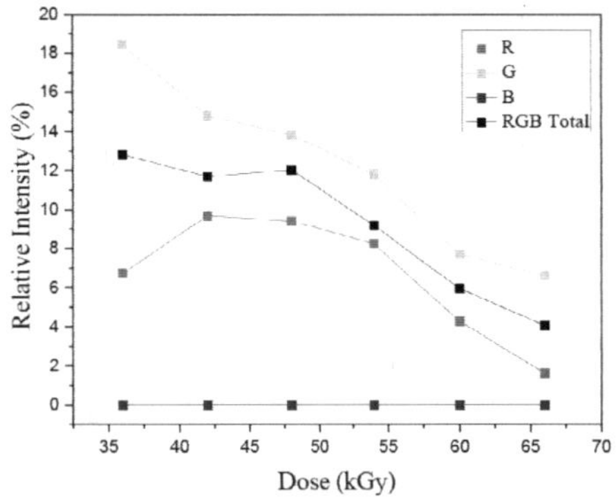

Figure 3. Δ Relative intensities of film labels containing *P. ferrugineum* in the dose range 36–66 kGy.

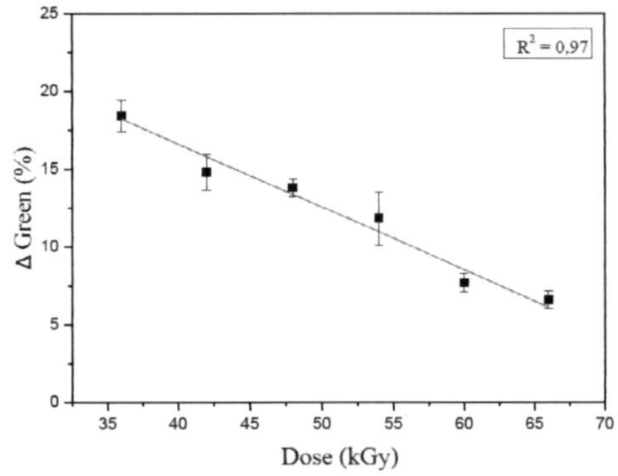

Figure 4. Linear fit of the green intensity (with standard deviation) for film labels containing *P. ferrugineum*.

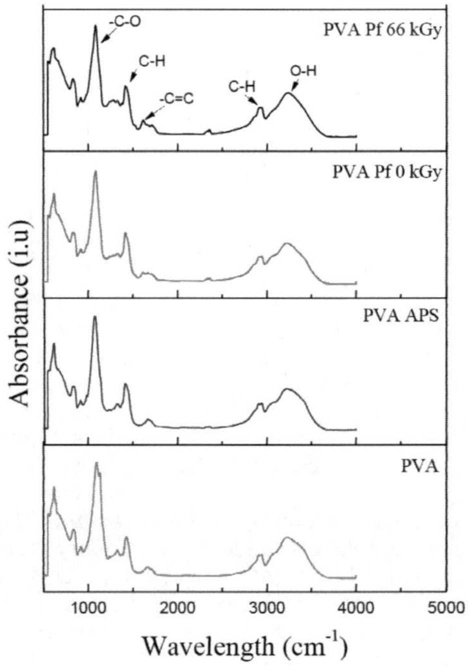

Figure 5. FTIR spectra of the PVA, PVA–ammonium peroxodisulfate (APS), and PVA–APS–*P. ferrugineum* samples irradiated at 0 kGy and that of a PVA–APS–*P. ferrugineum* sample irradiated at 66 kGy.

brown to brown), as indicated by the peak observed at 482 nm in the optical spectra of the samples. The spectrum intensity (RGB) of the colorimeter depicts that the intensity of G on the film labels reflects the sensivity to detect gamma-ray detection; the linear compatibility that is observed to be R^2 = 0.97 for the relative intensity of G at different doses also supports this analysis (figure 4). The standard deviation appears different for each dose, which is due to two factors: homogeneity differences in the film labels and the stability of the colorimeter.

3.3. FTIR characterization of film labels containing P. ferrugineum extract
The FTIR spectra of the PVA, PVA–APS, and PVA–APS–*P. ferrugineum* film labels taken before and

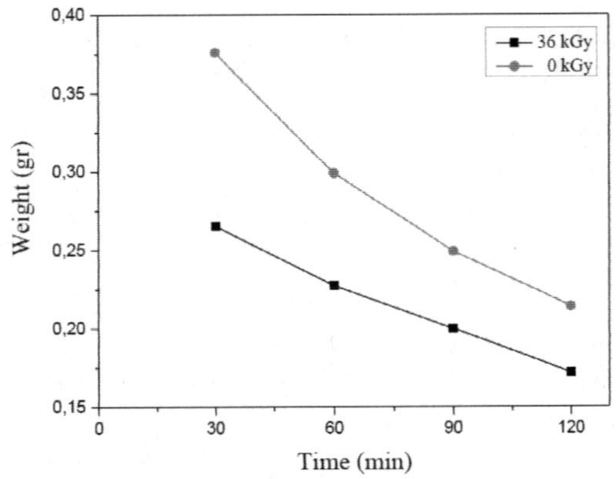

Figure 6. Swelling tests on film labels containing *P. ferrugineum* before (red) and after irradiation (black).

after gamma irradiation at 66 kGy are shown in figure 5. The absorbance spectra show the same pattern for all the film labels at 1089, 1410, 1611, 2918, and 3226 cm^{-1}, and they are identical to the spectra obtained by Sonker *et al.* [8] who studied the effects of the crosslinker on PVA. In both cases, the PVA pattern could be observed for all four-film labels. The signals observed at 1089, 1611, 2918, and 3226 cm^{-1} can be assigned to the stretching vibrations of the –C–O, –C=C, C–H, and O–H bonds, respectively. There also exists a band at 1410 cm^{-1}, which corresponds to the bending of the –CH bond. The wide absorbance band observed at 3226 cm^{-1} is due to O–H stretching from the hydroxyl group, which illustrates the hydrophilic properties of PVA [15]. This is also supported by the results of the swelling test. The FTIR characterization results observed from the film labels in this research are depicted in figure 5. No change is observed in the spectral form both before and after irradiation. Similarly, film labels containing PVA extract and *P. ferrugineum* exhibit the dominant PVA spectrum. This is strongly influenced by the composition of label making that is observed between PVA and *P. ferrugineum*.

3.4. Swelling tests

Results of the swelling tests for the film labels containing *P. ferrugineum* before and after gamma irradiation are depicted in figure 6. The weight of the film labels that is observed before and after irradiation is observed to decrease with the increasing test time. This was because some of the PVA molecules that was not well-crosslinked by APS dissolved in water. The weight loss for the film labels before irradiation was observed to be sharper than the weight loss observed in case of the film labels after irradiation. During the irradiation process, the film labels became hot so that the water molecules that were absorbed by the film would evaporate.

4. Conclusions

The plastic film labels made of PVA-APS and natural dye from P. *ferrugineum* extract has been successfully prepared as a simple dosimeter. The color of the film labels changed from yellowish brown to solid brown after gamma irradiation at a dose of 36–66 kGy, followed by a linear decrease in the relative intensity of the green color component (R^2 = 0.97). All the characterizations performed in this study showed positive results so that the PVA–APS–*P. ferrugineum* extract film labels could be used as a simple dosimeter label.

Acknowledgements

The authors would like to thank Einago (www.einago.com) for the English language review.

References

[1] International Atomic Energy Agency (IAEA) 2007 *Dosimetry in Diagnostic Radiology: An International Code of Practice*, Technical Report Series No. 457 (Vienna: IAEA)

[2] El Gohary M I, Soliman Y S, Amin E A, Gawad M H and Desouky O S 2016 *Appl. Radiat. Isot.* **113** 66–9
[3] Abdel-Fattah A, Beshir W B, Hassan H M and Soliman Y S 2017 *Radiat. Meas.* **100** 18–26
[4] Zayzafoon G and Alkassiri H 2008 *Radiat. Meas.* **43** 1550–3
[5] Kattan M, Al Kassiri H and Daher Y 2011 *Appl. Radiat. Isot.* **69** 377–80
[6] Al Zahrany A A, Rabaeh K A and Basfar A A 2011 *Radiat. Phys. Chem.* **80** 1263–67
[7] Rabaeh K A, Basfar A A and Moussa A A 2012 *Radiat. Phys. Chem.* **81** 479–83
[8] Sonker A K, Rathore K, Nagarale R K and Verma V 2018 *J. Polym. Environ.* **26** 1782–94
[9] Shahid M A K, Bashir B, Bashir H, Bashir H and Mubashir A 2014 *International Journal of Chemistry and Material Science* **2** 28–45
[10] deOliveira J C, David J M and David J P 2016 *Phytochem. Lett.* **16** 18–22
[11] Sivagnanasundaram P and Karunanayake K O L C 2015 *OUSL Journal* **9** 1–17
[12] Lavalle M, Corda U, Fuochi P G, Caminati S, Venturi M, Kovács A, Baranyai M, Sáfrány A and Miller A 2007 *Radiat. Phys. Chem.* **76** 1502–6
[13] Akhtar S, Hussain T, Shahzad A, Hussain M Y, Akhtar N 2013 *J. Basic Appl. Sci.* **9** 416–9
[14] El-Kelany M, Ebraheem S and Gafar S M 2012 *Chem. Mater. Res.* **2** 39–46
[15] Gonzalez J S, Ludueña L N, Ponce A and Alvarez V A 2014 *Mater. Sci. Eng. C Mater. Biol. Appl.* **34** 54–61

IOP Conf. Series: Materials Science and Engineering **496** (2019) 012042 doi:10.1088/1757-899X/496/1/012042

Synthesis and characterization of NaY zeolite from Bayat natural zeolite: effect of pH on synthesis

G S A Melaningtyas[1], Y K Krisnandi[1] and R Ekananda[2]

[1] Department of Chemistry, Faculty of Mathematics and Natural Sciences (FMIPA), Universitas Indonesia, Depok 16424, Indonesia

[2] Research and Development PT Pertamina (Persero), Jalan Raya Bekasi, Jakarta 13920, Indonesia

Corresponding author: yuni.krisnandi@sci.ui.ac.id

Abstract. Herein, NaY zeolite was synthesized by a hydrothermal method via seeding using Bayat natural zeolite with a Si:Al ratio of 6.26, which comprised the silica and alumina sources as well as sodium aluminate as the additional alumina source. Poorly crystalline natural zeolite contains several impurities; thus, activation and purification are crucial for removing some impurities and widening the zeolite pores. Before synthesis, Bayat natural zeolite was subjected to dealumination to remove some alumina and fragmentation to break the zeolite framework into their monomers. In addition, the effect of pH on synthesis was investigated. NaY zeolite was characterized via X-ray diffraction (XRD) and Fourier transform infrared (FTIR) spectroscopy equipped with energy-dispersive X-ray spectroscopy (EDX). The XRD patterns revealed that a pH of 11–12 is typical for synthesizing NaY zeolite. Different FTIR spectra were observed for the raw, fragmented, and as-synthesized NaY. The EDX data revealed that NaY zeolite synthesized at pH values of 11–13 exhibits Si:Al ratios of 5.35, 3.72, and 2.96, respectively.

Keywords: NaY zeolite, natural Bayat zeolite, fragmentation method, seeding method, pH effects

1. Introduction

NaY zeolite exhibits a faujasite-type framework with saturated sodium (Na^+) cations, which plays an important role in the petrochemical industry because it exhibits some favorable properties, including high catalytic activity, atomic surface ratio, specific surface area, and surface energy [1]. NaY zeolites have been widely used in separation and ion exchange as well as acid catalysts for size- and shape-selective catalytic reactions related to their molecular-sized microporosity [2]. In this study, Bayat natural zeolite was used as the main silica and alumina source with a dominant mordenite structure [2]. Poorly crystalline natural zeolite contains some minerals, which are unevenly distributed, with several impurities [4]. The presence of these impurities can decrease the activity of zeolite; hence, it is crucial to activate and purify zeolites. Before synthesis, Bayat natural zeolite was subjected to dealumination and fragmentation. pH is a crucial factor during synthesis [3]. At an extremely high pH, the solubilities of the Si and Al sources can increase, thereby possibly accelerating the polymerization rates of polysilicates and aluminate anions. This in turn leads to an increase in the amount of time required for shorter nucleation in order to accelerate the crystallization time. However, extremely high crystallization rates may lead to the formation of an imperfect zeolite framework or other zeolite frameworks [4]. Therefore, Bayat natural zeolite is synthesized before its conversion into NaY zeolite and the effect of pH on synthesis is investigated.

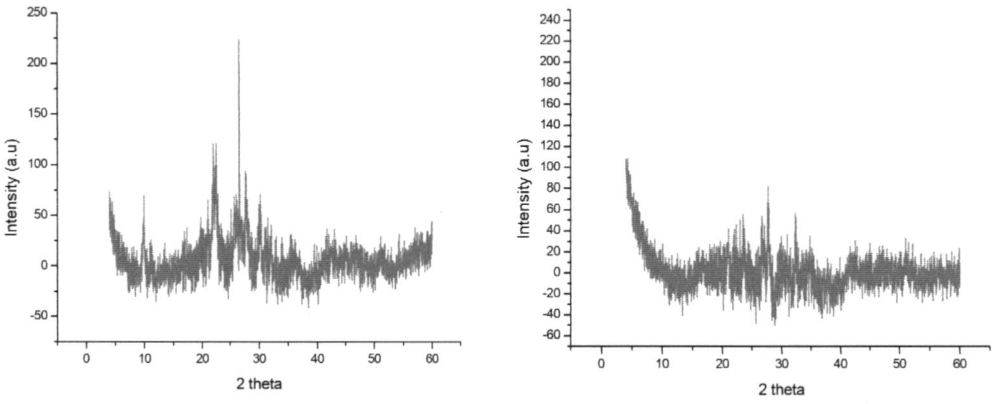

Figure 1. (a) Raw Bayat natural zeolite and (b) fragmented Bayat natural zeolite

2. Experimental

2.1. Synthesis of Bayat natural zeolite

2.1.1. Activation. Bayat natural zeolite was dispersed in deionized (DI) water at a ratio of 1:3 (w/v) and heated at 300 °C for 3 h to separate zeolite from impurities.

2.1.2. Purification. To dissolve carbonate salts, zeolite was dispersed into a pH 5 sodium acetate buffer at a ratio of 1:3 (w/v) at 80 °C for 1 h. Next, the solution was decanted and stirred again by the dropwise addition of DI water and a 30% H_2O_2 solution with simultaneous heating (in case of bubbles, 0.1 M of an acetic acid solution was added until bubbles were not observed) [5]. The mixture was decanted, and a precipitate was obtained by the addition of a Na–citrate–bicarbonate solution and sodium dithionite and stirring at 80 °C for 30 min. The precipitated zeolite was washed using DI water and then dried at 105 °C [6].

2.1.3. Dealumination. This step involves the reduction of the levels of alumina in zeolite. The purified zeolite was dissolved with 2-M HCl at a ratio of 1:10 (w/v) and stirred at 100 °C under reflux for 2 h.

2.1.4. Fragmentation. Zeolite and NaOH were homogenized at a ratio of 1:2, followed by the addition of DI water, stirring until homogeneity was attained, and heating at 250 °C for 2 h.

2.2 Synthesis of NaY zeolite

The seed gel was synthesized with a molar ratio of 10.67 Na_2O:Al_2O_3:10 SiO_2:180 H_2O [7]. A mixture of the seed gel was stirred in a polypropylene (PP) bottle and then aged for 24 h at room temperature. For synthesizing NaY zeolite, Bayat natural zeolite was used as the silica and alumina source and sodium aluminate was used as an additional alumina source. For this synthesis, the molar ratio was 6.25 Na_2O:Al_2O_3:10 SiO_2:180 H_2O. A mixture of Bayat natural zeolite, sodium aluminate, and DI water was stirred for 48 h, with the addition of varying volumes of glacial acetic acid (pH 11, 12, and 13). Next, the seed gel was inserted into a PP bottle and heated at 100 °C for 24 h. After heating, two phases were formed. Then, the precipitate phase was washed with DI water until a pH of 9 was attained, following by drying at 60 °C.

3. Results and discussion

3.1 Characterization of Bayat natural zeolite

The X-ray diffraction (XRD) patterns of the raw and fragmented Bayat natural zeolite (figure 1) revealed poor crystallinity for the raw zeolite; peak shifts at 19.10°, 20.74°, and 20.86° after fragmentation; and the loss of some peaks at 21.94°, 22.48°, and 26.60°. The XRD patterns also revealed the presence of a high amount of amorphous materials, indicating successful fragmentation.

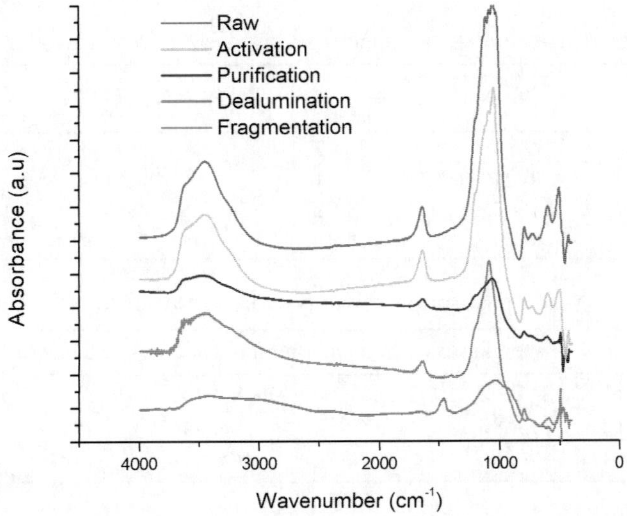

Figure 2. Fourier transform infrared (FTIR) spectrum of Bayat natural zeolite

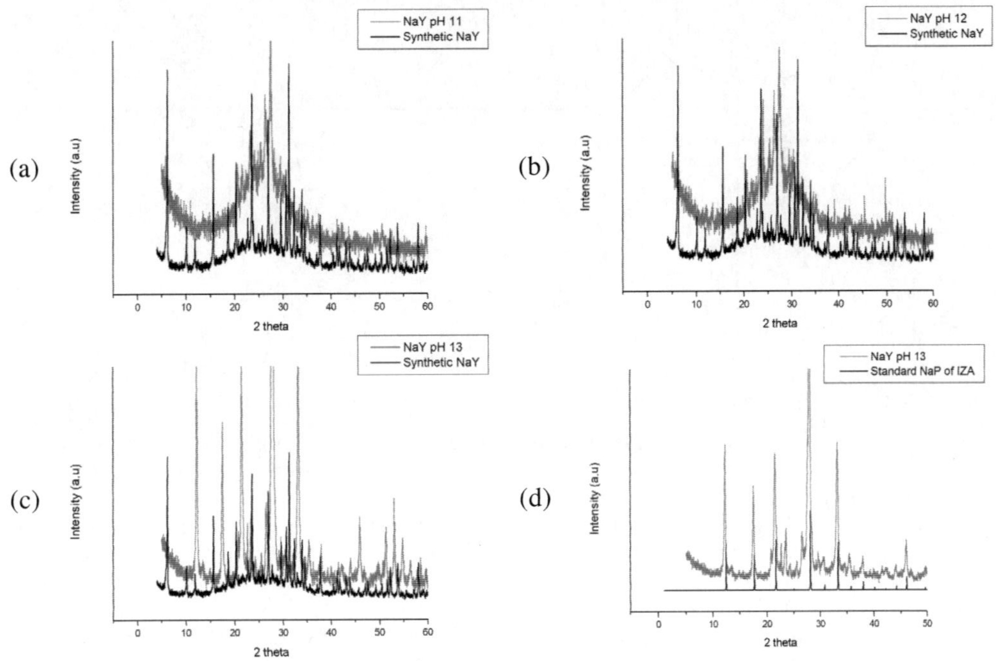

Figure 3. Comparison of NaY zeolite at (a) pH 11, (b) pH 12, (c) pH 13, and (d) pH 13 with that of standard NaP zeolite from the IZA database.

The Fourier transform infrared (FTIR) spectrum of the raw Bayat natural zeolite was compared with that of the fragmented natural zeolite (figure 2). Raw zeolite adsorbed a considerable amount of water, as can be observed from the peaks at 4000–3000 cm^{-1} and 1620 cm^{-1}, corresponding to the molecular water adsorbed on the zeolite surface. After fragmentation, a small peak was observed at 1500 cm^{-1}, corresponding to an aluminum oxide compound. This result indicated that the alumina-silicate framework collapses, affording smaller fragments. The FTIR analysis results supported the XRD data.

Table 1. Energy-dispersive X-ray spectroscopy (EDX) data of as-synthesized NaY.

Element (%wt)	Raw	NaY		
		pH 11	pH 12	pH 13
Si	37.84	38.02	35.09	30.87
Al	6.04	7.1	9.42	10.42
Na	0.42	4.02	6.4	6.63
Si/Al ratio	6.26	5.35	3.72	2.96

Table 2. Water adsorption capacity of as-synthesized NaY.

Zeolite	% Weight loss	Weight after adsorption of water (g)
NaY pH 11	3.92	0.5043
NaY pH 12	4.83	0.5244
NaY pH 13	13.82	0.5351

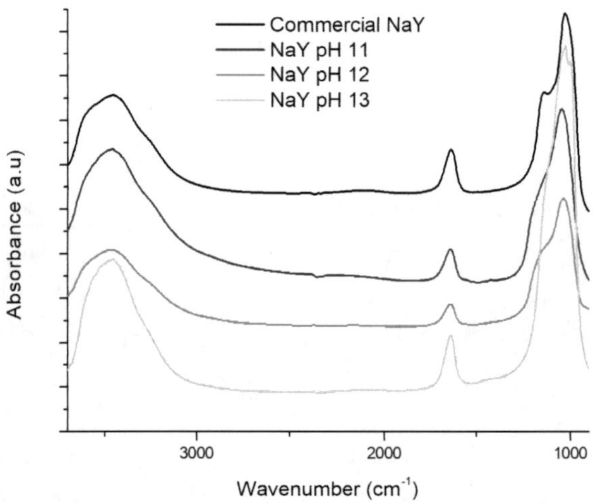

Figure 4. FTIR spectrum of as-synthesized NaY.

3.2. Characterization of NaY zeolite

The XRD patterns of NaY as-synthesized at pH 11–13 revealed the formation of crystalline NaY compared with that observed for synthetic NaY zeolite. This result was confirmed by comparing with the XRD pattern of standard NaY and NaP zeolite obtained from the International Zeolite Association (IZA) database (figure 3). NaY zeolite was not completely formed at pH 11 and 12, but some similarities to the synthetic NaY zeolite XRD pattern at 2θ values of 6.73°, 10.91°, 23.23°, and 27.47° were observed. This result is probably related to the presence of cations other than Na, such as K^+, Mg^{2+}, Ca^{2+}, and Fe^{2}, which were not removed during purification [8]. In addition, an uneven baseline was observed in the XRD patterns of NaY at pH 11 and 12, indicating a mixed amorphous fraction in zeolite. However, for NaY prepared at pH 13, the XRD pattern revealed an extremely high crystalline structure, but this structure was found to be similar to that of P zeolite based on its comparison with the XRD pattern of the standard NaP zeolite from the IZA database because of the typical peaks observed for P zeolite. Nevertheless, the XRD patterns revealed that the Bayat natural zeolite structure was successfully destroyed and converted into a different zeolite structure.

A comparison of the FTIR spectra of NaY prepared at pH values of 11, 12, and 13 with that of commercial NaY is shown in figure 4. From the band observed at 1600 cm^{-1}, NaY prepared at pH 13 exhibited the highest H–OH intensity. This result was in agreement with the energy-dispersive X-ray spectroscopy (EDX) data.

The EDX data (table 1) revealed that the lower the Si:Al ratio, the higher the Na ratio. Thus, NaY prepared at pH 13 exhibited high polarity and can adsorb a considerable amount of water.

The water adsorption capacity of as-synthesized NaY was investigated (table 2). From the % weight loss of zeolite after activation at 200 °C was in agreement with the EDX data.

4. Conclusions

In this study, NaY zeolite was prepared from Bayat natural zeolite, which was subjected to dealumination and fragmentation. A pH of 11–12 was successful for the synthesis of NaY zeolite, which was verified via the XRD pattern and the fact that Si:Al ratio was similar to that of synthetic NaY. However, at pH 13, multiple hydroxyl (–OH) groups were present, leading to the dissolution of silica and the formation of the NaP framework. Furthermore, a low Si:Al ratio in NaY leads to a high polarity, implying a high ability for adsorbing water.

Acknowledgements

The authors acknowledge funding from PT. Pertamina (Persero) with contract number No.004/E20400/2016-S0 and Einago (www.einago.com) for the English language review.

References

[1] Xu Y, Cai L, Shao L and Chen J 2012 *Pet. Sci.* **9** 106–9
[2] Liu Z, Shi C, Wu D, He S and Ren B 2016 *Journal of Nanotechnology* **2016** 1486107.
[3] Rohayati, Krisnandi Y K and Sihombing R 2017 *AIP Conf. Proc.* **1862** 030094
[4] Lechert H 1998 *Micropor. Mesopor. Mater.* **22** 519–23
[5] Ming D W and Dixon J B 1987 *Clays Clay Miner.* **35** 463–8
[6] Mehra O P and Jackson M L 1960 *Clays Clay Miner.* **7** 317–27
[7] Ginter D M, Went G T, Bell A T and Radke C J 1992 *Zeolites* **12** 742–9
[8] Matti A H and Surchi K M 2014 *Int. J. Innov. Res. Sci. Eng. Technol.* **3** 13333–42

Characterization of relaxor ferroelectrics from $BiFeO_3$ doped $(K_{0.5}Na_{0.5})NbO_3$ systems

U Nuraini, N A Triyuliana, M Mashuri and S Suasmoro

Departement of Physics, Faculty of Sciences, Institute of Technology Sepuluh Nopember, Surabaya 60111, Indonesia

Corresponding author: suasm@its.ac.id

Abstract. The additon of $BiFeO_3$ (BFO) to $(K_{0.5}Na_{0.5})NbO_3$ (KNN) systems and termed as (1-x)KNN-xBFO for x=0 and x=0.075 were successfully synthesised using a combination of solid-state reaction and auto-combustion methods. The single phase of KNN-BFO was formed by two step of heat threatments; it is calcined at 575 °C for 6h then sintered at 1100 °C for 6h. The phase structure of KNN material changes from the orthorombic symmetry to tetragonal symmetry after the addition of BFO. The relaxor ferroelectric properties were carried out using impedance analyzer at various temperature and frequency. The pure KNN shows a typical ferroelecrics that has two Curie temperature at 220 °C (orthorombic-tetragonal transition) and 440 °C (tetragonal-cubic transition). After the 7.5%mol.BFO doped in the KNN system, the maxima peak around Tc becomes broader and its transition temperature becomes invisible then has a relaxor behaviours. The Currie-Weiss law is used to prove the ferroelectric of KNN ($\gamma \leq 1$) and relaxor of KNN-BFO ($\gamma \geq 1$).

Keywords: KNN-BFO, relaxor, ferroelectric,

1. Introduction

$(K_{0.5}Na_{0.5})NbO_3$ (KNN) is one of the best candidates capable of replacing the role of $Pb(Zr,Ti)O_3$ (PZT) as a ferroelectric material. KNN has a good relative permittivity value, high Curie temperature, and environmental friendliness [1]. Zhou *et al.* [2] stated that pure KNN has an orthorhombic structure at room temperature and turn into a tetragonal symmetry structure at 200 °C then becomes a cubic symmetry structure at 400 °C. While Kumar *et al.* [3] revealed that this material can produce a dielectric constant of 648 at room temperature and the critical temperature of 372 °C. The transition of the phase structure in the material is marked by the change in the value of the dielectric constant at a certain temperature [4]. Pure KNN ceramics behave a good ferroelectric with Pr ~20 $\mu C/cm^2$ besides piezo ceramics with $d_{33} \approx 90$ pC/N [5].

Dopant that substituted into the KNN structure will affect the performance of the material either from its electrical value or piezoelectric constant. Agustinawati *et al.* [6] revealed that Li^{2+} ions that substitute on KNN system shifts the critical temperature from 440 °C to 510 °C. While Du *et al.* [7] reported that the 10% mol $Ba_{0.5}Sr_{0.5}TiO_3$ (BST) doped KNN indicates the dielectric loss value less than 4% (which is relatively low) for the temperature region of 100-250 °C and a broader dielectric peak with maximum permittivity close to 2500. The result also ensures that doped BST will modify KNN's dielectric properties from a normal ferroelectric to become a relaxor typical after x>0.05. $BiFeO_3$ (BFO) recently have attracted much attention for its high Neel temperature T_N of 643 K and its high Curie temperature T_C of 1103 K [8]. The amount of BFO doped into KNN structure is expected to provide better dielectric performance. In this study, addition of BFO with variation x=0.075 was used to substitute KNN materials. The influence of BFO on phase structure and electrical properties of KNN ceramics was investigated.

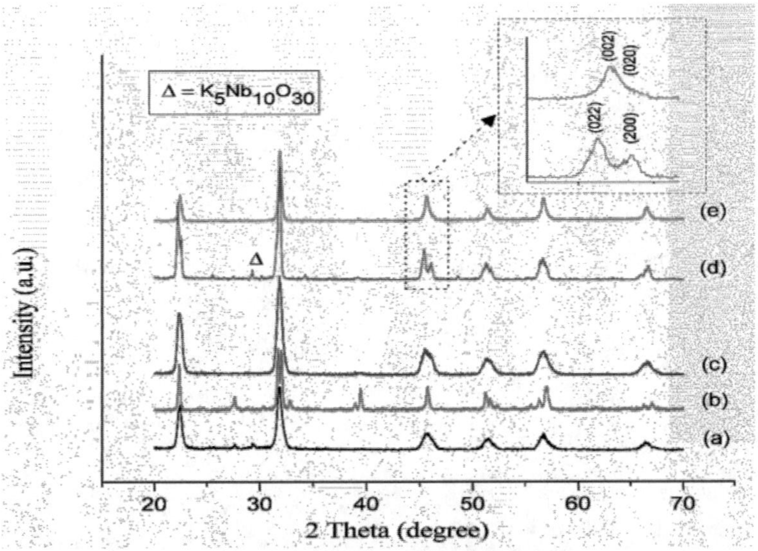

Figure 1. XRD spectra of samples calcined *ff*powders: (a) KNN, (b) BFO, (c) 0.925KNN-0.075BFO, and sintered pellet of (d) KNN, (e) 0.925KNN–0.075BFO

2. Methods

(1-x) $K_{0.5}Na_{0.5}NbO_3$ – x$BiFeO_3$ (KNN-BFO) materials (with variation x = 0 and 0.075) were prepared through two-step powder synthesis method. The first step was separately synthesized $K_{0.5}Na_{0.5}NbO_3$ (KNN) and $BiFeO_3$ (BFO) powders. The KNN powder was synthesized being similar from our work [9] while BFO powders was prepared by auto-combustion method by using $Bi(NO_3)_3.5H_2O$, $Fe(NO_3)_3.9H_2O$, and $C_6H_8O_7$ as raw materials [10]. The second step was mixed KNN and BFO powders in ethanol media and ball milled for 6 hours then evaporated to remove the alcohol content in the samples. After calcined at 575 °C for 6 hours, the mixture is placed a 13 mm diameter of disk then pressed at 55 MPa and sintered in air at 1100 °C for 6 hours.

X-Ray diffraction (XRD) characterization with monochromatic Cu Kα radiation (X'Pert Diffractometer, Philips) was used to determine the crystal structure of the samples. Then the phases exist in the samples were analysed using Match II software. The further analysis of structure was exploited using Rietvel's refinement method of Rietica software [11]. SI 1260 Impedance/Gain-Phase Analyzer (0.1 Hz-32 MHz) is used in electrical characterization as a function of temperature from room temperature to 500 °C). Silver paste was printed as electrodes on the pellet sample that has been sintered then it fired at 925 °C for 20 min. The relative permittivity that measured at 1.27 kHz, 12.7 kHz, and 127 kHz as function of temperature was used to obtain the dielectric properties from the samples. While, the characterization of relaxor was determined by using the modified of Curie-Weiss law.

3. Results and discussion

X-Ray Diffraction of sintered samples of (1-x) $K_{0.5}Na_{0.5}NbO_3$ -x$BiFeO_3$ (KNN-BFO) (with x = 0 and 0.075) in the range of 2θ (from 20° to 70°) is showed in figure 1. Similar with previous study [9], the minor phase which identified as $K_5Nb_{10}O_{30}$ was exist because alkali metals such as Na and K can volatile during the sintering process [12,13]. All ceramics poses typical perovskite (ABO_3) spectra while the peak intensity varies depending on the x value. At x=0, the XRD pattern shows orthorhombic symmetry at room temperature which can be evidenced by the (022/200) splitting of the diffraction peaks around 2θ ≈ 45°, as reported Zhou *et al.* [2]. When x=0.075, the split difraction peaks seem to merge into a single peak (002/020), indicating that the crystal structure was succesfully modified after BFO doped. The change of peak diffraction is explained by Xu *et al.* [14] as the success of BFO diffusion process to substitute KNN lattice. For complete this suggestion, the quantitative

IOP Conf. Series: Materials Science and Engineering **496** (2019) 012043 doi:10.1088/1757-899X/496/1/012043

Table 1. Lattice parameter of (1-x)KNN-xBFO analysed by Rietveld refinement using Rietica software

Material	Lattice Parameter (Å)			Crystal Structure
	a	b	c	
KNN	3.95523 ± 0.00148	5.63766 ± 0.00226	5.67845 ± 0.00215	Orthorhombic
0.925KNN-0.075BFO	3.97813 ± 0.00048	3.97813 ± 0.00048	3.99989 ± 0.00047	Tetragonal

Table 2. Density of KNN and KNN-BFO by Archimedes method.

No.	Sample	Density		
		Archimedes (g/cm^3)	Theoretic (g/cm^3)	Relative (%)
1	KNN	3,78	4,48	84,37
2	0.925KNN-0.075BFO	4,19	4,75	87.31

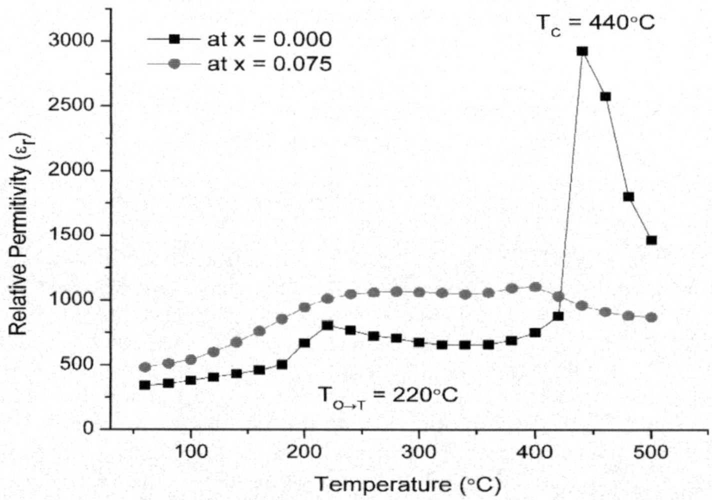

Figure 2. Temperature dependence of the relative permittivity for (1-x)KNNxBFO ceramics with x = 0 and 0.075 at 127 kHz

analysis is done by Rietveld's Refinement method using Rietica software. The result of refinement from all samples is described in table 1. It is presented that orthorhombic symmetry crystal structure is the initial structure of KNN. With the addition of BFO, its crystal structure begins to transform into a tetragonal symmetry (a=b≠c). The existance of minor phase ($K_3Nb_{10}O_{30}$) that appears in KNN caused the highest reliability factor, similar from our previous work [9]. The difference in the radius of the dopant and the main atom (Fe^{3+} on Nb^{5+} sites and Bi^{3+} on K^+/Na^+ sites) results in a rearrangement of the lattice where the a-axis becomes extended while c-axis will shrink. The radius of the Bi^{3+} (103 pm) ion is about the Na^+ (102 pm) and not far from the K^+ (138 pm) ions radii, while the radius of the Fe^{3+} (64.5 pm) ion is just about the Nb^{5+} (64 pm) ion radius, therefore it would be reasonable to speculate that both ions (Bi^{3+} and Fe^{3+}) could be easily substituted into KNN lattice, on site A and site B. The density of pellet samples is measured using Archimedes method and its result is written in table 2. BFO doped increase the density of KNN so it can be mean that BFO doped made the shrinkage process more optimal than pure KNN. The shrinkage process will influence the dielectric properties of materials.

Figure 2 presents the relative permittivity (ε_r) of (1-x) KNN- x BFO samples as a function of temperature measured at 127 kHz during heating process. When x=0, the material shows two-phase transition temperature, at 220 °C associated with the tetragonal-cubic (T_c) transition and the orthorhombic-tetragonal transition (T_{O-T}) at 440 °C. It is also have reported in previous study [9] which

IOP Conf. Series: Materials Science and Engineering **496** (2019) 012043 doi:10.1088/1757-899X/496/1/012043

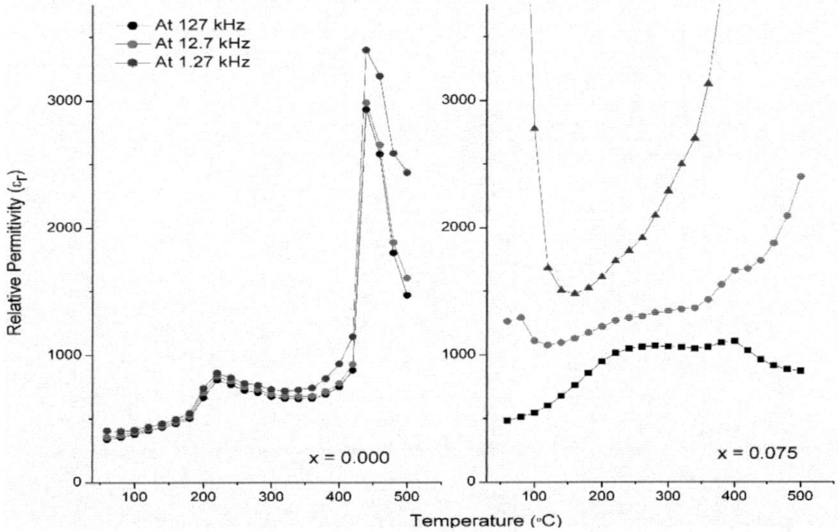

Figure 3. Temperature dependence of the relative permitivity for (1-x)KNN – xBFO ceramics with x 0 and 0.075 measured at different frequencies

corresponds to the result of Ahtee and Glazer project [4]. When x=0.075 mol.% BFO, no peaks that indicate the temperature of (T_{o-t}) or tetragonal-cubic (T_c) transition can be observed evidently. Xu *et al.* [14] has stated about the boundary between tetragonal phase and orthorhombic which called as MPB (morphotropic phase boundary). It appears when 0.01 mol.% BFO added into KNN. Moreover, a MPB also can be expected in the system through transition point for the structure (by diffraction pattern) when the split (022/200) peaks from an orthorhombic symmetry are gradually merged into a single peak [15].

The relative permittivity for (1-x)KNN-xBFO samples measured at different frequencies as a function of temperature is shown in figure 3. Increasing frequencies doesn't change the phase transition temperature for pure KNN, besides the fact that pure KNN ceramics are known to have two phase transition temperature, both orthorhombic-tetragonal and tetragonal to cubic [16]. These results indicate that pure KNN act as a normal ferroelectric which does not affected by frequencies changes. When x=0.075 the T_c point changed for those frequencies and at 1.27 and 12.7 kHz there are "tail curve" for high temperature. Our previous work [17] explained that there are two possible explanation; the existence of multiple ferroelectric structure and effect of defect dipoles. In this case, the defect dipole that created when BFO doped KNN should be:
On A sites,

$$Bi_2O_3 \xrightarrow{KNN} 2Bi_{K/Na}^{\bullet\bullet} + 4e' + O_O^X$$

On B sites,

$$Fe_2O_3 \xrightarrow{KNN} 2Fe_{Nb}'' + 2V_O^{\bullet\bullet} + O_O^X$$

The defect dipole creates an imbalance and produces local electric field in the KNN-BFO structure. Local electric fields and elastic fields will induce relaxor behaviour with the emerging of polar nanoregions (PNRs) [7].

The characteristic of ferroelectric-relaxor behavior can be proved with obey a modified Curie-Weiss law,

$$\frac{1}{\varepsilon} - \frac{1}{\varepsilon_m} = \frac{(T - T_m)^\gamma}{C}$$

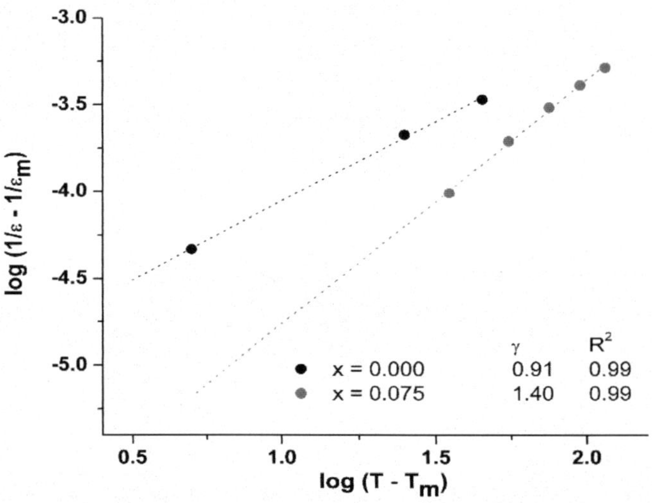

Figure 4. The plot of log ($1/\varepsilon$-$1/\varepsilon_m$) as function of log (T-T_m) at 127 kHz for (1-x) KNN–xBFO ceramics.

where ε and ε_m are relative permittivity and its maxima, T and T_m are temperature and its maxima. The value of γ is in the range between 1 and 2 which will indicates the normal ferroelectrics if its value approaches 1 and 2 for ideal relaxor. The value of γ can be determined from the results of fitting the logarithmic plots of the reciprocal permittivity versus temperature. Ignoring the influence of space charges at high temperature during frequency of 127 kHz results the γ value as shown in figure 4. The results confirms that KNN act as a normal ferroelectric with $\gamma \approx 1$, while BFO doped has a relaxor characteristic. It also means that the BFO doped modified the behaviour of KNN which initially pure ferroelectric to become relaxor.

4. Conclusions

The effect of $BiFeO_3$ (BFO) doping in the $K_{0.5}Na_{0.5}NbO_3$ (KNN) system has been investigated for its dielectric properties and its phase structures. The pure KNN that have orthorhombic symmetry transforms to a tetragonal symmetry after 0.075 mol.% BFO doped on the system. It is not also confirmed by quantitative analysis using Rietveld's refinement but also be compared to electrical analysis by cole-cole plot to obtain conductivity. The dielectric properties of KNN was successfully modified which initially act as a normal ferroelectric to become relaxor. The value of γ that be determined using modified Curie-Weiss law is approaches 1 (normal ferroelectric) for KNN and 2 (ideal relaxor) for KNN-BFO.

Acknowledgements

The research supported by Ministry of Research Technology and Higher Education of the Republic of Indonesia and LPPM ITS through program PMDSU 2017 No.489/PKS/ITS/2017.

References

[1] Chao X, Yang Z, Li Z and Li Y 2012 *J. Alloys Compd.* **518** 1–5
[2] Zhou J J, Li J F, Cheng L Q, Wang K, Zhang X W and Wang Q M 2012 *J. Eur. Ceram. Soc.* **32** 3575–82
[3] Kumar P, Pattanaik M and Sonia 2013 *Ceram. Int.* **39** 65–9
[4] Ahtee M and Glazer A M 1976 *Acta Cryst.* **A32** 434–46
[5] Qin Y, Zhang J, Yao W, Wang C and Zhang S 2015 *J. Am. Ceram. Soc.* **98** 1027–33
[6] Agustinawati D, Isnaini N L and Suasmoro S 2017 *AIP Conf. Proc.* **1788** 030129
[7] Du H, Zhou W, Luo F, Zhu D, Qu D and Pei Z 2009 *J. Appl. Phys.* **105** 124104
[8] Manzoor A, Afzal A M, Umair M, Ali A, Rizwan M and Yaqoob M Z 2015 *J. Magn. Magn. Mater.* **393** 269–72

[9] Nuraini U, Agustianawati D, Yahya E, Cahyono Y, Kidkhunthod P and Suasmoro S 2017 *Ceram. Int.* **43** 3664–9
[10] Layek S and Verma H C 2012 *Adv. Mater. Lett.* **3** 533–8
[11] Hunter B A 2000 *Rietica - a visual Rietveld program* (Menai: Australian Nuclear Science and Technology Organization)
[12] López-Juárez R, Novelo-Peralta O, González-García F, Rubio-Marcos F and Villafuerte-Castrejón M E 2011 *J. Eur. Ceram. Soc.* **31** 1861–4
[13] Kim J H, Kim J S, Han S H, Kang H W, Lee H G and Cheon C I 2016 *Ceram. Int.* **42** 5226–30
[14] Xu P, Jiang M H and Liu X Y 2011 *Adv. Mater. Res.* **335-336** 968–75.
[15] Zuo R, Ye C and Fang X 2008 *J. Phys. Chem. Solids* **69** 230–5
[16] Chen T, Wang H, Zhang T, Wang G, Zhou J, Zhang J and Liu Y 2013 *Ceram. Int.* **39** 6619–22
[17] Nuraini U, Triyuliana N A, Mashuri M, Kidkhunthod P and Suasmoro S 2017 *J. Mater. Sci. Mater. Electron.* **29** 1139–45

IOP Conf. Series: Materials Science and Engineering **496** (2019) 012044 doi:10.1088/1757-899X/496/1/012044

Electric and magnetic field influence on localization length G4 DNA molecule

L Apriyanti and E Yudiarsah

Department of Physics, Faculty of Mathematics and Natural Sciences (FMIPA) Universitas Indonesia, Depok 16424, Indonesia

Corresponding author: e.yudiarsah@sci.ui.ac.id

Abstract. Electric and magnetic field influence on localization length G4 DNA molecule has been studied. This research used G4 DNA model of 102 base pair. Localization length is calculated using the transfer matrix method. The electric field is applied in the direction parallel to DNA's symmetrical axis. The magnetic field is applied perpendicular to DNA symmetrical axis. The results of the calculation of localization length under influence of electric field showed that the localization length decreases with electric field strength, so more electron state becomes a localized state. Other results show magnetic field has no significant effect on the spectrum of the localization length, even at large magnetic field.

1. Introduction

In the development of nanoscale electronic devices, the electronic properties of DNA have attracted attentions of many scientists [1,2]. Electronic properties of DNA molecule could depend on its structure, temperature, and environment surround it. A special kind of DNA molecule is composed of guanine nucleotides (G), G4 DNA, with its distinctive structure and self-assembling properties [3]. The nature of self-assembly is the ability of molecules to make themselves into supramolecular. The structure of G4 DNA is very stable under physiological conditions because of the quadruple-G quartet that forms quadruple-helical [4].

Many experiments and theoretical studies focused on the nature of the transport of double-stranded DNA (ds-DNA) [5,6]. In this report, we studied the G4 DNA molecule. This research uses G4 DNA molecule to study electric and magnetic field influence on the localization length. G4 DNA molecule is placed in an applied electric and magnetic field. The electric fields is set parallel to DNA symmetrical axis [7]. Whereas, magnetic field is applied perpendicular to DNA symmetrical axis [8]. The parameters contained in this paper include the electric field, magnetic field, frequency, and temperature. Localization length is calculated using a transfer matrix method by operating electron wave function to tight-binding Hamiltonian. Calculation results are discussed in detail in the results and discussion. Then, the discussion will be summarized.

2. Theoretical method

DNA model used in this study is a G4 DNA model of 102 base pairs. The unique structure of G4 DNA consists of a stacked of quartet G. Each quartet G is a set of 4 G in planar structure with hydrogen bonds. The G4 DNA in the electric and magnetic fields is described as the scheme in figure 1. The system can be effectively described with a tight-binding Hamiltonian model. Tight-binding Hamiltonian model of G4 DNA is as follows [5,8].

In equation (1), L is the DNA chain length, 102 base pairs. Indices q, τ, and i represent backbone index, strand index, and site index, consecutively. Parameters ε_i^τ , B_i^q , and ϕ_i^τ respectively refer to base's onsite energy on site i-th and strand τ-th, backbone's onsite energy on site i-th and

$$
\begin{aligned}
H = \sum_{i=1}^{L} \Big[& \sum_{\tau=1}^{4} \left\{ (\varepsilon_i^{\tau} + \phi_i^{\tau}) |i,\tau\rangle\langle i,\tau| + t_{i,i+1}^{\tau,\tau} |i,\tau\rangle\langle i+1,\tau| \right\} \\
& + \sum_{\tau=1}^{4} \left\{ B_i^{\tau,q} |i,\tau\rangle\langle i,q| + t_{i,i}^{\tau,q} |i,\tau\rangle\langle i,q| \right\} \\
& + \sum_{\tau=1}^{4} \left\{ t_{i,i}^{\tau,\tau+1} |i,\tau\rangle\langle i,\tau+1| + t_{i,i+1}^{\tau,\tau+1} |i,\tau\rangle\langle i+1,\tau+1| \right\} \Big] + h.c.
\end{aligned}
\tag{1}
$$

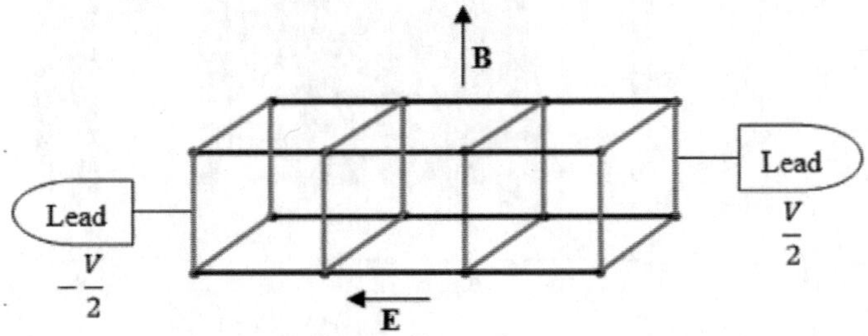

Figure 1. Scheme diagram of G4 DNA in the electric and magnetic fields. The black line is the strand, the blue line represent the boundary of the planar structure of G4 with a guanine base (black dot) at every corner.

backbone q-th, and onsite energy change on site i-th and strand τ-th due to twisting bases twisting motion [9]. Parameter t represents electron's hopping constant. It is dependent on temperature [10] as well as electric [7] and magnetic fields [11]. In the model, we assume that electron hopping between sites is influenced by internal electric field, which is equal in both magnitude and direction to applied electric field.

2.1. Localization Length

In this research, we calculated the localization length for studying the electrical properties of G4 DNA molecule. The matrix transfer method is used to calculate the length of localization based on the following references [12,13]. The matrix transfer method can be used to calculate the amplitude of electron wave function at a given energy from Hamiltonian equation (1). The amplitude of the wave function at site L and L-1 can be calculated using,

$$
\begin{pmatrix} \psi_L \\ \psi_{L-1} \end{pmatrix} = M_{L-1} \dots M_2 \; M_1 \begin{pmatrix} \psi_1 \\ \psi_0 \end{pmatrix} \tag{2}
$$

Multiplication matrices $M_{L-1} \cdots M_2 M_1$ is called a forward matrix multiplication. In addition, wave amplitude at site L-1 and L are calculated as follows,

$$
\begin{pmatrix} \psi_0 \\ \psi_1 \end{pmatrix} = M_1^b \dots M_{L-2}^b M_{L-1}^b \begin{pmatrix} \psi_{L-1} \\ \psi_L \end{pmatrix} \tag{3}
$$

Multiplication $M_1^b \cdots M_{L-2}^b M_{L-1}^b$ above is called a reverse matrix multiplication. The localization length is calculated by multiplying forward and reverse matrix multiplication several times. Every five multiplication, the matrix must be normalized to minimize errors. We used Gram-Schmidt orthonormalization to do that. The normalization coefficient is saved for calculating Lyapunov exponents. The Lyapunov exponent (γ_{cl}) is calculated by summing up the logarithm value of normalization coefficient and then dividing it with the total matrix multiplication,

$$
\gamma_{cl} = \left(\frac{Ln(\prod_m \alpha_{cl}^m)}{2K(L-1)} \right) \tag{4}
$$

The inverse of the smallest γ_{cl} gives the localization length (λ).

3. Results and discussion

The result of the calculations in this study is the electron localization length of G4 DNA molecule with 102 base pairs. The electron localization length when electric field is applied at voltage 2 mV, 4 mV, and 6 mV can be seen in figure 2, figure 3, and figure 4, respectively. The localization lengths are calculated at magnetic field of 1 Tesla, temperature of 4.2 Kelvin, and basis twisting motion frequency of 0.51 meV. The graph in the figures is the localization length as a function of electron energy. A break on Y axis in the graph from point 102 to 200, intended to mark the limit of localized state which

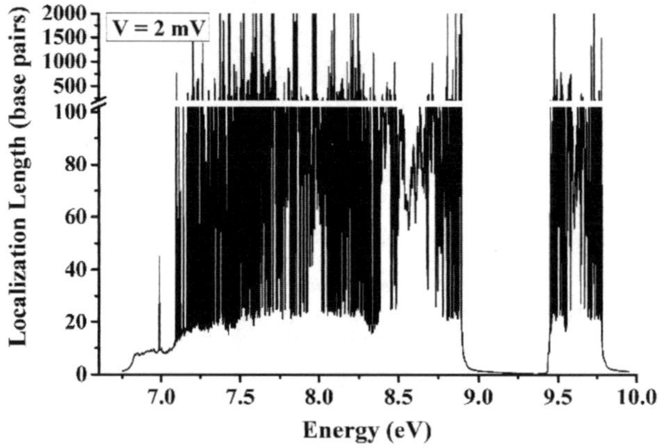

Figure 2. The localization length calculated at voltage of 2 mV, magnetic field of 1 Tesla, temperature of 4.2 Kelvin, and frequency of 0.51 meV.

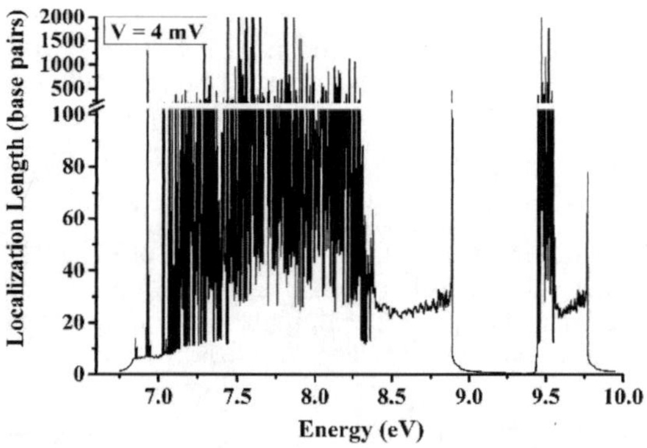

Figure 3. The localization length calculated at voltage of 4 mV, magnetic field of 1 T, temperature of 4.2 K, and frequency of 0.51 meV.

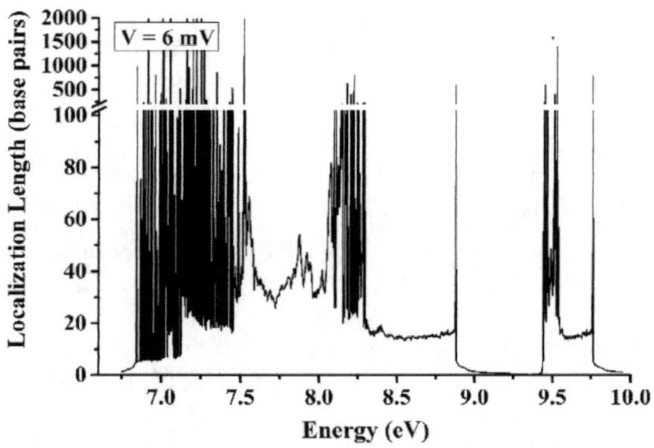

Figure 4. The localization length calculated at voltage of 6 mV, magnetic field of 1 Tesla, temperature of 4.2 K, and frequency of 0.51 meV.

has localization length less than the length of DNA (102 base pairs). Figure 2 shows many extended states in the energy range 6.75 eV - 8.9 eV and 9.4 eV - 9.8. The extended state can conduct electrical current since electron charge travel along the molecule. In these regions, there are also some localized states, where electron that starts traveling from one end of the DNA cannot reach the opposite end. In addition, there is a gap in the energy range of 8.9 eV - 9.4 eV. The states in the gap has really small or zero localization length. Figure 3 shows similar range of energy where the extended and localized states are. However, at this voltage more states become localized. Furthermore, as we increase electric field by increasing voltage, less extended states are observed, see figure 4. Even though the magnitude of localization length at certain energy decrease as the electric field increases, it is observed that the region of energy with non-zero localization length almost does not change. The effect of increasing the electric field changes the magnitude of localization length.

Figure 5, figure 6, and figure 7 show the effect of magnetic field on electron localization length on G4 DNA molecule. The localization lengths are calculated at applied electric field of 1 mV, temperature of 4.2 Kelvin, frequency of 0.51 meV, and three values of magnetic field. The localization length at the magnetic field of 4 Tesla, 6 Tesla, and 8 Tesla respectively can be seen in figure 5, figure 6, and figure 7. Figure 5 shows many extended states in the energy range of 7.3 eV–8.9 eV and 9.5 eV–9.8. In these two energy ranges, there are localized states such as in the energy range of 6.9 eV–

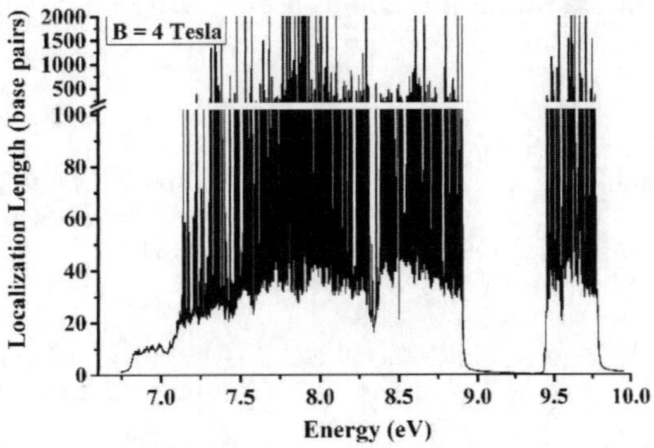

Figure 5. The localization length calculated at magnetic field of 4 Tesla, voltage of 2 mV, temperature of 4.2 Kelvin, and frequency of 0.51 meV.

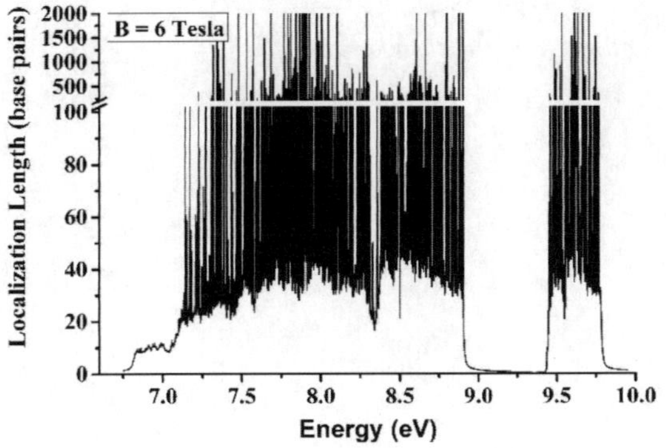

Figure 6. The localization length calculated at magnetic field of 6 Tesla, voltage of 2 mV, temperature of 4.2 Kelvin, and frequency of 0.51 meV.

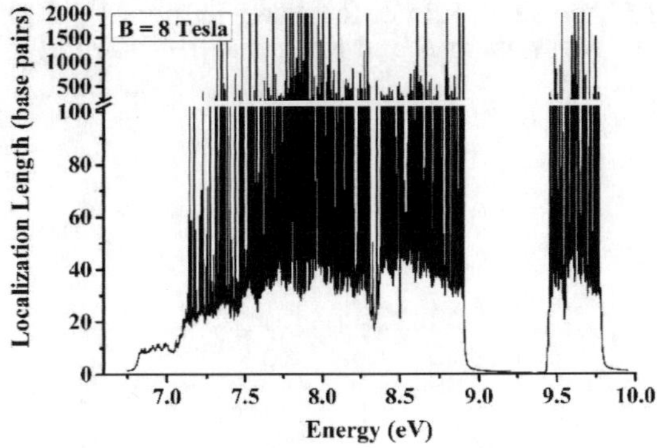

Figure 7. The localization length calculated at magnetic field of 8 Tesla, voltage of 2 mV, temperature of 4.2 Kelvin, and frequency of 0.51 meV

8.9 eV. The spectrum of localization length has a gap in the energy range of 8.9 eV – 9.4 eV. Figure 6 and figure 7 indicate similar feature of localization length to the one in figure 5. It can be said that the magnetic field influence on the electron localization length in G4 DNA molecule is minute. Even though, the magnetic field difference is quite large, magnetic field increases from 1 Tesla to 8 Tesla, the different in the localization length is hard to be observed. This is because the magnetic field in the molecule of G4 DNA only affects the phase of the electron hopping constant and this phase shift is small even for quite large magnetic field for short distance traveled by electron.

4. Conclusions

This research shows that the electric and magnetic field influence on the localization length of G4 DNA molecule. The electric reduces localization length. The greater the value of the applied electric field, then electron becomes more localized at some states. The magnetic field influence on the localization length is not significant. Although a large magnetic field has been applied, due to small change in the phase of electron hopping constant the change in the spectrum of localization length is hard to notice. There are states with the localization length greater than 102 base pairs upwards undergoes decrement in localization length.

Acknowledgements

The author would like to thank Universitas Indonesia for financing of this research through PITTA Grant 2017 with contract number: 649/U2.R3.1/HKP.05.00/2017 and Cisco Laboratory of Departemen

Fisika Universitas Indonesia for permission to use the laboratory, and also Einago (www.einago.com) for the English language review.

References

[1] Guo A M and Xu H 2007 *Phys. Lett.* A **364** 48–53
[2] Alvarez J R, Skachkov D, Massey S E, Lu J, Kalitsov A and Velev J P 2014 *Phys. Rev. Appl.* **1** 034001
[3] Guo A M and Xiong S J 2009 Phys. Rev. B **80** 035115
[4] Kang D W, Sun M L, Zuo Z W, Wang H X, Lv S J, Li X Z and Li L B 2016 *Phys. Lett.* A **380** 977–82
[5] Guo A M, Yang Z, Zhu H J and Xiong S J 2010 *J. Phys. Condens. Matter* **22** 065102
[6] Woiczikowski P B, Kubař T, Gutiérrez R, Cuniberti G and Elstner M 2010 *J. Chem. Phys.* **133** 035103
[7] Jadhav P J, Limketkai B N and Baldo M A 2010 *Adv. Polym. Sci.* **223** 29–44
[8] Kang D, Jiang H, Sun Z, Qu Z and Xie S 2011 *J. Phys. Condens. Matter* **23** 055302
[9] Suhendro D K, Yudiarsah E and Saleh R 2010 *Physica B Condens. Matter* **405** 4806–11
[10] Endres R G, Cox D L and Singh R R P 2004 *Rev. Mod. Phys.* **76** 195–214
[11] Peierls R 1933 *Z. Phys.* **80** 763–91
[12] Klotsa D, Römer R A, and Turner M S 2005 *Biophys. J.* **89** 2187–98
[13] Mackinnon A and Kramer B 1983 *Z. Phys. B. Condens. Matter* **53** 1-13

Study of $LaFeO_3$ perovskite material by high-pressure Raman scattering

D Triyono and H Laysandra

Department of Physics, Faculty of Mathematics and Natural Sciences (FMIPA)
Universitas Indonesia, Depok 16424, Indonesia

Corresponding author: djoko.triyono@sci.ui.ac.id

Abstract. We use high-pressure Raman-scattering spectroscopy to study $LaFeO_3$ synthesized by using the sol-gel method. The structural evolution of the samples under high pressure and at room temperature is monitored via the Raman-scattering spectrum in the pressure range of 0.50 to 22.1 GPa. When the pressure surpasses 12 GPa, the Raman spectra change significantly around 180 cm^{-1}, which indicates a structural phase transition. However, the Raman spectrum remains constant under 5.0 GPa, which indicates that the $LaFeO_3$ structure is preserved at these pressures.

Keywords: $LaFeO_3$, high pressure, Raman scattering, structural transition

1. Introduction

Perovskite orthoferrite is a distorted perovskite structure with the chemical formula $AFeO_3$, space group *Pnma* and/or its derived nonconventional space group, such as *Pbnm, Pmma, Ibnm*, etc. [1]. In this structure, the *A* cation sits in the interstitial area between the octahedral structure and is coordinated to 12 O anions. In addition, the Fe ion is located at the center of the octahedral and is coordinated to six O anions [2]. The magnitude of the distortion in the orthorhombic perovskite structure is closely related to tilting and rotation of the FeO_6 octahedral, where the tilt angle is very sensitive to substitution at the *A* and/or Fe sites, temperature, and pressure applied to the system [2-3].

$LaFeO_3$ is a typical orthoferrite perovskite compound with space group *Pbnm* [3]. At ambient pressure, it is an antiferromagnetic insulator [2,3]. The effect of pressure, temperature, and substitution on $LaFeO_3$ structure has been studied in previous work. For example, Mahapatra *et al.* [1] reported that Ho substitution at La sites causes the Fe-O-Fe angle to depart from 180° and the A–O and Fe–O bond lengths to shorten, which enhances the exchange interaction between A–A and Fe–A, thereby increasing the magnetization. Coutinho *et al.* [2] reported that Y substitution at the La site in $LaFeO_3$ shifts the vibrational modes to higher frequencies, indicating an increased octahedral distortion and decreased bond lengths. The high-pressure phase transition in $LaFeO_3$ was studied by Etter *et al.* [3], who prepared $LaFeO_3$ by using the solid-state method. The material first undergoes a second-order structural phase transition from *Pbnm* to *Ibmm* at 21.1 GPa, and then a first-order phase transition from high-spin to low-spin Fe^{3+} ions at 38.0 GPa.

Building on previous research [1–3], we use high-pressure Raman spectroscopy to understand in more detail the structural evolution of $LaFeO_3$ under high pressure. The vibrational characterization of $LaFeO_3$ in this work is detected by using Raman scattering in the pressure range of 0.50 to 22.0 GPa. The Raman shift reveals the atomic vibrations in the crystal lattice. $LaFeO_3$ was prepared by using the sol-gel method and the samples were used to study how high pressure affects the structural properties of the crystal (i.e., crystal structure and crystallography parameters of $LaFeO_3$). The

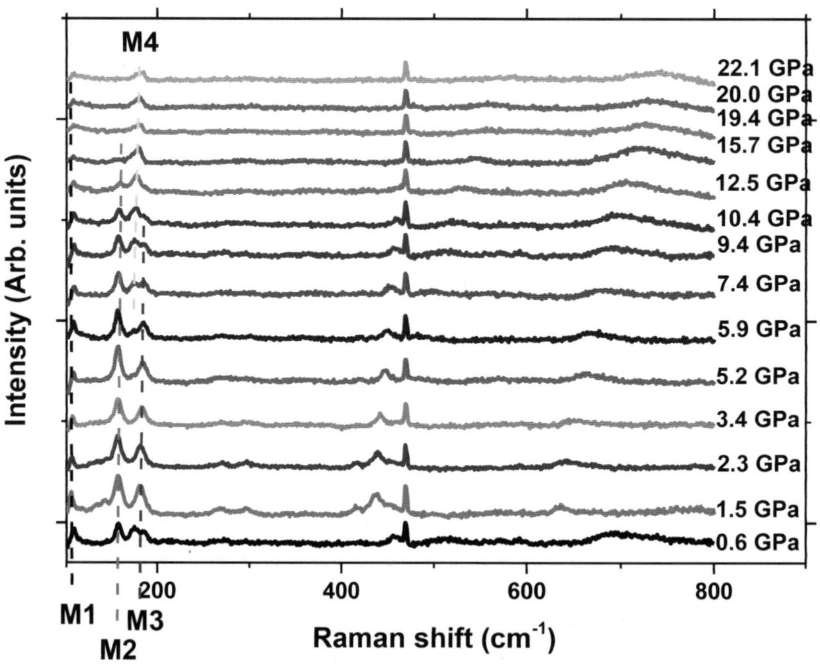

Figure 1. $LaFeO_3$ Raman-scattering spectra for several pressures.

vibrational modes obtained from Raman scattering spectra were fit by using Peakfit software, which allows us to estimate the pressure-induced changes in the structural properties of $LaFeO_3$.

2. Experimental details

By using the sol-gel method, $LaFeO_3$ was synthesized from the precursors lanthanum oxide and ferrite nitrate nonahydrate (all analytical grade). The precursors were balanced stoichiometrically and mixed with citric acid monohydrate and aquadest as solvents. Next, the solvent was stirred and heated with a magnetic stirrer (wisestir MSH-20D) until a brown gel was formed. The gel phase was dried to eliminate water and organic compounds, following which the resulting powder was calcined to obtain $LaFeO_3$ in powder form.

Raman spectroscopy measurements were made in an inVia Renishaw Raman spectrometer system with an excitation-laser wavelength of 512 nm. The powder was subjected to high pressure by using a symmetric diamond anvil cell consisting of a pair of 500 μm diamond anvil culets. The stainless steel gasket was pre-indented from 250 to ~40 μm thick and a 250-μm-diameter hole was laser drilled through the center of the indented area to serve as sample chamber. The powder sample was loaded into the sample chamber along with several ruby spheres for calibrating pressure [4]. Silicone oil was used as a pressure-transmitting medium.

3. Results and discussion

Figure 1 shows the Raman-scattering spectra of $LaFeO_3$ for several pressures at room temperature. $LaFeO_3$ has an orthorhombic perovskite structure with the *Pbnm* space group (not shown here). The increasing of pressure broadens the peaks and decreases of intensity, which indicates increased distortion of the crystal structure. In addition, several Raman modes below 200 cm^{-1} and between 400 and 500 cm^{-1} disappear with increasing pressure.

Based on previous research [2], the Raman modes between 100 and 160 cm^{-1} are attributed to an A_g mode associated with the La site vibration. The Raman mode around 180 cm^{-1} is attributed to the A_g mode associated with La rotation. The Raman mode between 400 and 500 cm^{-1} is attributed to the A_g mode associated with O octahedral bending vibration. However, the significant changes are in the Raman modes below 200 cm^{-1}, and these are the focus of the following discussion.

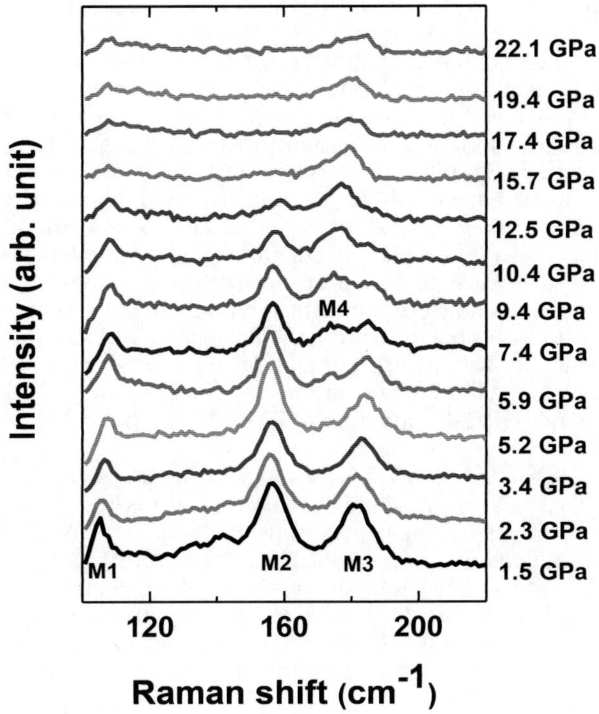

(a)

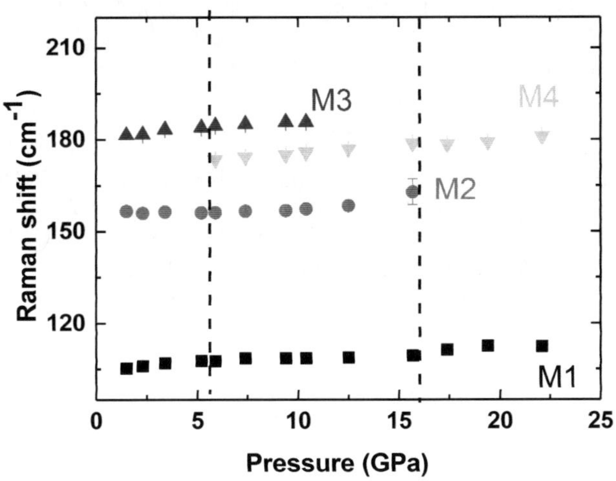

(b)

Figure 2. (a) Raman-scattering spectra in the range of 100 to 200 cm^{-1} and for several pressures. (b) Raman shift as a function of pressure based on fits to Raman modes below 200 cm^{-1}.

Figure 2a plots the Raman-scattering spectra from 100 to 200 cm^{-1}. The Raman modes near 110, 160, 180, and 170 cm^{-1} are labeled M1–M4, respectively. In accordance with previous studies [1,2], the Raman mode near 130, 150 and 180 cm^{-1} are associated with A_g mode. The band near 180 cm^{-1} is assigned to an E_g mode. The Raman modes below 200 cm^{-1} are closely related to rotation and vibration at La sites [3].

Based on previous research by Etter *et al.* [3], $LaFeO_3$ preserves its crystal structure at least up to 48.0 GPa. However, the space group changes from *Pbnm* to *Ibmm* around 21.1 GPa because the La *y* coordinate decreases from 0.029 to 0.0 and the Wyckoff factor changes from 4c to 4e [3].

Figure 2b shows the Raman shift of several modes as a function of pressure as determined by fitting the Raman modes below 200 cm^{-1}. The significant change in the Raman modes is the disappearance of M3 above 12.0 GPa and the appearance of the new Raman mode M4 around 6.0 GPa. As previously reported [2,3], the mode around 180 cm^{-1} indicates rotation at the La site. The disappearance of M3 and appearance of M4 reflect the change in the rotational state of the La site and its coordinates in the crystal lattice. This allows the possibility of the structural transition at the given pressure (12.0 GPa) from the *Pbnm* space group to the *Ibmm* space group [3]. Compared with previous research [3], the difference in transition pressure may be due to the method in which the samples were prepared and the sample environment. Thus, chemically synthesized $LaFeO_3$ has inferior stability under high pressure compared with $LaFeO_3$ synthesized by the conventional solid-state method. However, the characterization of the structural under high pressure should be confirmed to verify the changes in the crystallographic parameters. Finally, the M2 Raman mode completely disappears at 16 GPa. Under 5 GPa, the structure of $LaFeO_3$ does not change with pressure.

4. Conclusions

A vibrational analysis of $LaFeO_3$ reveals the disappearance of the M3 Raman-scattering mode at 180 cm^{-1} and the appearance of the M4 Raman-scattering mode at 170 cm^{-1} when the pressure surpasses 12.0 GPa, which indicates the possibility of a structural phase transition from the *Pbnm* space group to the *Ibmm* space group. The M3 and M4 Raman modes are closely related to the rotational state of the La site and its crystallographic coordinates. However, the structural changes should be confirmed by another structural characterization under high pressure to confirm this result. Under 5.0 GPa, the Raman modes do not change with pressure, which indicates that the stability of the $LaFeO_3$ structure is preserved at these pressures.

Acknowledgements

The authors acknowledge financial support from the Ministry of Research, Technology and Higher Education of Republic of Indonesia under grants N o.537/UN2.R3.1/HKP05.00/2018, R. A. Susilo from the Center for High Pressure Science & Technology Advanced Research (HPSTAR) for his help with the high pressure experiments and also would like to thank Einago (www.einago.com) for the English language review.

References

[1] Mahapatra A S, Mitra A, Mallick A, Ghosh M and Chakrabarti P K 2016 *Mater. Lett.* **169** 160–3
[2] Coutinho P V, Cunha F and Barrozo P 2017 *Solid State Commun.* **252** 59–63
[3] Etter M, Müller M, Handfland M and Dinnebier R E 2014 *Acta Crystallogr. B* **70** 452–8
[4] Mao H K, Xu J and Bell P M 1986 *J. Geophys. Res.* **91** 4673–6

Effect of Cr(VI) and bisphenol A on the formation of 8-hydroxy-2'deoxyguanosine DNA adduct from 2'-deoxyguanosine through a Fenton-like reaction

Budiawan, E Kusumowardani, S Handayani, I C Dani and R Bakri

Department of Chemistry, Faculty of Mathematics and Natural Sciences (FMIPA), Universitas Indonesia, Depok 16424, Indonesia

Corresponding author: dr.budiawan@gmail.com

Abstract. In this research, we study the effect of the addition of bisphenol A (BPA) and Cr(VI) on the formation of the DNA adduct, 8-hydroxy-2'-deoxyguanosine (8-OHdG). The formation of 8-OHdG, which is monitored using reversed phase high performance liquid chromatography (HPLC) with a Ultraviolet-Visible (UV-Vis) detector at 254 nm, can be achieved through the reaction of 2'-deoxyguanosine (dG) with BPA in the presence of Fenton-like reagents and vitamin C. The reaction is performed under different conditions, including several pH (7.4 and 8.4), temperatures (37 °C and 60 °C), and incubation times (7 and 12 h). To detect 8-OHdG, the minimum concentration was set as the limit of detection (LOD) value of 5.19 ppb, and the minimum concentration for its quantification was set as the limit of quantification (LOQ) value of 17.29 ppb. The results of this study indicate that the concentration of 8-OHdG increases upon addition of BPA and Cr(VI), and most of the concentration values obtained from the analysis are greater than the LOD value. The addition of Fenton-like reagent and vitamin C causes an increase in the concentration of 8-OHdG. For most samples, the conditions that cause the highest 8-OHdG concentration are a pH of 7.4, a temperature of 60 °C, and an incubation time of 12 h.

Keywords: BPA, Cr(VI), DNA adduct, 8-OHdG, Fenton-like reaction

1. Introduction

Bisphenol A (BPA; 4,4'-dihydroxy-2,2-diphenylpropane) is an extensively produced chemical that is normally used in goods for packaging foodstuffs, such as baby bottles, reusable plastic bottles, plates, cups, heat-resistant plates, and storage containers, and in food and beverage cans [1]. BPA is renowned to negatively impact the biological environment because it can trigger cellular oxidative stress, which, in turn, activates an inflammatory pathway that eventually leads to the formation of tumor cells [2].

However, Cr(VI) compounds, which is carcinogen to humans in accordance to classification by International Agency for Research on Cancer (IARC), are extensively observed in the environment, particularly in leather textiles, exhaust from cars or drainage, and cigarette smoke [3].

This research was conducted to analyze the influence of BPA during the formation of the deoxyribonucleic acid (DNA) adduct, 8-hydroxy-2'-deoxyguanosine (8-OHdG), which is a biomarker of the oxidative damage to DNA due to the exposure to chemical carcinogen. The formation of 8-OHdG was performed in vitro conducting a Fenton-like reaction of 2'-deoxyguanosine-monohydrate (dG) with BPA, Cr(VI), H_2O_2, and vitamin C, and the effect of the temperature, pH, and incubation time was evaluated.

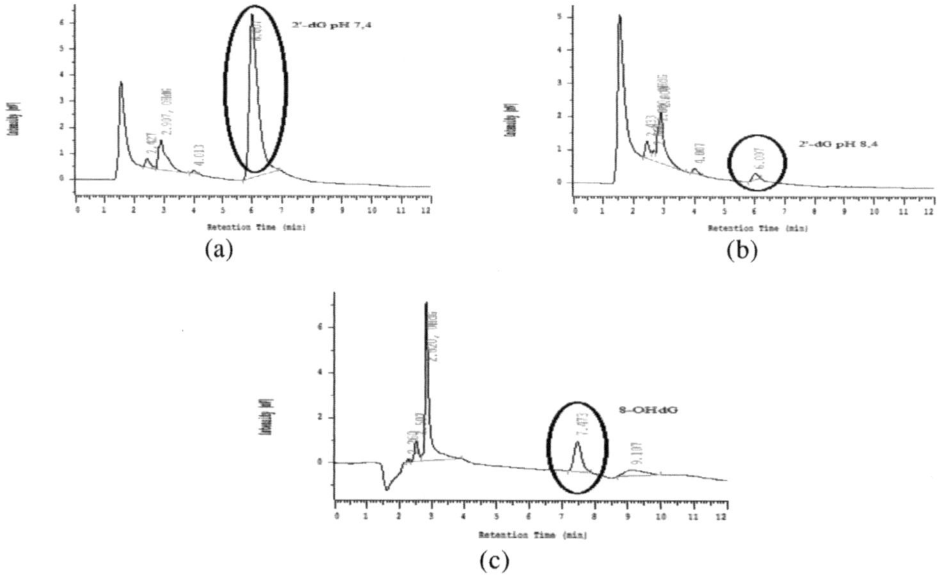

Figure 1. Standard chromatograms of (a) dG at a pH of 7.4 (6 ppm), (b) dG at a pH of 8.4 (6 ppm), and (c) 8-OHdG (500 ppb).

2. Materials and methods

2.1. Materials

To perform the high performance liquid chromatography (HPLC) analysis, a Hitachi Primaide HPLC instrument equipped with a YMC-TriartC18/S-5um/12 nm, a 250 × 4.6 mm reversed phase column, and a 254-nm UV detector was used. For the sample preparation, vials, a centrifugation instrument, centrifugation tubes, a pH meter, micropipettes, a spray bottle, a weighing bottle, and a scale were used. Bisphenol A (BPA), 2'-deoxyguanosine monohydrate (dG), and 8-hydroxy-2'-deoxyguanosine standard (8-OHdG) were purchased from Sigma Aldrich. Vitamin C, Cr(VI), hydrogen peroxide, KH_2PO_4 buffer (with pH of 7.4 and 8.4), Na-acetate buffers, Na-phosphate buffers, and LC-gradient grade methanol were purchased from Merck.

2.2. Methods

Upon incubation of 2'-deoxyguanosine (6 ppm) with BPA (60 ppm) under different pH (7.4 and 8.4), temperatures (37 °C and 60 °C), and incubation times (7 and 12 h), the formation of 8-OHdG was detected by HPLC analysis. To perform the analysis, the samples were first centrifuged for 15 min and then decanted. 20 μL of the decanted samples were injected into a reversed-phase HPLC column with a UV detector at a 254-nm wavelength. A solution of methanol and a 10 mmol/L sodium phosphate buffer (pH 6.7) in a ratio of 85:15 was used as the mobile phase (eluent) at a flow rate of 1 mL/min. The results of the measurements were compared with the 8-OHdG standard calibration curve. The quantification of 8-OHdG in the samples was calculated using the regression equation of y=81.184x–326.33 was obtained from the 8-OHdG standard calibration curve at various concentrations (10, 30, 50, 80, and 100 ppb with coefficient of correlation 0.998 [4]. Furthermore, similar tests with addition of Cr(VI) (120 ppm), vitamin C (60 ppm), H_2O2 (120 ppm), and other combinations of compounds were performed.

3. Results and discussion

3.1. Optimization of the HPLC conditions

In the optimum HPLC conditions, the peaks corresponding to standard dG at a pH of 7.4 and standard dG at a pH of 8.4 were observed at a retention time of 6,007 min, whereas that corresponding to 8-OHdG was detected at 7,473 min. The chromatogram profiles for standard dG at a pH of 7.4, dG at a pH of 8.4, and 8-OHdG can be observed in figure 1.

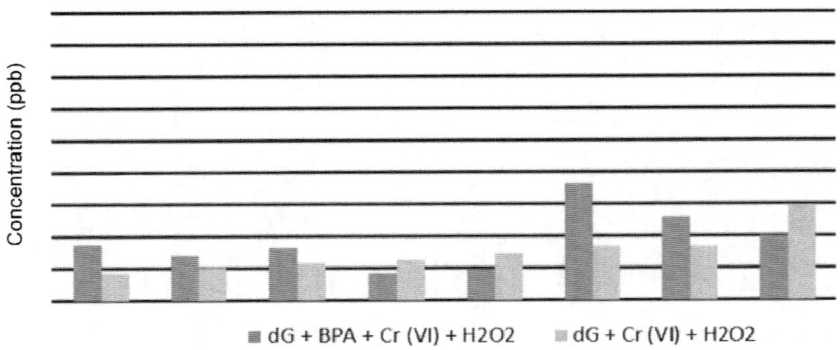

Figure 2. The effect of BPA addition on the 8-OdHG concentration.

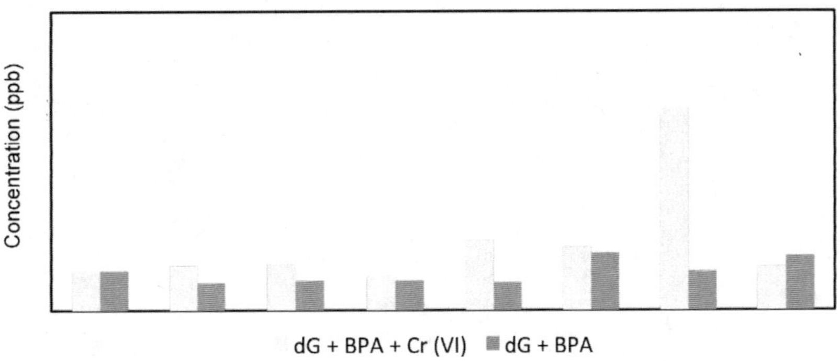

Figure 3. The effect of the addition of Cr(VI) on 8-OHdG concentration.

The regression equation obtained from the calibration curve was used to calculate the limit of detection (LOD) and limit of quantification (LOQ) values, which were estimated to be 5.19 and 17.29 ppb, respectively.

3.2. The effect of BPA addition
The metabolic conversion of BPA into a metabolite that can be bound to DNA is expected to be catalyzed by the microsomal cytochrome P450 [5]. Figure 2 depicts the effect of BPA incubation with 2'-dG on 8-OHdG concentration. For most of the samples, an increase in the concentration of 8-OHdG was observed upon the addition of BPA, which indicates the role of BPA as a prooxidant.

3.3. The effect of Cr(VI) and H_2O_2
Cr(VI) compounds can induce DNA damage because they produce several kinds of reactive intermediates and reactive oxygen species (ROS) in cells [6]. For most of the samples that were analyzed in this study, the addition of Cr(VI) was observed to cause an increase in the concentration of 8-OHdG (figure 3), and the highest 8-OHdG concentration of 30.32 ppb was obtained at a temperature of 60 °C, an incubation time of 12 h, and a pH of 7.4 in the presence of BPA and Cr(VI).

Hydrogen peroxide (H_2O_2) is also classified as ROS [7]. Although H_2O_2 is a stable compound, it can produce highly reactive hydroxyl radicals when it reacts with transition metals [7]. Figure 4 depicts the effect of adding H_2O_2 and Cr(VI) as Fenton reagents to the reaction. For most of the samples, a significant increase was observed in the 8-OHdG concentrations with the addition of Fenton-like reagents.

The interaction of vitamin C with metals is known to contribute to the oxidative damage through the production of hydroxyl radicals [8]. Thus, vitamin C can reduce Cr(VI) and interfere with the Fenton-like reaction, leading to the formation of hydroxyl radicals [9]. The effect of vitamin C on the

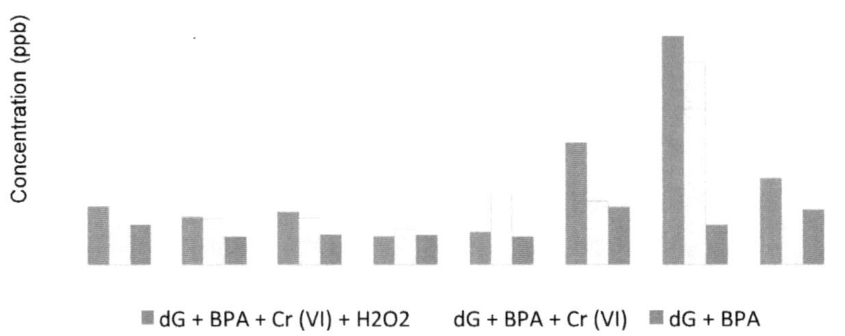

Figure 4. The effect of the addition of Fenton-like reagents on 8-OHdG concentration.

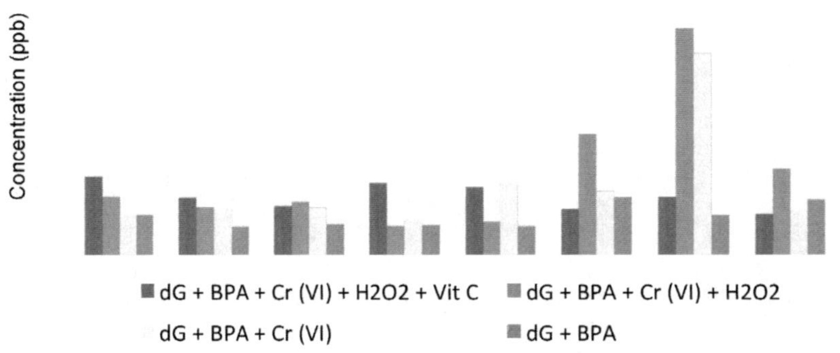

Figure 5. The effect of Fenton-like reagents and vitamin C on the 8-OHdG concentration.

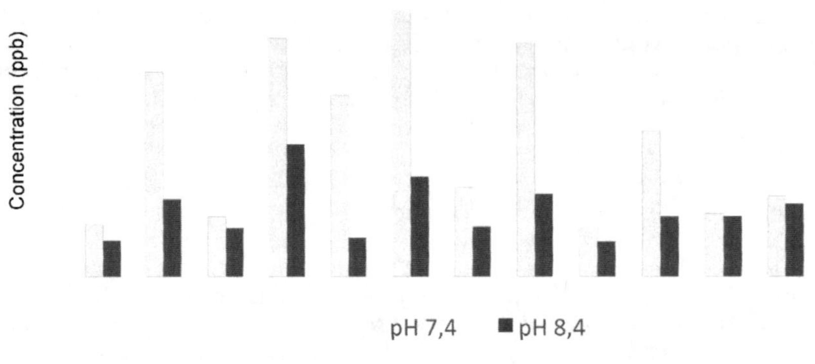

Figure 6. Effect of pH on the 8-OHdG concentration at 60 °C with a 12 h incubation time.

final concentration of 8-OHdG was also investigated in this study, and the results are depicted in figure 5. As can be observed, all the samples contained 8-OHdG in concentrations that were greater than the LOQ value in the presence of Fenton-like reagents and vitamin C. In some cases, the samples exhibited higher 8-OHdG concentration after the addition of these compounds.

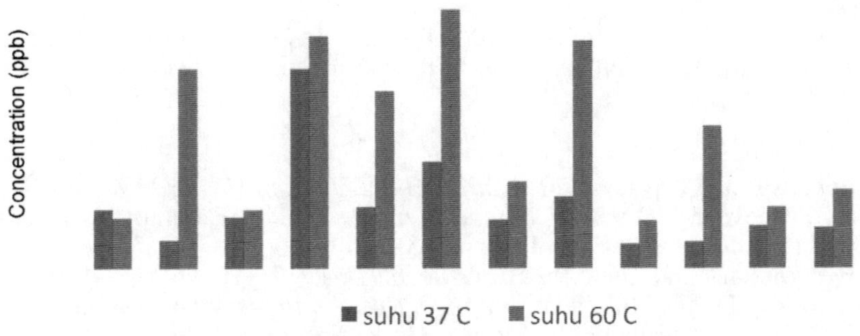

Figure 7. Effect of temperature on 8-OHdG concentration at a pH of 7.4 with 12 h of incubation time.

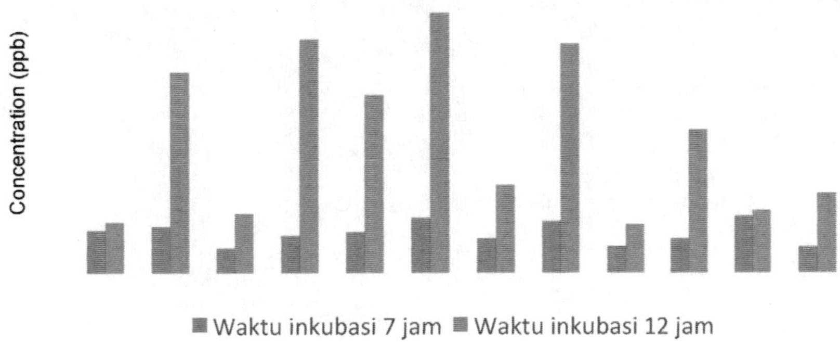

Figure 8. Effect of incubation time on 8-OHdG concentration at a pH of 7.4 with a temperature of 60 °C.

3.4. The effect of pH and temperature

Figure 6 depicts the effect of pH on the 8-OHdG concentration at 60 °C with an incubation time of 12 h. Most of the samples exhibited higher 8-OHdG concentration at a pH of 7.4 than that observed at a pH of 8.4. This is in accordance with a previous study conducted by Shi *et al.* [9] in which the production of free radicals due to the reduction of Cr(VI) by vitamin C, which can damage the double-stranded DNA, was demonstrated to occur at a physiological pH.

The temperature of the reaction also had an effect on the final concentration of 8-OHdG. Figure 7 displays the results that were obtained at a pH of 7.4 with an incubation time of 12 h at 60 °C and 37 °C. In most of the cases, higher 8-OHdG concentrations were obtained at 60 °C.

In this study, the effect of the incubation time was also evaluated. For most of the samples, higher 8-OHdG concentrations were obtained with 12 h of incubation time as compared with that observed at an incubation time of 7 h (Figure 8).

4. Conclusions

The formation of the biomarker 8-OHdG from 2'-deoxyguanosine was evaluated in the presence of BPA, Cr(VI), H_2O_2, vitamin C, and a combination of these compounds under different reaction conditions. Most of the samples exhibited 8-OHdG concentrations that were greater than that observed in the LOD and LOQ values. It was observed that higher 8-OHdG concentrations were obtained upon the addition of BPA, Cr(VI), Fenton-like reagent, and vitamin C at a pH of 7.4, a temperature of 60 °C, and an incubation time of 12 h.

Acknowledgements

Authors wishing to thank Universitas Indonesia for funding this project through PITTA Grant, Balai K3 for facilitating the instruments for analysis and also Einago (www.einago.com) for the English language review.

References

[1] Geens T, Goeyens L and Covaci A 2011 *Int. J. Hyg. Environ. Health* **214** 339–47
[2] Stoycheva T, Pesheva M, Dimitrov M and Venkov P 2012 Carcinogen ed Pesheva M and Dimitrov M (London: IntechOpen Ltd.) pp 83–110
[3] Henkler F, Brinkmann J and Luch A 2010 *Cancers (Basel)* **2** 376–96
[4] Atkinson A and Roy D 1995 *Biochem. Biophys. Res. Commun.* **210** 423–33
[5] Hu G, Zheng P, Feng H and Jia G 2017 *React. Oxyg. Species* **3** 1–11
[6] Bayir H 2005 *Reactive Oxygen Species Crit. Care Med.* **33**(12) S498–501
[7] Carr A and Frei B 1999 *FASEB J.* **13** 1007–24
[8] Valko M, Rhodes C J, Moncol J, Izakovic M and Mazur M 2006 *Chem. Biol. Interact.* **160** 1–40
[9] Shi X, Mao Y, Knapton A D, Ding M, Rojanasakul Y, Gannett P M, Dalal N and Liu K *Carcinogenesis* **15** 2475–8

IOP Conf. Series: Materials Science and Engineering **496** (2019) 012047 doi:10.1088/1757-899X/496/1/012047

Synthesis of $ZrSiO_4$ powders by a sol-gel method with varied calcination temperatures

G Elsandika, A D C Putri, M Musyarofah and S Pratapa

Department of Physics, Faculty of Science, Institute Technology Sepuluh Nopember, Surabaya 60111, Indonesia

Corresponding author: suminar_pratapa@physics.its.ac.id

Abstract. The effect of calcination temperature on phase formation of the powders prepared with starting materials $ZrCl_4$ and TEOS (tetraethylorthosilicate) was studied. Stoichiometric Zr/Si (1:1 mol%) mixture of samples were prepared by a sol-gel method after 5 wt.% of PVA (polyvinyl alcohol) addition as a catalyst and mechanical activation via ball milling for 5 h. The mixtures were then calcined at 900, 1000, 1100, 1200, and 1300 °C for 10 h. The phase composition was determined by Rietveld analysis of X-ray diffraction (XRD) data using *Rietica* software. Results showed that at calcination temperatures of 900, 1000, and 1100 °C, only tetragonal zirconia was observed, but retained silica in an amorphous state. Both phases then completely reacted to form zircon at 1200 °C. At 1300 °C, a little amount of zircon decomposed into tetragonal zirconia and cristobalite.

Keywords: $ZrSiO_4$(zircon) powder, sol-gel method, $ZrCl_4$, TEOS, decomposition.

1. Introduction

Ceramics are used to support human daily activities. One of functional ceramic materials is $ZrSiO_4$(zircon or zirconium silicate) that have been found to exhibit physical properties appropriate as sensors, circuit applications, semiconductor hetero junctions, ceramic pigments, nuclear waste disposal, and high dielectric materials [1]. Zircon has a high dielectric constant (≈15), strong corrosion resistance, low thermal expansion coefficient and conductivity, as well as excellent phase and chemical stability [2], and relatively high refractive index (1.92), so it can be used as a light-dyeing material [3]. Zircon has a tetragonal with I_4/amd space grup [4]. The unit cell parameters are a = 0.653 nm and c = 0.591 nm [5].

There were several ways to produce zircon powders: solid state reaction, low temperature hydrothermal, micro emulsion, co-precipitation, and sol-gel methods. Synthetic zircon was usually obtained from a high temperature process. Inspite of its effortlessness, the solid state reaction is difficult to provide high purity zircon at temperature lower than 1450 °C [6]. Pure zircon powder, can also be obtained at a lower calcination temperature 1300 °C but via a sol-gel method [7]. Itoh [8] began this method with mixing $ZrOCl_2$ with colloidal silica and heating at 500-1300 °C temperature. They got zircon and tetragonal zirconia phase at 1300 °C temperature. Du *et al.* [9] had 90% zircon at 1500 °C calcination temperature with addition 2.5 mol% Y_2O_3 and 3 wt.% zircon seeds to $ZrOCl_2$ and TEOS mixture before the heating treatment. In 2014, Cu *et al.* [10] used zirconium (IV) nitrate and TEOS as the precursors to obtain tetragonal zirconia at 1173-1373 K, also zircon at 1473 K.

The main role of PVA is a binder between some substance [11]. In a previous study, PVA was used as a binder between zirconia and silica powder [12]. Feng *et al.* [13] successfully synthesis monophases $LaFeO_3$ powers at low temperature by PVA addition.

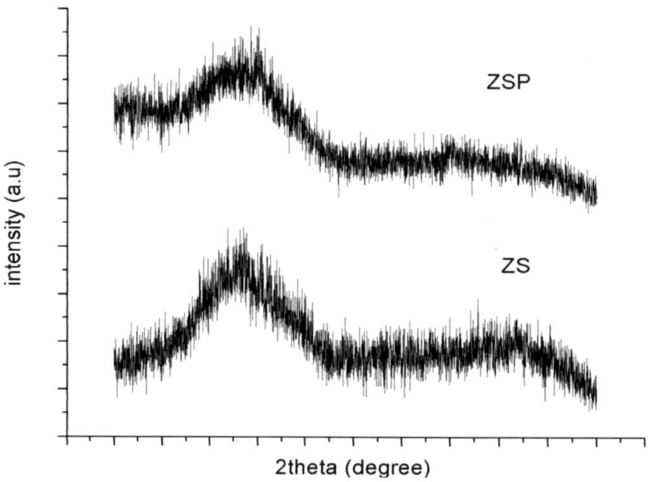

Figure 1. X-ray diffraction pattern (CuKα) of ZS and ZSP before calcination

In this research will be studied the effect of varied temperature in the zircon phase formation which is given by PVA as the additive material.

2. Experimental

The raw materials used in this study were zirconium chloride ($ZrCl_4$) powder from Merck and TEOS (tetraethylorthosilicate, $Si(OC_2H_5)_4$) solution from Sigma Aldrich. The precursors were mixed with 1:1 Si/Zr ratio. The mixing was done to obtain gel prior to washing with water. The washed gel was dried and then crushed to obtain powder. Two powders were prepared, i.e. the as-sol-gel route product, designated as ZS sample, and ZS sample with mechanical activation, named as ZSP sample. The ZSP powder were uniformly mixed with 3wt% polyvinyl alcohol (PVA) and then subjected to a mechanical activation process in a planetary ball-mill for 5 h and finally dried. These samples were calcined at various temperature following the DTA-TGA history of the samples.

The obtained zircon powders were characterized using a powder X-ray diffractometer (XRD). The XRD patterns of the samples were recorded using Cu-Kα (λ=1.54056 A) radiation, at 2-theta range of 15-65° and step size of 0.017°. The patterns were analysed using a Rietvield refinement method embedded in the *Rietica* software package [14] for phase composition calculation.

3. Results and discussion

Figure 1 presents the diffraction patterns for the uncalcined ZS and ZSP samples. What stands out in the pattern is no peak identified. This result indicates that both samples were in an amorphous state. The synthesis resulted in amorphous zirconia and silica powders. After precursor mixing and before calcination, crystallization did not occur. To obtain the crystalline structure, a driving force in the form of heat can be applied. The required heat can be predicted using TGA and DTA measurement.

The results of the TGA and DTA are set out in figure 2. The results of TGA measurements in the temperature range between 29 and 600 °C indicate a loss of 27%, which can be attributed to the release of alcohols, ammonium hydroxide, adsorbed water and organic species decomposition from the samples [15]. Confirmation of the loss of alcohol and water absorbed by the sample was also obtained from the DTA results with the presence of endothermic peak around 100 °C. There are two other exothermic peaks observed in the DTA result, one around 350 °C as a result of surfactant and acetate elimination and another one at approximately 892 °C as a result of crystallization of tetragonal zirconia [16]. The latter result is in accordance with the work by Veytizou *et al.* [7] who reported that crystallization process of amorphous zirconia becomes tetragonal-zirconia phase took place above 800 °C. Meanwhile, Du *et al.* [9] stated that the formation of low cristobalite phase of silica occurred at temperature range 1000-1200 °C [9]. We conclude that, for studying the zircon synthesis in our case, the calcination should be carried out at temperature range of 900-1300 °C.

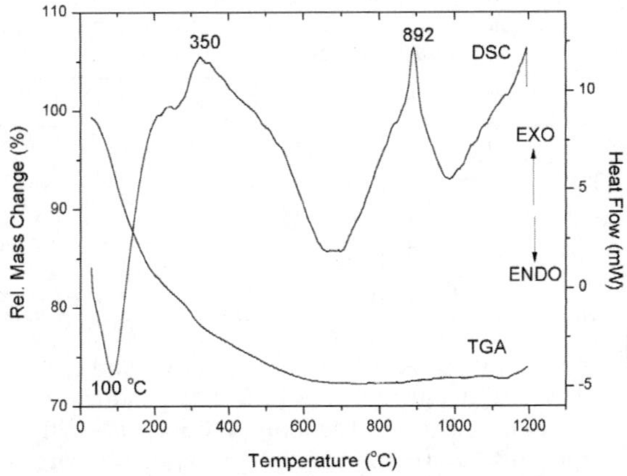

Figure 2. DTA-TGA analysis of ZSP powder

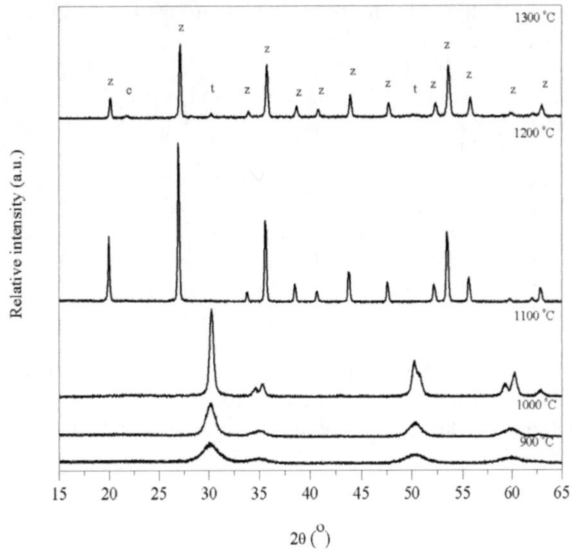

Figure 3. X-ray diffraction pattern (CuKα) of powders after calcination at 900-1300 °C for 10 hours (z=zircon, t=tetragonal zirconia, c=cristobalite)

Figure 3 compares the XRD patterns of the calcined samples. The patterns show that the samples have undergone crystallization as indicated by the appearance of some peaks. Three different groups of crystalline phases were obtained. The calcination temperature of 900-1100 °C gave only a tetragonal zirconia [No. PDF 079-1767], and that at 1200 °C produced only zircon [No. PDF 072-0402], while those at 1300 °C were zircon, t-zirconia, and low cristobalite [No. PDF 076-0940] phase. It appears that calcination below 1200 °C could not provide sufficient energy to allow zircon formation. The fact that only tetragonal was found at 1100 °C and only zircon at 1200 °C implies that below 1100 °C silica is in an amorphous state and unable to react with zirconia to form zircon. The reaction of zircon formation is shown in equation (1). The formation of a single-phase zircon at 1200 °C due to the formation Gibbs energy of the tetragonal zirconia and amorphous SiO_2 has been met, resulting in enough energy to obtain zircon crystal.

$$ZrO_2 + SiO_2 \rightarrow ZrSiO_4 \quad (1)$$

Table 1. Result of quantitative analysis of X-Ray diffraction pattern of samples calcined at varied temperatures using *Rietica* by Rietveld method

Sample (°C)	Phase Composition (mole%)		
	Zircon	t-Zirconia	Cristobalite
900	-	100.0 (0)	-
1000	-	100.0 (0)	-
1100	-	100.0 (0)	-
1200	100.0 (0)	-	-
1300	93.0 (16)	3.0 (2)	4.0 (4)

As shown in figure 3, the sample decomposes to zirconia and silica at 1300 °C. The decomposition of zircon results in tetragonal zirconia and cristobalite. The result was confirmed by the Rietvield analysis of XRD patterns as shown in table 1. Decomposition reaction is a chemical reaction on a single compound breaking down into two or more forms under a particular condition. This zircon decomposition is the result of thermal decomposition from high calcination temperature [17]. In this study, the amount of decomposed phase is 7%. Zircon phase usually form at 1600 °C, but in this study at 1200 °C has been done.

The role of PVA addition is an interesting aspect in our work. Basically, the purpose of PVA addition was to favor crystal growth [11]. However, we showed here that it can be used to accelerate the formation of high purity zircon at lower calcination temperature. Normally, the formation of zircon was at approximately 1350-1500 °C [6–10]. PVA melts at 228 °C [18]. When the ZSP mixture was calcined at 1200 °C, the PVA melted and then easily inserted through the samples to draw the zirconia and silica particles closer and allow them to react. Thus, in addition to its role as a crystal growth enhancer, PVA can also act as a catalyst.

On the other hand, the mechanical activation leads to formation of reactants with high surface energy [19]. As a result, the expected phase can be formed at the low calcination temperature. The formation of the phase is due to the increasing speed of diffusion and the higher particle homogeneity [20] . During the mechanical activation process, the sample exhibited plastic deformation to facilitate atomic mobility. Moreover, the milling time affects the number of collisions occurred between the particles in the sample. The collision led to a reduction of aggregate and grain size and hence improved the homogeneity of the mixture. The homogenization caused the reaction between amorphous zirconia and amorphous silica particles became easier. The surface area contact between zirconia and silica particles will be more widespread by reducing their grain size so that more particles react [19]. Therefore, another explanation for the formation of of zircon at 1200 °C is because of the increasing speed of diffusion and the higher particles homogeneity [20].

4. Summary

Zircon ($ZrSiO_4$) powders has been successfully synthesized using $ZrCl_4$ powders (zirconium chloride) and TEOS (tetraethylorthosilicate) solution by a sol-gel method followed by calcination. The application of mechanical activation by a planetary ball mill for 5 hours and the addition of 3wt% PVA significantly influenced the formation of zircon. The varied calcination temperature affected the phase formation. At 900-1100 °C tetragonal zirconia and silica amorphous were formed, while at 1200 °C, the single-phase zircon was obtained. Moreover, the zircon powder decomposed into multiphase, i.e. 93% zircon, 3% tetragonal zirconia and 4% low cristobalite after calcination at 1300 °C.

Acknowledgements

The authors would like to thank Ministry of Research Technology and Higher Education of the Republic of Indonesia to support finance through program Penelitian PMDSU (Pendidikan Magister Menuju Doktor untuk Sarjana Unggul) Tahun Anggaran 2017 No. 491/PKS/ITS/2017.

References

[1] Huang S, Li Q, Wang Z, Cheng X and Wen H 2017 *Ceram. Int.* **43** 875–9
[2] Hays D C, Gila B P, Pearton S J, Kim B J and Ren F 2016 *Vacuum* **125** 113–7
[3] Focke W W, Manhique A and Carter R 2002 *J. Am. Ceram. Soc.* **85** 1139–44
[4] Srikanth S, Devi V L and Kumar R 2016 *Hydrometallurgy* **165** 125–36

[5] Puthenkovilakam R, Carter E A and Chang J P 2004 *Phys. Rev. B* **69** 155329
[6] Wang F, Liu D W, Zhu J F and Li D 2011 *Adv. Mater. Res.* **295–297** 1485–8
[7] Veytizou C, Quinson J F and Douy A 2000 *J. Mater. Chem.* **10** 365–70
[8] Itoh T 1992 *J. Cryst. Growth* **125** 223–8
[9] Du C, Yuan Q and Yang Z 1999 *J. Mater. Sci. Lett.* **18** 965–6
[10] Cu H L, Hwang W S, Wang C L, Wang M C, Lee K C, Huang H H and Lee H E 2014 *J. Alloys Compd.* **616** 413–9
[11] Peko C, Groth B and Nettleship I 2010 *J. Am. Ceram. Soc.* **95** 115–20
[12] Musyarofah, Nurlaila R, Muwwaqor N F, Saukani M, Kuswoyo A, Triwikantoro and Pratapa S 2017 *J. Phys.: Conf. Ser.* **817** 012033
[13] Feng J, Liu T, Xu Y, Zhao J and He Y 2011 *Ceram. Int.* **37** 1203–7
[14] Pratapa S 2010 *Makara J. Sci.* **11** 85–90
[15] Tartaj P and de Jonghe L C 2000 *J. Mater. Chem.* **10** 2786–90
[16] Aguilar D H, Torres-Gonzales L C, Fuentes A F, Torres-Martinez L M, Lopez T and Quintana P 2000 *J. Solid State Chem.* **158** 349–57
[17] Abdelkader A M, Daher A and El-Kashef E 2008 *J. Alloys Compd.* **460** 577–80
[18] Tubbs R K 1965 *J. Polym. Sci. A* **3** 4181–9
[19] Emadi R, Ashrafi H and Zamani Foroushani R 2015 *Ceram. Int.* **41** 14400–5
[20] Ahmadzadeh M, Ataie A and Mostafavi E 2015 *J. Alloys Compd.* **622** 548–56

IOP Conf. Series: Materials Science and Engineering **496** (2019) 012048 doi:10.1088/1757-899X/496/1/012048

Utilization of natural compounds (chlorophyll and carotene extracts) as an octane-boosting additive in gasoline

E Yuliarita[1] and A Zulys[2]

[1] Research and Development Centre for Oil and Gas Technology, Jalan Ciledug Raya No.109, Jakarta 12230, Indonesia

[2] Department of Chemistry, Faculty of Mathematics and Natural Sciences (FMIPA), Universitas Indonesia, Depok 16424, Indonesia

Corresponding author: zulys@ui.ac.id

Abstract. Natural materials should be substituted for organometallic and oxygenated compounds in octane-boosting additives for gasoline because of environmental considerations. Chlorophyll and carotene extracts can be used as octane-boosting additives in gasoline. Maceration and adsorption methods are used to extract them. In this work, chlorophyll and carotene extracts were mixed with fatty acid methyl esters (FAMEs) and a solvent (xylene and gasoline) in a particular composition. The extracted pigment was characterized using ultraviolet–visible (UV–Vis) spectrophotometry. Three different additive formulations with chlorophyll/carotene extract ratios of 0 : 1, 3 : 1, and 2 : 3 were investigated. Additive performance tests included measurements of physical properties, octane number, exhaust hydrocarbon (HC) and carbon monoxide (CO) emissions. The results showed that additives with a chlorophyll/carotene extract ratio of 2 : 3, when added to premium-grade gasoline, increased the octane number by 0.3%. The increase of the octane number was further demonstrated by analysis of the HC composition of the gasoline, where the additive with a 2 : 3 ratio resulted in the greatest increase in the olefin content (22.9%), and the emission tests showed that it could reduce HC emissions by 30.9%.

Keywords: octane boosting additive, gasoline, octane number, chlorophyll and carotene extracts

1. Introduction

The limited ability of refineries to produce gasoline with a high octane number has led to the introduction of octane boosters as gasoline additives into the market. Such fuel additives are a common approach for increasing the octane number of gasoline [1]. The octane number is an important parameter in distinguishing the quality of a fuel [2]. Octane boosters with a considerable improvement effect are generally based on organometallic compounds such as tetraethyl lead (TEL), methylcyclopentadienyl manganese tricarbonyl, and ferrocene. The results of the research showed that TEL is very effective in increasing the octane number of gasoline. However, the use of TEL causes environmental pollution by the lead content of vehicle exhaust emissions. When used as a gasoline additive, TEL may also damage the catalytic converter in the vehicle [3]. TEL is therefore no longer used in gasoline.

Other compounds used as additives to increase the octane number of gasoline include oxygenated compounds such as methyl *tert*-butyl ether (MTBE). This compound can also reduce carbon monoxide

IOP Conf. Series: Materials Science and Engineering **496** (2019) 012048 doi:10.1088/1757-899X/496/1/012048

(CO) and NO_x emissions from motor vehicles. However, these compounds are carcinogenic and harmful to human health, and MTBE can contaminate groundwater [4].

In addition to MTBE, oxygenated compounds of the alcoholic groups, such as ethanol, can be used to increase the octane number of gasoline [5]. Their octane-boosting ability is supported by the presence of oxygen atoms that enable fuel to burn more completely; alcohol thus improves combustion efficiency and reduces air pollution [2]. However, the presence of alcohol in fuel can lead to the corrosion of metal components of the fuel system; thus, engines that operate on alcohol-containing fuels must be modified to run on alcohol or run on fuel with a low alcohol concentration [6]. These oxygenated compounds (ethers and alcohols) each have considerable side effects. Thus, the development of natural materials as a source for the manufacture of octane-boosting gasoline additives is needed to enable the production of high-quality fuel with both good engine performance and low environmental impact.

Carotenoid and chlorophyll compounds have been speculated to act as octane boosters in gasoline additives. Carotenoid compounds, as antioxidants, participate in the energy transfer process through electron transfer, thereby protecting the chlorophyll in the main reaction of the combustion process from autooxidation. By contrast, chlorophyll is a complex compound with a high molecular weight, a very high energy content, and an oxygen-rich composition; it thus promotes combustion in the engine's combustion chamber.

The richest natural source of carotenoids is crude palm oil (CPO), in which carotenoids exist primarily as retinol compounds (vitamin A). CPO contains approximately 15–300 times more carotenoid compounds than carrots and is a byproduct of the palm oil industry. Despite its high economic value, CPO has not been optimally utilized [7, 8].

β-Carotene, along with chlorophyll and jojoba oil, has been successfully used as a gasoline additive to enhance combustion characteristics. In this case, carotenoids act as scavengers of pollutants or react with the pollutants present in the fuel [9]. Previous research on carotenoid compounds has been carried out in the energy field, where carotenoid compounds extracted from carrots were used as a combustion-boosting additive in gasoline. The results of the aforementioned study indicate that such additives effectively reduce both CO and hydrocarbon (HC) emissions [10].

In the present study, CPO is used as a source of carotenoids, and *eceng gondok* (water hyacinth, *Eichhornia crassipes*) leaves are used as a source of chlorophyll. CPO was selected because of its high carotenoid content, and water hyacinth was selected as a source of chlorophyll because it is often regarded as a weed or a plant that disrupts the aquatic ecosystem; thus, the development of an application will provide added value to the water hyacinth. In this paper, we report the manufacture of octane-boosting additives for gasoline through the utilization of carotene along with chlorophyll and fatty acid methyl esters (FAMEs) and compare their optimum utilization conditions.

2. Experimental

2.1. Materials and methods

All solvents were used as obtained from commercial suppliers. Isopropanol (IPA), acetone, *n*-hexane, and xylene were of industrial grade (Pro Analysis, PA). Silica gel 60 (0.063–0.2 mm) was used as an adsorbent. Premium gasoline was purchased from the market (a gas station). The water hyacinth leaves (*Eichhornia crassipes*) were acquired from Cipondoh Lake, whereas CPO was acquired from PT Wilmar International Jakarta, rotary evaporator, CFR engine, UV–Vis spectrometer, GC.

2.2. Extraction methods

2.2.1. Chlorophyll extraction. Chlorophyll extraction was carried out by a maceration method from water hyacinth plants. Leaves of water hyacinth were first cleaned and dried. They were then blended until smooth and transferred to a clean container. As much as 500 g of blended water hyacinth leaves was added to 2 L of solvent consisting of a mixture of acetone and hexane (80 : 20, v/v). The maceration process was conducted in dark-colored glass bottles for 24 h in a refrigerator, where the samples were not exposed to direct light. The contents of the bottles were then filtered and concentrated at 40 °C using a vacuum rotary evaporator. The concentration of the extract of chlorophyll obtained was calculated, and it was characterized with an ultraviolet–visible (UV–Vis) spectrophotometer.

IOP Conf. Series: Materials Science and Engineering **496** (2019) 012048 doi:10.1088/1757-899X/496/1/012048

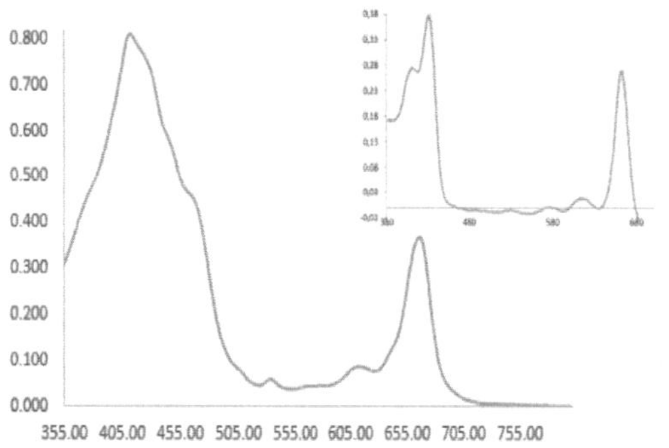

Figure 1. UV–Vis absorption spectra of chlorophyll compared to that of a chlorophyll-*a* standard from Sigma-Aldrich (inset).

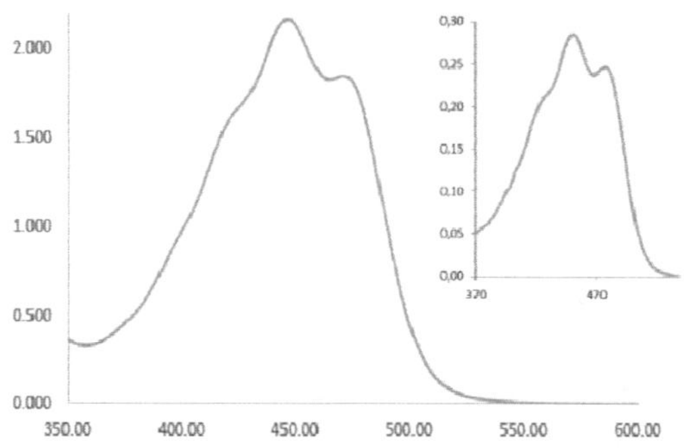

Figure 2. UV–Vis absorption spectra of the carotene extract compared to that of the β-carotene standard from Sigma-Aldrich (inset).

2.2.2. Carotene extraction. Carotene extraction was carried out by adsorption from CPO. 20 g of adsorbent was agitated with an IPA solvent until the adsorbent was submerged. The agitation process was conducted for 15 min at a speed of 1,500 rpm. After the agitation process was completed, the adsorbent was decanted. The CPO was then dissolved in IPA with a ratio of 1 : 3 m/v. Soluble CPO was added to the dried adsorbent, and the IPA was added until the adsorbent was submerged. The adsorption process was carried out at 40 °C for 30 min, and the liquid was then decanted. The adsorbent containing CPO was further extracted with hexane for 3 h and then filtered and concentrated. The concentration of the obtained carotene extract was calculated, and the extract was characterized using UV–Vis spectrophotometry with wavelength range 400–500 nm.

2.3. Manufacture of octane-boosting additives and formulations with gasoline

An octane-boosting additive was made from a mixture of additive A and additive B, which are based on chlorophyll extracts and carotene extracts, respectively. Additives A and B were each added to a FAME and a solvent (xylene and gasoline, respectively) in certain compositions, and they were also mixed with premium-grade gasoline, where the carotene/chlorophyll ratio was 0 : 1, 3 : 1, or 2 : 3. The performance of the additives was investigated by testing the main chemical properties of the gasoline according to the gasoline specifications in Indonesia and by testing the emissions of CO and HCs.

3. Results and discussion

3.1. Chlorophyll and carotene extraction

The percentage yields from the chlorophyll pigment and carotene extraction processes were 23% and 1.4%, respectively.

3.2. Characterization by UV–Vis spectrophotometry

In order to ensure that the extracts obtained were chlorophyll and carotene pigments, characterization was performed by UV–Vis spectrophotometry and the resulting spectra were compared with the absorption spectra of chlorophyll and carotene standards [7]. The spectra show that the extract of chlorophyll exhibits a visible-light absorption spectrum with two main peaks at wavelengths of 413 nm and 666 nm, which is typical for the absorption spectrum of chlorophyll-*a* (figure 1). The spectrum of the carotene extract is similar to that of the β-carotene standard obtained from Sigma-Aldrich (figure 2). The spectrum of the carotene pigment shows three typical absorption peaks in the wavelength region near 400 nm, as observed in both the spectrum of the carotene rough extract CPO and that of the β-carotene standard. The typical peaks in the spectrum of the crude extract of carotene CPO are at wavelengths of 424, 447, and 472 nm.

Table 1: Test results of the main physical and chemical properties of fuel formulations.

Number	Parameter	Gasoline without additive	Gasoline + additive (0 : 1)	Gasoline + additive (3 : 1)	Gasoline + additive (2 : 3)	Specification Limits	
						Min.	Max.
1	Octane number (RON)	88.0	88.0	88.0	88.3	88	-
2	Sulfur content (%wt.)	0.0099	0.0094	0.0102	0.0110	-	0.05
3	Density (kg/m³)	718.50	719.21	719.41	719.52	715	770
4	Manganase/Iron (mg/L)	0/1.7	0/1.2	0/2.4	0/2.5		
5	Olefine content (%vol.)	13.6115	16.4397	16.7221	17.7964		
6	Aromatic content (%vol.)	6.7023	6.6071	6.7047	6.8610		
7	Wash gum (g/100 mL)	2.4	1.2	1.4	1.6	-	5

3.3. Test results of the main physical and chemical properties

The physical/chemical properties of the gasoline formulations are summarized in table 1. The results of the mercaptan sulfur content analysis showed that the addition of the additive caused a decrease in the mercaptan sulfur value comparable to the concentration of the additive added to premium gasoline. This similarity is attributed to the formation of silver mercaptide at the equivalent point during the analysis. Additives in gasoline affect the formation of silver mercaptide, enabling the equivalent point to be reached faster and lowering the mercaptan sulfur value.

The results of the iron content analysis show that the addition of additives tended to increase the iron content in proportion to the increase in the additive concentration in the premium gasoline. This result is reasonable because the pigment originates from secondary metabolites of plants, where most of the plant cells contain various minerals as nutrients supporting plant metabolism. The presence of iron as an additive will directly improve the performance of the fuel [11].

The results of the gum content analysis gave varying results, although the gum content tended to increase with increasing concentration of the additives in the gasoline. This behavior is strongly influenced by the antioxidant properties of the carotene-pigment components of the additives as well as by the stabilizers of the chlorophyll pigments, which beneficially affect gasoline by decreasing its gum content. This effect is evidenced by the gum content before the addition of additives being higher than that in the premium gasoline after the addition of additives. The gum content is directly proportional to the additive concentration in the premium gasoline [12].

The research octane number (RON) determined by a Cooperative Fuel Research (CFR) engine is useful for stating the fuel's performance and is strongly related to the physical and chemical properties of premium gasoline, which are improved by the addition of chlorophyll and carotene in a 2:3 ratio. The octane number of gasoline is also highly dependent on the composition of its HC compounds, especially its aromatic and olefin contents. Greater aromatic and olefin contents in gasoline lead to higher octane numbers. The test results of the physicochemical properties of the premium gasoline with and without additives showed that no parameter exceeded the maximum limit for gasoline, as specified by the government through the Decree of the Director General of Oil and Gas number 933.K/10/DJM.S/2013 [13].

3.4. Emission testing

Emission testing was carried out on a two-wheeled motor vehicle that was manufactured in 2013. Emission tests were performed under idle machine conditions. The emission test results for premium gasoline with and without additives are shown in table 2.

3.4.1. HC emissions. The successful performance of the chlorophyll and carotene additives is evident from the substantial reductions in HC emission levels. Premium gasoline containing chlorophyll and carotene additives in a 2 : 3 ratio exhibited the best performance. The additives enhanced HC combustion in the machine, resulting in less HC emissions by increasing the combustion heat and

Table 2. Emission test results for premium gasoline under idle machine conditions.

Parameter (ppm)	Premium without additive	Premium + additive A (1 : 0)	Premium + additive B (3 : 1)	Premium + additive C (2 : 3)
Hydrocarbon, HC	259	229	196	180
Carbon monoxide, CO	0.82	1.93	1.55	1.94
Carbon dioxide, CO_2	7.68	6.84	7.41	6.49

maintaining a stable temperature in the test machine [3]. Furthermore, the additive with a 3 : 1 ratio resulted in emissions with the lowest CO content, whereas the additive with a 2 : 3 ratio resulted in emissions with the highest CO content. This increase in CO emissions is derived from the effect of secondary metabolite macromolecules having carbon double bonds and low oxygen content, thereby requiring a high combustion energy and increasing CO emissions [14].

3.4.2. Carbon dioxide (CO_2) emissions. The CO_2 emission for premium gasoline without additives was 7.68 %. Premium gasoline with additive B resulted in CO_2 emissions almost equal to those for premium gasoline without additives. The CO_2 emissions in this case were predominantly unaffected because the ratio of carbon to HC atoms in the fuel was relatively unchanged by the additives. Additives A and C resulted in a slight decrease in CO_2 emission levels. In these cases, the additives also functioned as stabilizers, where the effect of carotene on premium gasoline is more pronounced because these additives have higher concentrations of carotene than chlorophyll. Larger chlorophyll compositions provide more constant combustion, whereas CO_2 emissions are equal to emissions when no additives are used [14]. Emissions of CO_2 from gasoline will increase if there is complete combustion of fuel.

4. Conclusions

From the results presented here, we concluded that the crude extract of carotene from CPO has a visible-light absorption spectrum similar to that of a β-carotene standard, demonstrating that the crude extract contains a carotene pigment. The visible-light absorption spectrum of the chlorophyll extract from water hyacinth leaves is similar to that of chlorophyll-*a*. The blending of these additives with premium gasoline resulted in no substantial change in the physicochemical properties of the gasoline; thus, these gasoline additives can be marketed domestically in accordance with the Standard and Quality of Fuel Type Petrol 88 specified in the Decree of the Director General of Oil and Gas Number 933.K/10/DJM.S/2013 Year 2013. The use of an additive with a chlorophyll/carotene ratio of 2 : 3 in premium gasoline improves the performance of the gasoline compared with the other two investigated additive compositions, especially in terms of increasing the gasoline octane number and decreasing the emissions of HCs. The additive with a chlorophyll/carotene ratio of 2 : 3 can therefore be classified as an octane-boosting additive for gasoline.

Acknowledgements

The author would like to thank Einago (www.einago.com) for the English language review.

References

[1] Rankovic N, Bourhis G, Loos M and Dauphin R 2015 *Fuel* **150** 41–7
[2] Wang C, Zeraati-Rezaei S, Xiang L and Xu H 2017 *Appl. Energy* **191** 603–19
[3] Elfasakhany A 2015 *Eng. Sci. Technol. Int. J.* **18** 713–9
[4] Kiadehi A D, Ebadi A and Aghaeinejad-Meybodi A 2017 *Sep. Purif. Technol.* **182** 238–46
[5] Alramadan S, Sarathy S M, Kurshid M and Badra J 2016 *Fuel* **180** 175–86
[6] Prado J M, Veggi P C and Meireles M A A 2014 *Curr. Anal. Chem.* **10** 29–66.
[7] Hairunnisa 2013 *2nd Natural Pigments Conference for South-East Asia (Malang)* (Malang: Ma Chung University)
[8] Kupan S, Kulkarni A D, Hamid H, Yusoff M M 2015 *International Journal of Engineering Research and Technology* **4** 662–7.

IOP Conf. Series: Materials Science and Engineering **496** (2019) 012048 doi:10.1088/1757-899X/496/1/012048

[9] Jordan S 1998 *Chlorophyll based fuel additive for reducing pollutant emissions* U.S. Patent 5826369A

[10] Adriany R, Fibria M, Yuliani C and Hanifuddin M 2011 *Pembuatan Aditif Combustion Booster dan Pack Kolom Untuk Mengurangi Emisi CO dan CO_2 pada Kendaraan Bermotor* (Jakarta: Puslitbang Lemigas)

[11] Stratiev D and Kirilov K 2009 Pet. Coal **51** 244–8

[12] Ackerman L 1968 *Period. Polytech. Chem. Eng.* **13** 29–39

[13] Dirjen Migas Kementrian ESDM 2013 *Keputusan Nomor 933.K/10/DJM.S/2013* (Jakarta: Dirjen Migas Kementrian ESDM)

[14] Zhu R, Hu J, Bao X, He L, Zu L 2017 Energy Policy **101** 185–93

IOP Conf. Series: Materials Science and Engineering **496** (2019) 012049 doi:10.1088/1757-899X/496/1/012049

The extraction of methanol via the partial oxidation of methane over a hierarchical Co/ZSM-5 catalyst

D A Nurani, D U C Rahayu, K F Prayudha and Y K Krisnandi

Department of Chemistry, Faculty of Mathematics and Natural Sciences (FMIPA), Universitas Indonesia, Depok 16424, Indonesia

Corresponding author: yuni.krisnandi@sci.ui.ac.id

Abstract. Methanol is one of the main products obtained from the partial oxidation of methane over a mesoporous Co/ZSM-5 catalyst. The cobalt species are present as cobalt oxides. Issues associated with the extraction of methanol from the zeolite surface include the carbon imbalance caused by the transformation of methane in the catalyst pores as well as low extraction efficiency. The developed extraction method involved the use of a dry solvent and a mixture of the solvent with water for the reliable quantitative measurement and enhancement of product recovery. Furthermore, extraction parameters such as solvent type and volume, extraction time, and number of extraction cycles were investigated. A mixture of 1:1 ethanol and water afforded a methanol recovery of 94.2%. Optimum extraction was achieved using 1 mL of solvent/0.15 g of catalyst, an extraction time of 30 min, and three extraction cycles. As reuse was crucial for catalyst performance, Fourier transform infrared spectra revealed that the extraction solvent did not significantly affect the catalyst structure.

Keywords: hierarchical Co-ZSM5, partial oxidation of methane, methanol recovery, solvent extraction

1. Introduction

It is desirable to develop a method for the direct conversion of methane to methanol owing to environmental issues. Heterogeneous catalysts (ZSM-5) are promising support materials for examining the oxidation of methane to other products [1–6]. Bezniz *et al.* [1] have reported the low-temperature partial oxidation of methane to methanol and formaldehyde over Co–ZSM-5 agglomerates. The catalytic activity as well as the product selectivity for methane conversion depend on cobalt species that is impregnated into the catalyst. Hence, it is crucial to modify the ZSM-5 pores by the double-template method because of the high pore volume distribution and large crystal size [7]. Furthermore, Krisnandi *et al.* [8] and Karim *et al.* [9] have successfully reported the high conversion for methanol as the product obtained from the partial oxidation of methane over a heterogenous catalyst (Co/ZSM-5). The results indicated that the zeolite porosity and Co loading method affect the yield of the partial oxidation reaction.

Previous studies have reported certain issues associated with the quantitative measurement of methanol as the conversion product [3,5,10]. The main issue is that the methanol recovery is not as high as the theoretical value, thereby indicating that inefficient extraction is a crucial factor in catalysis. Extraction involves the transfer of the target component from one phase to another second phase. Some factors affecting extraction include the extraction solvent, extractant composition, temperature, extraction time, and extraction cycle number.

Sobolev *et al.* [10] have reported the extraction of methanol by the partial oxidation of methane. Water and an acetonitrile–water mixture were used as solvents. After four extraction cycles, 80% of

Table 1. Extraction parameters

Type of solvent	Solvent volume (mL)	Extraction time (min)	Extraction cycle
Ethanol Propanol Water Ethanol:propanol v/v (1:1) Ethanol:water v/v (1:1)	0.5; 1.0; 1.5	15; 30; 45	1–5

methanol is recovered [10]. Knops-Gerrits and Goddard [3] have reported the preference of water over the acetonitrile–water mixture during extraction, whereas the reaction between methanol and water releases heat, which tends to facilitate the release of any extraneous dissolved air as bubbles. Starokon *et al.* [5] have compared methods for the extraction of methanol over an Fe/ZSM-5 surface using dry solvent as well as a mixture of organic solvents with water. Seventy-five percent of methane is recovered; however, the complete extraction of the reaction products is not achieved [5].

In this study, the recovery of methanol over Co/ZSM-5 was examined. In addition, the extraction method was modified. Furthermore, the products were quantitatively analyzed on account of the additional experiments conducted for the catalytic performance of zeolites.

2. Materials and methods

2.1. Materials

$NaAlO_2$, tetrapropylammonium hydroxide (TPAOH, 25 wt%), tetraethyl orthosilicate (TEOS), dimethyl diallyl ammonium chloride acrylamide copolymer (PDDA, 10 wt%), NaOH (Merck, >99%), pellet and an ammonia aqueous solution (Merck, 29%), HCl aqueous solution (Merck, 36%), $Co(NO_3)_2.6H_2O$ (Aldrich, 98%), CH_4 gas (99,99%, BOC), N_2 gas (UHP), methanol (99.8%, Merck), ethanol (99.8%, Merck), propanol (99.9%, Merck), and distilled water were used.

2.2 Methods

2.2.1 Regeneration of the Co/ZSM-5 catalyst. The synthesis of the Co-ZSM-5 catalyst and its reaction with methane in a pressurized batch reactor have been reported elsewhere [8]. The catalyst was calcined at 550 °C for 2 h prior to regeneration.

2.2.2. Product extraction from Co/ZSM-5. First, 0.15 g of the regenerated catalyst was mixed with 0.05 mL of methanol, followed by transfer into a vial (2 mL). Second, the corresponding solvent (volumes: 0.5, 1.0 and 1.5 mL) was added into the vial. Afterward, the vial was sealed and placed in a shaker at 200 rpm for 30 min to ensure appropriate mixing. After extraction, the vial was subjected to centrifugation at 7500 rpm for 10 min for the separation of the solvent from catalyst particles. Extraction parameters such as solvent type and volume, extraction time, and extraction cycle number are summarized in table 1. The ethanol content was analyzed by gas chromatography (GC-FID Shimadzu 2010), and the catalyst was characterized by FTIR spectroscopy (Shimadzu Prestige 21).

3. Results and discussion

The double-template method was employed to synthesize the catalyst. TPAOH was used as the directing agent in the microporous MFI zeolite structure, and mesoscale cationic PDDA was used as the mesoporous template. Cobalt species was impregnated into the hierarchical ZSM-5 catalyst. The partial oxidation of methane was affected by the cobalt species. Cobalt oxide (Co_3O_4 and CoO) is selective to methanol production, while Co(II) affords formaldehyde [1]. As the product (methanol) is strongly attached on the catalyst surface and only being released with solvent extraction, modified extraction parameters, i.e., solvent type and volume, extraction time, and extraction cycle number, were utilized to increase the extraction efficiency. This experiment was carried out using the regenerated catalyst. The characteristics of the catalyst used in this experiment have been described previously [8].

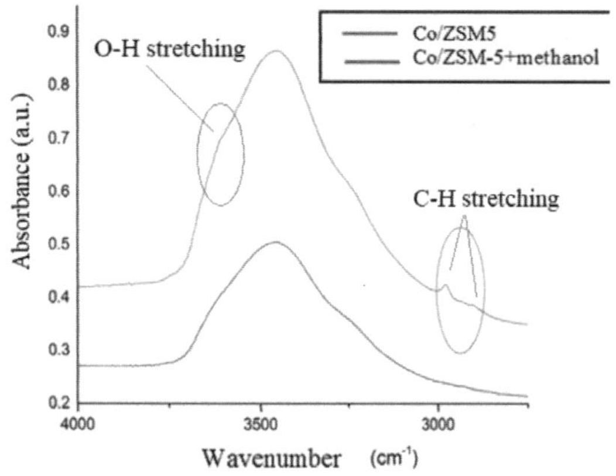

Figure 1. FTIR spectra of Co/ZSM-5 catalyst before and after regeneration, providing evidence for the methanol absorbed into Co/ZSM-5.

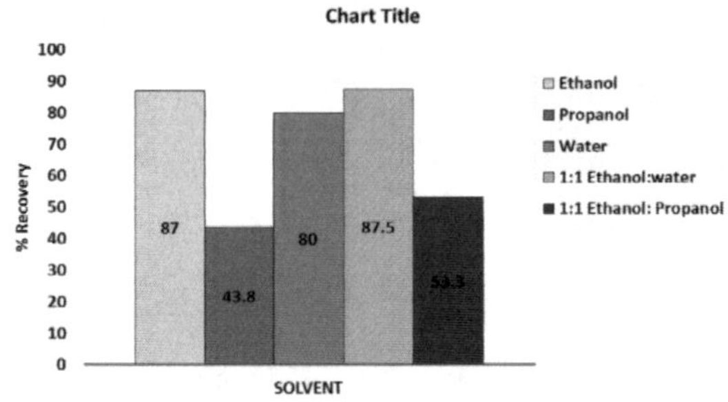

Figure 2. Effect of solvent type on the percentage recovery of methanol

After regeneration, FTIR spectra were recorded to examine whether methanol and other products were still adsorbed onto the Co/ZSM-5 at room temperature. The infrared spectrum of the methanol adsorbed on the Co species is shown in figure 1. C–H symmetric stretching and C–H asymmetric stretching bands were observed at 2855 cm^{-1} and 2958 cm^{-1}, respectively. Starokon *et al.* [11] have reported that methanol molecules are adsorbed on the catalyst surface. Oxygen atoms from methanol and cobalt species exhibited electrostatic interactions (i.e., dipole–dipole interaction).

3.1. Solvent type

Solvents were selected on the basis of their physical and chemical properties, such as dielectric constant, polarity, and solubility. Two organic solvents, their mixtures, and a mixture of water with the solvents were compared. The percentage recovery of methanol as a function of solvent type is shown in figure 2.

The use of ethanol afforded a high methanol recovery (87.0%), while the use of propanol afforded the lowest methanol recovery (43.8%). This result is related to solvent polarity, which plays a key role in increasing methanol recovery. Furthermore, methanol is formed by the substitution of methoxy groups by ethoxy groups [5]. A mixture of these solvents with water afforded higher methanol recoveries of 87.5% (1:1 v/v of water/ethanol) and of 53.3% (1:1 v/v water/propanol). This result can be explained by the fact that the interaction between methanol and the solvent is affected by the formation of hydrogen bonds between oxygen atoms from methanol and hydrogen atoms from water. Thus, methanol is released from the catalyst surface. During partial oxidation, methoxy species are initially formed from the chemisorption of methane on the oxide species, leading to the strong absorption observed for the product on the catalyst surface [1]. The hydrolysis of methoxy groups assisted by water led to the high recovery of methanol.

FTIR spectra were recorded to observe the effect of solvent on the catalyst structure. Five solvents were compared. An additional peak corresponding to Co/ZSM-5 was not observed during extraction using ethanol, water, and their mixtures, while extraction using propanol and 1:1 ethanol:propanol revealed differences (figure 3). C–H stretching peaks were observed at 2960–2850 cm^{-1} in the FTIR spectra of these solvents. Propanol is still thought to be present within the catalyst pores because of its molecular size and boiling point. Hence, regeneration (i.e., heating at 550°C) leads to the evaporation of the remaining solvent. Hence, the solvent used herein does not significantly affect the catalyst structure.

3.2. Extraction time

The effect of extraction time on methanol recovery was investigated using ethanol as the solvent. From figure 4, the extraction time (15, 30, and 45 min) affects methanol recovery. The methanol

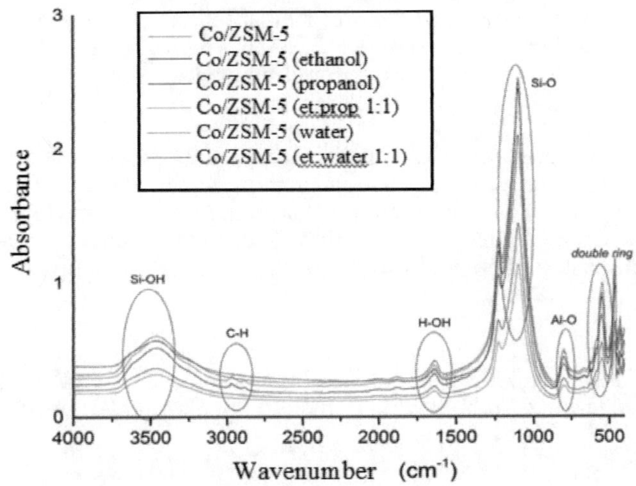

Figure 3. FTIR spectra of Co/ZSM 5 after extraction using different solvents.

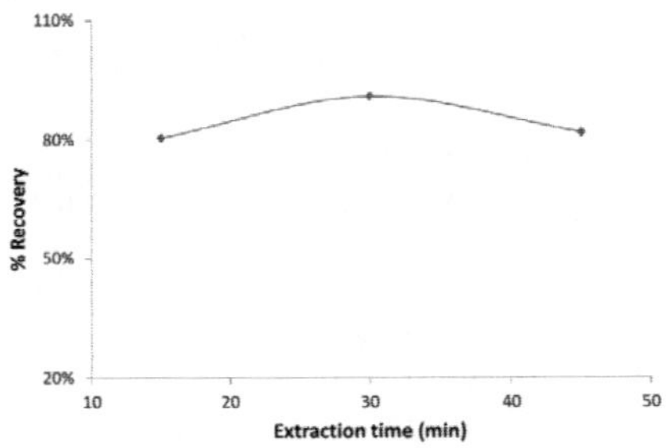

Figure 4. Effect of extraction time on the methanol recovery from the Co/ZSM-5 surface.

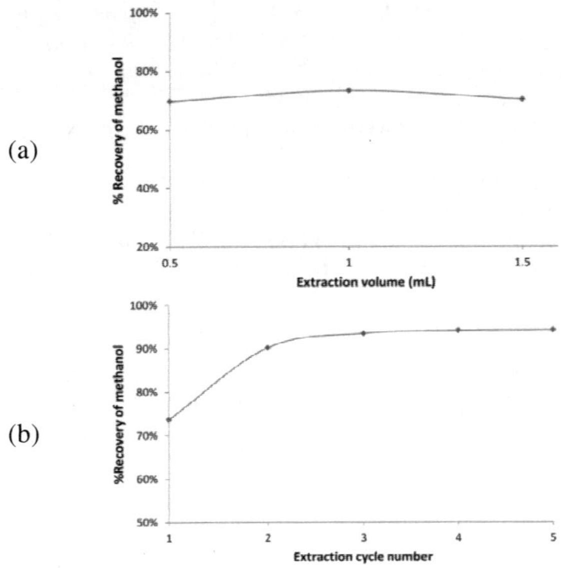

Figure 5. (a) effects of extraction volume and (b) extraction cycle number on the recovery of methanol from the Co/ZSM-5 surface.

content increased with time, i.e., for extraction times of 0.5 and 1 min, methanol recoveries of 80.5% and 91% were obtained, respectively. Based on the product composition, a longer extraction time of 45 min led to decreased methanol recovery (82%). The increase in the extraction time leads to the increase in solvent evaporation. Hence, it is recommended to utilize an extraction time of less than 1 min for methanol.

3.3 Volume and extraction cycle number

The ethanol volume did not significantly affect the methanol recovery. As figure 5 shows, 1 mL of ethanol led to a methanol recovery of 73%. In addition, the effect of the extraction cycle number using ethanol as the solvent was investigated under the above-mentioned optimum conditions. The first extraction cycle recovered a majority of the methanol from Co/ZSM-5, while some methanol was recovered from the second extraction cycle. Moreover, the third and fourth extraction cycles led to the recovery of low amounts of methanol. Furthermore, methanol was not obtained from the 5th cycle. In total, 94% of methanol was recovered from Co/ZSM-5. Notably, the extraction efficiency was obtained using a low volume of solvent and three extraction cycles.

4. Conclusions

Methanol, which was the main product obtained from the partial oxidation of methane, was extracted from the Co/ZSM-5 catalyst surface. An extraction method was developed herein. A high methanol recovery was achieved using a 1:1 v/v of ethanol:water as the solvent. An appropriate reaction time, solvent volume, and extraction cycle led to a high extraction efficiency. Despite the extraction conditions, Co/ZSM-5 demonstrated potential for reuse in the partial oxidation of methane to methanol as the extraction solvent did not alter its structure.

Acknowledgements

This work was partially funded by Hibah Kolaborasi International UI, 2014 and OGFICE Grant 2016 from OSAKA GAS Foundation. Authors would like to thank Einago (www.einago.com) for the English language review.

References

[1] Beznis N V, van Laak A N C, Weckhuysen B M and Bitter J H 2011 *Micro. Meso. Mater.* **138** 176–83
[2] Grundner S, Markovits M A C, Li G, Tromp M, Pidko E A, Hensen E J M, Jentys A, Sanchez-Sanchez M and Lercher J A 2015 *Nat. Commun.* **6** 7546
[3] Knops-Gerrits P P and Goddard W A 2001 *J. Mol. Catal. A: Chem.* **166** 135–45
[4] Sheppard T, Daly H, Goguet A and Thompson J M 2016 *Chem. Cat. Chem.* **8** 562–70
[5] Starokon E V, Parfenov M V, Pirutko L V, Abornev S I and Panov G I 2011 *J. Phys. Chem. C* **115** 2155–61
[6] Vanelderen P, Hadt R G, Smeets P J, Solomon E I, Schoonheydt R A and Sels B F 2011 *J. Catal.* **284** 157–64
[7] Narsimhan K, Iyoki K, Dinh K and Román-Leshkov Y 2016 *ACS Cent. Sci.* **2** 424–9
[8] Krisnandi Y K, Putra B A P, Bahtiar M, Abdullah I, and Howe R F 2015 *Procedia Chem.* **14** 508–15
[9] Karim T M, Nuzuliyah R, Nurani D A and Krisnandi Y K 2017 *AIP Conf. Proc.* **1862** 030086
[10] Sobolev V I, Dubkov K A, Panna O V and Panov G I 1995 *Catal. Today* **24** 251–2.
[11] Starokon E V, Parfenov M V, Arzumanov S S, Pirutko L V, Stepanov A G and Panov G I 2013 *J. Catal.* **300** 47–54

A paper-based visual indicator for detection of ammonia using *Ruellia simplex*

W Sholihah, A Listyarini, R Fitriana and C Imawan

Department of Physics, Faculty of Mathematics and Natural Sciences (FMIPA) Universitas Indonesia, Depok 16424, Indonesia

Corresponding author: cuk.imawan@sci.ui.ac.id

Abstract. Ammonia is one of the most important chemicals for agriculture, chemical, and food industries. This condition prompts the need for a research on detecting ammonia gas with a simple and easy method in daily practices. One of the methods for detecting ammonia gas is done through the use of visual indicators. Aiming for making the visual indicators, *Ruellia simplex* flower extract is used in this research as an indicator, which visually detects ammonia gas. The extract functions as a smart material, which changes its colour according to the extract's pH condition. At pH 2 – pH 3 the extract colour is pink and then turn to purple at pH 6 and becoming green at pH 10 – pH 11. The results of the UV-Vis spectra for the extract solution at pH 2 – pH 3 show the absorbance peak at 535 nm wavelength, then shifted to 537 nm for pH 6 and shifted to 621 nm for pH 10 – pH 11. For developing paper-based indicators, cellulose paper (1cm x 6cm) is used for substrates. The cellulose paper's tip has been coated with extracts whose pH have been preconditioned to 2 correspondingly. Using ammonia gas with a concentration between 0% and 0.1%, a test on the visual indicator paper is performed. Colour changes that occur in the indicator paper are observed as a function of detection time and then analyzed through RGB image analyzer. The result of the experiment shows that the indicator paper undergoes a change of colour upon the detection of ammonia gas. The colour change depends on the detected gas concentration. The paper indicators with pH value of 2 has the most optimum contrast in colour change as the paper's original pink colour changes into green upon the detection of ammonia. This research may be able to be developed further for the purpose of detecting food freshness.

Keywords: ammonia detection; visual indicator; *Ruellia simplex*

1. Introduction

Meat is a source of protein composed of free amino acids. Proteins in meat can brake down into amino acid through process of hydrolysis. The amino acid itself can be further decomposed partly or entirely into simple compounds such as CO_2, NH_3, H_2S and H_2O, [1]. The degree of damage in a meat can be obtained by measuring its chemical changes. Alternatively, it can be done by reacting the substances released during meat degradation with freshness indicator based on acid-base [2]. Therefore, whether a food is fresh or spoiled rotten can be determined by detecting the emitted gas, including ammonia (NH_3).

Ammonia can be detected by using a paper of visual indicator, which contains colouring agent with specific stability influenced by factors such as acid-base. One of the colouring agents is anthocyanin, which is soluble in water and alcohol. This substance can be found in various plants including flowers or any part of plants, which are coloured red, blue, and purple [3]. Use Utilization of anthocyanin as natural colouring agent has been done in several researches such as utilization of red cabbage extract (*Brassica olereceae*) [4], flowers of *Bauhinia blakeana Dunn* [5], and grape skin [6].

In this research, flowers of *Ruellia simplex* are used as natural source for anthocyanin to create a paper visual indicator capable of detecting ammonia gas. Then, RGB values of the visual indicator which has passed been tested and underwent changed in colour changes are characterized. Characterization is done to obtain the RGB intensity of each spectrum.

2. Experimental

2.1. Preparation of Ruellia simplex flower extract

Flowers of *Ruellia simplex* were gathered from the vicinity of Universitas Indonesia, Depok. The extract of *Ruellia simplex* was made by maceration method which is soaking 80 g of *Ruellia simplex* flower which was previously crushed first by using a blender into ethanol water solution 7: 3 v/v [4]. Then the extract solution was conditioned at pH 2 with the addition of 1 M hydrochloric acid (HCl) from Merck. At pH 2, anthocyanin becomes stable and durable as well as long lasting [7]. The extract of *Ruellia simplex* was then stored for 24 hours in a refrigerator. After that, the extract was filtered using Whatman paper No 1 and the extract was tested for colour changes against pH by adding buffer solutions of pH 2–11. Colour change test against pH was also conducted using UV-Vis Spectrophotometer to read the absorbance spectrum of the *Ruellia simplex* extract. The *Ruellia simplex* extract used in this research functions as a visual indicator to detect ammonia gas. The extract itself is a smart substance which changes colour according to its pH levels.

Cellulose paper (1cm x 6cm) of brand Whatman no. 1 is used as substrate. The tip of the celluloid paper (1 cm x 4 cm) is coated with *Ruellia simplex* extract which was conditioned at pH 2. After being coated, the cellulose paper was then dried. Test was conducted on the dried visual indicator to detect ammonia.

2.2. Ammonia detection test

Test on the visual indicator paper was conducted on ammonia with concentration of 0.000%, 0.005%, 0.010%, 0.040%, 0.070%, and 0.100%, put into a 100 ml bottle at 20 mL. Test was conducted on various temperature levels, refrigerator temperature at ±13 °C, room temperature at ±25 °C, and oven temperature at 40 °C which were observed for 3 hours. Colour changes on the visual indicator were observed to determine its detecting time and then analysed using RGB visual analyzer.

2.3. Characterization of RGB

The colour change of the indicator label on ammonia was observed visually using a digital camera. The photo results are then quantified with ImageJ software using RGB values. Scale 0 to 255 (8 bit) indicated the intensity of the analyzed spectrum colour, where dark colour (black) was represented by 0 while bright colour (white) was represented by 255. Combined RGB of all three that surpassed 255 shall result in white colour. Standard Deviation showed the degree of intensity change (View Histograms and Pixels Value, 2017). The more homogeneous its visuals, the lower its standard deviation becomes. Total RGB is the sum of each R, G, and B value on the visual. Equation (1) showed how to calculate total RGB:

$$RGB\ total = \sqrt{0.299(R^2) + 0.587(G^2) + 0.114(B^2)} \tag{1}$$

Equation (2) showed how to calculate total standard deviation.

$$Std\ total = \sqrt{Std_{red}^2 + Std_{green}^2 + Std_{blue}^2} \tag{2}$$

3. Results and discussion

3.1. Colour change test on the extract against pH and UV-Vis spectrum

Results of colour change test on *Ruellia simplex* extract with buffer pH 2–11 can be seen in figure 1. From visual analysis, colours resulted from the extract ranged from pink to green. Extract of *Ruellia simplex* flower, which contains anthocyanin, clearly changed colour at different pH. As shown in figure 1, pink solution is obtained at pH 2–3, and then yellowish red at pH 4–5, purple at pH 6, blue at pH 7 and pH 8–9, and becoming green at pH 10–11. Based on the colours of *Ruellia simplex* flower extract at buffer solution pH 2–11, this extract possesses a wide range of colours (from pink to green), which makes *Ruellia simplex* a good natural material candidate for the sensor.

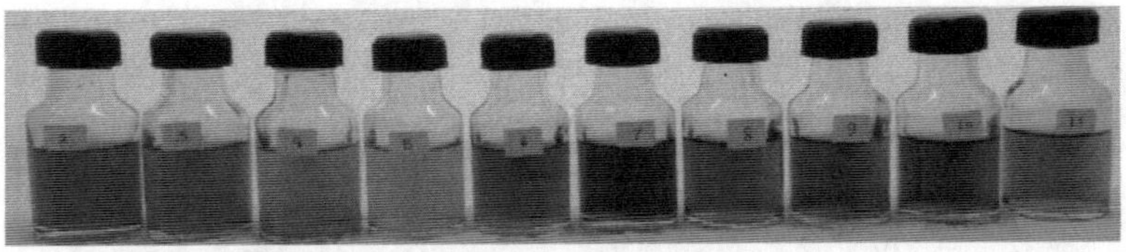

Figure 1. Extract of *Ruellia simplex* flower added with pH 2-11 buffers

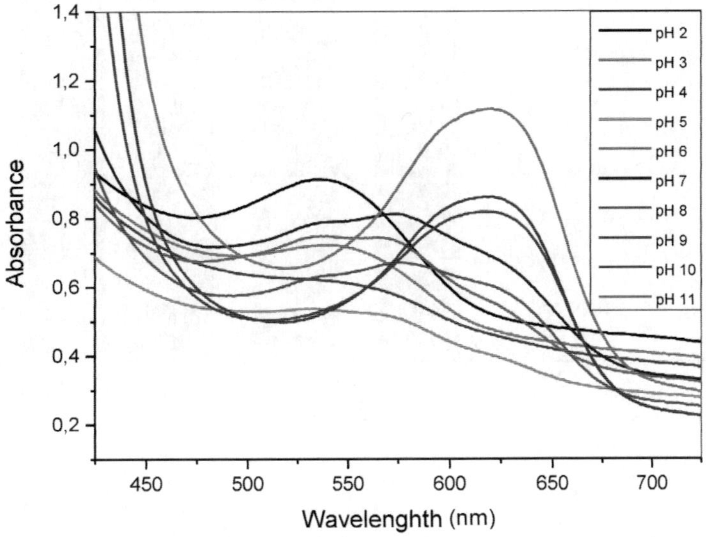

Figure 2. UV-Vis spectrum of *Ruellia simplex* flowers

UV-Vis spectrums of the solution at similar pH level were tested using spectrophotometer in range 400-750, as shown in figure 2. UV-Vis spectrum of *Ruellia simplex* extract at pH 2-11 show a colour change from pink to green approximately at wavenumber 535 to 621 nm. At pH 2-3, showing pink colour, clear absorbance peaks can be observed at 535 and 539 nm. The intensity decreases along with increase in pH, where the peak almost disappears when the solution is at pH 5. Absorbance intensity at pH 5 is the lowest compared to other pH levels. At higher pH solution, absorbance peak starts increasing around 537 nm from pH 6, with colour change from cream to purple. The respective absorbance peaks for pH 9, 10, and pH 11 are at 618 nm, 621 nm, and 621 nm, where in this pH range, the colour changes into green with further wavelength peaks.

3.2. Results of test and characterization of indicator labels

Table 1 shows the colour changes on the indicator paper after being exposed to ammonia gas of varying concentrations. These colour changes are compared between the original indicator paper and post-ammonia gas exposure. There was no visual colour change on the indicator paper when the temperature reaches 25°C at 0.000% concentration after 3 hours. At 0.005% and 0.010% concentration after 3 hours the colour changed into blue and at 0.040% concentration after 3 hours the colour starts changing into yellow. It becomes complete yellow after 2 hours at 0.100% and 0.070% concentration. The colour changes showed that the greater the ammonia concentration, the faster indicator labels changes their visual colours. This is due to the fact that solid concentration of ammonia results in larger size of ammonia gas.

Colour changes on the indicator labels at exposure of ammonia gas are quantified based on intensity of green colour as shown in figure 3. In figure 3, it can be seen that the higher concentration of ammonia, the higher intensity of green colour on the paper.

IOP Conf. Series: Materials Science and Engineering **496** (2019) 012050 doi:10.1088/1757-899X/496/1/012050

Table 1. Visual indicators with varying duration and concentration at 25 °C room

NH_3 concentrations (%)	Time (hours) 0	1	2	3
0.000				
0.005				
0.010				
0.040				
0.070				
0.100				

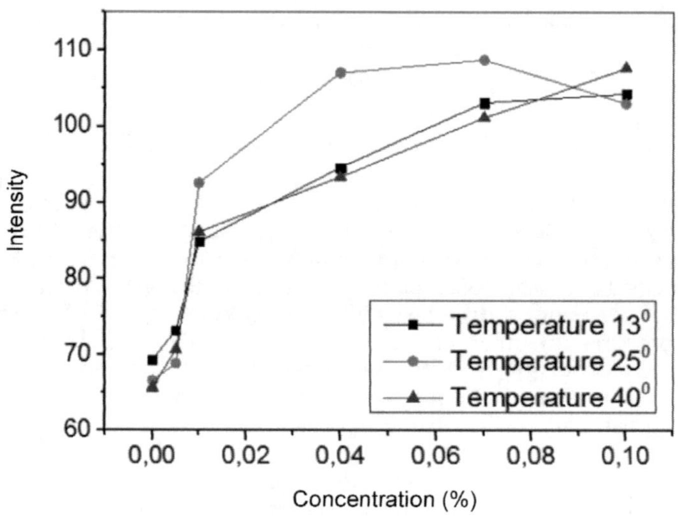

Figure 3. Concentration and intensity of green colour after 1 hour of test for three variation of temperature

Ideally, both temperature and duration factors should not affect colour changes on the visual indicators since the concentration stays the same. However, during the test, variations on temperature and duration as shown in table 2 also affect colour changes. To observe the influence of temperature on these colour changes, tests are conducted at 13 °C, 25 °C, and 40 °C. Table 2 shows the colour changes at these temperatures. These colour changes are quantified with RGB system using ImageJ program. Change in intensity of green colour is illustrated in figure 4 shows how the intensity rises alongside the temperature.

Table 2. Visual indicators from varying duration and concentration after 1 hour of test

Temperature (°C)	Ammonia concentrations (%)		
	0.000	0.005	0.100
initial condition			
13			
25			
40			

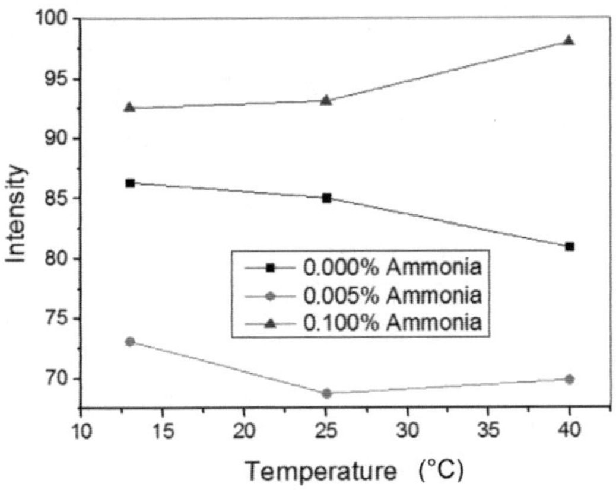

Figure 4. Temperature and intensity of green colour after one hour of test for three variation of ammonia concentration

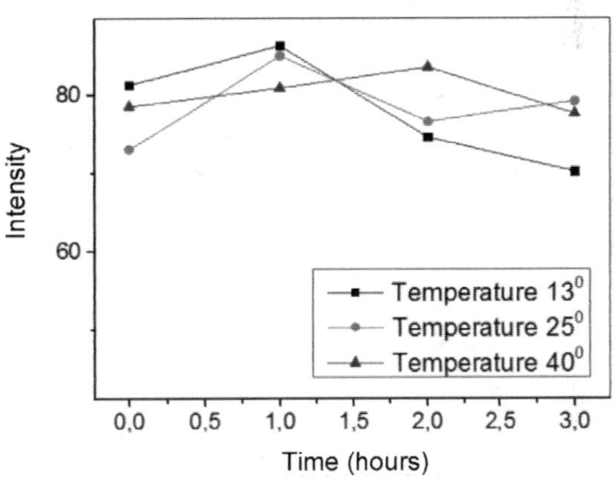

Figure 5. Time and intensity of total RGB at 0,000% concentration for three variation of temperature

To observe the stability of indicator labels due to environmental factors, test was conducted on indicator labels exposed to the environment (without ammonia gas) as shown in figure 5. Graph on 0.000% concentration with varying duration and temperature in figure 5 shows that 0.000% concentration made no significant change on total RGB value of the paper (less than 15 pixel). As seen from table 1 and table 2, at 0.000% concentration there is no visual change from the original colour. This proves that the visual indicator labels are stable against environmental influence and will not change colour at the absence of ammonia gas.

4. Conclusions

UV-Vis spectrum results from the *Ruellia simplex* flower extract at pH 2-11 show colour changes from pink to green with wavelength peaks from 535 nm to 621 nm. Visual indicator labels at pH 2 are the most applicable due to ammonia being an alkaloid, so that visual indicator labels at pH 2 have optimum contrast in the colour change. Concentration of ammonia, temperature, and duration of test can affect the colour changes on the indicator labels in parallel relation.

References

[1] Lawal A T and Adeloju S B 2012 *Talanta* **100** 217–28
[2] Smolander M, Alakomi H L, Ritvanen T, Vainionpää J and Ahvenainen R 2004 *Food Control* **15** 217–29
[3] Rodriguez-Amaya D B 2016 *Curr. Opin. Food Sci.* **7** 20–6
[4] Silva-Pereira M, Teixeira J A, Pereira-Júnior V A and Stefani R 2015 *LWT - Food Sci. Technol.* **61** 258–62
[5] Zhang X, Lu S and Chen X 2014 *Sens. Actuators B Chem.* **198** 268–73
[6] Ma Q and Wang L 2016 *Sens. Actuators B Chem.* **235** 401–7
[7] Farahmandazad H 2015 *Recovery and Purification of Anthocyanins from Purple-blue Potato* Master Thesis (Lappeenranta: Lappeenranta University of Technology)

Electrochemical study of nickel-cobalt deposited on boron-doped diamond as working electrodes for urea fuel cells

Y M T A Putri, J Gunlazuardi and T A Ivandini

Department of Chemistry, Faculty of Mathematics and Natural Sciences (FMIPA), Universitas Indonesia, Depok 16424, Indonesia

Corresponding author: ivandini.tri@sci.ui.ac.id

Abstract. Urea as one of the most abundant compounds that can be used as a fuel in fuel cells. It contains a nitrogen–hydrogen bond that can easily be broken by electrochemical processes to produce two molecules of hydrogen. The hydrogen released can generate electricity as in a fuel cell. In this research, boron-doped diamond (BDD) was modified using nickel and/or cobalt electrodeposition to provide Ni-Co-BDD, Ni-BDD, and Co-BDD working electrodes for enhanced urea electro-oxidation. The highest power density and potential was achieved using Ni-Co-BDD (96.4 mW cm^{-2} and 0.92 V, respectively), while the highest current density was achieved using Co-BDD (165.34 mA cm^{-2}). Overall, the results indicate that the best and most stable electrode for fuel cells is that based on Ni-Co-BDD.

Keywords: boron-doped diamond, fuel cell, surface modification, nickel, cobalt

1. Introduction

Electricity demand in Indonesia continues to increase in line with increasing national economic growth. Electrical energy is currently mainly produced from fossil fuels, which are a limited resource and produce large quantities of pollutants. Therefore, alternative energy sources that can satisfy increasing demands while remaining efficient and environmentally friendly are essential.

A fuel cell is a device that can instantly convert chemical energy into electricity with high efficiency [1]. Fuel cells can be used for various purposes, including as power sources for portable devices and electric vehicles, as well as stationary energy storage devices. Therefore, the use of fuel cells is expected to replace fossil-fueled electric energy.

Urea is considered as a potentially excellent source of hydrogen owing to its abundance, stability, and non-toxicity [2]. Furthermore, electricity can be generated if the hydrogen gas is released via electrochemical processes [3]. Accordingly, Guo *et al.* [2] used urea as a fuel in a fuel cell using a nickel–cobalt electrode in 2016 [2]. Previously, Boggs *et al.* [4] had used urea as a fuel in a fuel cell using a nickel electrode in 2009. Nickel is known to be a good catalyst and has a high current density and low oxidation potential in urea electro-oxidation. The catalytically active component of Ni electrodes for urea electro-oxidation is NiOOH [4]. The formation potential of NiOOH from $Ni(OH)_2$ can reach ≈0.25 V (vs. Ag/AgCl) in 5.0 M KOH. Accordingly, a large over potential is required to operate direct urea fuel cell [4]. Theoretically, urea fuel cells produce water, electric energy, and heat by the following processes [2]:

Overall reaction:

$$CO(NH_2)_2 + 8KOH + 3H_2O_2 + 3H_2SO_4 \leftrightarrow N_2 + 12H_2O + K_2CO_3 + 3K_2SO_4 \ (E^0 = 2.51 \text{ V})$$

IOP Conf. Series: Materials Science and Engineering **496** (2019) 012051 doi:10.1088/1757-899X/496/1/012051

(a)

(b)

(c)

(d)

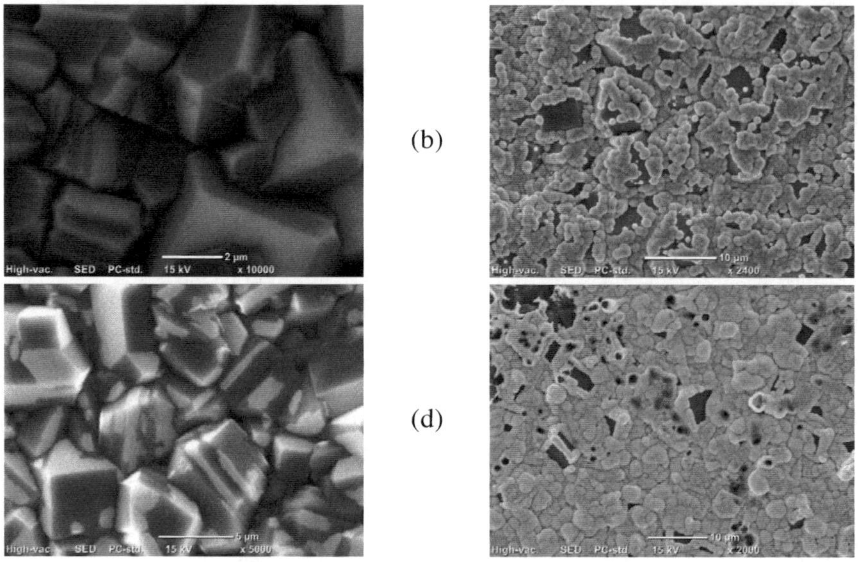

Figure 1. SEM images of (a) bare BDD, (b) Ni-BDD, (c) Co-BDD, and (d) Ni-Co-BDD.

On the contrary, boron-doped diamond (BDD) is known as an electrode with high stability and wide potential window [5]. Furthermore, Asgari *et al.* [6] reported that the charge acceptance of nickel can be increased by doping with cobalt, lowering the NiOOH formation potential [6]. Therefore, in the present study, BDD was modified with nickel and/or cobalt to synthesize three different electrodes (Ni-BDD, Co-BDD, and Ni-Co-BDD), which were compared and assessed for urea fuel cells [1].

2. Experimental

2.1. Preparation of Ni-BDD, Co-BDD, and Ni-Co-BDD

BDD film was obtained from Einaga Laboratory, Keio University, Japan. To prepare Ni-BDD, $NiSO_4.6H_2O$ (0.53 g) was dissolved in 0.1 M acetate buffer solution (pH 5). This solution was used to electrodeposit onto a BDD at a potential of −1.2 V for 600 s using an amperometric electrodeposition technique. The same procedure was performed using $CoCl_2.6H_2O$ (0.951 g), followed by electrodeposition at −1.0 V to prepare Co–BDD. Ni–Co–BDD was prepared by dissolving $NiSO_4.6H_2O$ (0.84 g) and $CoCl_2.6H_2O$ (0.19 g) in acetate buffer solution (pH 5), followed by electrodeposition at −1.2 V for 600 s. The modified electrodes were then placed in 1 M KOH, and cyclic voltammetry (CV) was performed 300 times at 100 mV s^{-1} to generate $Ni(OH)_2$ layers [7]. The electrodes were characterized using scanning electron microscopy (SEM).

2.2. Fuel cell test

For the fuel cell tests, Nafion 115 (Dupont) was placed between the anodic and cathodic chambers. Prior to use, the Nafion was boiled in 5% H_2O_2 and deionized water, respectively. The membranes were immersed in deionized water before use. The anolyte contained urea and KOH while the catholyte contained H_2O_2 and H_2SO_4 [2]. Direct urea/H_2O_2 fuel cell tests were carried out with a potentiostat using the eDAQ Chart application.

3. Results and discussion

3.1. Characterization of Ni-BDD, Co-BDD, and Ni-Co-BDD

For Ni–BDD, Ni(OH) layer was perfomed by cyclic voltammetry for 300 times in 1M KOH at the scan rate of 100 mV s^{-1} [7]. The same procedure was applied to Co–BDD and Ni–Co–BDD. Then all the electrodes were characterized using SEM. Representative SEM images are presented in figure 1.

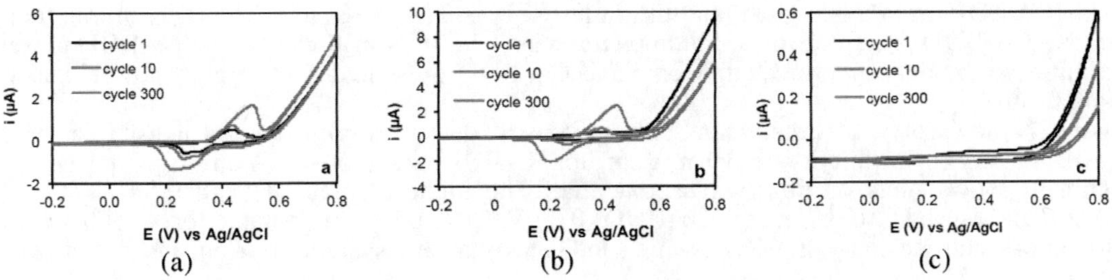

Figure 2. Cyclic voltammograms of (a) Ni–BDD, (b) Ni–Co–BDD, and (c) Co–BDD in 1 M KOH at a scan rate 100 mV s^{-1} over 300 scans.

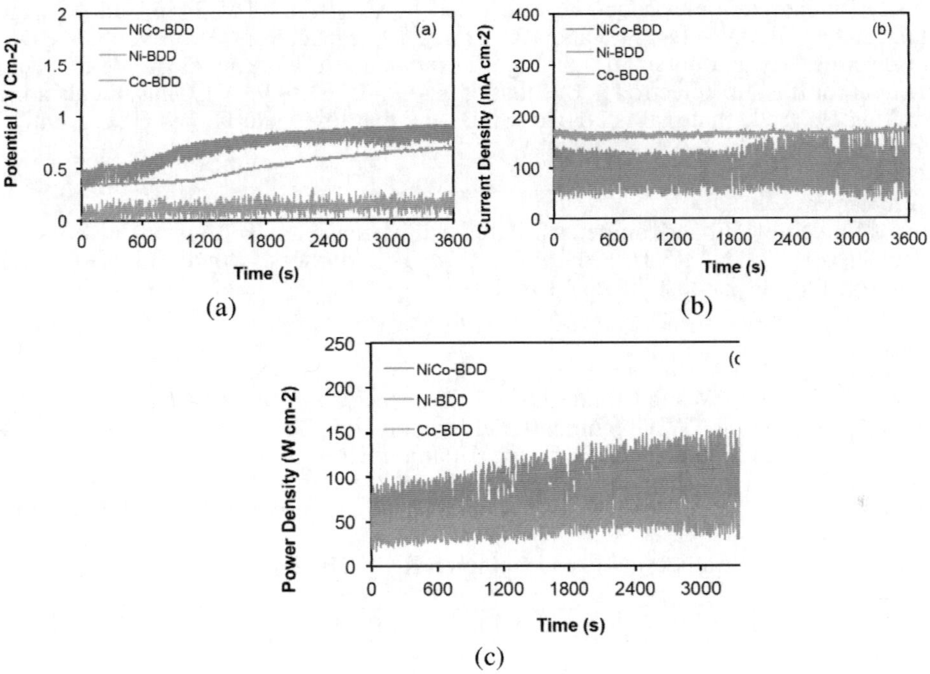

Figure 3. Plots of time vs. power density, current density, and potential for urea/H_2O_2 fuel cells using 0.001 M KOH

This treatment is required to form two crystallographic species, i.e., hydrated α-$Ni(OH)_2$ and anhydrous β-$Ni(OH)_2$ [7]. β-$Ni(OH)_2$ is more stable than α-$Ni(OH)_2$ and is thus formed by performing 300 cycles.

The CV results for Ni–BDD, Co–BDD, and Ni–Co–BDD are shown in figure 2. The oxidation peak of the last cycle for both Ni-BDD and Ni-Co-BDD is shifted to the right of the previous oxidation peaks as shown in figure 2a and figure 2b. This indicates the formation of the more stable β-$Ni(OH)_2$ structure. However, as shown in figure 2c, this phenomenon is not observed for Co–BDD because it has no $Ni(OH)_2$ layer.

3.2. Fuel cell tests

Plots of time versus power density for direct urea/H_2O_2 fuel cells operating for 1 h at room temperature are presented in figure 3a. Ni–Co–BDD, Ni–BDD, and Co–BDD were used as the working electrodes with a platinum spiral as the counter electrode and a saturated Ag/AgCl electrode as the reference electrode. Nafion was used to transfer protons from the anode to the cathode chambers. The anolyte was 0.001 M KOH and 0.33 M urea, and the catholyte was 2 M H_2SO_4 and 2 M H_2O_2. In this fuel cell system, Ni–Co–BDD presented a maximum power density of 96.4 mW cm^{-2}, while Ni–BDD presented a maximum power density of 73.9 mW cm^{-2} and Co–BDD presented a maximum power density of 52.74 mW cm^{-2}.

Figure 3b shows plots of time versus current density. The maximum current density for Ni–Co–BDD is 104.3 mA cm^{-2}, while it is 103.1 mA cm^{-2} for Ni–BDD and 165.34 mA cm^{-2} for Co–BDD.

Figure 3c shows plots of time versus potential. The highest potential is that for Ni–Co–BDD (0.92 V), then Ni–BDD (0.71 V), and Co–BDD (0.32 V). The power was stable for ~3,500 s with the best working electrode (Ni–Co–BDD), which indicates that our system is promising for direct urea fuel cells.

4. Conclusions

Application of Ni–Co–BDD, Ni–BDD, and Co–BDD as electrode materials in fuel cells containing urea provided maximum power densities of 96.4, 73.9, and 52.74 mW cm^{-2}, respectively. The highest current density for the three electrodes was provided by Co–BDD (165.34 mA cm^{-2}), while those of Ni–Co–BDD, and Ni–BDD were 104.3 and 103.1 mA cm^{-2}, respectively. However, Co–BDD provided the lowest power density and potential, making it unsuitable as a working electrode for fuel cells. The highest potential for the three electrodes was that of Ni–Co–BDD (0.92 V). Thus, the data indicate that the best working electrode material is Ni–Co–BDD, and that this material has great promise as a urea fuel cell electrode material.

Acknowledgements

This work was funded by Hibah Kompetensi Publikasi Internasional Terindeks Universitas Indonesia, with contract No.1822/UN2.R3.1/PPM.00.01/2017. The authors would like to thank Einago (www.einago.com) for the English language review.

References

[1] Lan R and Tao S 2001 *J. Power Sources* **196** 5021–6
[2] Guo F, Cheng K, Ye K, Wang G and Cao D 2016 *Electrochim. Acta* **199** 290–6
[3] Walter X A, Stinchcombe A, Greenman J and Ieropoulos I 2017 *Appl. Energy* **192** 575–81
[4] Boggs B K, King R L and Botte G G *Chem. Commun.* **32** 4859–61
[5] Shibano S, Ivandini T A, Terashima C, Nakata K and Einaga Y 2014 *Chem. Lett.* **43** 1292-3
[6] Asgari M, Maragheh M G, Davarkhah R, Lohrasbi E and Golikand A N 2012 *Electrochim. Acta* **59** 284–9
[7] Toghill K E, Xiao L, Stradiotto N R and Compton R G 2010 *Electroanalysis* **22** 491–500

IOP Conf. Series: Materials Science and Engineering **496** (2019) 012052 doi:10.1088/1757-899X/496/1/012052

Catalytic performance of CuO@SiO_2 for one-pot synthesis of pyrimidine derivative, ethyl 6-methyl-2-oxo-4-phenyl-1,2,3,4-tetrahydropyrimidine-5-carboxylate

A H Cahyana, S P Sari and B Ardiansah

Department of Chemistry, FMIPA, Universitas Indonesia, Kampus Depok, 16424, Jawa Barat, Indonesia

Corresponding author: herrykim@ui.ac.id

Abstract. We have prepared a three-component reaction of ethyl acetoacetate, urea and benzaldehyde for synthesis of a pyrimidine derivative, ethyl 6-methyl-2-oxo-4-phenyl-1,2,3,4-tetrahydropyrimidine-5-carboxylate using CuO@SiO_2 nanocatalyst. The catalyst was prepared by using glycerol as green solvent, and then it was characterized by FT-IR, EDAX spectroscopy and TEM analysis. The copper oxide nanoparticles are dispersed as fine and amorphous phase and made nanoclusters with average size of 14 nm. In pyrimidine derivative synthesis, the catalyst shows good activity for one-pot three-component reaction in mild reaction condition with the highest yield of 85%.

1. Introduction

Pyrimidines and their derivatives play a key role in some biological processes, as they are part of nucleic acid DNA and RNA [1,2]. They continue to attract great attention because many pyrimidine derivatives are reported to exhibit numerous bioactivities such as antimicrobial [3], anti-inflammatory [4], anticancer [5], antiviral [6] and antimycobacterial [7]. Thus, facile and applicable methods for the synthesis of pyrimidine derivatives are of considerable research interest.

Supported copper and nickel catalysts are frequently used in a variety of organic transformation and fine chemical production, such as methane reforming, hydrogenolysis, hydrogenation, dehydrogenation and methanol steam reforming [8–14]. Particularly, CuO@SiO_2 has been prepared by several methods such as sol-gel, hydrothermal and precipitation [15–17]. It has been applied in high performance electrochemical sensing and multifunctional coating in building stone [18,19]. However, CuO@SiO_2 has not been reported for pyrimidine synthesis yet. Therefore, in the present work, we will explore catalytic ability of CuO@SiO_2 for the one-pot three-component reaction in pyrimidine derivative synthesis.

2. Materials and methods

2.1. Materials

All chemicals were purchased from Merck in high purity, used without further purification. Element composition of the catalyst was determined by EDAX spectroscopy. Size and shape of the CuO@SiO_2 was observed using Tecnai D2350 S-Twin Transmission Electron Microscope (TEM) with the accelerated voltage of 200 kV. FT-IR spectrum of the catalyst was recorded on Shimadzu 8400 spectrophotometer with KBr background. The UV-Vis spectrum of pyrimidine derivative was recorded on Shimadzu UV mini type 1240 spectrophotometer, whereas mass spectrum was obtained from Shimadzu GC-MS QP Mass Spectrometer 2010A.

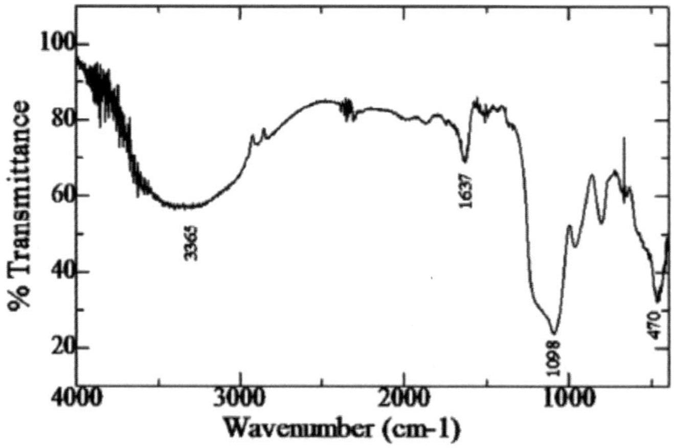

Figure 1. IR spectrum of $CuO@SiO_2$

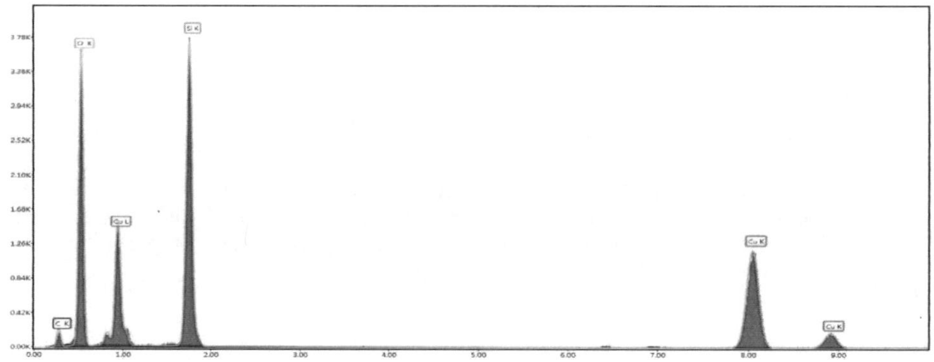

Figure 2. EDAX spectrum of $CuO@SiO_2$

2.2. Preparation of $CuO@SiO_2$ nanocatalyst

In a 50 mL Beaker glass, $Cu(NO_3)_2.3H_2O$ (1 mol/L), TEOS (10 mL) and glycerol (5 mL) were mixed and pH of solution was adjusted to 1.5 by adding 0.125 mol/L nitric acid. The mixture was stirred at room temperature for 3h. Then, ammonia solution was added dropwise until pH reached 9. Resulting gel was filtered and washed with ethanol:water (1:1), then it was aged for 3 days. Gel was calcined at 500 °C for 5h. the formed catalyst was characterized by FT-IR, EDAX and TEM.

2.3. Synthesis of a pyrimidine derivative

In a 50 mL round-bottom flask equipped with condenser, mixture of benzaldehyde (2 mmol), ethyl acetoacetate (2 mmol), urea (3 mmol) and 5 mL ethanol as solvent was stirred in the presence of certain amount of catalyst in various reaction condition (table 1), for example, 7.5% mol catalyst at 70 °C. After specified time, the obtained solid was filtered and then washed with deionized water. The crude product was recrystallized using hot ethanol. The product was identified by FT-IR, UV-Vis and GC-MS instruments.

3. Results and discussion

3.1. Catalyst characterization

Functional groups identification of $CuO@SiO_2$ was done using FTIR spectroscopy at 400–4000 cm^{-1} wavenumber range (figure 1). Stretching vibrations of the Cu-O group at 470 cm^{-1} and Si-O-Si groups at 1098 cm^{-1} indicated that $CuO@SiO_2$ were successfully prepared.

Table 1. Different conditions in the synthesis of pyrimidine derivative

Entry	Catalyst (% wt.)	Time (h)	Temperature (°C)	Yield (%)
1	Free	4	70	31
2	2.5	4	70	52
3	5	2	50	67
4	5	3	50	75
5	5	4	50	80
6	5	5	50	78
7	5	6	50	78
8	5	4	RT	36
9	5	5	70	80
10	5	4	90	83
11	7.5	4	70	85
12	10	4	70	82

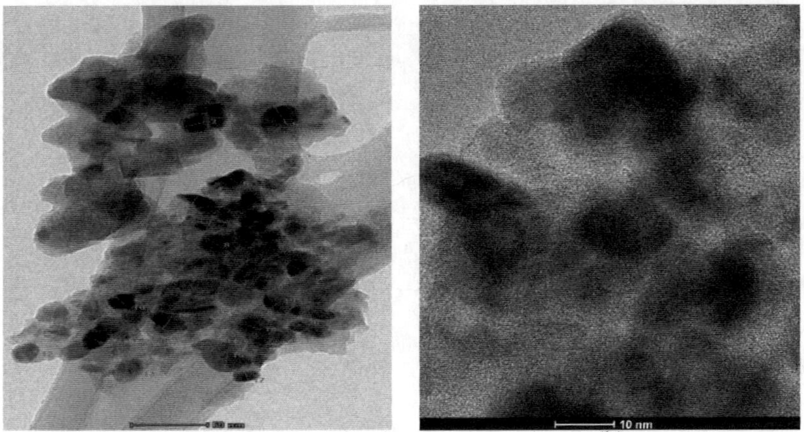

(a) (b)

Figure 3. (a) TEM and (b) HRTEM images of $CuO@SiO_2$

Composition of $CuO@SiO_2$ was analysed by using EDAX (figure 2). The material contains several atoms such as Cu (38.05%), Si (33.24%), O (27.16%), and C (1.54%). Metal impurities are not observed in the spectrum indicating the successful of catalyst preparation. Meanwhile, trace of carbon may come from background upon EDAX analysis. Morphology and size of $CuO@SiO_2$ were characterized by TEM (figure 3). From analysis, $CuO@SiO_2$ is in spherical shape with size of approximately 14 nm based on HRTEM result.

3.2. Synthesis of pyrimidine derivative

In catalysts-free system, only 31% yield of pyrimidine derivative was isolated. The optimum result of obtained pyrimidine derivative was in 4 hours at 70 °C and 7.5% weight of catalyst. The best yield of pyrimidine derivative was 85% (table 1, Entry 11). Increasing amount of catalyst to 10% will not increase the pyrimidine yield in significant amount. Also, prolonging reaction time up to 6 hours cannot increase the yield of product significantly.

FTIR data in figure 4 shows the presence of secondary N-H vibration at 3325 cm^{-1}, whereas at 2950 cm^{-1} indicating the presence of a C-H sp^3 group. Sharp peak at 1653 cm^{-1} is corresponds to vibration of C=O ester group. The maximum wavelength of this pyrimidine derivative was 284 nm (figure 5). Mass spectra data showed the molecular weight of pyrimidine derivatives with value of 260 g/mol, suitable with the molecular formula of $C_{14}H_{16}N_2O_3$ (figure 6).

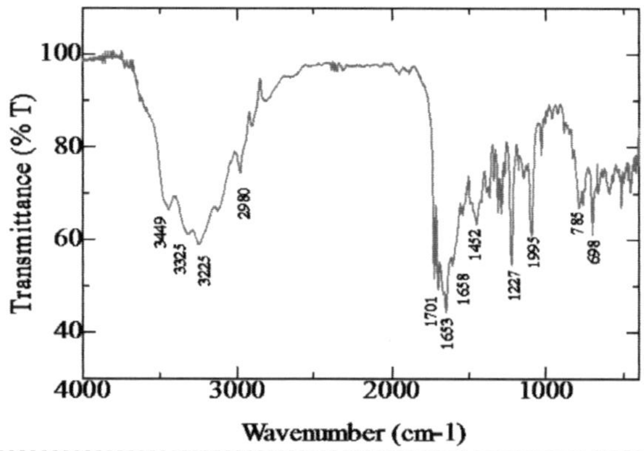

Figure 4. FTIR spectra of pyrimidine derivatives compound

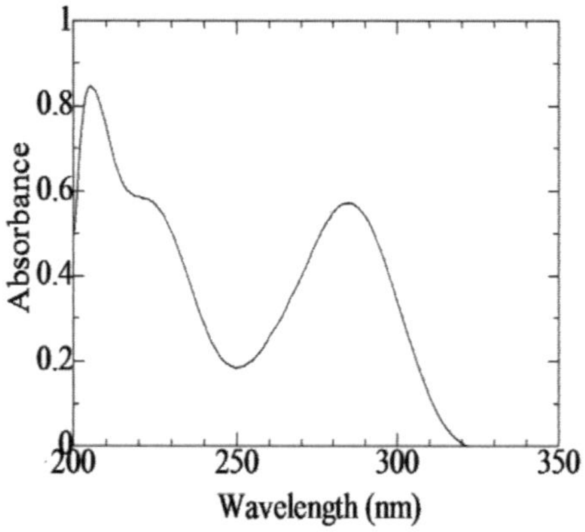

Figure 5. Uv-Vis spectra of pyrimidine derivatives compound

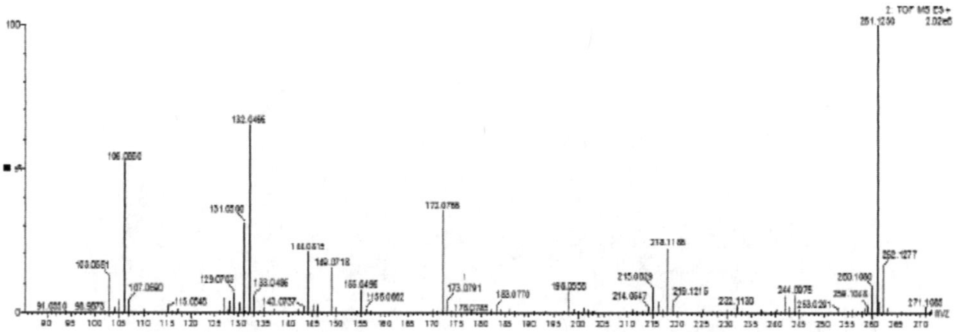

Figure 6. Mass spectral of pyrimidine derivatives compound

Reaction scheme for the synthesis of 3,4-dihydropyrimidine-2-ones was showed in scheme-1 following below:

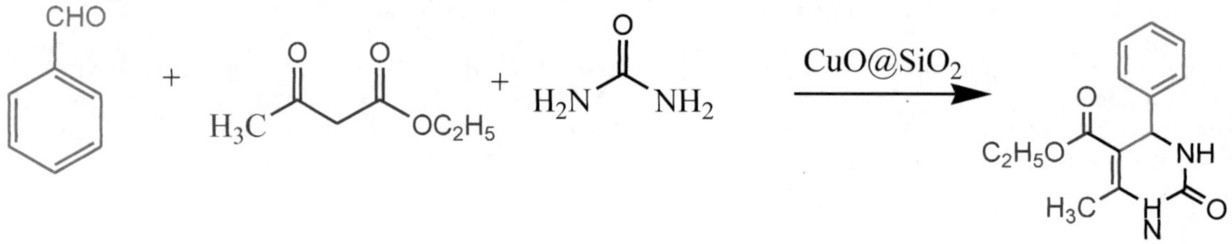

Scheme-1: Reaction scheme for the synthesis of 3,4-dihydropyrimidine-2-ones from benzaldehyde, ethyl acetoacetate and urea using $CuO@SiO_2$ nanocatalyst.

4. Conclusions

The $CuO@SiO_2$ nanocatalyst was successfully prepared using glycerol as stabilizer and capping agent with silica from TEOS and characterized by FTIR, EDX, and TEM. Pyrimidine derivative was obtained from the reaction of benzaldehyde, ethyl acetoacetate, and urea using $CuO@SiO_2$ heterogeneous catalysts. According to mass spectra data, molecular weight of the product was 260 g/mol.

Acknowledgements

The authors would like to thank Einago (www.einago.com) for the English language review.

References

[1] Gore R P and Rajput A P 2013 *Drug Invent. Today* **5** 148–52
[2] Xia S, Yin S, Tao S, Shi Y, Rong L, Wei X and Zong Z 2012 *Res. Chem. Intermed.* **38** 2435–42
[3] Kumar N, Singh G and Yadav A K 2001 *Heteroatom Chem.* **12** 52–6
[4] Noval Z M, Fahmy H H, Zarea E S and El-Eraky W 2011 *Acta Pol. Pharm.* **68** 507–17
[5] Al-Issa S A 2013 *Saudi Pharm. J.* **21** 305–6
[6] Kimura H, Katoh T, Kajimoto T, Node M, Hisaki M, Sugimoto Y, Majima T, Uehara Y and Yamori T 2006 *Anticancer Res.* **26** 91–7
[7] Kumar A, Sinha S and Chauhan P M 2002 *Bioorg. Med. Chem. Lett.* **12** 667–9
[8] Keane M A 2004 *Appl. Catal. A* **271** 109–18
[9] Couttenye R A, de Vila M H and Suib S L 2005 *J. Catal.* **233** 317–26
[10] Fridman V Z and Davydov A A 2000 *J. Catal.* **195** 20–30
[11] Sekizawa K, Yano S, Eguchi K and Arai H 1998 *Appl. Catal. A* **169** 291–7
[12] Brands D S, Poels E K and Bliek A 1999 *Appl. Catal. A* **184** 279–89
[13] Murthy K V, Patterson P M, Jacobs G, Davis B H and Keane M A 2004 *J. Catal.* **223** 74–85
[14] Ermakova M A and Ermakov D Y 2003 *Appl. Catal. A* **245** 277–88
[15] Tenkyong T, Bachan N, Raja J, Naveen Kumar P, Shyla J M 2015 *Mater. Sci. Poland* **33** 826–34
[16] Niu X, Zhao T, Yuan F and Zhu Y 2015 *Sci. Rep.* **5** 9153
[17] Huang Z, Cui F, Kang H, Ceng J and Xia C 2009 *Appl. Catal. A* **366** 288–98
[18] Gan T, Shi Z, Hu D, Lv Z, Sun J and Liu Y 2016 *Food Chem.* **190** 544–51
[19] Zarzuela R, Carbű M, Almoraima Gil M L, Cantoral J M and Mosquera M J 2016 *Mater. Design* **114** 364–72

Dependence on the length of localization of G4 DNA molecule to temperature and electric field

E N Sinurat and E Yudiarsah

Department of Physics, Faculty of Mathematics and Natural Sciences (FMIPA)
Universitas Indonesia, Depok 16424, Indonesia

Corresponding author: e.yudiarsah@sci.ui.ac.id

Abstract. The charge transport property of G4 DNA molecule that represented mathematically by using Hamiltonian tight binding approach has been studied. Charge transport property in the model of 102 base pairs long G4 DNA is studied from localization length under influence of electric field and temperature. Electric field affects electron hopping constants in G4 DNA molecule, while temperature leads to DNA structural disorder. Localization length is calculated by using transfer matrix method and using gram-schmidt orthonormalization method to reduce calculation errors. The result show, the increment of electric field and temperature decreases localization length, so that G4 DNA molecules becomes less conducting.

Keywords: DNA, DNA G4, Localization Length, Electric Field, Temperature

1. Introduction

The research progress has been bringing physicists to examine and study electrical properties of various biological materials, for instance DNA molecule. DNA molecule is an active element candidate that interesting to be studied because DNA can be passed by electric current along π-way [1]. DNA molecule has self-assembly and recognition properties. Self-assembly property is the ability of molecules to spontaneously assemble themselves. Recognition property is the ability of molecule to form selective bond with other molecules. These ability make DNA to be a promising candidate for used in molecular electronics [2].

DNA structure has been frequently modified in order to improve its effectiveness in transporting charges. Ones found structures such as single stranded DNA (ss-DNA), double stranded DNA (ds-DNA) and G-quadruplexes DNA (G4 DNA). Recently, G4 DNA has been investigated extensively because it has unique structure and self-assembly property [3]. Theoretical and experimental results show G4 DNA is efficient for long distance charge transport and more stable than ss-DNA and ds-DNA [3,4]. G4 DNA structural disorder is smaller than ds-DNA so that its conduction property is better [4]. Structural disorder can be caused by thermal agitation such as twisting motion [5].

In this report, DNA G4 model is studied in the presence of electric field along its symmetrical axis. This report will be focused on electron localization length which measures the distance spanned by an electron wave function. The localization length of DNA model was calculated by considering that electric field affect electron hopping constants of DNA molecule [6]. In the model, environment temperature is also taken into account. Temperature influences charge transport so that localization length decreases.

2. Theoretical model

G-quadruplexes also known as G4-DNA is a helix structure containing guanine tetrad where four

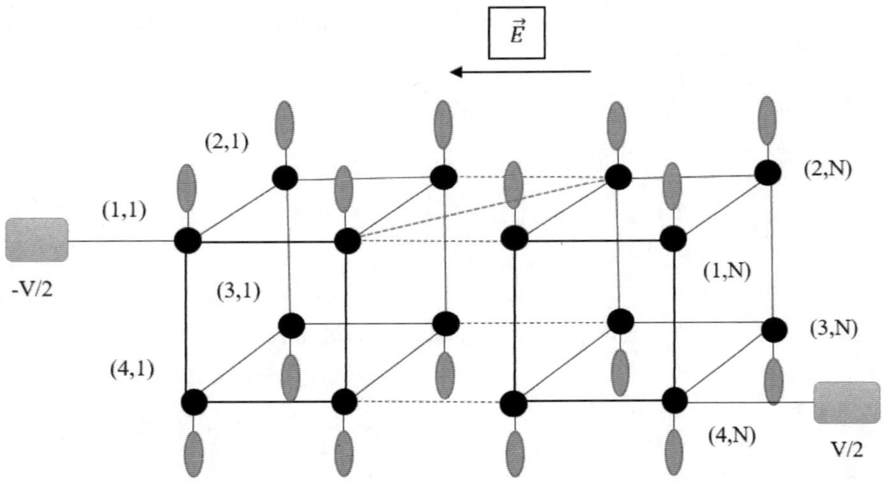

Figure 1. G4 DNA illustration scheme with dark circles describing guanine bases and orange circle depicting backbone with direction of electric field from V/2 to -V/2

guanine bases are linked by hydrogen bonds forming square planar structure. This planar structure is called guanine tetrad. Two or more guanine tetrad stack on top one another forming a G-quadruplex. The presence of cations, especially potassium, stabilizes the quadruplex structure [7]. The static G4-DNA molecule schematic diagram is shown in figure 1.

The charge transport of G4 DNA molecule was studied using tight binding Hamiltonian model as follows [7]

$$H = \sum_{n=1}^{N}\sum_{\tau=1}^{4}\left\{(\varepsilon_n^{\tau} + \emptyset_n^{\tau})|n,\tau\rangle\langle n,\tau| + t_{n,n+1}^{\tau,\tau}|n,\tau\rangle\langle n+1,\tau|)\right\} + \sum_{q=u,d}\sum_{\tau=1}^{4}\left\{B_n^{\tau,q}|n,\tau\rangle\langle n,q| + t_{n,n}^{\tau,q}|n,\tau\rangle\langle n,q|\right\} + \sum_{\tau=1}^{4}\left\{t_{n,n}^{\tau,\tau+1}|n,\tau\rangle\langle n,\tau+1| + t_{n,n+1}^{\tau,\tau+1}|n,\tau\rangle\langle n+1,\tau+1|\right\} + \text{h.c} \tag{1}$$

The parameter in equation (1) has been described in paper [8]. Electric field is applied parallel to DNA symmetrical axis (axis-Z). The direction of electron flows is opposite to the direction of electric field. Electric field leads to changes in the value of electron hopping constant from site *i* (initial state) to site *i+1* (final state) according to Miller-Abrahams formula [6]. In this report, the electron hopping constant can be written as

$$t_{i,i+1} = t_0 \exp\left\{-eE\left(\frac{Z(i+1) - Z(i)}{K_B T}\right)\right\} \tag{2}$$

$$t_{i,i-1} = t_0 \tag{3}$$

In equation (1) and equation (2), $t_{i,i+1}$ is electron hopping constant from initial state $(i+1)$ to final state (i) and $t_{i,i-1}$ is electric field affected electron hopping constant from initial state $(i-1)$ to final state (i). t_0 is electron hopping constant in the absence of electric field, *e* is electron charge, $Z(i)$ is the position of site-*i* along the *z* axis of DNA. k_B is Boltzmann constant and *T* is temperature. The effect of temperature is considered in this study following procedure in reference [5].

2.1. Localization length

Localization length is the size of room in length that might be traveled by charge in the system. Localization length can be calculated using transfer matrix method. Tight binding Hamiltonian approach can be used to calculated transfer matrix method using Schrödinger equation $H\psi = E\psi$. In 1-dimensi, the equation can be written as:

$$-t_{i-1,i}\psi_{i-1} + (\varepsilon_i - E)\psi_i - t_{i,i+1}\psi_{i+1} = 0 \tag{4}$$

Equation (4) can be transform into matrix equation

$$\begin{pmatrix} \psi_{i+1} \\ \psi_i \end{pmatrix} = T(i) \begin{pmatrix} \psi_i \\ \psi_{i-1} \end{pmatrix}, \tag{5}$$

where $T(i)$ is transfer matrix [9] and in 2x2 matrix can be written as

$$T(i) = \begin{pmatrix} \frac{\varepsilon_i - E}{t_{i,i+1}} & -\frac{t_{i-1,i}}{t_{i,i+1}} \\ 1 & 0 \end{pmatrix} \tag{6}$$

Equation (5) can be used to calculate wave function amplitude at one sites as long as the wave functions at two previous sites are known. Even more, through multiplication of consecutive transfer matrices ones obtain the relation of wave function amplitude at both ends of a molecule. To reduce error due to many multiplicative operation, in the calculation to types of multiplication are employed which are forward transfer matrix multiplication and, then, backward transfer matrix multiplication, gram-schmidt orthonormalization method are employed after certain number of matrix multiplication.

Following reference [9], after several times of matrix multiplication, the Lyapunov exponents γ_{cl} is calculated using

$$\gamma_{cl} = \left[\frac{Ln(\prod_r \alpha_{cl}^r)}{2K(L-1)}\right], \tag{7}$$

where cl is order of matrix columns, α is normalization coefficient obtained from orthonormalization process, K is repetition of multiplication, L is DNA length, and r is number of orthonormalization. The inverse of smallest γ_{cl} gives the localization length (λ)

$$\lambda = \frac{1}{\gamma} \tag{8}$$

3. Results and discussion

Electron localization lengths of G4-DNA molecule under the influence of electric field and temperature have been calculated at twisting oscillations frequency of 0.51 meV. The localization length in 102 base pair long DNA as function of electron energy has been calculated for voltages of 0, 1 mV, 3 mV, and 7 mV at temperature of 4.2 K and 30 K, respectively.

Electron localization length of G4 DNA molecule at temperature 4,2 K and various voltages are shown in figure 2. Spectrum of localization length shows the existence of extended states and localized states, which form bands. Extended state has localization length longer than 102 base pair, while states with electron localization length smaller than 102 base pair are called localized state. Figure 2a shows localization length when the electric field is absent. It shows two bands of states with localization length greater than zero separated by one gap which is in the energy range of 8,9 eV to 9,4 eV. In the energy range of 7.35 eV to 8.9 eV and 9.5 eV to 9.7 eV many states are extended states. However, there are no extended states in the energy below 7.35 eV. The localization lengths in the presence of electric field are shown in Fig 2(b-c). It is shown that the electric field decreases the localization length at certain energy. With the increment of electric field, more states changes from extended to be localized states. It can be seen clearly at energy greater than 8.4 eV. However, at energy below 7.35 eV some states experience an increment of localization length and become extended states. In addition to that, energy gap is slightly getting wider as voltage increases. This trend

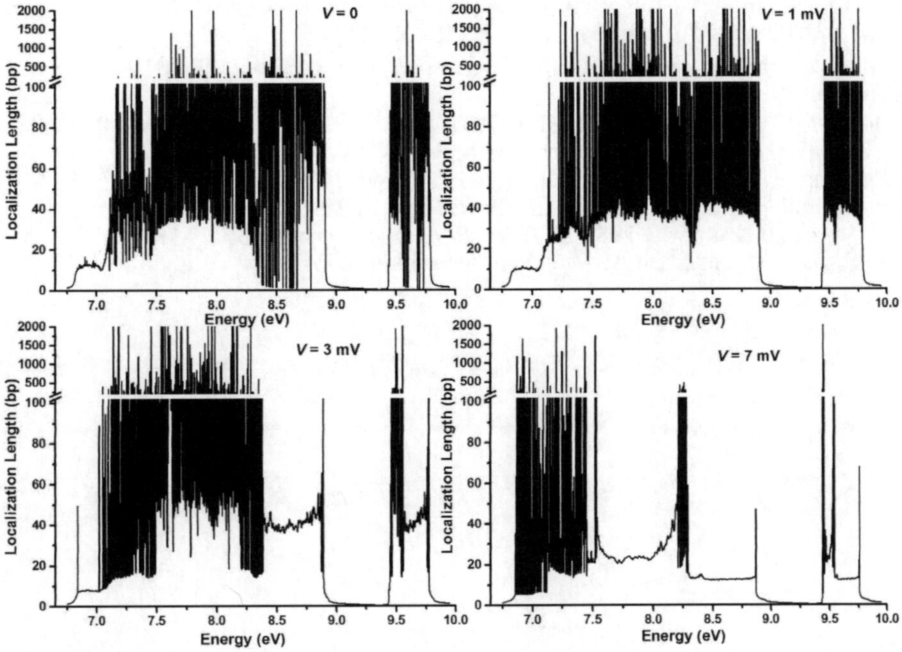

of localization length changes may be the result of the decrement of electron hopping constant in the direction of electric field. Sergiy nokhrin *et al.* find threshold voltage of 1 volt at a distance of 5 nm leads to decrease conductivity of M-DNA by at least 10-fold [10].

In figure 3 shows localization length calculated at temperature 30 K for several values of voltage. The increase in temperature leads to localization length decreases such that at this temperature all states become localized. Thermal fluctuations can affect the twist angle of DNA molecule so that the distribution of the twist angle of bases becomes wider. It causes the DNA structure becomes more irregular as temperature increases, thus disrupting electrons to jump. The result shows that temperature increases disorder in system so that inhibit charge transport in G4-DNA molecule. This result of calculation are in agreement to Suhendro *et al.* [5] research, which showed that localization length of double strand DNA molecule decreases due the increase in temperature, especially at low frequencies because of the distribution of DNA rotating angle widen. All states are localized is observed at all voltages used in the calculation. The effect of electric field is not as dramatic as at temperature 4.2 K in the sense of the decrease in localization length is much smaller. There is competing effect of twisting motion and electric field in the localization length.

4. Conclusions

Electron localization length in G4-DNA molecule under the influence of external electric field and at temperature 4.2 K and 30 K has been studied. The increment of electric field leads to lesser number of extended states. The electric field affects the electronic state more at higher energy where electron states become localized. This results shows that, in the scheme of electric field dependent hopping constant model we used, the electric field decreasing electron localization length and it leads to the deterioration of G4-DNA transport property. This trend is observed at both temperature used in the study. However, electric field influence on electron localization length becomes weaker as the temperature increases since the temperature effect is dominant.

Acknowledgements

Authors would like to thank D K Suhendro for parameters in process of calculation, Cisco Lab in Department of Physics FMIPA UI for computer lab facility and also Einago (www.einago.com) for the English language review. This work is funded by Universitas Indonesia through PITTA Grant 2017 with contract number 649/U2.R3.1/HKP.05.00/2017.

References

[1] Ketabi S A and Khouzestani H F 2014 *Iranian J. Phys. Research* **14** 67–72
[2] Malakooti S, Hedin E and Joe Y 2013 *J. Appl. Phys.* **114** 014701
[3] Guo A M and Xiong S J 2009 *Phys. Rev. B* **80** 035115
[4] Woiczikowski P B, Kubař T, Gutiérrez R, Cuniberti G and Elstner M 2010 *J. Chem. Phys.* **133** 035103
[5] Suhendro D K, Yudiarsah E and Saleh R 2010 *Physica B Condens. Matter* **405** 4806–11
[6] Li L, Meller G and Kosina H 2007 *Microelectron. J.* **38** 47–51
[7] Kang D, Jiang H, Sun Z, Qu Z and Xie S 2011 *J. Phys. Condens. Matter* **23** 55302
[8] Wiliyanti V and Yudiarsah E 2016 *AIP Conf. Proc.* **1729** 020033
[9] Mackinnon A and Kramer B 1983 *Z. Phys. B. Condens. Matter* **53** 1–13
[10] Nokhrin S, Baru M and Lee J S 2007 *Nanotech.* **18** 95205

Fe_3O_4/eggshells composite as green catalyst for Hantzsch condensation in acridine synthesis

A H Cahyana, E Saepudin, M Surya and B Ardiansah

Department of Chemistry, FMIPA, Universitas Indonesia, Kampus Depok, 16424, Jawa Barat, Indonesia

Corresponding author: endang.saefudin@sci.ui.ac.id

Abstract. Acridine derivatives are a class of nitrogen heterocycles which have broad spectrum of important biological properties and pharmaceutical applications. Low-cost eggshells waste material was used as source of $CaCO_3$ to modified magnetic Fe_3O_4 particles prepared from co-precipitation to serve as reusable heterogeneous catalyst. The synthesized eggshells based catalyst is characterized by X-ray diffraction (XRD) and TEM studies. The activity of the catalyst has been investigated in Hantzsch condensation reaction with good yields and simple work-up procedure. The synthesis of acridine derivative was achieved through a one-pot, three-component condensation of dimedone, aryl amine and aryl aldehydes utilizing Fe_3O_4/eggshells composite as catalyst under mild conditions. Maximum yield obtained is 91.3% with reflux condition, amount of catalyst 0.0254 g, and 1.5 h reaction time. Catalyst used can be separated magnetically and can be reused up to three times without significant decrease of yield. The synthesized compound was characterized using IR, UV-Vis and LC-MS.

Keywords: acridine, Hantzsch condensation, Fe_3O_4/eggshell

1. Introduction

Indonesia is one of the world largest cinnamon producer. Cinnamon production in Indonesia reached 91,400 ton in 2014, surpassing China (71,146 ton) and Sri Lanka (16,766 ton) [1]. However, Indonesia sold cinnamon at low price because it is often marketed or exported as unprocessed cinnamon bark stick [2]. Cinnamon powder or cinnamon bark oil usually used in culinary as an flavoring or food additive and medicine. Exploration of cinnamon potential is important to rise the price value of Indonesia's cinnamon.

Main constituent of cinnamon bark oil is (E)-cinnamaldehyde. Cinnamon bark oil can be obtained using Soxhlet extraction and steam distilation, and its content ranging from 62 to 90% depend on extraction technique [3]. Utilization of (E)-cinnamaldehyde as precursor in organic synthesis can be a way to increase its value.

One of the simple reaction that can be done using (E)-cinnamaldehyde as precursor is Hantzsch condensation reaction. The reaction involves aldol reaction of aldehyde and 1,3-dicarbonyl, followed by imine formation and cyclization forming pyridine skeleton [4]. Acridine, a tricyclic compound with similiar structure with anthracene with pyridine ring in center, can be obtained from Hantszch condensation reaction using dimedone as 1,3-diketone. Acridine possess many biological activities, from antitumor [5], anticancer [6], antifungal [7], cytotoxic [8], and antibacterial [9]. Research on synthesis of acridine has been widely done by researchers around the world, mainly focused on improving catalyst, such as using ionic liquid [10], amberlyst [11], functionalized silica [12], and refluxing water [13]. Improvement of catalyst mainly focused around green chemistry concept. This article reports our research in synthesis of 1,8-dioxoacridine using Fe_3O_4 coated on

Figure 1. SEM Image of Fe_3O_4/eggshell using
(a) 1000x magnification and (b) 5000x magnification.

eggshell particles. Fe_3O_4 is reported to have catalytic activity on acridine synthesis [14], meanwhile eggshell particle as Fe_3O_4 support has been proven to increase yield of xanthene, where the proposed synthetic mechanism is similiar with Hantzsch synthesis [15].

2. Materials and methods

All chemicals were on analytical grade, purchased from Fluka Chemical Company, Merck and Sigma-Aldrich. Silica gel 60F-254 with a mobile phase consist of ethyl acetate and n-hexane (1:1) solvents was used in thin layer chromatography (TLC) analysis. Fourier Transform Infrared Spectroscopy (FTIR) characterization was performed by using Shimadzu 8400 spectrometer with KBr background. UV-Vis absorption recorded with Shimadzu UV mini type 1240. Mass spectroscopy analysis was conducted with Xevo G2-S Qtof with J12CHA 602G serial.

2.1. General procedure for the Fe_3O_4/eggshell synthesis

Fe_3O_4/eggshell was prepared using modified method of previous studies [14,16]. Eggshell was collected from domestic waste. Collected eggshell was rinsed, separated from the membran, and boiled in NaOH 1 mol.L^{-1}. After boiling, eggshell dried and crushed, to further screened through a sieve to get < 100 µm size. Fe_3O_4/eggshell as a catalyst was prepared using coprecipitation method. $FeCl_3$ (0.02 mol) and $FeCl_2$ (0.01 mol) was dissolved in 1.5 M HCl solution. Both solution was added dropwise to NaOH solution (1.5 mol.L^{-1}) that was heated to 65 °C and stirred vigorously at 60 °C. 20 g of previously prepared eggshell powder was added and then stirring continues for 0.5 hour. Black precipitate then collected by filtering and washed by distilled water to neutralize excess acid.

2.2. General procedure for the synthesis of 1,8-dioxoacridine synthesis

A mixture of aniline (0.5 mmol), dimedone (1.0 mmol), and cinnamaldehyde (0.5 mmol) was dissolved in ethanol in the presence of Fe_3O_4/eggshell as catalyst. Reaction mixture stirred and heated to 80 °C (reflux). After completion of reaction as monitored by TLC, reaction mixture was separated from catalyst, and cooled to room temperature. Distilled water was added to mixture in 1:1 portion, and left overnight. Precipitated compound then separated, heated to evaporate excess ethanol and water, and then recrystallized and washed by hot water. Obtained product were further characterized by FT-IR spectroscopy and LC-MS with mobile phase acetonitrile and water using gradient concentration technique.

3. Results and discussion

3.1. Characterization of Fe_3O_4/eggshell catalyst

Fe_3O_4/eggshell particle formed as black precipitate. The particle showed magnetic properties and can be separated when it is suspended in liquid. SEM-EDX analysis on catalyst powder shown irregular pattern and nonuniform size of the eggshell particle (figure 1). Eggshell particle is succesfully attached by Fe_3O_4 particle, as confirmed by attracting suspended catalyst with magnet. Elemental composition of one random point in catalyst by EDX is shown in the table 1.

Table 1. Chemical composition of Fe_3O_4/eggshell catalyst as analyzed by EDX

Elements	Fe	O	C	Ca	Al
% wt.	40.6	37.3	12.3	6.7	1.5

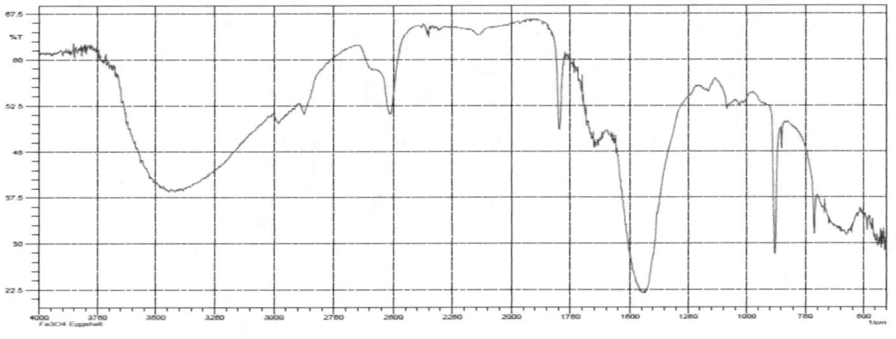

Figure 2. FTIR spectra of Fe_3O_4/eggshell catalyst

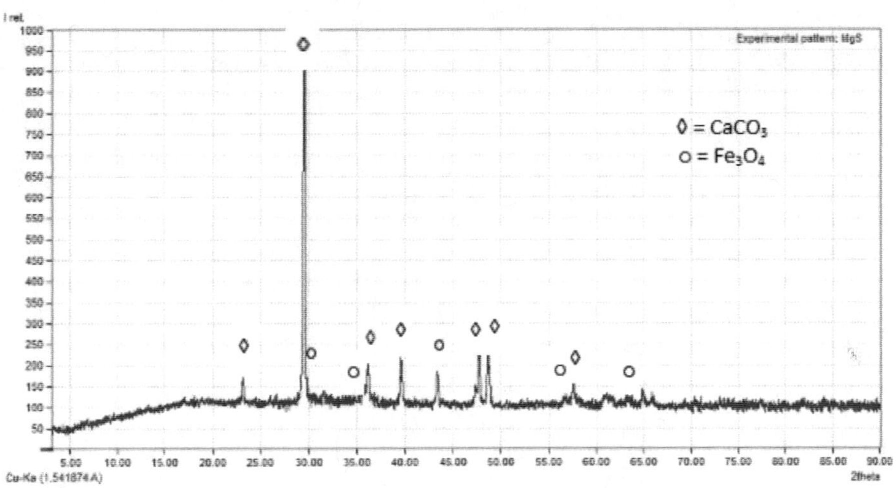

Figure 3. X-ray diffractogram of Fe_3O_4/eggshell catalyst

FTIR analysis of catalyst (figure 2) showed a sharp peak at 1400–1500 cm^{-1} indicating presence of carbonyl functional group contained in eggshell as $CaCO_3$. Meanwhile, Fe-O absorption recorded at 500–600 cm^{-1} [17]. Broad peak at 3300–3600 cm^{-1} appeared due to water vapor contamination during analysis.

XRD analysis of Fe_3O_4/eggshell showed peak at 2θ = 23.18; 29.56; 36.14; 39.58; 47.70; 48.68; and 57.66 indicated presence of $CaCO_3$, and 2θ = 30.60; 35.66; 43.36; 56.88; 61.78 indicated presence of magnetite (figure 3) [18,19].

3.2. Synthesis and Characterization of 1,8-dioxoacridine

Catalytic activity of Fe_3O_4/eggshell catalyst was studied using one pot three component synthesis of 1,8-dioxoacridine (figure 4). Condensation reaction occurs between three components; cinnamaldehyde, aniline, and dimedone; forming 1,8-dioxoacridine. Resulting product is yellow solid, soluble in ethanol but insoluble in water and n-hexane. Reaction conditions were optimized by varying reaction time, temperature, and catalyst amount. In addition, reusability of Fe_3O_4/eggshell catalyst also examined using optimized reaction conditions. Reaction mechanism is likely to follow Hantzsch condensation reaction mechanism, starting with aldol product formation. Few seconds after mixing all reactant with catalyst, yellow insoluble product obtained. Further examination by comparing reaction with dimedone and cinnamaldehyde alone in base condition, followed by TLC analysis, it is concluded that both reaction formed same product.

Table 2. Optimization of reaction condition results

No	Time (min)	Catalyst (mg)	Temp. (°C)	Yield (%)
1	60	16.6	80	76.2
2	90	16.6	80	88.2
3	120	16.6	80	87.8
4	150	16.6	80	78.8
5	90	16.6	30	77.8
6	90	16.6	50	78.1
7	90	0	80	69.0
8	90	8.8	80	79.9
9	90	25.4	80	91.3

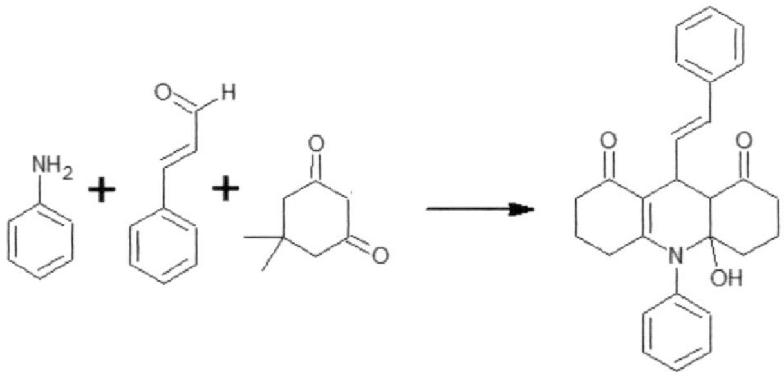

Figure 4. Reaction scheme of 1,8-dioxoacridine synthesis and obtained product

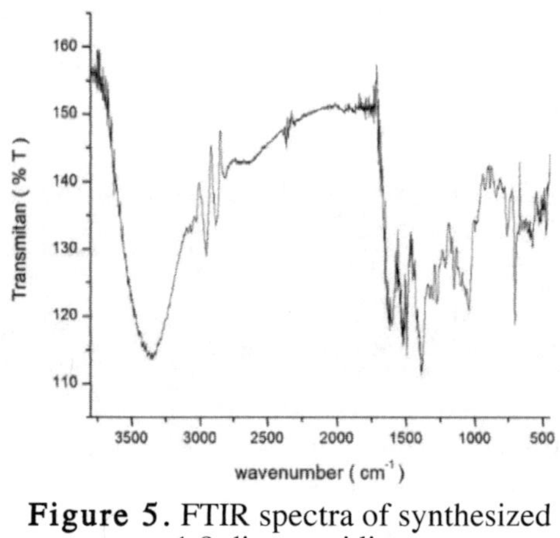

Figure 5. FTIR spectra of synthesized 1,8-dioxoacridine

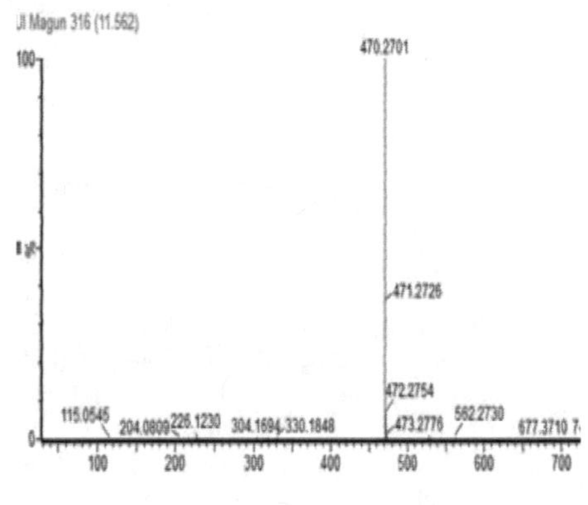

Figure 6. Mass spectrum of synthesized 1,8-dioxoacridine

FTIR analysis of products shows broad absorption at 3300-3500 cm^{-1}, indicating presence of hydroxyl group in the compound (figure 5). Other absorption occurs at ~1450 cm^{-1}, possibly indicating presence of carbonyl group in acridine. Absorption occured at lower wavenumber than normal carbonyl shift (1670-1820 cm^{-1}) due to effect of conjugation. Medium absorption at 1000-1200 cm^{-1} indicating C-N strecth in acridine skeleton.

LCMS analysis of product resulted a single dominant peak (as base peak) in mass spectrum at m/z = 470, fit with the $[M+H]^{+}$ value with the reaction product. Mass spectrum analysis resulted single dominant peak (figure 6) and only tiny other peak because energy of electron used in LCMS analysis is the lowest one in the instrument, in order to get the M^{+} value as a base peak and minimizing fragmentation.

Optimization result shown on table 2. Optimum reaction time obtained is 90 min. Longer reaction time resulted in lower yield obtained, possibly due to side reaction of product decomposition. Optimization of reacton temperature was done with optimum time obtained. Results shown that yield

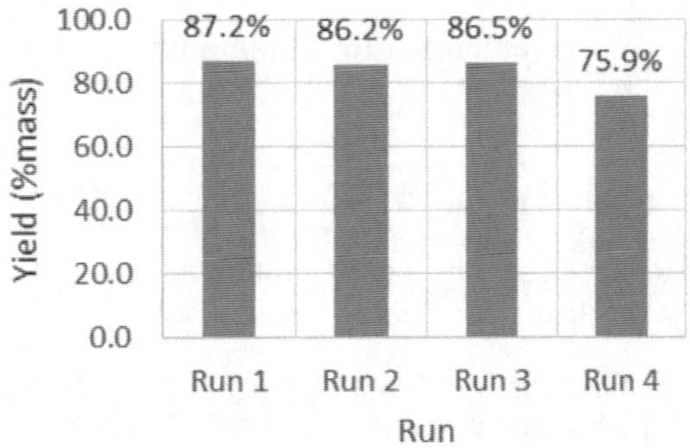

Figure 7. Reusability tes result on synthesis of 1,8-dioxoacridine

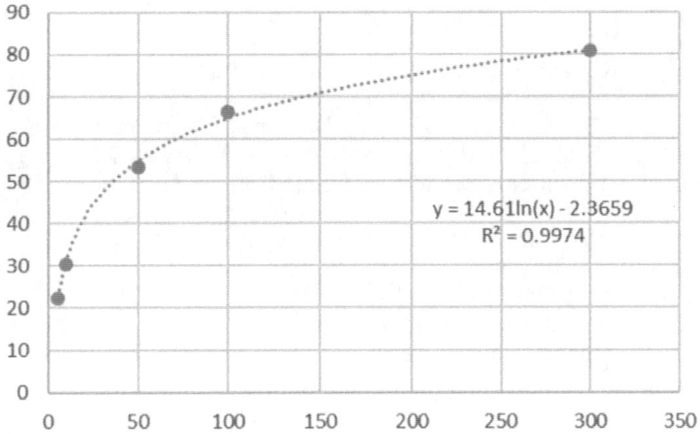

Figure 8. Plot of acridine concentration vs. inhibition percent from DPPH assay

obtained is proportional to reaction temperature. Condensation reaction needs heat to occur, and higher temperature favor condensation reaction. Meanwhile, variation of catalyst amount also directly proportional to yield obtained. However, reaction still occured when no catalyst added. It is important to note that normally Hantzsch condensation reaction catalyzed by base, and presence of aniline acted as catalyst for the reaction itself, so with no catalyst added, reaction still occured. Fe_3O_4/eggshell catayst reusability was tested by running 4 reaction using same catalyst, washed by ethanol between usage. Obtained yield indicated the catalyst can be reused, first three run showed no significant decrease in yield, and fourth run the yield decreased by 10,6% (figure 7).

3.3. Antioxidant assay result

Series of 1,8-dioxoacridine solution in ethanol with concentration of 5; 10; 50; 100; and 300 ppm used for this experiment. Inhibition percent after reaction were calculated from blank absorbance (A_0) and test solution absorbance (A) as follow [20]:

$$\% \text{ inhibition} = \frac{A_0 - A}{A_0} x\ 100\%$$

From the calculation, IC_{50} value was obtained using regression analysis of the curve (figure 8). IC_{50} value obtained for the compound is 36.03 ppm. This result is far lower than cinnamaldehyde and dimedone that posses IC_{50} value above 500 ppm [21].

4. Conclusions

A low-cost, reusable and harmless catalyst, Fe_3O_4/eggshell had been synthesized using simple precipitation method. The catalyst showed excellent catalytic activity in 1,8-dioxoacridine synthesis with yield up to 91.3%. In addition, using multi-component reaction techniques maximizes atom economy, utilization of renewable and waste material (eggshell waste), and reusable property of catalyst is an important properties that is in accordance with green chemistry principle. Catalyst used can be reused up to three times without significant decrease in yield.

Acknowledgements
The authors are grateful to Universitas Indonesia for funding this research through PITTA Grant No. 2068/UN2.R3.1/PPM.00/2017 and Einago (www.einago.com) for the English language review.

References
[1] Food and Agriculture Organization of the United Nation 2014 FAOSTAT data: Production-Crops – production quantity – cinnamon (Rome: FAO)
[2] Ferry Y 2013 Sirkuler Inovasi Tanaman Industri dan Penyegar (SIRINOV) **1** 11–20
[3] Nabavi S F, Di Lorenzo A, Izadi M, Sobarzo-Sánchez E, Daglia M and Nabavi S M 2015 *Nutrients* **7** 7729–48
[4] Muscia G C, Buldain G Y and Asis S E 2009 *Monatsh. Chem.* **140** 1529–32
[5] Mikata Y, Yokoyama M, Mogami K, Kato M, Okura I, Chikira M and Yano S 1998 *Inorg. Chim. Acta.* **279** 51–7
[6] Gamega S A, Spicer J A, Atwell G J, Finlay G J, Baguley B C and Denny W A 1999 *J. Med. Chem.* **42** 2383–93
[7] Ngadi L, Galy A M, Galy J P, Barbe J, Cremieux A, Chevalier J and Sharples D 1990 *Eur. J. Med. Chem.* **25** 67–70
[8] Antonini I, Polucci P, Kelland L R, Menta E, Pescalli N and Martelli S 1999 *J. Med Chem.* **42** 2535–41
[9] Amini M M, Fazaeli Y, Yassaee Z, Feizi S and Bazgir A 2009 *Open Catal. J.* **2** 40–4
[10] Zolfigol M A, Bahrami-Nejad N and Baghery S 2016 *J. Mol. Liq.* **218** 558–64
[11] Das B, Thirupathi P, Mahender I, Reddy V S and Rao Y K 2006 *J. Mol. Cat. A Chem.* **247** 233–9
[12] Ramesh K B and Pasha M A 2014 *Bioorg. Med. Chem. Lett.* **24** 3907–13
[13] Xia J J and Zhang K H 2012 Molecules **17** 5339–45
[14] Ghasemzadeh M A, Safaei-Ghomi J and Molaei H 2012 *C. R. Chimie* **15** 969–74
[15] Mosaddegh E, Hosseininasab F A and Hassankhani A 2015 *RSC Adv.* **5** 106561–7
[16] Ren J, Bopape M F, Setshedi K, Kitinya J O and Onyango M S 2012 *Chem. Ind. Chem. Eng.* Q. **18** 221–31
[17] Khalil M I 2015 *Arab. J. of Chem.* **8** 279–84
[18] Rahman M A, Halfar J and Shinjo R 2013 A*dvances in Materials Physics and Chemistry* **3** 120–5
[19] Hirmas D R, Platt B F, Hasiotis S T 2012 *Soil Sci. Soc. Am. J.* **76** 1100–6
[20] Shekhar T C and Goyal A 2014 *Am. J. Ethnomed.* **1** 244–9
[21] Cahyana A H, Nadila A and Ardiansah B 2017 *AIP Conf. Proc.* **1862** 030097

A hierarchical MnOx/ZSM-5 heterogeneous catalyst for the conversion of cellulose from mahogany wood to levulinic acid

D U C Rahayu, D A Nurani, L N H Rochmah, N F Antra and Y K Krisnandi

Department of Chemistry, Faculty of Mathematics and Natural Sciences (FMIPA), Universitas Indonesia, Depok 16424, Indonesia

Corresponding author: yuni.krisnandi@sci.ui.ac.id

Abstract. At present, lignocellulose holds significant potential as a renewable biomass resource. Lignocellulose can be converted to levulinic acid, one of the most important platform chemicals used in several applications. The forestry of mahogany wood produces abundant waste in the urban areas of Indonesia. This waste is a source of biomass, containing the major chemical constituents of holocellulose, α-cellulose, and lignin in proportions of 53.37%, 33.05%, and 31.82%, respectively. Pretreatment processes of mahogany wood were conducted through dewaxing processes followed by delignification using NaOH and NaOCl to pre-concentrate the cellulose substrate. The delignification processes using NaOH and NaOCl reduced the lignin content to 15.58% and 6.40%, respectively, from 31.82% of dewaxed-mahogany wood. The cellulose conversion from mahogany wood to levulinic acid was performed using the MnOx/Hi_ZSM-5 heterogeneous catalyst in the presence of H_2O_2 and H_3PO_4 solutions. The parameters of cellulose conversion, including conversion time and concentration of H_3PO_4 and H_2O_2, were also studied, and HPLC analysis revealed the highest cellulose conversion to levulinic acid to be 2.44% with 20% H_2O_2 (v/v) and 30% H_3PO_4 (v/v) after 8 h of reaction time.

Keywords: hierarchical Mn/ZSM-5, cellulose, mahogany wood, levulinic acid

1. Introduction

The conversion of biomass into valuable chemicals or biofuel is a field that has captured the attention of many researchers [1–3]. It is of interest to discover improved uses for residual biomass and endeavour to increase its value and profitability. Levulinic acid is an excellent chemical platform and holds great potential as a product derived through biomass conversion. It has two functional groups, i.e., a ketone and a carboxylic acid, which can be used for further functionalization in synthesizing organic compounds such as 2-methyl-tetrahydrofuran, levulinic ester (biodiesel additive), and δ-aminolevulinate (herbicide) [4,5].

In Indonesia, mahogany wood (*Swietenia macrophylla* King) is one of the most harvested timber forest products that are currently used in industry. Based on the statistics of forestry production from Statistics Indonesia (*Biro Pusat Statistik*, BPS); 295,739.28 m^3 of mahogany wood was produced in Indonesia during 2015 [6]. The greater the production of mahogany wood, the greater the mahogany wood waste will be. The waste should therefore be used so as to maximize its economic value and utility. This abundance of waste and chemical composition of mahogany wood, including holocellulose, α-cellulose, and lignin, allow it to be used as feedstock in biomass conversions [7].

Biomass conversion into levulinic acid occurs in mild conditions during the acid hydrolysis of lignocellulose leading to the cleavage inter and intramolecular hydrogen bonds, activation of a Fenton-like system, and the dehydration of glucose to yield the product [8,9]. This research intends to initiate a Fenton-like system by optimizing the concentrations of H_3PO_4 and H_2O_2 in the cellulose conversion of mahogany wood as the biomass source into levulinic acid. The heterogenous catalyst of MnOx/ZSM-5 will be used for its pore structure and high activity. It was proven that the impregnated Mn^{2+} in Hi_ZSM-5 pores lead to higher yields of levulinic acid than that observed in Fe/ZSM-5 during the conversion of glucose to levulinic acid [10]. This catalyst is also expected to exhibit good product selectivity and should allow for the conversion in a single-step reaction.

2. Materials and methods

2.1. Materials

Mahogany wood was obtained from Depok, West Java, Indonesia. The chemicals used were of technical grade. NaOCl (pro-analytical grade) and levulinic acid (standard grade) were obtained from Wako Pure Chemicals (Osaka, Japan). Glucose (standard grade), 5-HMF (standard grade), 96-97% H_2SO_4, NaOH, $NaClO_2$, glacial CH_3COOH, 89% H_3PO_4, 30% H_2O_2, n-hexane and 95% ethanol were obtained from Merck. Hierarchical 2% Mn/ZSM-5 (MnOx/ZSM-5) was synthesized based on the protocol by Krisnandi *et al.* [11].

2.2. Preparation of mahogany wood

Mahogany wood was finely ground and filtered with a particle size diameter of 315-710 µm. The feedstock was then dried overnight and stored in an airtight plastic box at 25 °C.

2.3. Chemical content analysis of mahogany wood

The chemical constitution of mahogany wood was determined according to methods AOAC 1984 [12] for the moisture and ash content; according to ASTM D 1107-56, reapproved in 1972 [13] for the extractive content; according to Dance *et al.* [14] and Yoshihara *et al.* [15] for the lignin content; according to Browning [16] for the holocellulose content and ASTM D 1103-60, reapproved in 1978 [17] for the α-cellulose content.

2.4. Pretreatment processes of mahogany wood

The pretreatment processes of mahogany wood were conducted by dewaxing and delignification processes using NaOH and NaOCl. The dewaxing process was conducted twice by Soxhlet extraction using n-hexane/ethanol (1:2) for 6 h at 80 °C. Delignification processes were conducted using a 10% NaOH solution for 1.5 h at 55 °C, followed by NaOCl at 70 °C. The reaction contents were further neutralized using glacial CH_3COOH. The remaining cellulose was then analyzed for its resulting lignin, holocellulose, and α-cellulose content.

2.5. Cellulose conversion from mahogany wood to levulinic acid

The pretreated mahogany wood (1 g) was mixed with MnOx/Hi_ZSM-5 (1 g) and H_3PO_4 (20 mL) at various concentrations ranging from 30%–60% (v/v) and 30% H_2O_2 (v/v) (0.5 mL) in a three-neck round-bottom flask. The mixture was heated to reflux at 100 °C for 2, 4, 6, and 8 h. A similar reaction was also conducted with a 10% and 20% concentration of H_2O_2 (v/v).

2.6. Product analysis

An HPLC HITACHI L2000 equipped with an Aminex HPX-87H column and an RI-Detector was used to analyze the conversion products. The detector and column temperatures were set to 40 °C and 60 °C, respectively, and H_2SO_4 (0.005 M) was used as the mobile phase at a flow rate of 0.55 mL min^{-1}. Equation (1) was used to calculate the yield of the cellulose conversion, where C_{LA} is the concentration of levulinic acid (mg L^{-1}), V_R is the volume of reactant (L), and m_s is the mass of substrate (mg).

$$\% \, yield = \frac{C_{LA} \times V_R}{m_S} \times 100 \, \% \tag{1}$$

3. Results and discussion

The chemical constituents of raw mahogany wood are tabulated in table 1. Pretreating the mahogany wood with NaOH decreased the lignin content wood from 51.03% to 15.58%. According to Chang *et al.* [1], effective pretreatment should decrease the lignin content by more than 10% from its

Table 1. The chemical constitution of raw mahogany woodsare.

Chemical Content	Content (wt. %)
Water	5.79
Ash	1.31
Extractives	2.73
Holocellulose	53.37
α-Cellulose	33.05
Lignin	31.82

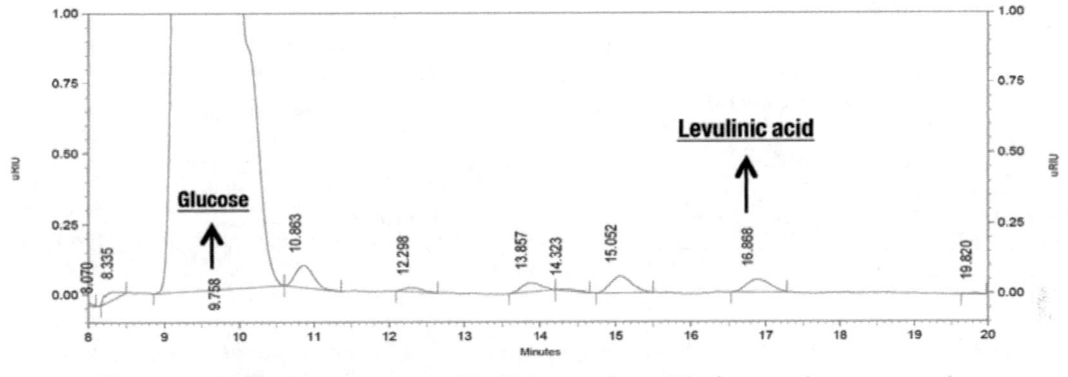

Figure 1. The chromatogram of cellulose conversion from mahogany wood to levulinic acid with 40% H_3PO_4 and 30% H_2O_2 after 8 h of reaction time.

raw material lignin content [1]. After alkaline pretreatment with NaOH, the biomass was pretreated using NaOCl, decreasing the lignin content from 15.58% to 6.4%. The ClO• radical and ClO- anion attacked the phenolic group resulting in an interaction between the ClO- anion and the phenolic compound in lignin. The ClO- anion acts as a strong oxidant causing cleavage of the C-C bonds in lignocellulose. The remaining bonds between lignin and cellulose in lignocellulose were then cleaved causing lignin to dissolve, rendering white samples. This result was similar to the result reported by Lee *et al.* [18].

The conversion of cellulose to levulinic acid is facilitated by the acidic pH from the H_3PO_4 addition, which causes the cleavage of hydrogen bond intra- and intermolecular bonds [8]. The conversion occurs in a Fenton-like system due to the presence of H_2O_2. HO• radicals from H_2O_2 caused the depolymerization of cellulose into its monomer (glucose) and further decomposed glucose to 5-HMF as intermediate before producing levulinic acid as the target product [8]. The chemical reactions are shown in equations (2) and equation (3).

$$Mn^{2+} + H_2O_2 \rightarrow Mn^{3+} + HO^- + HO^{\bullet} \quad (2)$$

$$Mn^{3+} + H_2O_2 \rightarrow Mn^{2+} + H_2O^{\bullet} + H^{+\bullet} \quad (3)$$

The reaction was conducted at various concentrations of H_3PO_4 ranging from 30-60% (v/v). In the early stages of the reaction, charcoal formed when using H_3PO_4 at concentrations higher than 40%, so the product could not be analyzed using HPLC. Based on figure 1, two key products were formed i.e., glucose (rt. 9.7 min) and levulinic acid (rt. 16.8 min). The absence of 5-HMF (rt. 31.2 min) indicated that 5-HMF might be further converted into levulinic acid. This analysis served to prove that MnOx/ZSM-5 had good selectivity for levulinic acid. Figure 2a shows that the highest biomass conversion to levulinic acid was after 8 h of reaction time where a 1.89% yield of levulinic acid for 40% H_3PO_4 (v/v) was obtained. The higher concentration of H_3PO_4 might dominate the reaction causing the catalytic activity of MnOx/ZSM-5 to decrease, leading to charcoal formation because of faster cellulose depolymerization into glucose. An investigation into the effect of the H_2O_2 was therefore conducted using 30% H_3PO_4 (v/v).

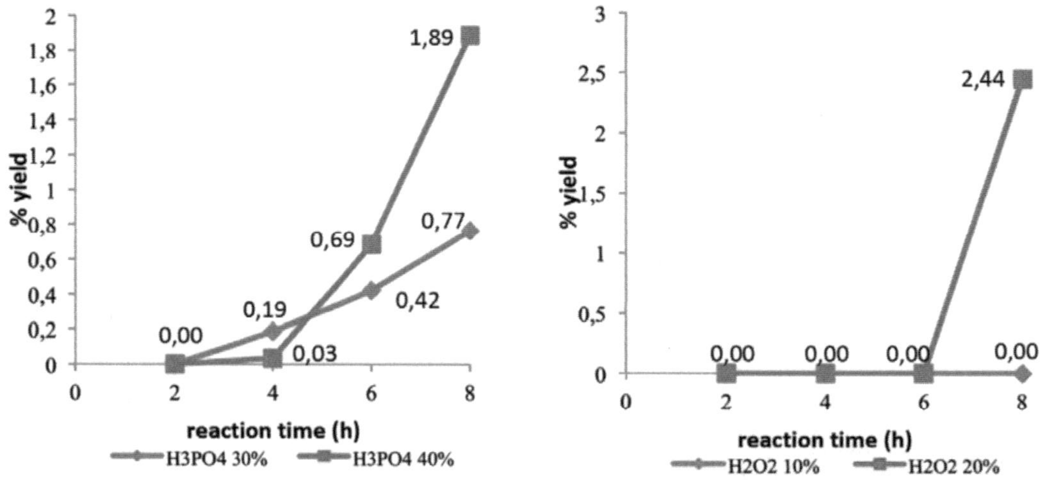

Figure 2. The percentage yield of levulinic acid as a function of reaction time for pretreated mahogany wood using (a) 30-40% H_3PO_4 (20 mL, v/v), 30% H_2O_2 (0.5 mL, v/v), and MnOx/ZSM-5 (0.1 g) refluxed at 100 °C, and (b) 10-20% H_2O_2 (0.5 mL, v/v), 30% H_3PO_4 (20 mL, v/v), and MnOx/ZSM-5 (0.1 g) refluxed at 100 °C.

To elucidate the role of H_2O_2 in cellulose conversion, various concentrations of H_2O_2 ranging from 10-20% (v/v) were used and the effects thereof were studied. Levulinic acid was not obtained when 10% H_2O_2 was used in this reaction (figure 2b). This could be due to the limited formation of HO• radicals which were therefore unable to initiate the conversion of cellulose into levulinic acid. Moreover, according to figure 2b, the conversion reaction using 20% H_2O_2 (v/v) with 30% H_3PO_4 (v/v) resulted in the highest percentage yield of levulinic acid of up to 2.44% after 8 h of reaction time. The MnOx/ZSM-5 catalyst has an inherent mesoporous and microporous character that is needed to catalyze the conversion of cellulose into levulinic acid. The active sites of MnOx/ZSM-5 and H_2O_2 facilitated the formation of HO• radicals that could cleave glycosidic bonds in the cellulose. Furthermore, the mesoporous ZSM-5 increased the ability of cellulose to convert to levulinic acid.

4. Conclusions

In this work, the optimization reaction of cellulose conversion from mahogany wood into levulinic acid was studied using a MnOx/ZSM-5 heterogeneous catalyst. Cellulose from mahogany wood was obtained through alkali and oxidative delignification. Alkali delignification using NaOH decreased the lignin content from 31.82% to 15.58%, and further oxidative delignification using NaOCl decreased the lignin content to 6.4%. HPLC analysis showed the highest cellulose conversion to levulinic acid was 2.44% with 20% H_2O_2 (v/v) and 30% H_3PO_4 (v/v) after 8 h of reaction time. With the absence of the 5-HMF peak in the HPLC analysis, this work demonstrated that the MnOx/ZSM-5 catalyst exhibits good selectivity for levulinic acid.

Acknowledgements

This work was supported by International Collaboration Research Grant 2014 from Universitas Indonesia and PUPT BOPTN 2016 research grant No. 1136/UN2/R12/HKP/05.00/2016 from Ministry of Research, Technology and Higher Education (Kemenristekdikti) Republic of Indonesia. Authors would like to thank Einago (www.einago.com) for the English language review.

References

[1] Chang C, Cen P and Ma X 2007 *Bioresour. Technol.* **98** 1448–53
[2] Peng L, Lin L, Li H and Yang Y 2011 *Appl. Energ.* **88** 4590–6
[3] Menon V and Rao M 2012 *Prog. Energ. Combust. Sci.* **38** 522–50
[4] Bozell J J, Moens L, Elliott D C, Wang Y, Neuenscwander G G, Fitzpatrick S W, Bilski R J and Jarnefeld J L 2000 *Resour. Conserv. Recycl.* **28** 227–39

[5] Chalid M 2012 *Levulinic Acid as a Renewable Source for Novel Polymers* (Groningen: University of Groningen) p 109
[6] Statistics Indonesia 2015 *Statistics of Forestry Production* (Jakarta: Biro Pusat Statistik) p 164
[7] Kumar P, Barrett D M, Delwiche M J and Stroeve P 2009 *Ind. Eng. Chem. Res.* **48** 3713–29
[8] Chen Y, Li G, Yang F and Zhang S M 2011 *Polym. Degrad. Stab.* **96** 863–9
[9] Girisuta B, Janssen L P B M and Heeres H J 2007 *Ind. Eng. Chem. Res.* **46** 1696–1708
[10] Saskhina K A, Labko V S, Rudina N A, Parmon V N and Parkhomchuck E V 2013 *J. Catal.* **299** 44–52
[11] Krisnandi Y K, Anggraini A R, Azaria A P, Agnes A, Pertiwi R, Antra F, Nurani D A, Suwardiyanto and Howe R F 2017 *Indones. J. Chem* (in review)
[12] AOAC 1984 *Official Methods of Analysis (*Arlington: Association of Official Analytical Chemists)
[13] ASTM 1972 *Standard Test Method of Test for Alcohol-toluene Solubility of Wood* (West Conshohocken: ASTM International) ASTM D 1107-56 (reapproved 1972)
[14] Lin S Y and Dance C W 1992 *Methods in Lignin Chemistry* (Berlin: Springer-Verlag)
[15] Yoshihara K, Kobayashi T, Fujii T and Akamatsu I 1984 *Japan Tappi J.* **38(4)** 466–75
[16] Browning B L 1967 *Methods of Wood Chemistry* (New York: Wiley & Sons)
[17] ASTM 1978 *Standard Test Method for α-Cellulose* (West Conshohocken: ASTM International) ASTM D 1103-60 (reapproved 1978)
[18] Lee Y J, Chung C H and Day D F 2009 *Bioresour. Technol.* **100** 935–41

Modification of boron-doped diamond with gold nanoparticles and its preliminary study for electrosynthesis of vanillin

R P Putranto, H Mubarok, K Umam, E Saepudin and T A Ivandini

Department of Chemistry, Faculty of Mathematics and Natural Sciences (FMIPA), Universitas Indonesia, Depok 16424, Indonesia

Corresponding author: ivandini.tri@sci.ui.ac.id

Abstract. Vanillin is generally synthesized from coniferin, cresol, ferulic acid, guaiacol, lignosulphonates, and vanillyl mandelic acid using oxidation-reduction methods with various chemicals and consecutive stages. These methods are less environmentally friendly due to their use of harmful compounds and the ineffective multistage process. In this study, we investigated electrochemical oxidation of isoeugenol to vanillin. A boron-doped diamond (BDD) modified by gold nanoparticles (AuNPs) was prepared for use as the working electrode, with sulfuric acid used as the electrolyte. Peak oxidation and reduction of BDD-AuNPs occurred at 1.20 V and 0.40 V, respectively, and a voltammetric range of −1.50 V to 1.50 V was applied to determine the potential oxidation of isoeugenol. The results indicated that the optimal pH for electrooxidation was 8, with potential oxidation occurring at 0.50 V.

Keywords: boron-doped diamond, electrosynthesis, gold nanoparticles, isoeugenol, vanilin

1. Introduction

The growing demand for vanillin exceeds its availability. Sources of vanillin production worldwide originate from the petroleum industry and are nonedible, whereas the natural edible vanillin is derived from the extraction of vanilla plant seeds only (*Vanilla plantifolia*). One extraction method involves the catechol pathways, which requires continuous oxidation stages and the use petroleum-based phenol [1].

Vanillin is the most abundant pure monoaromatic phenol currently produced at an industrial scale from lignin (at ~20,000 tons annually), with 3000 of those tons originating from lignin. Vanillin has many advantages as an aromatic compound, including its safety, and harbors two functional groups that can be modified chemically [2]. In addition, vanillin has potential as a difunctional feedstock for the preparation of thermoplastic polymers [1].

Alternative methods for vanillin conversion involve the use of lignin, guaiacol, coniferin, eugenol, and vanillyl mandelic acid [3], which utilize vanillyl alcohol oxidase from microorganisms, such as *Bacillus subtilis*, *Pseudomonas sp*., and *Pseudomonas aeruginosa* [4]. Adilina *et al.* [5] converted isoeugenol to vanillin using a cobalt(II) porphyrin heterogeneous catalyst, and electrochemical vanillin synthesis from lignin derivatives was reported by Reichert *et al.* [6]. Additionally, Mubarok *et al.* [7] performed vanillin electrosynthesis from isoeugenol using a platinum electrode in solvent.

In this study, we described the preparation of a gold-nanoparticle (AuNP)-modified boron-doped diamond (BDD-AuNPs) advantageous as a functional material because of its durability and catalytic characteristics [8,9]. We applied this material for a preliminary study of vanillin electrosynthesis from isoeugenol using pH variation.

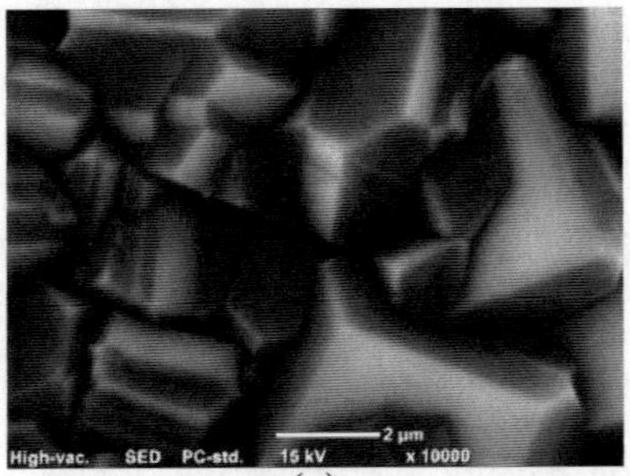

(a)

(b)

Figure 1. SEM image of surface (a) BDD (1%) and (b) BDD–AuNPs.

2. Materials and methods

2.1. Materials and instruments

We used a mixture of *cis*- and *trans*-isoeugenol (98%; Wako Pure Chemical Co., Dalian, China), methanol p.a. (Sigma Aldrich, St. Louis, MO, USA), phosphoric acid, boric acid, acetic acid, tetrabutyl ammonium perchlorate (Tokyo Chemical Industries, Tokyo, Japan), allylamine (Wako Pure Chemical Co.), deionized ultrapure water (Merck Millipore, Billerica, MA, USA), tetrachloroauric(III) acid (Tokyo Chemical Industries), sulfuric acid p.a (Sigma Aldrich), and vanillin (Wako Pure Chemical Co.). The pH meter and potentiostat were obtained from (EDAQ Instrument Corp., Taipei, Taiwan), while the scanning electron microscopy (SEM) was from JEOL/EO JCM-6000Plus, Japan

2.2. Experimental

2.2.1. Preparation of BDD-AuNPs. The BDD was cleaned by sonication in isopropanol and deionized ultrapure water for 5 min, respectively. Tetrachloroauric(III) acid diluted in sulfuric acid was used, as described previously [8]. Electrodeposition was conducted by amperometry at potential −200 mV (vs. Ag/AgCl) for 100 s. The results were characterized by SEM and cyclic voltammetry. Cyclic voltammetry of BDD-AuNPs was conducted in phosphate-buffered saline (PBS; pH 7) at a scan rate of 100 mV/s.

2.2.2. Electrochemical study of isoeugenol. Electrooxidation of isoeugenol was performed by cyclic voltammetry in one compartment. Variation of the pH was performed using Britton–Robinson (B–R) buffer solution as described previously [9], with slight modifications. The electrooxidation system contained a volumetric ratio of isoeugenol:tetrabutyl ammonium perchlorate: pH buffer (2:2:1, v/v) in a total volume of 5 mL, as described previously and with slight modifications [7]. Isoeugenol and tetrabutyl ammonium perchlorate (TBAP) were diluted in methanol to final concentrations of 0.1 M, and B–R buffer was diluted in ultrapure water.

3. Results and discussion

SEM images revealed the differences between the surfaces of BDD (figure 1a) and BDD-AuNPs (figure 1b). AuNPs electrodeposited on the surface of the BDD enables increases in surface activity as a catalytic site of chemical reactions. In addition to the BDD, AuNPs improve the formation of hydroxyl radical and chemically absorb and substrates with d-orbitals, which are preserved. Therefore, AuNPs increase electron conductivity to increase the reaction rate and selectivity based on the natural affinity of AuNPs. Electrodeposition of AuNPs onto the BDD was performed at molecular ratio of tetrachloroauric(III) acid as a gold precursor and sulfuric acid as an electrolyte of 1:100, respectively [8]. A constant potential at 200 mV for 100 s allowed the formation of AuNPs ordered with nucleation, which continued with increasing layers of AuNPs. This process is affected by the applied potential, programmed step deposition, deposition time, mass transport of the electrolyte, pH system, and handling material [10]. In this work (figure 1b), the distribution size of the nanoparticles on the BDD surface was quite uniform.

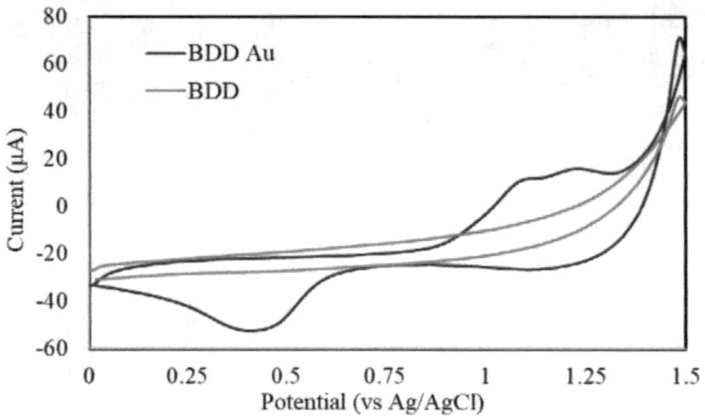

Figure 2. Cyclic voltammogram of BDD and BDD-AuNPs in PBS (pH 7).

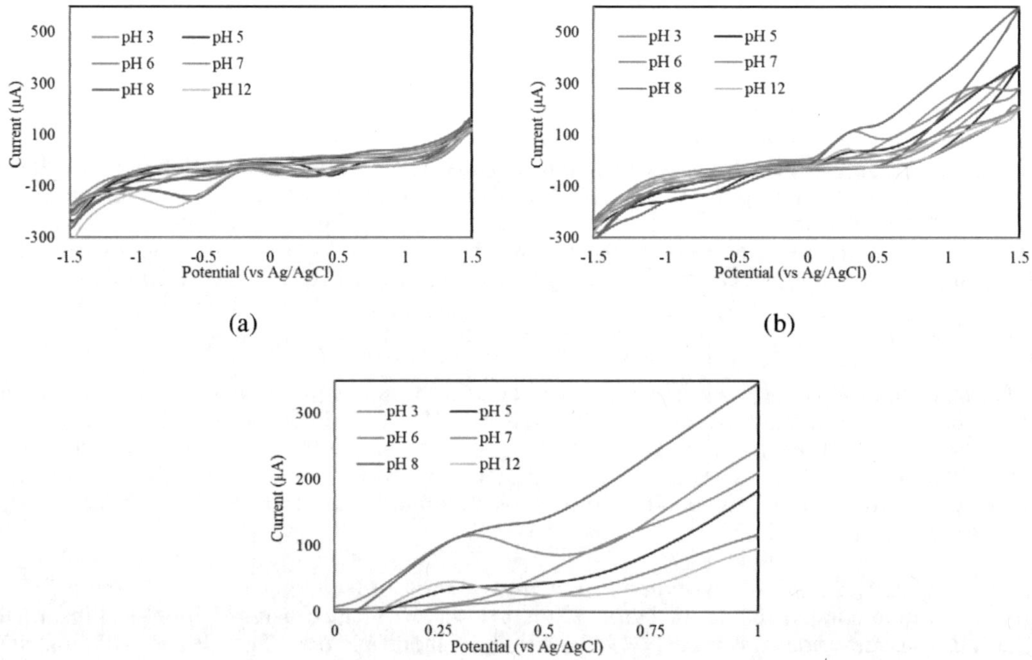

(c)

Figure 3. Cyclic voltammogram in the (a) absence and (b) presence of isoeugenol (c) magnification of the result in the presence of isoeugenol at an oxidation range of 0 mV to 1000 mV. The experiment was performed at room temperature without stirring and at a scan rate of 100 mV/s.

Cyclic voltammograms from experiments in PBS (pH 7) containing BDD–AuNPs showed three peaks representing peak Au oxidation at 1115 mV and 1231 mV along with a reduction peak at 408 mV. The reduction peak indicated the acquisition of Au(0) from Au(III). Figure 2 shows that the reaction initiated at oxidation and continued through reduction [11]:

$$Au^{0}{}_{(s)} \rightarrow Au^{+}{}_{(aq)} + e^{-} \quad (1)$$

$$Au^{0}{}_{(s)} \rightarrow Au^{3+}{}_{(aq)} + 3e^{-} \quad (2)$$

$$Au^{3+}{}_{(aq)} + 3e^{-} \rightarrow Au^{0}{}_{(s)} \quad (3)$$

$$O_{2(aq)} + 2H_2O_{(aq)} + 4e^{-} \longrightarrow 4OH^{-}_{(aq)} \quad (4)$$

Table 1. Potential oxidation and current-based oxidation of isoeugenol with pH variation.

pH	3	5	6	7	8	12
Potential oxidation (vs. Ag/AgCl)	Unidentified	324	322	428	500	282
Current oxidation (μA)	Unidentified	69.47	45.21	23.03	177.94	45.58

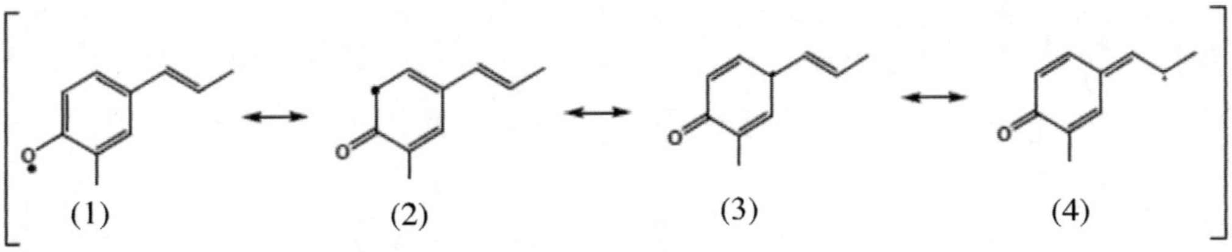

Figure 4. Resonance of radical isoeugenol

Figure 3 shows a cyclic voltammogram for an experiment using blank solvent in the presence and absence of isoeugenol, which resulted in different oxidation and reduction peaks. B–R buffer contains boric acid (H_3BO_3), acetic acid (CH_3COOH), phosphoric acid (H_3PO_4), and sodium hydroxide (NaOH), which are suitable in an electrochemical system, especially one involving polarography, to stabilize ionic strength in solvent [12,13].

The pKa of vanillin 7.4 affects its stability in product solutions [14]. The pH values of isoeugenol and TBAP in methanol are 6.6 and 7.0, respectively. The reactivity of isoeugenol and vanillin under extreme alkaline conditions caused more rapid deprotonation of the phenolic substituent. In a more acidic system, isoeugenol became less electroactive, because the solvent began losing its ability to facilitate formation of the phenolic transition state [7,15,16].

The resonance of the phenolic transition state was initiated by the homolytic proton from the hydroxyl group [7,17]. Therefore, electron nonpairing freely mobilized around four possible transition states, representing the majority transition state in for this species (1). This transition state represented the rate-determining step of the formation of various products.

4. Conclusions

A BDD-AuNP electrode was successfully prepared using electrodeposition methods and sulfuric acid as an electrolyte at −200 mV for 100 s. Electrooxidation of isoeugenol was optimized at pH 8 with a potential oxidation of 500 mV. These results demonstrated the potential for vanillin electrosynthesis using an AuNP electrode with isoeugenol in methanol.

Acknowledgements

This work was funded by PITTA Grant Universitas Indonesia with contract number 1822/UN2.R3.1/PPM.00.01/2017. The authors would like to thank Einago (www.einago.com) for the English language review.

References

[1] Caillol S, Boutevin B and Fache M 2016 *ACS Sustainable Chem. Eng.* 4 35
[2] Møller B L and Gallage N J 2015 Molecular Plant 8 40–57
[3] Fu W, Duan X, Yue L, Dai S, Xu L, Lu G and Mao D 2015 *Reac Kinet Mech Cat* 116 191–204
[4] Rana R, Mathur A, Jain C K, Sharma S K and Mathur G 2013 *Int. J. Biotechnol. Bioeng. Res.* 4 227–34
[5] Adilina I B, Hara T, Ichikuni N and Shimazu S 2012 *J. Mol. Catal. A: Chem.* 361-362 72–9
[6] Reichert E, Wintringer R, Volmer DA and Hempelmann R 2012 *Phys. Chem. Chem. Phys.* 14 5214–21
[7] Mubarok H, Hilyatudini, Saepudin E, and Ivandini T A 2017 *IOP Conf. Ser.: Mater. Sci. Eng.* 188 012050
[8] Ivandini T A and Einaga Y 2012 *Chem. Comm.* 53 1338–47

IOP Conf. Series: Materials Science and Engineering **496** (2019) 012056 doi:10.1088/1757-899X/496/1/012056

[9] Suzuki A, Ivandini TA, Kamiya A, Nomura S, Yamanuki M, Matsumoto K, Fujishima A and Einaga Y 2007 *Sens. Actuators B Chem.* **120** 500–7
[10] Macpherson JV 2015 *Phys. Chem. Chem. Phys.* **17** 2935–49
[11] Nicol M J 1980 *Gold Bull.* **13** 46–55
[12] Fry A J 1989 *Synthetic Organic Electrochemistry, 2nd edition* (New Jersey: John Wiley & Sons, Inc.) p 185
[13] Kyriacou D 1981 *Basics of Electroorganic Synthesis* (New Jersey: John Wiley & Sons, Inc.) p 260
[14] Frenkel C and Havkin-Frenkel D 2006 *Perfumer and Flavorist* **31** 28–36
[15] Wu X, Davis A P and Fry A J 2007 *Org. Lett.* **9** 5633–6
[16] Qi Y B, Wang X L, Shi T, Liu S, Xu Z H, Li X, Shi X, Xu P and Zhao Y L 2015 *Phys. Chem. Chem. Phys.* **17** 29597–607
[17] Sumi T, Saitoh T, Natsiu K, Yamamoto T, Atobe M, Einaga Y and Nishiyama S 2012 *Angew. Chem. Int. Ed.* **51** 5553–6

Quantification of phosphate ions by diffusive gradients in thin films technique using ferrihydrite as a binding gel and application to marine sediments

A Maimulyanti[1], Budiawan[1*], A Saefumillah[1] and H Suseno[2]

[1] Department of Chemistry, Faculty of Mathematics and Natural Sciences (FMIPA), Universitas Indonesia, Depok 16424, Indonesia

[2] Marine Radioecology Group, Center Technology for Radiation Safety and Metrology, National Nuclear Energy Agency (BATAN), Jl. Lebak Bulus Raya No. 49 Jakarta 12440, Indonesia

Corresponding author: dr.budiawan@gmail.com

Abstract. A diffusive gradient in thin films (DGT) is a newly developed analytical technique that can isolate and quantify the phosphate ions. A DGT unit comprises two components: the diffusion and binding layers. The diffusion layer comprises polyacrylamide hydrogel, whereas the binding layer from the oxide gel of ferrihydrite is impregnated with polyacrylamide gel. The technique employs a thin film of diffusive hydrogel in contact with a binding phase (ferrihydrite) that can bind phosphate ions. The diffusive and binding gels were characterized via infrared spectroscopy. The main peaks of the diffusive gel were observed at 3336 and 1636 cm^{-1}, whereas those of the binding gel were observed at 3332 and 1638 cm^{-1}. The DGT technique was evaluated as a way to quantify the amount of phosphate ions present in an aqueous KH_2PO_4 solution. After a contact time of 24 h, ferrihydrite exhibited a high affinity for phosphate, with an uptake efficiency of 98.17%. A linear relationship (R^2 = 0.9938) was found to exist between the contact time (3–24 h) and uptake efficiency. Within the 0.2–1.2-mgL^{-1} KH_2PO_4 concentration range, a linear relationship existed between analyte concentration and the mass of phosphate salt accumulated in the binding gel (R^2 = 0.9748). The phosphate ion was eluted from the ferrihydrite hydrogel using 0.25-M H_2SO_4 over 30 min. The DGT unit was used to measure the phosphate level in six marine sediment samples from Jakarta Bay. The results revealed mass accumulation of P-containing compounds in the 1.009–2.4676-μg range.

Keywords: phosphate, DGT, ferrihydrite, marine sediment

1. Introduction

A diffusive gradient in thin films (DGT) serves as a passive sampling method, which was primarily promoted as an instrument to examine the species of trace metal contained in water, sediments, and soil. This technique enables us to determine the solute content diffuses in an established concentration gradient of hydrogel layer isolated from the analyte solution by a protective filter membrane [1]. Lately, the DGT method was employed for measuring the dissolved P content in environmental samples. DGT is an *in situ* passive sampling technique that enables to measure the time-weighted average concentration of an analyte in surface and interstitial water. This technique has been applied to the measurement of the concentration of phosphate [2], trace metals [3], radionuclides [4,5], sulfide [6], and nitrate [7].

DGT instrument comprises two sections: a diffusion layer and a binding layer. Regardless of the analyte being measured, the diffusion layer has been reported to comprise the diffusive gel of polyacrylamide and cover membrane, which individually characterized by a specigied thickness [8]. In the DGT-based approach, a solute diffuses through a diffusive layer and is immobilized within a binding layer (often hydrogel-based) containing a high affinity-binding agent against a specified analyte. The analyte content in the bulk solution is measured by analyte's mass accumulated in the binding layer [9]. The DGT instrument is a layer set, which with a sink (iron oxide layer) -where P is contained- serve as diffusive medium simultaneously. The device is characterized by a diffusion-driven analyte resupply from a kinetically unstable solid-phase in respect to the formed concentration gradient in the solution by DGT [10]. Acrylamide gel combined with agarose-derived cross-linker (patented by DGT Research), allow the target analyte present in solution to diffuse across the membrane and reach the binding gel [11]. The analyte thus binds to the binding phase and is removed from the liquid. The analyte content accumulated in the binding-phase in the described form is regarded as equal toward the one passed thru the diffusive layer. In our study, ferrihydrite was used as the analyte-binding species in the binding layer because it has a strong adsorption capacity for phosphorus-containing compounds. These compounds are some of the most important nutrients for biological growth. Additionally, they are nutrients, which mostly limit the biological productivity in the land and marine environments [12]. Observing the concentration of phosphate is considerably important to avoid and manage the eutrophication, which has a serious ecological impact, as it decreases species biodiversity and abundance [13]. The study aimed to determine phosphate concentration in analyte samples using a DGT device having ferrihydrite as a binding gel.

2. Materials and methods

2.1. Reagents and materials
Plastic wares were washed in 5% HNO_3, followed by rinsing thoroughly with deionized water prior to use. Reagents were analytical grade. A stock solution (100 mgL^{-1}) of phosphate was prepared by dissolving the relevant amount of KH_2PO_4 in deionized water. The concentrated solution was kept at 4 °C and immediately added with solvent to the desired concentration before use. The DGT samples were purchased from DGT Research.

2.2. Characterization of the diffusive and binding gels
The functional groups of the binding and oxide gels were analyzed via Fourier transform infrared (FTIR) spectroscopy.

2.3. Analyte uptake and elution
Triplicate experiments were conducted wherein ferrihydrite binding gel discs were exposed to known volumes of phosphate solutions for varying contact time and with varying phosphate concentration. The samples of the solution were collected before and after the solution's exposure to the binding gel discs for colorimetric analysis to determine the mass of phosphate accumulated on the binding gel. The elution of the phosphate ions from the binding gels was performed using H_2SO_4. The eluent thus obtained was diluted and analyzed via UV–Vis spectrophotometry to determine the efficiency of the elution procedure.

2.4. Contact time and concentration dependence of phosphate uptake by ferrihydrite
To determine whether the sorption capacity of phosphorus by ferrihydrite was linear, three sets of triplicate oxide gel were exposed to a well-stirred solution containing 0.2 mgL^{-1} of KH_2PO_4 for 3, 6, 9, 12, and 24 h. In particular, the oxide gel containing ferrihydrite was exposed to a 10-mL solution of 0.2 mgL^{-1} KH_2PO_4. The phosphate ion accumulated in the binding gels was eluted with 10 mL of H_2SO_4 0.25 M. The amount of phosphate ion eluted was determined through the spectrophotometric method. To study the concentration dependence of the adsorption efficiency (sorption capacity) of the oxide gel, KH_2PO_4 solutions with concentrations ranging between 0.2 and 1.2 mgL−1 were utilized. In these experiments, three sets of triplicate oxide gel were exposed to a well-stirred KH_2PO_4 solution for 24 h. Phosphorus in the binding gel was eluted with 10 mL of H_2SO_4 0.25 M, and its amount determined through the spectrophotometric method.

2.5. Acid concentrations used to effect phosphate elution
Sulfuric acid concentrations used for the elution of the phosphate salt from the binding gel were 0.05,

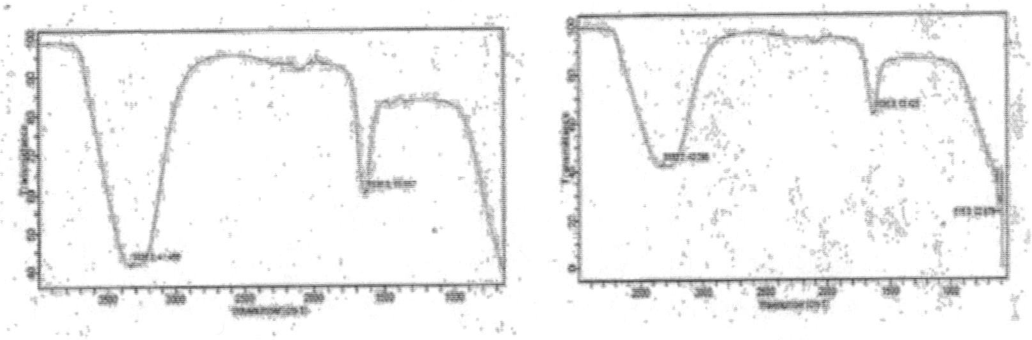

(a) (b)

Figure 1. Fourier transform infrared (FTIR) spectra of (a) the diffusive gel and (b) the binding gel.

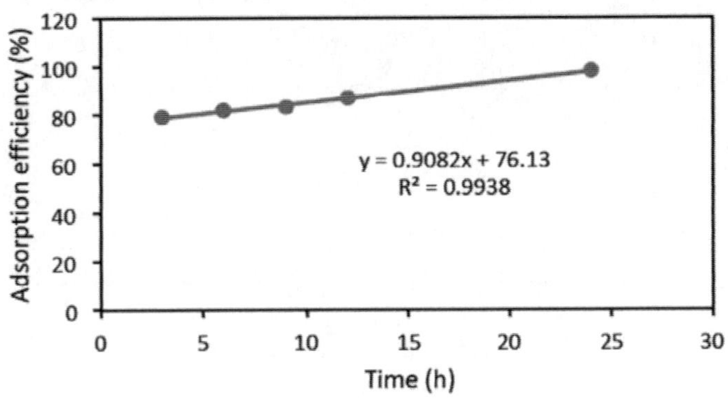

Figure 2. Effect of contact time on the uptake of phosphate ions by ferrihydrite

0.10, 0.15, 0.20, 0.25, and 0.30 M. The eluent was then analyzed using the spectrophotometric method.

3. Results and discussion

3.1. Infrared analysis

The characteristics of the diffusive and binding gels were analyzed via infrared spectroscopy. In particular, the FTIR spectra of the diffusive gel and the ferrihydrite-containing binding gel are shown in figure 1.

The FTIR spectrum of the diffusive gel is characterized by the presence of two main peaks at wavelengths of 3336 and 1638 cm^{-1}. The peak at 3336 cm^{-1} is related to the content of a primary amine ($-NH_2$), whereas that at 1638 cm^{-1} is attributed to the presence of a C–N bond. The peak at 615 cm^{-1} is because of the presence of ferrihydrite. The functional groups in the diffusive and binding gels were similar and the process of impregnation of the hydrogel with ferrihydrite did not cause substantial changes in the structure of the hydrogel. Based on these spectra, we showed that the interactions between ferrihydrite and polyacrylamide gel are physical in nature and do not involve any reactivity.

3.2. Phosphate ion uptake and elution

The optimization of the binding gel to ferrihydrite was performed to show the ability of ferrihydrite gel into phosphorus (KH_2PO_4) with a concentration of 0.2 mgL^{-1}. Figure 2 shows that increasing the contact time between the phosphate-containing solution and the ferrihydrite binding gel increases the

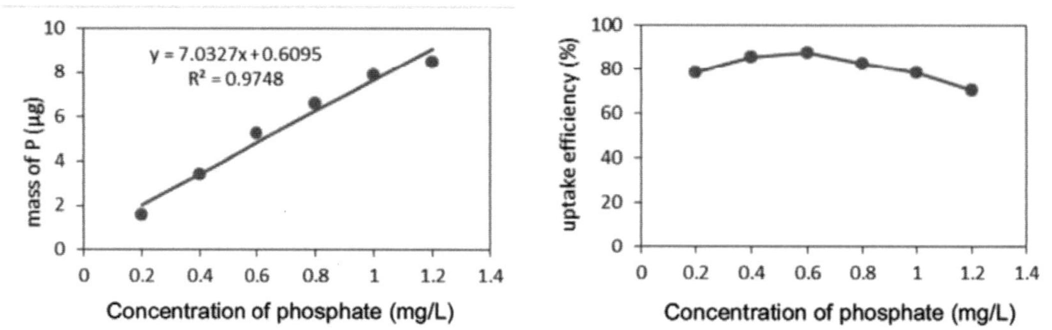

(a) (b)

Figure 3. (a) Effect of the phosphate concentration on the accumulated mass of KH_2PO_4 and (b) the efficiency of the uptake of phosphate ferrihydrite

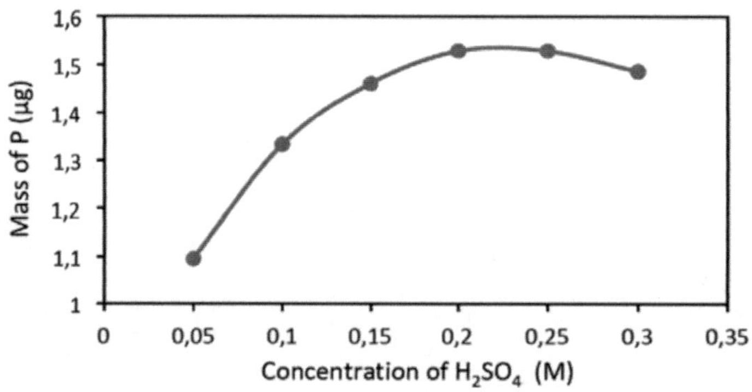

Figure 4. Effect of eluent concentration on the removal of analyte ions from the binding phase

efficiency of phosphate absorption. In particular, a linear correlation was observed between the uptake of phosphate ions and the contact time between the analyte solution and the oxide gel, with an R^2 value of 0.9938. The efficiency of the uptake of phosphate ions by ferrihydrite was 79.31 % over 3 h, an observation that provided the initial verification of the the high affinity of ferrihydrite binding gels against phosphate, hence compatible to be applied in further experiment. The binding phase's affinity for the target ion and its binding capacity were sufficiently high for the solute to accumulate within the oxide gel, thereby maintaining the solute concentration to be zero on the gel's surface contacted with the diffusive layer [14]. The accumulation of the phosphate ion on ferrihydrite was tested by the deployment samples of each type in triplicates for 3, 6, 9, 12, and 24 h. As can be seen in the graph reported in figure 2, the P uptake rapidly increased across the chosen contact time interval. The P uptake efficiency after a contact time of 24 h displayed its maximum value, i.e., 98.17%.

The effect of the concentration of the phosphate ion in solution on this ion's accumulation on ferrihydrite was studied by putting the binding gel into contact with KH_2PO_4 aqueous solutions with concentrations ranging between 0.2 and 1.2 mgL^{-1}. The contact time was 20 h, and the mixture was agitated at 150 rpm. The data reported in figure 3 indicate that there was a linear relationship between phosphate ion concentration in solution and the accumulated mass of the corresponding phosphate salt. In other words, an increase in the concentration of the phosphate ion in solution was associated with an increase in the amount of phosphate adsorbed onto the binding gel. Specifically, after a 20-h contact with a 1.2 mgL^{-1} solution of KH_2PO_4, the amount of this salt accumulated in the binding gel was 8.5525 grams. The mass of P's linear regression accumulated by binding gel samples confirmed the linearity of the uptake process after sometimes and was characterized by an R^2 value of 0.9748. On the other hand, data show that the relationship between phosphate ion concentration and binding efficiency was such that, as the KH_2PO_4 concentration increased from 0.8 to 1.2 mgL^{-1}, the binding

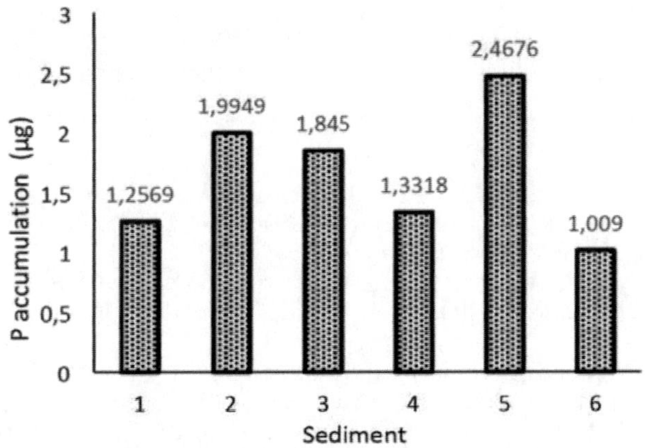

Figure 5. Results of the implementation of the DGT approach to perform the analysis of six marine sediment samples.

efficiency decreased from 83.85% to 71.72%. In particular, the highest binding efficiency was measured to be 87.37% at a KH_2PO_4 concentration of 0.6 mgL^{-1}. Notably, the near-infinite sink no longer applies under the condition of accumulated analyte's amount is close to the binding layers' highest capacity.

Elution of the phosphate ion from the binding gel was investigated placing the oxide gel in sulfuric acid solutions at concentrations ranging from 0.05 to 0.3 M. The effect of the concentration of sulfuric acid on the mass of KH_2PO_4 accumulated in ferrihydrite was studied after putting into contact with the ferrihydrite binding gel a 10-mL solution of 0.2-mgL^{-1} KH_2PO_4. This processes, with the amount of phosphorus of binding to oxide-gel was determined for the mass of P that was released from oxide gel. The mass of KH_2PO_4 released from the binding gel increased with the concentration of H_2SO_4, up to a maximum at an acid concentration of 0.25 M. When the concentration of the acid increased further to 0.3 M, the mass of KH_2PO_4 released decreased. The optimum concentration of sulfuric acid to elute of KH_2PO_4 from the binding gel was thus determined to be 0.25 M (see figure 4), in which case mass of released KH_2PO_4 was 5298 µg. Elution is an important step when performing a DGT measurement, as an efficient removal of analyte ions from the binding phase ensures the accuracy of the measurement.

3.3. Use of a DGT device to analyze marine sediments

In this study, DGT analysis was performed on marine sediments collected in six stations located in Teluk Jakarta (Jakarta Bay). In particular, DGT units were placed into sediments to observe the phosphate migration process from sediments to the ferryhydrite hydrogel. The DGT unit responds by binding to phosphate in the sediment. The results obtained indicated that the amount of phosphate accumulated ranged between 1.0090 and 2.4676 µg P (see figure 5). Evidence thus suggests that a DGT unit could be used to measure the amount of unstable phosphate in the sediments. Should it be deployed in sediment, DGT device might be applied for quantitative measurement, either porewater SRP concentration as well as resupply of kinetically unstable phosphate present in the sediment, as the porewater SRP has already depleted by the instrument. In the aquatic system, P bound to amorphous iron oxides (note that ferrihydrite is an amorphous iron oxide) are considered some of the most active inorganic phosphorous pools [15].

4. Conclusions

The isolation and quantification of phosphorus was performed by implementing the DGT technique using a ferrihydrite-based binding gel. The efficiency of ferrihydrite for the uptake of phosphate ions was measured to be 98.17% when the DGT device was kept in contact with the analyte solution for 24 h. The elution of the binding gel with 0.25-M H_2SO_4 was determined to be optimal for the recovery of bound phosphate (1.5298 µg of isolated KH_2PO_4). The performance of the DGT unit was found to depend on the time of deployment, the concentration of phosphate, and the concentration of acid used in the elution step.

Acknowledgements
We would like to thanks to Pusdiklat Industry (Ministry of Industry, RI), who provided full support through all the stages of this study. The authors would like to thank Einago (www.einago.com) for the English language review.

References
[1] Santner J, Prohaska T, Luo J and Zhang H 2010 *Anal. Chem.* **82** 7668–74
[2] Zhang H, Davison W, Gadi R and Kobayashi T 1998 *Anal. Chim. Acta* **370** 29–38
[3] Zhang H and Davison W 1995 *Anal. Chem.* **67** 3391–400
[4] Six L, Pypers P, Degryse F, Smolders E and Merckx R 2012 *Plant Soil* **359** 267–79
[5] Murdock C, Kelly M, Chang LY, Davison W and Zhang H 2001 *Environ. Sci. Technol.* **35** 4530–5
[6] Teasdale P R, Hayward S and Davison W 1999 *Anal. Chem.* **71** 2186–91
[7] Huang L, Li Z, Bai X, Li R Y, Wu H, Wei D and Yu L 2016 *Ecol. Eng.* **95** 438–46
[8] Li W, Zhao H, Teasdale P R, John R and Zhang S 2002 *Anal. Chim. Acta* **464** 331–9
[9] Bennett W W, Arsic M, Welsh D T and Teasdale P R 2016 *Environ. Sci. Process Impacts* **18** 992–8
[10] Davison W and Zhang H 2012 *Environ. Chem.* **9** 1–13
[11] Zhang C, Ding S, Xu D, Tang Y and Wong M H 2014 *Environ. Monit. Assess.* **186** 7367–78
[12] Ding S, Wang Y, Xu D, Zhu C and Zhang C 2013 *Environ. Sci. Technol.* **47** 7821–9
[13] Burns E E, Comber S, Blake W, Goddard R and Couldrick L 2015 *Environ. Sci. Pollut. Res.* **22** 9816–28
[14] Ding S, Xu D, Sun Q, Yin H, Zhang C 2010 *Environ. Sci. Technol.* **44** 8169–74
[15] Gao Y, Liang T, Tian S, Wang L, Holm P E and Hansen H C B 2016 *Environ. Pollut.* **219** 466-74

Quality improvement of water from post Tin mining based on a wasted cockle shell

Y Tiandho

Department of Physics, Universitas Bangka Belitung, Jl. Kampus Peradaban, Bangka 33172, Indonesia

Corresponding author: yuant@ubb.ac.id

Abstract. The high Tin mining activity in the Bangka Belitung Islands caused a considerable amount of environmental damage in the province. One of the most common impacts of contamination is the poor quality of water. *Kulong* is a former excavation of Tin mines that may be used as a reservoir of water. But most *kulong*, especially young ones, store acidic water and dispersed-pollutant water so that it does not meet the minimum requirements of good water. In this study, we developed a simple method to improve the quality of water from post Tin mining using wasted cockle shell. The utilisation of wasted cockle shell because of its abundant amount in this province. Based on the results it is known that calcined cockle shell powder is sufficient to increase the pH level of the water and can coagulate the pollutants that contained in it. Through the water clarity level and the number of total dissolved particles in the water after testing, it is known that the cockle shell powders have a significant effect on improving quality water.

Keywords: water treatment, cockle shell, green mining.

1. Introduction

The Bangka Belitung Islands are the largest Tin-producing province in Indonesia and are among the top three Tin producers in the world [1,2]. The Tin mining sector remains the primary commodity in the region even though in recent years there has been a decline in production [3]. High Tin mining activities in the province cause severe environmental damage. The lack of wastewater treatment in the Tin mining process, especially by unconventional miners, has created a lot of water sources in this province that is polluted and does not meet the Ministry of Health Decree No.907/Menkes/SK/VII/ 2002 on water quality requirements [4,5]. One of the effects of poor water quality in the Bangka Belitung Islands, regardless of the daily needs of the community, is the lack of water supply to meet agriculture and freshwater fisheries needs while the freshwater aquaculture development program is one of the flagship programs for this province.

When land has undergone excessive exploration, it will produce basins, which are then called *kulong*. *Kulong* can be a source of water in the dry season, or it can be used for agricultural and fisheries purposes. However, as ex-mining land, it is feared that there are remnants of heavy metals that make *kulong* water unsuitable to meet these needs. One indication of the presence of heavy metals in water is its turbidity of colour and its acidity. Therefore, to be applied in agriculture and freshwater fisheries, this water requires further treatment.

The maritime sector is another potential natural resource from the Bangka Belitung Islands. One of them is cockle. The cockles can be easily found in the province's almost all coastal areas. The primary composition of cockle shell is calcium carbonate, which is usually present in the aragonite phase [6]. The calcination process can decompose calcium carbonate into coagulated calcium oxide

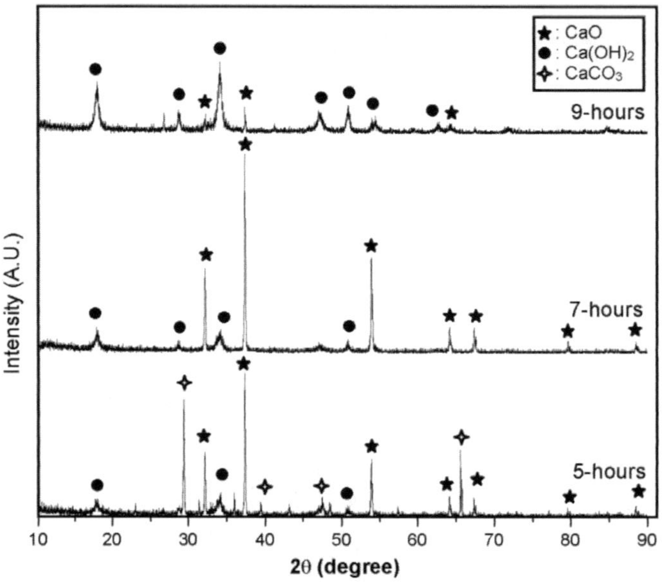

Figure 1. XRD diffraction pattern of cockle shell after calcinations

and can increase its pH to form hydroxide group in water. This hydroxide group can be created because calcium oxide powder, when encountered with acidic water, which has H^+ ions, will form calcium hydroxide [7]. Moreover, the shell powder if interacting with Ca^{2+} ions, it can coagulate the pollutants dispersed in water [7,8]. The superiority of shell powder as a coagulator is that it is non-toxic, readily biodegradable and easily interacts with other organic substances such as proteins. Even shell powder can be more effective than alum. Thus the use of shell powder as a coagulator is environmentally friendly [9]. Coagulation of pollutants in post Tin mining water is the first step in filtering heavy metals because heavy metals in water will be coagulated into larger particles so that they can be separated further.

2. Methods

Prior to use, the cockle shell waste is rinsed then dried under the sun for two days. After the drying process, cockle shell waste is then made into powder through a milling process followed by a calcination process at a temperature of 700 °C for 5, 7, and 9 hours. The phase of shell powder was identified using X-Ray Diffraction (XRD, Rigaku Smartlab 3 KV, Cu - K_α = 1.5406 Å) method. XRD pattern analysis is performed using X'pert HighScore Plus software. Calcined cockle shell powder is then mixed into turbid and acidic water from the post Tin mine. The post Tin mining water used comes from Tanjungratu area in Bangka Regency. The effectiveness of water quality improvements that will be observed in this research include the aspect of turbidity, total of dissolved particles, and increased in pH. These three aspects are the basic parameters in determining water quality. The measurement of the total dissolved particles in the water is carried out using a TDS meter (TDS Testers-139) that has been calibrated with a calibration solution (1382 ppm, 25°C). The turbidity of the water is measured by a calibrated turbidimeter (HACH-2100N) using standard liquids of <0.1, 20, 200, 1000, and 4000 NTU. SEM/EDX (Hitachi, SU-3500) is used to observe the microstructure of the results of coagulation and its elements. Meanwhile, to find out the presence of heavy metal elements in water is done by atomic absorption spectroscopy (AAS) for aluminium and iron elements in water.

3. Results and discussion

The results of X-Ray Diffraction pattern from cockle shell powder after undergoing the calcination process are shown in figure 1. From these patterns, it can be concluded that calcination process effectively decomposes $CaCO_3$ (calcite, Ref. Code: 01-081-2027) to CaO (Ref. Code: 00-037-1497). $CaCO_3$ compounds remaining when the calcination carried out for 5 hours, while the other calcination

Table 1. EDX analysis of coagulated pollutants

Elements	Weight (%)	Atomic (%)
C	13.80	20.03
O	62.13	67.72
Al	4.73	3.06
Si	4.28	2.66
Ca	14.89	6.48

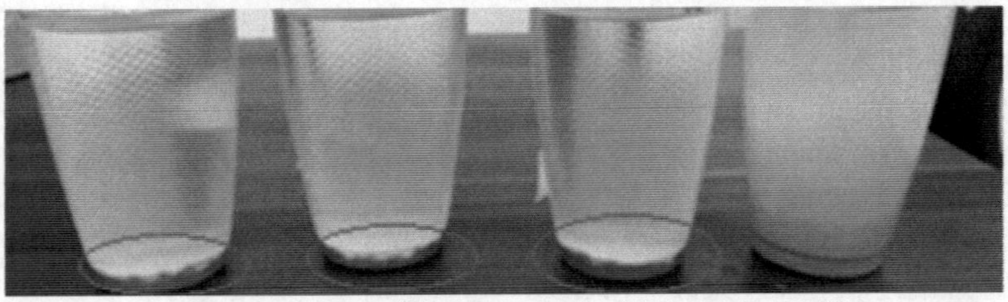

(a) (b) (c) (d)

Figure 2. The coagulation result (red circle) with the addition of the cockle shell: (a) without calcination, (b) 5 hours calcination, (c) 7 hours calcination, and (d) 9 hours calcination.

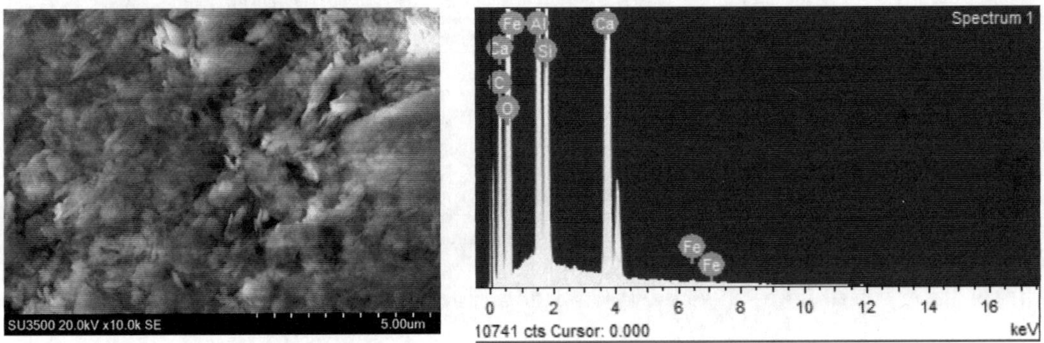

(a) (b)

Figure 3. (a) coagulated pollutants micrographs and (b) EDX analysis results (spectrum)

treatments, when the calcination carried out for 5 hours, while the other calcination treatments, which is 7 hours and 9 hours, $CaCO_3$ compound completely degradable. The presence of $CaCO_3$ for 5 hours calcination time shows that the calcination process is too short. The diffraction patterns also reveal the presence of $Ca(OH)_2$ compound (Ref. Code: 00-044-1481) after calcination process. This compound is obtained because CaO is highly reactive to moisture to form $Ca(OH)_2$.

After passing the mixing process of the calcined cockle shell powder into the ex-mining water and the mixture is aging overnight until the coagulation is obtained as shown in figure 2. These results indicate that the pollutants coagulation at the bottom of the container occurs because of calcined cockle shell powder. The figure also shows the effectiveness of the coagulation of pollutants with the calcination process. Without the calcination process, pollutants in water are still scattered and only partially coagulate. The calcination process converts the phase of the cockle shell powder which initially takes the form of $CaCO_3$ to CaO. $CaCO_3$ compounds are not reactive compounds in water so that their presence in water is less able to coagulate dispersed pollutants. Unlike CaO which is a

Table 2. Turbidity measurement of water

Type of the cockle shell powder or water	Turbidity (NTU)
Post tin mining water (raw)	1806
Without calcination	147
5 hours calcination	9.60
7 hours calcination	3.98
9 hours calcination	0.421

Table 3. Effect of the addition cockle shell powder on the number of dissolved particles in the water from post Tin mining

Type of the cockle shell powder or water	Total dissolved particles (ppm)
Post Tin mining water (raw)	87
Without calcination	118
5 hours calcination	122
7 hours calcination	125
9 hours calcination	121

Table 4. The results of spectrophotometric analysis of post Tin mining water before and after the addition of cockle shell powder.

Elements	Before treatment (mg/l)	After treatment (mg/l)
Al	0.41	0.09
Fe	0.95	0.27

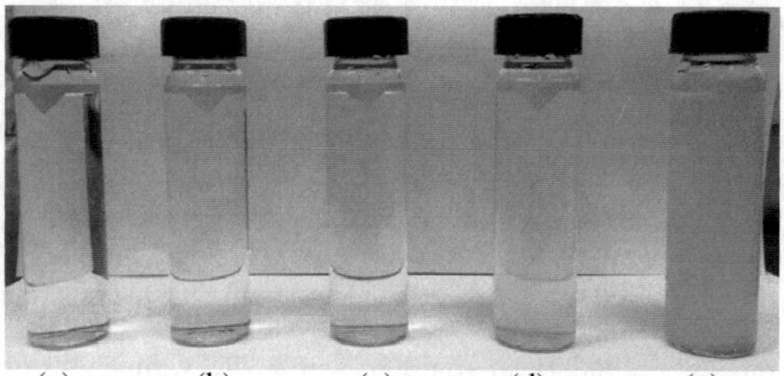

(a) (b) (c) (d) (e)

Figure 4. Results of water treatment after the separation between water and coagulation results of (a) post Tin mining water (raw), addition of cockle shell powder (b) without calcination, (c) calcination 5 hours, (d) 7 hours, and (e) 9 hours)

reactive compound in water which produces Ca^{2+} ions which bind pollutants in water, this compound also has good solubility in water and can form calcium hydroxide compounds in water. Microscopy of pollutant coagulation is shown in figure 3. The results of SEM micrographs showed that the results of coagulation had a microstructure in the form of coagulum. This coagulum is formed due to the binding of pollutant elements by Ca^{2+} ions. While the results of EDX analysis of the obtained coagulation show the presence of dominant elements of pollutants such as C, Fe, Al, and Si with details as shown in table 1. These results indicate that calcined cockle shell can bind various heavy metal elements like Fe and Al and organic elements like C.

Table 5. Effect of the addition cockle shell powder on pH level of water from post Tin mining

Type of the cockle shell powder	pH level
Without calcination	8.4
5 hours calcination	10.2
7 hours calcination	9.0
9 hours calcination	11.0

Comparison of the appearance of water after going through the coagulant separation process is shown in figure 4. The quantitative results of water turbidity using turbidimeters are presented in table 2. Based on water turbidity after the separation from coagulant, it was found that 9 hours calcined cockle shell powder produced the highest turbidity compared to the other two. This result is also supported by the smallest water turbidity value of 0.421 NTU. Also, it can be said that the use of calcined cockle shell powders effectively reduces the turbidity of the post Tin mining water and the longer the calcination process will be more effective in coagulating pollutants in the post Tin mining water. The TDS measurement results used to analyse the number of dissolved particles in water are presented in table 3. The table shows that the highest number of dissolved particles was found for cockle shell powder calcined for 7 hours. However, in general, the calcination process of cockle shell powder increases the number of dissolved particles. This can be understood because of the dominance of high solubility CaO in water.

The atomic absorption spectrophotometric (AAS) results for water before and after mixing with cockle shell powder (9 hours of calcination) are shown in table 4. The table shows that before mixing with cockle shell powder, the water contains Al and Fe in sufficiently high amounts, and the amount is above the limit set by the Indonesian Ministry of Health. However, after being mixed with cockle shell powder, the content of Al and Fe in post Tin mining water is reduced, and the amounts meet the requirements set by the Indonesian Ministry of Health. This result corresponds to the results of EDX measurements, which obtain Al and Fe metal elements in the results of coagulation.

The pH measurement results showed an increase for post Tin mining water after being mixed with calcined cockle shell powder as shown in table 5. The table shows that the addition of cockle shell powder (mass of the shell powders: 0.05 gram and the volume of water: 240 mL) has a very significant effect in changing pH. However, it was the calcination for 7 hours that gave an increase in pH, which was close to the water conditions suitable for use. Overall, the increase in pH indicates a value above the water level that can be used, so that additional processes are still needed to filter dissolved cockle shell powder.

4. Conclusions

Based on the results obtained it can be concluded that the calcination process at temperatures of 700 °C for 7 and 9 hours can decompose $CaCO_3$ as a whole from cockle shell powder to CaO. The presence of $Ca(OH)_2$ in calcined cockle shell powder is caused by water vapour contamination from the environment. After the process of separation of water with pollutant coagulants, it can be concluded that the use of calcined cockle shell powder can significantly reduce turbidity, reduce of Al and Fe metals, and organic C elements, increase the number of dissolved particles and pH values from post Tin mining water substantially.

Acknowledgements

This research was funded by Universitas Bangka Belitung through PDTJ UBB 2017 funding scheme.

References

[1] Ibrahim 2016 GSTF Journal of Law and Social Sciences **5**, 1–7
[2] Rohendi R and Aryanto N C D 2012 Bulletin of the Marine Geology **27** 7–18
[3] Aspinall C 2001 Mining, Minerals, and Sustainable Development Report No. 79 pp. 1–30 (London: International Institute for Environmental Development)
[4] Syarbaini, Warsona A and Iskandar D 2014 Atom Indonesia **40** 27–32
[5] Sari S P and Rosalina D 2016 Procedia Environ. Sci. **33** 436–42
[6] Mohamed M, Yusup S and Maitra S 2012 Journal of Engineering Scinece and Technology **7** 1-10, 2012

[7] Tiandho Y, Aldila H, Mustari, Megiyo and Afriani F 2018 J. Phys.: Conf. Ser. **1013** 012181
[8] Moideen S N F, Din M F M, Ponraj M, Yusof M B M, Ismail Z, Songip A R and Chelliapan S 2015 Desalin. Water Treat. **57** 1–9
[9] Daud Z, Abubakar M H, Kadir A A, Latiff A A A, Awang H, Halim A A and Marto A 2017 International Journal of Geomate **12** 46–52

IOP Conf. Series: Materials Science and Engineering **496** (2019) 012059 doi:10.1088/1757-899X/496/1/012059

PANi/ZrO_2-composite coating for corrosion protection in 3.5 M NaCl solution

N Nadliriyah, A L Putri and T Triwikantoro

Department of Physics, Faculty Science, Institute of Technology Sepuluh Nopember, Jalan Arief Rahman Hakim, Surabaya 60111, Indonesia

E-mail: triwi@physics.its.ac.id

Abstract. This study was designed to investigate PANi/ZrO_2 as an anti-corrosion coating material. An extraction of Zirconia (ZrO_2) from natural zircon sand of Kereng Pangi has been conducted by means hydrothermal process and alkali fusion. In this composite system consist of Polyaniline (PANi) as a matrix and ZrO_2 as a filler material for an anti-corrosion coating. PANi/ZrO_2 composites were synthesized by an in-situ polymerization method. The structure of synthesized ZrO_2 was characterized using X-Ray Diffraction (XRD). The samples for corrosion testing were original steel (SS), paint coated steel (PS), paint/ZrO_2 coated steel (ZS), paint/PANi/ZrO_2 coated steel (PZ) with dimensions of 2 cm x 2 cm. Corrosion rate was tested in 3.5M NaCl using polarization method. XRD result showed that the structure of obtained ZrO_2 was amorphous. The corrosion rates for the four samples SS, PS, ZS, and PZ were $2{,}36\times10^{-2}$, $1{,}3\times10^{-5}$, $1{,}35\times10^{-7}$, and $5{,}97\times10^{-7}$ mpy, respectively. The result of corrosion rate showed that the lowest corrosion rate was steel coated with ZrO_2 (ZS). It was $1{,}35 \times 10^{-7}$ mpy. The result indicated that the zirconia coating is a prospective material for anti-corrosion.

1. Introduction

Zirconia, ZrO_2 is refractory ceramic that has low heat conductivity, high hardness and density, and good polymorphs [1]. Zirconia is a mineral that is rarely found in nature in its form as a ZrO_2 crystal. It is found naturally as ZrO_2 crystal in baddeleyite minerals, and has a monoclinic crystal structure. It is rarely found with zircons because it forms on rocks with unsaturated silicates. If the silica in the rock is saturated or oversaturated, what is found in the rock is no longer a zirconia crystal but zircon crystals. Zircon ($ZrSiO_4$) crystal is the principal mineral for producing zirconia. Zircon is found everywhere in the earth's crust with a content of around 25 %. [2]. It is a common element of most sand so it usually found in sediment deposits. Many zircon minerals are found in nature as zircon sand. This sand mining, legally and illegally, mostly found in the Kereng Pangi area (Central Kalimantan).

Various techniques for zirconia extraction from zircon have been proposed in the literature [3], including extraction with basic oxides, chlorination reductive smelting and alkali digestion. The present work has processed zircon (Puya sand) by alkali fusion. The principle of the alkali fusion method is to disassemble chemical bonds in the material by using alkali compounds such as KOH, NaOH, and Na_2CO_3 [4,5].

Corrosion is a material degradation due to electrochemical reaction with the surrounding environment where electron transfer process occurs from metal to environment. Corrosion is one of the problems that often occur in everyday life, especially on materials made of metal. Corrosion cannot be eliminated at all but can only be slowed down the process.

There has been a lot of research that investigated the corrosion protection method, including with cathodic protection, the application of a thick coat of paint, and the application of anti-corrosion paint that can inhibit corrosion rate [1]. The application of this anti-corrosion paint usually contains pigments which can inhibit corrosion rate, one of them by using conductive polymer. Basically, polymer is an insulating material but can be used as conductive material if it is doped with a strong acid.

IOP Conf. Series: Materials Science and Engineering **496** (2019) 012059 doi:10.1088/1757-899X/496/1/012059

Polyaniline (PANi) is one of the most conductive polymers used as a protection against corrosion due to some advantages compared to other conductive polymers. The advantage of using PANi is that its chemical and electrochemical polymerization process is simple, easy to be doped and dedoped by treatment using strong acid [1,4]. But its melting point is low, so it needs to be combined with heat-resistant material and become a composite with that material so that the composite will not easily cracked.

In this research, zirconia was composited with polyaniline polymer by an in-situ polymerization method. The PANi/ZrO_2 composite was coated on stainless steel 304. This work focuses on the effect of the PANi/ZrO_2 composite coating for corrosion protection in the 3,5M sodium chloride solution.

2. Experimental

2.1. Material preparation

In the present work, zirconia was obtained from the natural zircon sand from Kereng Pangi, Central Kalimantan. Before being processed, magnetic elements are separated from zircon sand using a magnet. The sand is then milled using wet milling in alcohol, after that leached with chloride acid (HCl). After the leaching process, the sand is extracted by hydrothermal method using 7M NaOH solutions at 250 °C. The extracted zircon powder was then processed into amorphous zirconia (*a*-ZrO_2) powder using NaOH solution with alkali fusion method and then continued with co-precipitation method using NH_4OH solution. The zirconia powder and polyaniline (PANi) are then processed into composites. Polyaniline solution was prepared for the process using ammonium peroxy disulfate, aniline and dodecyl benzene sulfonic acid. Then the in-situ polymerization of the composite process was carried out for 8 hours.

Stainless steel used in this research is SS 304. Before coated, it smoothed first by using sandpaper numbered mesh 80 - 1000. The paint used in this research is Agatha Paint Ship. The paint is then mixed with ZrO_2 and the composite PANi/ZrO_2 separately. The steel coating process is carried out using spraying techniques on all steel surfaces but previously the steel is coated with epoxy to improve the quality of coating adhesion.

2.2. Characterization

Characterization of samples in this study were carried out using *X-Ray Diffractometer* (XRD) to determine the phase in the sample, *Fourier Transform Infrared* (FTIR) to find out the bond or functional group formed through the process of synthesis PANi/ZrO_2 composite, and polarization test to measure the corrosion rate.

3. Results and discussion

3.1. X-Ray Diffraction spectroscopy

Natural zircon sand was extracted by several processing stages: wet milling, leaching, and hydrothermal processes. Each process generates a material with a different phase. From these processes, we obtained zircon ($ZrSiO_4$) with tetrahedral structure [6]. The process of producing amorphous zirconia (ZrO_2) powder from the extracted zircon powder involves the alkaline fusion and co-precipitation methods.

The chemical reaction to the alkaline fusion process with NaOH as a fluxing agent with various molar variations is: [7]

$$ZrSiO_4 + 4NaOH \rightarrow Na_2ZrO_3 + Na_2SiO_3 + 2H_2O$$

$$ZrSiO_4 + 6NaOH \rightarrow Na_2ZrO_3 + Na_4SiO_4 + 3H_2O$$

$$ZrSiO_4 + 2NaOH \rightarrow Na_2ZrSiO_5 + H_2O$$

The X-Ray diffraction pattern of the zirconia powder shows that it has an amorphous structure as shown in figure 1.

3.2. FTIR spectroscopy

The purpose of Fourier Transform Infrared (FTIR) spectroscopy characterization is to find out the functional groups of zirconia (ZrO_2), polyaniline (PANi), and PANi/ZrO_2 composites. The result of FTIR spectroscopy characterization is shown by figure 2.

Table 1. Functional group absorption of the sampels

Wavenumber (cm^{-1})	Functionality	ZrO_2	PANi	PANi/ZrO_2
400-500	Zr-O Bending Vibration [8]	420.5	-	443.64
600-900	C-H Bending	-	665.08	663.53
1250-1360	Aromatic C-N Stretching	-	1284.46	1338.64
1400-1500	N-H Stretching	-	1442.81	1485.24
1430-1600	C=C Aromatic rings	-	1557.23	1535.39

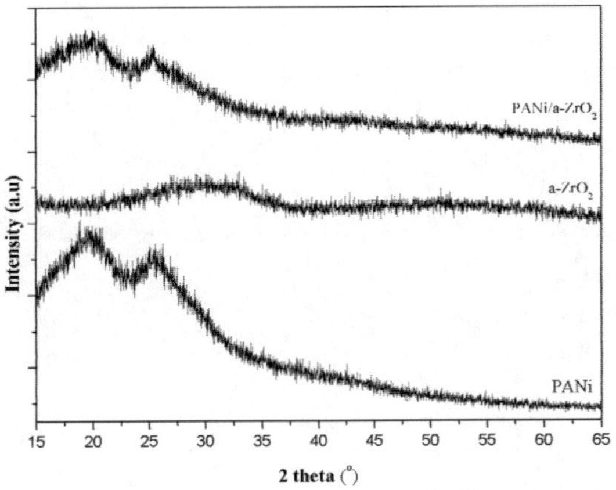

Figure 1. X-ray diffraction patterns of the composites

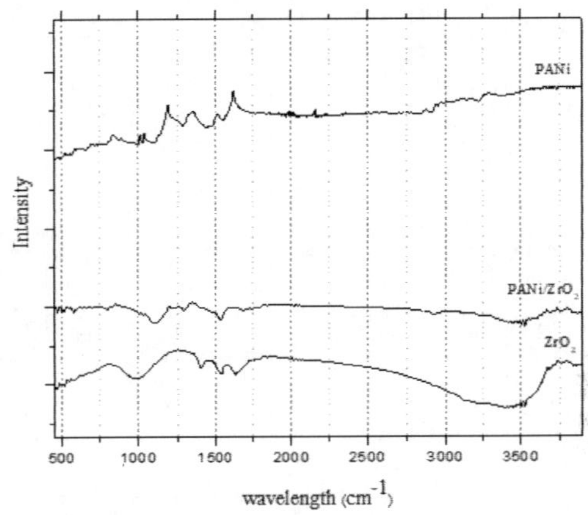

Figure 2. FTIR spectra of the samples

There are several wave absorption points, which represent the type of bond or a functional group formed. The results are then analyzed by using reference wave numbers to find out what functional groups are present in the material under test. Table 1 discusses the type of bond or functional group read from figure 2.

The absorption pattern of the PANi-ZrO_2 composites in figure 2 generally tends to resemble the PANi absorption pattern. It shows that the addition of ZrO_2 in the composite PANI-ZrO_2 synthesis process does not alter the matrix structure of polyaniline but will bring up a new functional group. The existence of absorption pattern in the wavelength range of 400-450 cm^{-1} shows the absorption pattern of functional group Zr-O that is characteristic of ZrO_2.

3.3. Corrosion performance test

In this work corrosion rate was tested by the polarization technique with a NaCl 3,5M electrolyte solution was used. The result of the polarization test is a Tafel diagram, which can be explained in figure 3. The Tafel diagrams give the value of corrosion current (I_{Corr}) and corrosion potential (V_{corr}) of the tested materials. The value of the corrosion rate is obtained from the meeting point between the slope I_{Corr} and V_{corr} in the Tafel diagrams.

The polarization test gives the corrosion rate of the stainless steels in mpy as shown in table 2. The highest corrosion rate is steel without coating (SS). The steel is exposed directly to the electrolyte solution so that nothing is resistant to negative ions in the 3.5 M NaCl solution resulting in an easily corroded steel. The PS sample shows protection against negative ions. The corrosion rate in this PS sample is as small as 1.3×10^{-5} mpy since the paint is a high-performance anti-corrosion paint for ships. The steel coated with PANi/ZrO_2 composite has a corrosion rate value in-between, i.e. lower than the the paint-coated steel but higher than the ZrO_2 coated steel.

4. Summary

In this work, the extraction of zirconia from the natural zircon sand using alkali fusion method has been performed and the extracted zirconia was used as corrosion inhibitor component with stainless

IOP Conf. Series: Materials Science and Engineering **496** (2019) 012059 doi:10.1088/1757-899X/496/1/012059

Table 2. Corrosion rate of the samples

Samples	Corrosion rate (mpy)
Stainless steel (SS)	2.36×10^{-2}
Stainless Steel +paint (PS)	1.30×10^{-5}
Stainless Steel+paint+ZrO_2(ZS)	1.35×10^{-7}
Stainless Steel+paint+PANi/ZrO_2 (PZ)	5.97×10^{-7}

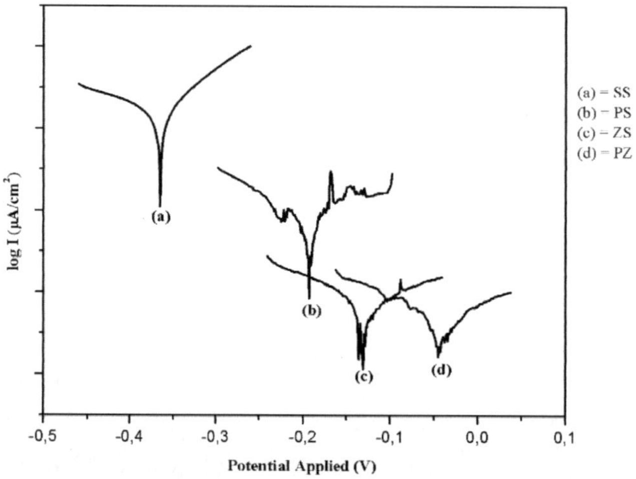

Figure 3. Tafel diagram of the samples

steel as the tested metal. The corrosion testing showed that the lowest corrosion rate was steel coated with ZrO_2 (ZS), i.e., as low as 1.35×10^{-7} mpy. The result indicated that zirconia coating is a prospective material for anti-corrosion.

Acknowledgements

The authors would like to thank Ministry of Research Technology and Higher Education of the Republic of Indonesia and LPPM ITS to support finance through program Non PNPB Funds 2017 No. 833/PKS/ITS/2017 given to SP.

References

[1] Affatato S, Goldoni M, Testoni M and Toni A 2001 *Biomaterial* **22** 717–23.
[2] Putri A K, Kirom M R, Abrar and Syarif D G 2016 e-Proceeding of Engineering **3** 2062–9
[3] Hancock J D 1977 *Miner. Sci. Eng.* **9** 25–30
[4] Kwela Z N 2007 *Alkali Fusion Processes for Recovery of Zirconia and Zirconium from Zircon Sand* (Pretoria:University of Pretoria) MSc. dissertation
[5] Mori T, Hoshino H, Ishikawa Y, Yamaguchi T, Yamamura H, Kobayashi H and Mitamura T 1991 *J. Ceram. Soc. Jpn.* **99** 227–32
[6] Musyarofah, Nurlaila R, Muwwaqor N F, Saukani M, Kuswoyo A, Triwikantoro and S Pratapa 2017 *J. Phys.: Conf. Ser.* **817** 012033
[7] Abdelkader A M, Daher A and El-Kashef 2008 *J. Alloy Compd.* **460** 577–80
[8] M. Behzadnasab, Mirabedini S M, Kabiri K and Jamali S 2011 *Corros. Sci.* **53** 89–98

The effect of antioxidant additives on the characteristics of food-grade grease using castor oil (*Ricinus communis L.*) as the base oil

E J Mayasari[1,2], R Bakri[1] and M Fibria[2]

[1] Department of Chemistry, Faculty of Mathematics and Natural Sciences (FMIPA), Universitas Indonesia, Depok 16424, Indonesia

[2] Research and Development Centre for Oil and Gas Technology (LEMIGAS), Jalan Ciledug Raya No.109, Jakarta 12230, Indonesia

Corresponding author: juitamayasari@gmail.com

Abstract. Grease has three basic components: base oil, thickening agents, and additives. Castor oil has a potential role as a lubricant base oil in grease, but it has lower oxidative stability than mineral oil. Antioxidant additives can delay oxidation reactions in food-grade grease. Li-12-hydroxystearate soap is used as a thickening agent. The food-grade grease is formulated via saponification, dilution, recrystallization, and homogenization. This research aimed to study the effects of different types (BHT, TBHQ, and high molecular weight phenolic) and concentrations (0%, 0.5%, 1%, 1.5%, and 2%) of antioxidants on food-grade grease characteristics. Several characteristics of greases, such as dropping point, pH, consistency, and corrosion resistance were observed. The results showed that different antioxidants induced differences in dropping points, and different antioxidant concentrations induced differences in dropping points and consistency. The best performance was obtained from food-grade grease formulated with high molecular weight phenolic antioxidant. This had the following characteristics: dropping point 193 °C, pH 7, consistency NLGI 2, and corrosion resistance group 1a.

Keywords: grease, food-grade grease, castor oil, antioxidant additive

1. Introduction

Grease is a semi-fluid form, a compound contains of liquid lubricant, thickening agent and auxiliaries. The liquid lubricant can be in the form of petroleum (mineral), oil derived from plants or synthetic oil [1]. Therefore, grease has three basic components: base oil, thickening agents, and additives [2]. Grease lubricants are regarded as colloidal dispersions of thickening agent contained in liquid lubricant. In environmentally friendly products, the main components of greases must be rapidly biodegradable [3]. Biodegradable greases synthesized from renewable resources such as vegetable oils offer a convenient alternative for petroleum-based greases which known to carry a toxic property toward the environment as well as non-biodegradable. Grease contains of vegetable oil mostly found in semi-solid form of colloidal dispersion from thickener (metal soap) in a liquid lubricant matrix (vegetable oil). Therefore, the current urge in tribology is an experiment to discover renewable, biodegradable, as well as highly thick oil, which meet the requisite of a lubricant [4]. Biological lubricants are particularly applied in food or pharmaceutical industry, namely in an oven or kiln in a bakery, in which the risk of contamination should be diminished. Sources of biological oils are vegetables and animals. Vegetable oils include palm, rapeseed, and castor oils, while sperm and fish oils and lanolin are sourced from animals [5]. Vegetable oils and synthetic ester-based greases give better performances and are eco-friendly and biodegradable [6]. The neat castor oil is in lower quality compared to mineral oil in terms of its tribological properties, even though an intense review against the experimental data indicated that both are comparable. Commercial mineral oil contained additives and thus holds a better performance than neat castor oil. In this case, a more thorough experiment is to be performed by utilizing a varied of additives with castor oil to obtain a suitable formulation that can substitute mineral oil for various usages and therefore, reduce the leverage toward the environment [7].

Lubrication is essential for ensuring engine performance and efficiency. Therefore, effective and safe lubricants are necessary. Lubricants that are widely used commercially are mineral-based and synthetic oils that are harmful when in contact with food products. Food-grade grease is a lubricant based on vegetable oils that is safe for health as well as environmentally friendly, so it is beneficial to the pharmaceutical, animal feed, cosmetics, and food industries in particular [8]. Standard instrument complies with the requirement of biodegradable grease manufacture; however, the equipment needs to be completely cleaned to ensure that conventional greases and their components will not contaminate [3]. Machinery used for preparing and processing the edible stuffs, make-up products, tobacco products, and pharmaceutical products require all type of lubricants, i.e. grease, oil etc. to lubricate, for heat transfer purpose, as well as transmission of power, and protection against corrosion in a machine, including its part, equipment, and instruments. Sometimes, lubricant and food contact cannot be avoided, which will cause contamination to the food products. One way to overcome this issue is by using a food-grade lubricant [9].

Castor oil is inedible and is used almost entirely for pharmaceutical and industrial applications [10]. Lubricating grease based on castor oil is suitable for use on metal surfaces. Castor oil is not consumed in foodstuffs so would not impact food security [11]. Research performed on crude castor seed oil confirmed the presence of ricinoleic, oleic, palmitic, stearic, and dihydroxystearic acid; this indicated that it could be modified for use as food additives as well as in transportation, cosmetic, and pharmaceutical industries [12]. Castor oil was found to be a rich source of ricinoleic acid, as a major fatty acid, and has significant potential as industrial oil [13].

Castor oil has a potential role as lubricant base oil in grease, but it has lower oxidative stability than mineral oil. According to Mohammed [14], natural antioxidants are among the most important additives used in greases. Selected antioxidants were found to be effective in improving the oxidative stability of coconut oil [15]. Antioxidant additives can delay oxidation reactions in food-grade grease [8].

This research aimed to study the effects of different types (BHT, TBHQ, and high molecular weight phenolic (HMWP)) and concentrations of antioxidants on food-grade grease characteristics.

2. Materials and methods

Lithium hydroxide and 12-hydroxystearic acid were used to prepare a lithium 12-hydroxystearate soap lubricating grease (thickening agent) containing 17% soap. BHT (butylated hydroxytoluene), TBHQ (tert-butylhydroquinone), and HMWP were used as antioxidant additives at percentages of 0%, 0.5%, 1%, 1.5%, and 2% by weight of the total formulation. Castor oil and all the components were kindly supplied by PPPTMGB "LEMIGAS."

A food-grade grease was formulated by a saponification–dilution–recrystallization–homogenization process. The greases were prepared by mixing lithium 12-hydroxystearate soap, castor oil, and the additive, heating to 80 °C under constant stirring at 400 rpm for 90 min, and then allowing the hot mixture to cool to room temperature. Desirable characteristics of greases relate to their dropping point, pH, consistency, and corrosion resistance, as well as oxidation stability for castor oil, and these properties can be measured. The experimental greases produced were characterized using ASTM (American Society for Testing and Materials) test methods.

3. Results and discussion

True grease consists of oil or another fluid lubricant that is mixed with a thickener such as a soap to form a solid. Lithium grease has many useful properties such as load carrying ability, high-temperature tolerance, and shear stability; thus, it is considered valuable and is commonly used in most parts of the world [14]. Table 1 lists the tribological properties of food-grade greases.

Table 1. Characteristics of Food-Grade Greases.

	Dropping Point (°C)	pH	Worked Penetration (0.1 mm)	Consistency NLGI	Corrosion Resistance
G-0	191	7	271	2	la
G-BHT0,5	189	7	290	2	la
G-BHT1	189	7	287	2	la
G-BHT1.5	190	7	285	2	la
G-BHT2	190	7	281	2	la
G-TBHQ0.5	188	7	295	2	la
G-TBHQ1	190	7	291	2	la
G-TBHQ1.5	192	7	275	2	la
G-TBHQ2	192	7	270	2	la
G-HMWP0,5	192	7	289	2	la
G-HMWP1	193	7	285	2	la
G-HMWP1.5	193	7	265	2	la
G-HMWP2	194	7	258	3	la

3.1. Dropping point
Grease's heat resistance indicator is represented by the dropping point parameter. When the grease's temperature rises, the penetration property increases to the point it liquefies and lost its desirable consistency. The dropping point refers to the temperature where the grease is adequately fluid to drip, suggested the upper temperature boundary in which the structure of grease remains, instead of maximum temperature which can be applied [16]; this is usually 20 °C–30 °C below the dropping point [4].

Food-grade grease with added BHT antioxidants has a lower dropping point than conventional grease with and without added TBHQ and HMWP antioxidants. This is because BHT has a lower melting point than TBHQ and HMWP, which have melting points ranging from 125 °C to 129 °C, while that of BHT is only 70 °C.

3.2. pH
The expected pH of food-grade grease is neutral (7). Addition of antioxidants did not affect the pH. All the food-grade greases produced had a pH of 7.

3.3. Consistency
Grease's physical and chemical properties of are greatly regulated by its consistency or hardness [17]. The thickening agent provides consistency characteristic in grease and keeps the oil properly [1]. Rigidity or consistency is the grease's most significant characteristic. Effective lubrication depends on the consistency of grease, which in turn relies on the applied thickening agent's type and quantity as well as the base oil's viscosity. The consistency of grease reflects in the resistance against deformation by a specified force. Penetration is a parameter for measuring the consistency of grease. Penetration of the grease that works and does not work is measured by ASTM D 217 and D 1403 methods. The Consistency or grade numbers, ranging from 000 to 6 has been established by NLGI related to a stated range of penetration number [16].

The greases with BHT had a lower consistency than the others, while those with HMWP had a higher consistency. Increasing the concentrations of antioxidants tended to increase the consistency of food-grade grease.

3.4. Corrosion resistance
Corrosion in air is caused by oxygen reactions, generally starting by the formation of peroxides and hydroperoxide, followed by the breakdown of fatty acids and the conversion into aldehydes or ketones as well as the formation of free fatty acids (rancidity). Oxidation process can increase the peroxide

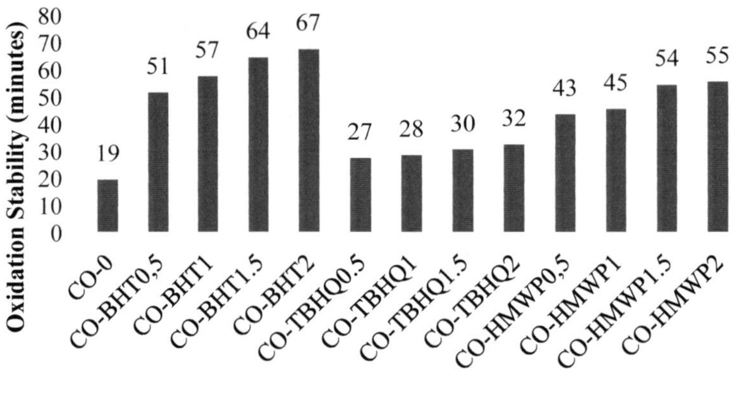

Figure 1: Impact of three different antioxidants on a castor oil

number [8]. Addition of antioxidants did not affect corrosion tests. All the food-grade greases produced showed good resistance to corrosion (group 1a). Group 1a indicates that there is no corrosion.

3.5. Oxidation stability
Oxidation causes breakdown of the chemical and physical properties of materials. For example, samples containing fats and oils can become rancid, exhibiting increased acidity that can lead to corrosion and rusting. Oils can thicken, becoming more viscous, which affects their lubricating properties. Sludge can also form, blocking nozzles and filters, and further compromising lubrication. Lacquering can affect cooling and friction properties, and dyes can undergo discolouration. To slow these processes, antioxidants are added to inhibit oxidation reactions [18]. The vegetable oil's oxidation stability decreases toward the double bond number, and the poor low-temperature properties is even worse as the number of double bonds decreases [19].

Figure 1 shows the performance of three different antioxidants at various concentrations in a castor oil in the RBOT test at 150 °C. It can be seen that all the antioxidants are effective in inhibiting oxidation, although BHT and HMWP are more effective than TBHQ. BHT is the most effective antioxidant.

4. Conclusions
A food-grade grease, NLGI Grade 2 (depth of penetration 265–295 mm (0.1 mm)), can be manufactured using 17% lithium soap thickener. Increased concentrations of antioxidants tend to increase the dropping point and consistency of food-grade grease. Food-grade grease with added BHT antioxidants has a lower dropping point. HMWP antioxidant additives increased the dropping point. Addition of antioxidants did not affect the pH or corrosion test. All the food-grade greases produced have a neutral pH (7) and good corrosion resistance (group 1a). All the antioxidants effectively inhibit oxidation, but BHT and HMWP are more effective than TBHQ. BHT is the most effective antioxidant. The best performance was shown by food-grade grease formulated with HMWP antioxidant. This food-grade grease had the following characteristics: dropping point 193 °C, pH 7, consistency NLGI 2 (multipurpose grease), and corrosion resistance group 1a.

Acknowledgements
This research was supported by the Research and Development Centre for Oil and Gas Technology (LEMIGAS). We thank our colleagues from the Department of Chemistry, Universitas Indonesia, who provided valuable insight and expertise that greatly assisted this research and Einago (www.einago.com) for the English language review.

References
[1] Yakuub N F 2008 *B. Sc. Final Report*, University Malaysia Pahang, Pahang

[2] Yanto T and Septiana A T 2012 *Jurnal Teknologi Pertanian* **13** 34–42
[3] Florea O, Luca M and Steliean C 2004 *Tribol. Ind.* **26** 52–57
[4] Adamu MS, Bilal SAA and Mashi AH 2016 *Journal of Scientific and Engineering Research* **3** 39–42
[5] Shahnazar S, Bagheri S and Hamid S B A 2016 *Int. J. Hydrog. Energy* **41** 3153–70
[6] Nagendramma P and Kumar P 2015 *Lubricants* **3** 628–36
[7] Bhaumik S and Pathak S D 2016 *Tribol. Ind.* **38** 33–44
[8] Yanto T, Naufalin R and Erminawati E 2013 *Jurnal Teknologi Pertanian* **14** 1–8
[9] Moens E 2003 *Trends Food Sci. Technol.* **14** 157–62
[10] Severino LS *et al.* 2012 *Agron. J.* **104** 853–80
[11] Ulfiati R, Malik M R and Bangun P A 2013 *Majalah M&E* **11** 15–19
[12] Nangbes J G, Nvau J B, Buba W M and Zukdimma A N 2013 *The International Journal of Engineering and Science* **2** 105–9
[13] Danlami J M, Arsad A and Zaini M A A 2015 *J. Taiwan Inst. Chem. Eng.* **47** 99–104
[14] Mohammed M A R 2013 *Iraqi Journal of Chemical and Petroleum Engineering* **14** 11–21
[15] Jayadas N H and Nair K P 2006 *Tribol. Int.* **39** 873–8
[16] US Army Corps of Engineer 2016 Engineer Manual Number EM 1110-2-1424: Lubricants and Hydraulic Fluids (Washington, USACE Publication)
[17] Delgado M A, Sánchez M C, Valencia C, Franco J M and Gallegos C 2005 *Chem. Eng. Res. Des.* **83** 1085–92
[18] Nolan S J and Savin R 2016 *The Evaluation of Oxidation Resistance of Lubricating Greases Using the Rapid Small Scale Oxidation Test (RSSOT)*, ELGI Paper pp 1–10
[19] Dresel W H 1994 *Ind. Crops Prod.* **2** 281–8

IOP Conf. Series: Materials Science and Engineering **496** (2019) 012061 doi:10.1088/1757-899X/496/1/012061

Electric field-dependent charge transport in a poly (dA)-(dT) DNA molecule: localization length study

F Rahmani and E Yudiarsah

Department of Physics, Faculty of Mathematics and Natural Sciences (FMIPA) Universitas Indonesia, Depok 16424, Indonesia

Corresponding author: e.yudiarsah@sci.ui.ac.id

Abstract. Electronic transport in DNA molecules could depend on many factors, one of them is external electric field influence. In this research we analyse localization length of poly (dA)-(dT) DNA molecule to study its electronic transport properties. Tight binding Hamiltonian model is used along with electric field and temperature influence. Electric field is set to be parallel to the helix axis of DNA. Transfer-matrix method is required to calculate localization length, while gram-schmidt orthonormalization method is used to reduce errors. The results show that electric field changes localization length in all energy considered in calculation.

1. Introduction

In many decades researcher is constantly trying to make smaller and more advance electronics circuits. Silicon based microprocessors has been widely used in today electronics industry, but trial in shrinking the size of it has been started to reach its limits. This challenges scientist to find another alternative materials to be the next generation of nano-device. In 1962 it was proposed that evidently DNA can conduct electrical current [1], this outcome may have inspired scientist to create biomolecular nano-device based on DNA molecule. The process of electron transport is believed because of the π-π stacking interactions along the base pairs and also the phosphate backbone sugar molecules [2]. Charge hopping mechanism is used to describe long range electron transport in DNA where it also occurs between hydrogen bonds in either Guanine-Cytocine or Adenine-Thymine [3].

There are many discoveries regarding DNA conductivity, and some of experimental results indicate the behavior of DNA as insulator [4] and even its behavior as superconductor [5]. This variation in result could happens because DNA is very sensitive to changes due to factors, such as electric field [6], magnetic field [7], temperature [8], and base motion [9] that will alter its structure. Alteration on DNA structure could mainly determine its quality for charge transport. This research will be discussed external electric field influence on electron localization length in poly(dA)-(dT) DNA molecule. In our calculation tight binding hamiltonian model is used to describe interaction between each base and backbone molecule, with adding electric field and temperature contribution. Electric field is set to be parallel to the helix axis of DNA. Transfer-matrix method is also required to calculate localization length, while gram-schmidt orthonormalization method is used to reduce errors. Miller-Abraham formula is used to modify hopping contant that had been affected by electric field. We also consider electron hopping between backbone in our calculation. The results will be represented in terms of localization length that will be elaborated in third section. Finally, the paper will be concluded in the final section.

2. Theoretical model

Double strand DNA molecule, which consist one strand of Adenine (A) base sequence and Thymine (T) base sequence as its complement, is used as medium for electron to transfer. Both ends of DNA strands are attached to electrode molecules which give different potential in between. In this calculation, we assume that electron could travel from base to nearest bases within the same strand, its pair, and also nearest backbone or vice versa. Along with that, electron also could travel from backbone to nearest backbone. A schematic diagram of a DNA model can be found in the reference [10], and the tight binding Hamiltonian model is determined by: [10]

$$H = \sum_{i=1}^{N}\left[\sum_{\tau,\sigma=1}^{4}(\varepsilon_i^{\tau} + \phi_i^{\tau})|i,\tau\rangle\langle i,\tau| + t_{i.i}^{\tau,\sigma}|i,\tau\rangle\langle i,\sigma| + t_{i,i+1}^{\tau,\sigma}|i,\tau\rangle\langle i+1,\sigma|\right] + h.c \quad (1)$$

Symbol N, i, and τ (σ) in equation (1), consecutively, are the length of the molecule sequence (102 base pairs), site index, strand index. Parameter ε_i^{τ} is on-site energy of base (9.06 eV for T and 8.22 eV for A), while $\emptyset_i^{\tau}$ is phonon effect on DNA base on-site energy. Parameter $t_{i.j}^{\tau,\sigma}$ represents electron hopping constant between sites. Medium and temperature are considered to be able to make backbone on-site energy randomly distributed around 9.36 eV with a deviation of 0,1 eV. External electric field is set to be parallel to the helix axis of DNA (z-axis). When we give electric field to the system, it will change electron hopping constant from site i to j that described by the following formula [11],

$$t_{ij} = t_0\, exp\left(\frac{qV}{k_BT} cos\theta\, Z_{ij} - \left(\frac{\varepsilon_j - \varepsilon_i}{k_BT}\right)\right) \quad (2)$$

where Z_{ij} is the distance between site i to j, q is electron charge, θ is the angle between Z_{ij} and electric field, and k_BT is energy unit. V is voltage difference across the molecule, higher voltages means higher electric field.

Localization length is giving information to what extent electrons could travel in DNA molecule. By analyzing localization length we could choose which DNA model and sequence is better to be molecular wire. The localization length is calculated by employing transfer matrix method. Due to large number of matrices multiplication, Gram-schmidt orthonormalization method is used to reduce errors. This method will produce coefficient value that is used to calculate Lyapunov exponents using equation:

$$\delta_{cl} = \left\{\frac{Ni(\prod_m \beta_{cl}^m)}{2R(N-1)}\right\} \quad (3)$$

where cl is sequence of matrix columns, β is normalization coefficient, R is repetition of multiplication, N is the length of DNA and m is ortonormalization number. Smallest value of δ is used in obtaining localization length (λ) through relation [12]

$$\lambda = \frac{1}{\delta} \quad (4)$$

3. Results and discussion

The results of the calculation are showed in figure 1 and figure 2. Figure 1 shows electron localization length in DNA poly(dA)-(dT) calculated at temperature 4.2 K. In the spectrum, it appears three regions of energy where the localization length larger than zero, they are around the energy of 7.5 eV – 8 eV (band I); 8.75 eV – 9.25 eV (band II); and 9.5 eV – 9.9 eV (band III). Electronic state in band I are localized states where the value of localization length less than N (102 pairs). According to Suhendro et al, localized states occurs because of disorder in backbone onsite energy along with thermal agitation caused disorder in effective bases onsite energy, however the difference between backbone onsite energy and bases onsite energy in few electronic energy states are quite small to obtain extended states [13]. Band II and III show some extended states, a condition when the value of localization length more than N. In extended states, the mobility of electron is higher than in localized states, it means that electron can transfer from first base pair to the next 102 pairs. This circumtanse in quantum mechanics occurs due to the wavefunction of electron in one quatum state significantly overlap with the neighboring quantum states.

Electric field applied parallel to the helix axis of the poly(dA)-(dT) DNA could change energy states which accordingly change its localization length. It can be seen from figure 1, as we set the value of electric field higher (0.005–5 V), the localization length is decreasing. The width of energy gap also seems increasing, especially in high electric field (5 V), whereas the extended states region looks narrowed. It could be translated that charge trasport in DNA poly(dA)-(dT) will be compromised by the presence of electric field. This phenomenon occurs because, in our model, electric field changes the electron hopping constant such that it decreases when electron moves in the direction of electric field and it is constant otherwise.

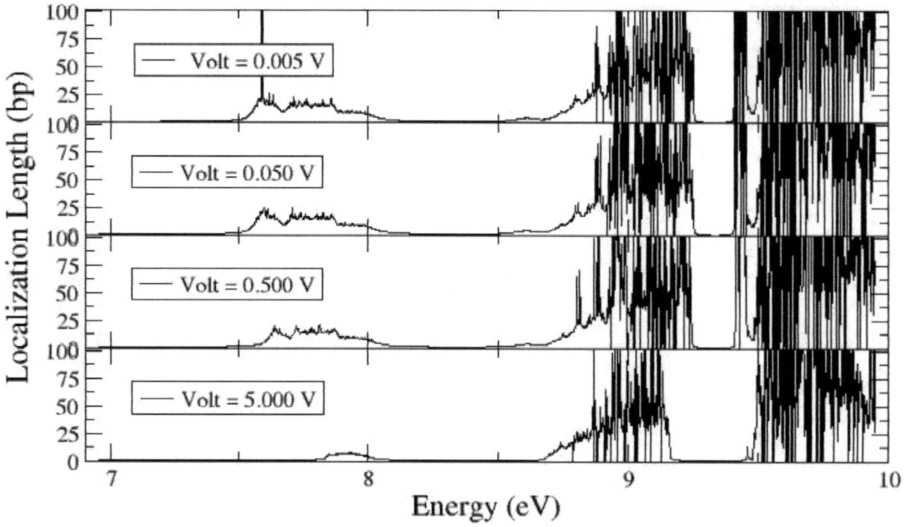

Figure 1. Localization length of poly(dA)-poly(dT) DNA in various voltage at 4.2 K

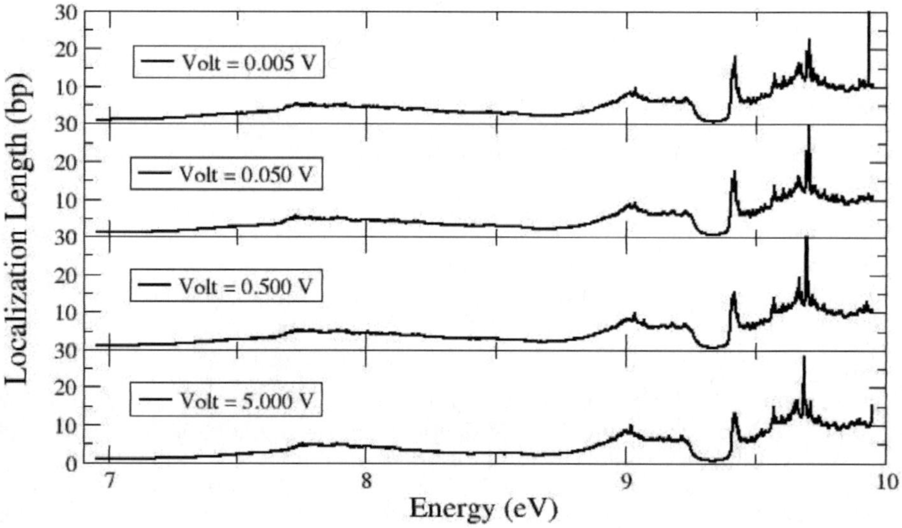

Figure 2. Localization length of poly(dA)-poly(dT) DNA in various voltage at 30 K

Figure 2 shows electron localization length calculated at higher temperature, 30 K. We still observed three energy region with localization length larger than zero separated by two gap with zero localization length. It appears that the localization length decreases significantly with the temperature, thus all three energy region become localized. Higher temperature will affect standard deviation of twist angle distribution to become wider. In consequence, it would causes variation in intrastrand and interstrand hopping amplitudes among bases, and induce onsite energy disorder that block charge transport process in the DNA molecule [14]. Thermal energy changes the onsite energy of bases which produces negative impact on energy structure of DNA molecule. The energy structure of DNA molecule becomes more disorder as the temperature increases. Since we set small magnitude of twisting motion frequency (0.51 meV), the system is furthermore disordered. The results show that thermal fluctuation causes bigger changes in the localization length than the electric field. The same tendency as in figure 1 in which terms of localization length decrement due to voltage increment is observed, but the voltage increase could only slightly change the localization length.

4. Conclusions

The electric field influence of electron localization length in a poly(dA)-(dT) DNA molecule have been studied. The localization length had been calculated by using transfer matrix method. Miller-Abraham formula is used to modify hopping constant under influence of electric field. The results show that higher electric field results in lower localization length in some energy. Higher electric result in wider gap between two energy regions (energy band) with nonzero localization, in consequences the energy band appears narrower. Under the model used in this study, at higher voltage the poly(dA)-(dT) DNA molecule would be less conductive. This tendency is observed in both temperature used in the study. In addition, higher temperature also decreases the mobility of electron to transfer.

Acknowledgements

Our gratitude to D K Suhendro for calculation parameters. We would also give thanks to Cisco Laboratory, Department of Physics FMIPA UI for computational facility. This work is supported by PITTA Grant 2017 with contract number: 648/UN2.R3.1/HKP.05.00/2017.

References

[1] Eley D D and Spivey D I 1962 *Trans. Faraday Soc.* **58** 411–5
[2] Cuniberti G, Craco L, Porath D and Dekker C 2002 *Phys. Rev. B* **65** 241314
[3] Dekker C and Ratner M 2001 *Phys. World* **14** 29--33.
[4] Braun E, Eichen Y, Sivan U and Ben-Yoseph G 1998 *Nature* **391** 775–8
[5] Kasumov A Y, Kociak M, Gueron S, Reulet B, Volkov V T, Klinov D V and Bouchiat H
[6] 2001 *Science* **291** 280–2
[7] Malyshev A V 2007 *Phys. Rev. Lett.* 98 096801
[8] Wong J R, Lee K J, Shu J J and Shao F 2015 *Biochemistry* 54 3392–99
[9] Guo A M and Xu H 2007 *Physica B Condens. Matter* **391** 292–8
[10] Roche S 2003 *Phys. Rev. Lett.* **91** 108101
[11] Rahmi K A and Yudiarsah E 2016 *AIP Conf. Proc.* **1729** 020037
[12] Li L, Meller G and Kosina H 2007 *Microelectron. J.* **38** 47–51
[13] Mackinnon A and Kramer B 1983 *Z. Phys. B. Condens. Matter* **53** 1–13
[14] Suhendro D K, Yudiarsah E and Saleh R 2010 Physica B Condens. Matter **405** 4806–11

IOP Conf. Series: Materials Science and Engineering **496** (2019) 012062 doi:10.1088/1757-899X/496/1/012062

Desorption behaviour of MgH_2-Ni-SiC prepared by ball milling for hydrogen storage material application

Z Jalil[1*], A Rahwanto[1], A Akhyar[2] and E Handoko[3]

[1]Department of Physics, Syiah Kuala University, Banda Aceh 23111, Indonesia

[2]Department of Mechanical Engineering, Syiah Kuala University, Banda Aceh 23111, Indonesia

[3]Department of Physics, Jakarta State University, Jakarta, Indonesia

Corresponding author: zjalil@unsyiah.ac.id

Abstract. Magnesium (Mg) is one potential candidate that can be used as a hydrogen absorbent. Theoretically, Mg can absorb large amounts of hydrogen up to 7.6 wt.%. However, the kinetic reaction of Mg in absorbing hydrogen takes at least 60 minutes when it occurs at high temperatures (300–400 °C). The purpose of this study is to increase hydrogen desorption capacity and the working temperature of the MgH_2-Ni-SiC-based storage system. The materials preparation was done with mechanical alloying technique. The milling process was varied for three different times of 5 h, 10 h and 15 h, with the two catalyst ratios of five wt.% and ten wt.% and the speed of 350 rpm. X-Ray Diffraction (XRD) measurement results revealed that the sample has successfully become nano-sized crystals with the dominant phase of MgH_2 and the minor phases of Ni and SiC. The irregular shape of the particles and their non-homogeneous size are known form the Scanning Electron Microscopy (SEM) results, which are caused by agglomeration. The best hydrogen desorption capacity of up to 2.6 % and the lowest temperature desorption of was 365 °C was obtained for samples that were milled for 10 hours with five wt.% Ni-SiC catalyst.

Keywords: Magnesium, hydrogen storage, reactive mechanical alloying, catalyst, silicon carbide.

1. Introduction
Magnesium (Mg) and Nickel (Ni) are hydrogen absorbent materials that are potential for applications in fuel cell vehicles. Theoretically, Mg is capable of absorbing large amounts of hydrogen (7.6 wt%), which exceeds the target of the International Energy Agency (IEA) for 5 wt.% absorption capacity and temperature of absorption-desorption of ≤100 °C [1]. Moreover, Mg is a lightweight and easily obtained element at affordable prices, making it considered to be studied today. But the drawback is that Mg has a prolonged kinetic reaction and high operating temperature in absorbing hydrogen, it takes at least 60 minutes and a temperature of 300 °C. The provisions of the IEA state that the material to be used as a hydrogen absorbent must be able to show the performace of hydrogen absorption occurring at temperatures below 100 °C in a short time [2–5].

Several researchers have studied magnesium alloys with various manufacturing methods and variations in the alloying elements. Ranjbar *et al.* [6] have carried out research using MgH_2 as a matrix with SiC particles dispersed in the matrix. After a small amount of SiC (5 wt.%) has been inserted in the matrix, the materials are capable of absorbing hydrogen as much as 6.3 wt.% at a temperature of 350 °C in a very short time (5 minutes).

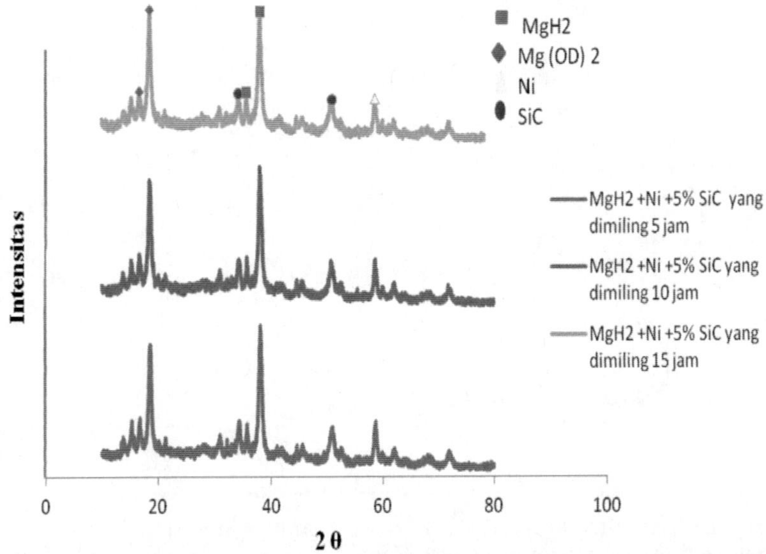

Figure 1. XRD profile on MgH_2 + Ni + SiC material with a catalyst of 5 wt% SiC and three milling time variations of 5 hours, 10 hours and 15 hours

Other studies were done using reactive ball milling technique for double catalyzed Ni and SiO_2 inserted in MgH_2 material with four variations of SiO_2 wt.% of 3, 5, 10 and 15. The result of Differential Scanning Calorimetry (DSC) showed that the lowest temperature for hydrogen desorption to occur is reached at 361 °C for 15 wt. % of SiO_2 catalyst [7–8]. The temperature of the hydrogen desorption temperature of is too high for its application in the vehicle industry. Several researchers have carried out further material and involving nano-sized catalysts [9,10] and double nano-sized catalyst [11] to improve hydrogen absorption/desorption characteristics of MgH_2 material to be applied as a hydrogen storage material.

2. Materials and methods

The materials used in this study were MgH_2 (99.99%, 50 μm size, Sigma Aldrich production), Ni (Hongwu Nano, 99%, 50 nm particle size), and silica carbide (SiC). These three reagents were combined and then split into two parts, namely MgH_2 + Ni + 5 wt.% SiC; and MgH_2 + Ni + 10 wt% SiC. Both parts then went through the milling process with three variations of time for 5, 10 and 15 h. The ratio of ball mills to samples is 10:1 and each process is carried out for 1 g of sample at 350 rpm. The phases present in the sample were determined by X-ray diffraction (XRD) (Shimadzu D6000, CuKα, λ = 1,54060 Å). The hydrogen desorption characteristics of the samples were analyzed by using Thermogravimetric Analyzer (TGA).

3. Results and discussion

3.1. Phase identification results with X-Ray Diffraction (XRD)

Identification results using XRD for samples with five wt.% catalysts for three variations of the milling process times of 5 hours, 10 hours and 15 hours are shown in figure 1, while figure 2 shows the XRD results for ten wt.% catalysts. The phases in the sample can be identified by comparing experimental data with standard data published by the Joint Committee for Powder Diffraction Standards (JCPDS). The comparison results are shown in both figures (figure 1 and figure 2).

The result of the XRD pattern comparison indicate that the MgH_2 phase is the dominant phase at 2θ of 37.9° and 16.6° whereas the peak at 2θ of 35.6° belongs to Ni phase. The presence of the SiC phase is represented by the peaks at 2θ = 58.6° and 31.1°. The XRD results also indicate the presence of additional $Mg(OD)_2$ phases at 2θ = 50.7° and 18.4° which is estimated to be the result of oxidation that occurs during the weighing process until the transfer of samples to the milling container. The presence of this phase indicates that the material is contaminated with air so that the oxygen enters the sample.

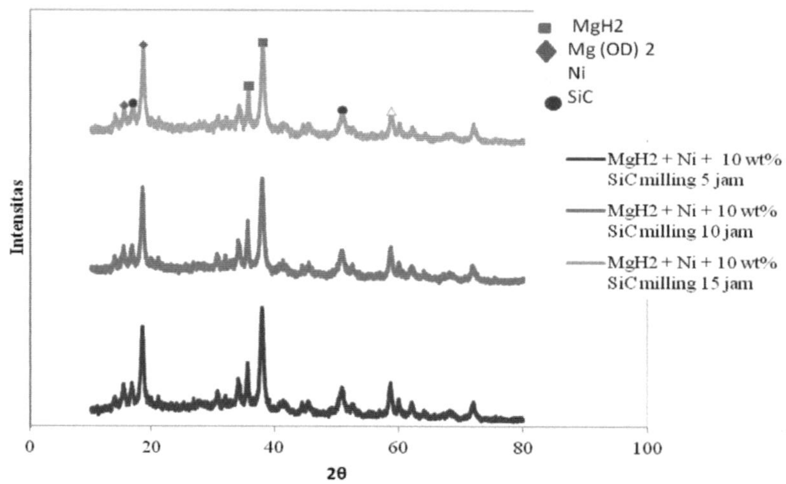

Figure 2. XRD profile on MgH_2 + Ni + SiC material with a catalyst of 10 wt% SiC and three milling time variationx of 5 hours, 10 hours and 15 hours

This impurity may affect the hydrogen desorption process. Ni and SiC serve as a catalyst and is used in very small amounts, where the catalyst itself help to accelerate the reaction but actually does not react. That is why both the phases of the catalyst Ni and SiC are present in the XRD spectrum.

The width of the peaks of the XRD spectrum is represented by the FWHM value of each peak. The relatively similar shape and width of the diffraction spectra in figure 1 show that there is no significant change in the value of FWHM with the difference in milling process time. FWHM can also be used to get the crystal size from the sample using the Scherrer equation. The MgH_2 peaks from the samples with a 5% SiC catalyst have smaller FWHM value compared to a 10% SiC catalyst. Thus it can be said that the addition of the SiC catalyst decreases the crystal size of the sample. However, the XRD spectrum shows an increase in FWHM from the MgH_2 peaks by increasing the milling time of both the 5% and 10 % SiC catalysts.

Ranjbar *et al.* [6] have found that silicon carbide (SiC) in addition to its role in the reactive milling process; it can also accelerate grain refinement due to the characteristics of the hard SiC as well as other carbide metals. The absorption capacity of hydrogen by MgH_2 network will increase with finer grain size since the hydrogen atoms can be easier when infiltrating to the hydrogen magnesium network. Furthermore, as reported by Ranjbar *et al.* [6] that the decrease in the grain size of the ductile Mg during the milling process of both Mg only or MgH_2 is also observed if the process is carried out in hydrogen environment. Furthermore, with its hardness characteristics, SiC also plays a role in reducing the size of the grain that makes the MgH_2 surface larger. The addition of this surface area will have a positive impact on the hydrogen desorption characteristics of MgH_2

3.2. Surface morphological observation results with SEM

The surface morphology observation using SEM provides information that the grain dimensions for samples with a 10-hour milling time are smaller than grain dimensions for samples with a 5-hour milling time as shown in figure 3. However, as reported earlier by Ranjbar *et al.* [6] that during the milling process is believed that the hardness of Ni elements plays a role in reducing the grain size of MgH_2 so that its surface area becomes more substantial, which then affects the absorption and desorption of MgH_2. However, the amount of Ni catalysts in MgH_2 must also be limited, because excessive doping Ni will cause reduce the hydrogen absorption capacity, the reaction kinetics will be slow and can also hinder the hydrogen diffusion process [6].

3.3. Thermal investigation

Gravimetric testing using Thermogravimetric Analyzer (TGA) is required in studying hydrogen release capacity. Figure 4 shows the desorption behaviour of MgH_2 + Ni + 5wt% SiC with a milling process for 10 hours. These results indicate that the milling process for 10 hours can produce hydrogen

(a) (b)

Figure 3. The morphological structure of the sample MgH_2 + Ni + 5 wt% SiC for (a) 5 hours and (b) 10 hours milling time

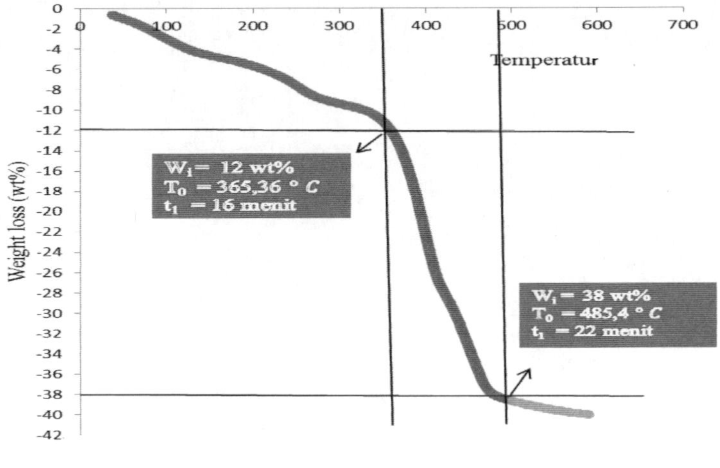

Figure 4. Desorption behaviour of MgH_2 + Ni + 5wt% SiC with 10 hours of milling process.

as much as 12 wt.% in 16 minutes. These results correspond to Jalil *et al.* [10] which get smaller particle sizes can produce a larger contact surface area between the samples and hydrogen, which will increase hydrogen absorption rate.

4. Conclusions

From the results of the use of Ni and SiC catalysts on MgH_2 solid hydrogen storage material with two variations of wt.% SiC catalysts and three variations of milling time process, it can be concluded that the desorption properties of MgH_2 material can be improved, and both Ni and SiC catalysts are suitable catalysts.

Acknowledgements

Authors would like to express the gratitude to Andia Fatmaliana, M.Si. (PPs Unsyiah) who is responsible for XRD measurements.

References

[1] Schlapbach L and Züttel, *Nature* **414** 353–8
[2] Denis A, Sellier E, Aymoniera C and Bobet J L 2009 *J. Alloys Compd.* **476** 152–9
[3] Barkhordarian G, Klassen T and Bormann R 2006 *J. Phys. Chem. B* **110** 11020 – 4

IOP Conf. Series: Materials Science and Engineering **496** (2019) 012062 doi:10.1088/1757-899X/496/1/012062

[4] Song M Y, Kwon S N, Bae J S and Hong S H 2009 *J. Alloys Compd.* **478** 501–6
[5] Oelerich W, Klassen T and Bormann R 2001 *Mater. Trans.* **42** 1588–92
[6] Ranjbar A, Guo, Z P, Yu X B, Wexler D, Calka A, Kim C J and Liu H K *Mater. Chem. Phys.* **114** 168–172
[7] Sun L and Gong K 2001 *Ind. Eng. Chem. Res.* **40** 5861–77
[8] Akihiro Y, Makoto S and Junichi K 1998 *J. Jpn. Soc. Powder Powder Metall.* **48** 61–6
[9] Larsson P, Moysés-Araújo C, Larsson J A, Jena P and Ahuja R 2008 *PNAS* **105** 8227–31
[10] Jalil Z, Rahwanto A, Ismail I, Sofyan H and Handoko E 2018 *Mater. Res. Express* **5** 064002
[11] Jalil Z, Rahwanto A, Akhyar H, Razali R and Handoko E 2018 *IOP Conf. Ser.: Earth Environ. Sci.* **105** 012098

Preparation of boron–doped diamond electrodes modified with copper–gold nanoparticles (CuAuNP) as oxygen sensors

A D Kurniawan, E Saepudin and T A Ivandini

Department of Chemistry, Faculty of Mathematics and Natural Sciences (FMIPA), Universitas Indonesia, Depok 16424, Indonesia

Corresponding author: ivandini.tri@sci.ui.ac.id

Abstract. Boron-doped diamond (BDD) electrodes are promising because they offer a wide electrochemical potential window, low background current, and good physical and chemical stability. In this study, to make them sensitive to dissolved oxygen, BDD electrodes were modified with copper–gold nanoparticles (CuAuNPs). A one-pot chemical reduction method was used to synthesize the CuAuNPs. Transmission electron microscopy revealed that the average diameter of the CuAuNPs was around 11 nm. The electrochemical behavior during oxygen reduction was characterized by cyclic voltammetry in a potential range from −1.0 to −1.5 V in a pH 7 phosphate-buffered saline solution containing various concentrations of the dissolved oxygen. The cyclic voltammograms showed that peaks corresponding to oxygen reduction occurred at −0.13 V. An approximately linear calibration curve was derived to correlate the peak current to the concentration of dissolved oxygen, indicating that the CuAuNP–BDD electrodes are promising for use as oxygen sensors. These electrodes can be readily applied to environmental monitoring, industrial safety, fuel cells, and automotive applications.

Keywords: Boron-doped diamond, surface modification, CuAuNP, oxygen sensors, cyclic voltammetry

1. Introduction

Over the years, the water quality in Indonesian rivers has degraded and the rivers are now classified as heavy polluted [1]. Thus, over the last 50 years, the detection of dissolved oxygen (DO) in water has been extensively studied because of its significant role to control the environmental, as well as to maintain safety in industry, in technology for fuel cell, and in the automotive industry [2].

Boron-doped diamond (BDD) is a carbon-based electrode that has many advantages over other electrodes, because of its wide electrochemical potential window, low background current, physical and chemical stability, and biocompatibility [3–5]. However, it does not provide catalytic activity for the reduction-oxidation reaction with oxygen.

Recently, nanomaterials have become integral to the development of sensors since their properties can be adjusted based on the size and shape of material and can be applied to various techniques. The use of nanoparticles, such as Au, Ag, and Cu nanoparticles, in sensors has been explored recently. Moreover, because of their optical, magnetic, and catalytic properties, a combination between multimetallic nanoparticles and an alloy or core-shell structures shows an interesting feature. Presently, advance trials are conducted to synthesis the nanoparticle alloys for applications in sensor, devices for photonic, and catalysis, because they often exhibit better catalytic properties than the monometallic forms of the metal itself do [6]. The BDD's electrochemical characteristics can be upgraded through the modification of its surface by the metal nanoparticles deposition, to enhance the activity of its electrocatalytic property, and thereupon, increase the process in terms of its sensitivity and selectivity.

This study focused on coating the surfaces of BDD electrodes with copper–gold nanoparticles (CuAuNPs) for this reason. Copper (Cu) is known to have a higher conductivity than gold and silver but is cheaper than both. Gold (Au) is known to have good sensitivity as an oxygen sensor since the introduction of the Clark electrode in the mid-1950s; in addition, it is favored because it is biocompatible, stable, and has high electrical and corrosion resistance [7].

We hypothesize that modifying the surfaces of BDD electrodes with shell-core CuAuNPs will provide better sensitivity to oxygen reduction. Thus, in this study, BDD electrodes modified with Au nanoparticles (AuNP–BDDN electrodes), copper nanoparticles (CuNP–BDD electrodes), and copper-gold nanoparticles (CuAuNP–BDD electrodes) were compared. Furthermore, the oxygen sensing capabilities of these electrodes were validated against measurements obtained using a conventional sensor designed for detecting DO in water.

2. Experimental

2.1. Materials and instruments

BDD was obtained from Keio University in Japan on a (1 0 0) silicon wafer template. The BDD and the silica had a total thickness of 1 mm and had a 0.1% boron/carbon content. Analytical-grade $CuSO_4$, $HAuCl_4$, K_2HPO_4, KH_2PO_4, $NaBH_4$, L-ascorbic acid, citric acid, and other chemicals were obtained from Merck and WAKO.

An EDAQ Potentiostat (from the Bioelectrochemistry main laboratory of Universitas Indonesia) was used to conduct electrochemical experiments. A three-electrode system was applied with Ag/AgCl as a reference and a Pt wire as the counter electrode. The working electrode was either a bulk–Au BDD electrode or an AuNP–BDD, CuNP–BDD, or CuAuNP–BDD electrode.

2.2. Preparation of AuNP–BDD and CuAuNP–BDD electrodes

Before modification with nanoparticles, the BDD was modified into nitrogen-terminated diamond based on an adaptation of a previously developed method [7,8]. Briefly, the BDD electrodes were ultrasonically washed for 15 min in isopropanol and followed by 15 min ultrasonic washing in ultra-pure water. Then, they were dried with nitrogen gas. Next, the washed BDD electrodes were modified by a reaction with allylamine molecules under 254 nm ultraviolet light irradiation for 6–8 hours.

Gold nanoparticles (AuNPs) and copper–gold nanoparticles (CuAuNPs) were prepared based on an adaptation of the Zhang method [9] with a few adjustments. 100 μL of 0.1 M $CuSO_4$ and 150 μL of 0.1 M sodium citrate ($Na_3C_6H_5O_7$) were added to 20 mL of water at room temperature. Then, 1 mL of fresh $NaBH_4$ (3.8 mg in 4 mL water) was added to the solution under rapid stirring followed by continued stirring for 15 min. A 0.1 M $HAuCl_4$ solution was added to this mixture followed by stirring for 20 min. CuAuNPs were stored at room temperature (14 ± 2) for 24 h before use. The nanoparticles were characterized by UV-Vis spectroscopy and transmission electron microscopy (TEM). Then, the nanoparticles were deposited on the nitrogen-terminated BDD by immersing the BDD in a colloidal nanoparticle solution for 12 hours followed by drying.

2.3. Electrochemical studies using cyclic voltammetry

The bulk-Au electrode and CuAuNP–BDD electrodes were used as working electrodes, as the supporting electrode, Pt-spiral was employed, while for the reference electrode, Ag/AgCl (KCl–saturated) system was applied. The analysis was performed in 7 mL 0.1 M electrolyte solution of phosphate buffer solution (PBS) with 100 mV/s scan rate, within range of potential from –1.0 V to 1.5 V (vs. Ag/AgCl).

3. Results and discussion

3.1. Preparation and characterization of BDDN-CuAuNP electrodes

The preparation of CuAuNPs was based on a reduction reaction of $AuCl_4^-$ ions at Cu nanoparticles by the remaining $NaBH_4$ after the formation of CuNPs. The yellow CuNP solution did not exhibit a surface plasmon resonance (SPR) peak and the CuNPs had diameters of less than 2 nm. With the addition of $HAuCl_4$, the solution color changed to red wine, which indicates the growth of the Au shell

IOP Conf. Series: Materials Science and Engineering **496** (2019) 012063 doi:10.1088/1757-899X/496/1/012063

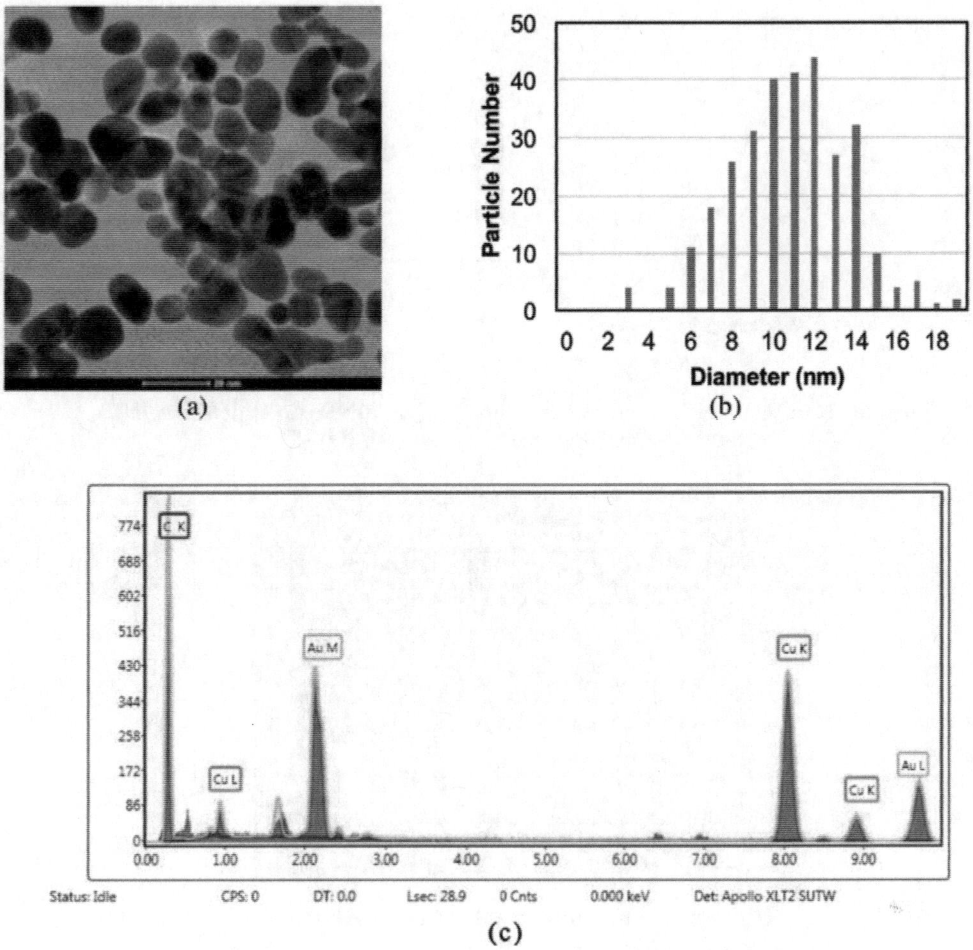

Figure 1. (a) Representative TEM image of the CuAuNPs, (b) the size distribution, (c) and EDX data.

on the Cu core. In addition, the increase in the SPR of the solution at 534 nm confirms the formation of AuNPs with a size increase of over 2 nm, which confirms that the formation of an Au shell on the Cu core [9]. The formed CuAuNPs were found to have diameters around 11 nm based on TEM observations. Moreover, the presence of the Cu core was confirmed by the EDX results, showing the presence of both Cu and Au (figure 1).

Figure 2 shows a comparison of the UV-Vis spectra for the three nanoparticles. Whereas the CuNPs did not show any peak, the AuNPs and CuAuNPs exhibited SPR peaks around 522 and 534 nm, respectively. This slight shift of the peak indicates an interaction between the two metals in the nanoparticles. Furthermore, the stability of the CuAuNPs was characterized by UV-Vis spectrometer readings over 10 days; these results showed a decrease in the intensity, indicating agglomeration of the nanoparticles.

Previous study showed the better affinity of the nitrogen–modified BDD toward AuNPs compared to the one modified with oxygen or hydrogen [7,8]. Because the surfaces of the CuAuNPs were shown to be covered with an Au shell, these nanoparticles were expected to interact strongly with the nitrogen-terminated BDD. Thus, before modification with nanoparticles, the BDD surfaces were pre-treated via a photochemical reaction under ultraviolet light (λ = 254 nm) with allylamine dropped at the surface; by this process, the allylamine's double bonds will be broken to form a new covalent bond between the allylamine's C ions and BDD's C ions [7,8]. This process induced nitrogen termination to facilitate the deposition of the nanoparticles. Figure 3 shows an SEM image of the CuAuNP–BDD.

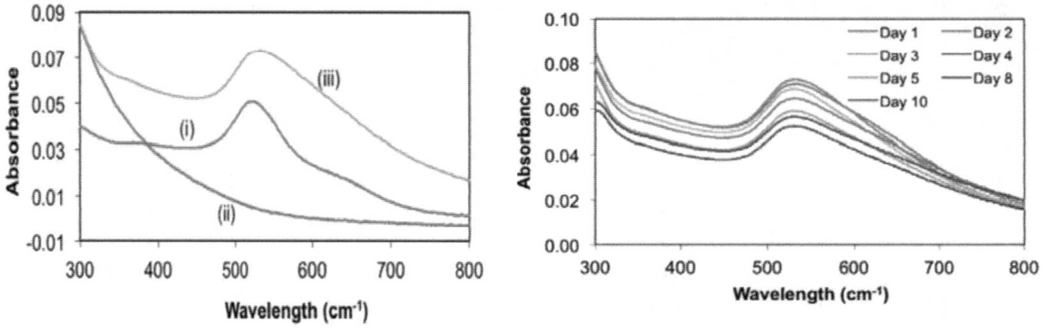

(a) (b)

Figure 2. (a) UV-VIS spectra of (i) CuNPs, (ii) AuNPs and (iii) CuAuNPs and (b) CuAuNP stability over 10 days.

Figure 3. SEM image of CuAuNP–BDD.

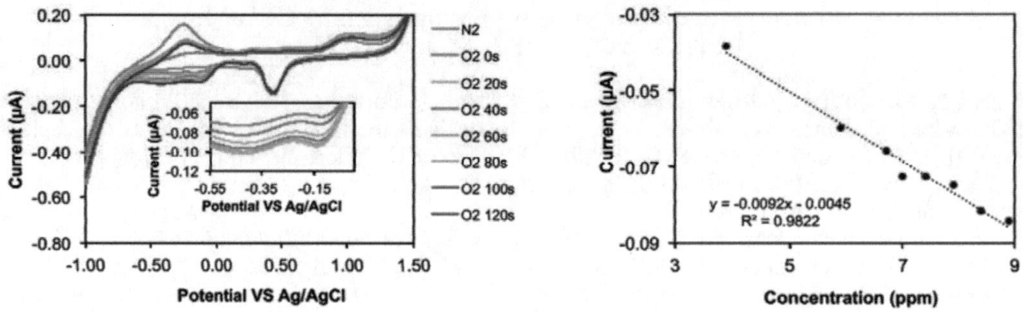

(a) (b)

Figure 4. (a) Cyclic voltammograms obtained with the CuAuNP–BDD electrode for 0.1 M PBS (pH 7) with different DP concentrations and (b) a linear curve representing the relationship between the peak oxygen reduction current and the DO concentration.

The observed white spots covering the BDD surface indicate that the BDD was modified with CuAuNPs successfully.

3.2. Electrochemical properties of oxygen at CuAuNP–BDD electrode

Cyclic voltammetry was used to examine the electrochemical behavior of the electrodes in the presence of oxygen. Figure 4 shows cyclic voltammograms obtained with the CuAuNp–BDD electrode in 0.1 M PBS (pH 7) with different DO concentrations. The results show a reduction peak at −0.13 V (vs. Ag/AgCl). The voltammograms obtained with the bulk-Au electrode were similar to those obtained with the CuAuNP–BDD electrode, indicating that the gold shell influences the electrode properties and the reduction peak in the CuAuNP–BDD electrode. The peak current associated with the oxygen reduction increased linearly as the DO concentration increased. Based on the results, the limit of detection was estimated to be 0.168 ppm, indicating that the proposed electrodes can be used as oxygen sensors.

4. Conclusions

In this study, a one-pot chemical reduction method was used to synthesize CuAuNPs were successfully. Nitrogen-terminated BDD surfaces were modified with CuAuNPs successfully as confirmed by the SEM image. The obtained CuAu–BDD electrodes resulted in similar voltammograms to those obtained using a bulk-Au electrode, indicating that the gold shell affects the CuAu–BDD electrode properties in terms of the reduction peak. In addition, the electrochemical results showed that the reduction peak current increased as the oxygen concentration increased, suggesting that the CuAu–BDD electrodes are promising for applications in oxygen sensing.

Acknowledgements

This research was supported by PITTA Grant Universitas Indonesia, Contract No. 1822/UN2.R3.1/PPM.00.01/2017. The authors would like to thank Einago (www.einago.com) for the English language review.

References

[1] Ministry of Environment and Foresty 2014 Ministry of Environment and Forest Statistic 2014 pp 1–418.
[2] Hutton L, Newton M E, Unwin P R and Macpherson J V 2009 *Anal. Chem.* **81** 1023–32.
[3] Macpherson J V 2015 *Phys. Chem. Chem. Phys.* **17** 2935–49.
[4] Ivandini T A and Einaga Y 2012 *Chem. Comm.* **53** 1338–47.
[5] Medina-Sánchez M, Mayorga-Martinez C C, Watanabe T, Ivandini T A, Honda Y, Pino F, Nakata K, Fujishima A, Einaga Y and Merkoçi A 2016 *Biosens. Bioelectron.* **75** 365–74.
[6] Lou T, Qiang H and Chen Z 2017 *Talanta* **163** 132–9.
[7] Tian R H and Zhi J F 2007 *Electrochem. Commun.* **9** 1120–6.
[8] Ivandini T A, Saepudin E and Einaga Y 2015 *Anal. Sci.* **31** 643–9.
[9] Zhang J, Xu X, Yang C, Yang F and Yang X 2011 *Anal. Chem.*, **83** 3911–7.